SPRINGER
LAB MANUALS

Springer-Verlag Berlin Heidelberg GmbH

Ilse Kranner Richard Beckett Ajit Varma (Eds.)

Protocols in Lichenology

Culturing, Biochemistry, Ecophysiology and Use in Biomonitoring

With 88 Figures, 4 in Color

Springer

Dr. ILSE C. KRANNER
Institut für Pflanzenphysiologie
Karl-Franzens-Universität Graz
Schubertstraße 51
8010 Graz, Austria
e-mail: ilse.kranner@kfunigraz.ac.at

Prof. Dr. AJIT K. VARMA
School of Life Sciences
Jawaharlal Nehru University
110067 New Delhi, India
e-mail: ajitkrvarma@hotmail.com

Prof. Dr. RICHARD P. BECKETT
School of Botany and Zoology
University of Natal
PBag X01
Pietermaritzburg 3209, Republic of South Africa
e-mail: beckett@nu.ac.za

ISBN 978-3-540-41139-0

Library of Congress Cataloging-in-Publication Data
Protocols in lichenology : culturing, biochemistry, ecophysiology, and use in biomonitoring / Ilse Kranner, Richard Beckett, Ajit Varma, eds.
p. cm. – (Springer lab manuals)
Includes bibliographical references and index.
ISBN 978-3-540-41139-0 ISBN 978-3-642-56359-1 (eBook)
DOI 10.1007/978-3-642-56359-1
1. Lichenology – Laboratory manuals. I. Kranner, Ilse. II. Beckett, Richard P., 1958–
III. Varma, A. (Ajit), 1939– IV. Springer lab manual.

http://www.springer.de

Originally published by Springer-Verlag Berlin Heidelberg New York in 2002

Production: PRO EDIT GmbH, 69126 Heidelberg, Germany
Cover design: design & production GmbH, 69121 Heidelberg, Germany
Typesetting: Mitterweger + Partner, 68723 Plankstadt, Germany
Printed on acid free paper SPIN 10698986 31/313So 5 4 3 2 1 0

Foreword

Biology is a complex and multifarious science, so much so that one could legitimately wonder whether this term actually designates a well-defined scientific discipline. The operations involved in giving a name to a beetle, in estimating the primary production of an ecosystem, in sequencing a string of DNA are so different, and require such different cultural and technical backgrounds, that the term "biology" appears today to be an umbrella which is not large enough to cover them all. In fact, it is hardly possible to find concepts, methods and epistemological backgrounds that are common to all the diverse branches of biology, and that can characterise it with respect to other sciences. As a scientific discipline, biology is only defined by its very general object of investigation: life on earth.

On the one hand, modern Biology tends towards specialisation, on the other hand the recognition of the basic similarity of processes common to all living beings is making the boundaries between traditional sub-disciplines less precise. Within the same Department of Biology, a "botanist" studying the mitochondria in a cell of a weed can communicate more easily with a "zoologist" working on the ribosomes of a snail than with another botanist studying the colonisation of railway tracks by weeds. Most of the traditional subdivisions of biology were based – as modern biology itself – on the respective subjects of study, but for many of them this criterion has become a sort of straightjacket. "Lichenology" is no exception.

The index of the present book well reflects this state of affairs: it lists a spectrum of methodologies encompassing an exaggeratedly broad range of techniques, disciplines, and levels of organisation. Furthermore, very few of the methods presented here are typical of lichenology: most of them are adaptations of much more general techniques to the peculiarities of a relatively narrow group of organisms, the lichenized fungi. Such an impressive heterogeneity could be considered by some as the weak point of the present book: many of its chapters could be better thought of as appendices in specialised monographs dealing with methodologies in physiology, ecology, genetics, etc.

However, there is a reason for this book, an important one, one which has to do with the recent history of lichenology.

The period between 1930 and 1950 marked the nadir of Lichenology world-wide: there was only a handful persons still interested in lichens, several of which found refuge in the lichenological fortress of Scandinavia. However, after the War, and almost suddenly, lichenology experienced a veritable boom. The first generation of post-war lichenologists were the pupils of a few masters – mostly classical taxonomists – who knew each other well and who had a common cultural background. Fortunately for modern lichenology, they were good masters: they pushed many of their pupils beyond the borders of taxonomy into all possible directions. The third or fourth generation is showing up today, but the common roots, and hence the common language, are still here. Many years ago, as a newcomer in a meeting of the International Association of Lichenology, I experienced something which completely changed my scientific interests. Coming from a different background, what I was familiar with were either highly specialised meetings – where everybody is studying more or less the same thing – or huge international congresses – which to me felt like an indigestible arithmetic sum of specialised meetings. My late arrival in the small Austrian village – just after dinner – directed me immediately, with a glass of beer, to a table where several people were engaged in a discussion on which characters could best distinguish *Hypogymnia physodes* from crooked forms of related species. I proudly believed I had understood at once what tribe of biologists was hosting me. After a few minutes, however, I started feeling confused: a man with a beard was explaining to us how low the photosynthetic performance of *Hypogymnia* in polluted areas is; a bold man replied that this could not be due only to sulphur dioxide, but also to heavy metal contamination, while a lady with dark glasses maintained that her SEM and TEM data showed that most metals were concentrated outside the algal layer. My confusion became total when they all began a lively discussion on the evolution of this genus, its possible centres of origin, and the main mechanisms of evolution with relation to reproductive strategies.

The present book is a mirror of my early experiences in the world of lichenology: it puts together a bright and diversified palette of methods, one that is unlikely to fit in any subdivision of modern biology. I am certain, however, that most lichenologists – even those from the fourth generation – will be able to appreciate it. Methods are subject to rapid change, this book might have overlooked several important details and become outdated soon, but here it is! It will not become a "bible" for lichenologists: it only offers an important basis for discussion, criticism, and, hence, progress. I don't know how many from the fourth generation are still aware of their common historical roots, but I am certain that such roots provide an enormous "added value" to their apparently outdated discipline. The cross-fertilisation among different branches of biology is a rare event today, which, however, lichenologists can still enjoy. Losing it will be their fault, if – following fashion – they decide to follow other paths.

Trieste, July 2000

PIER LUIGI NIMIS
Dipartimento di Biologia
Università di Trieste
Via L. Giorgieri 10
34127 Trieste, Italy

phone: +39+040+6763885
fax: +39+040+568855
e-mail: nimis@univ.trieste.it

Preface

As an intricate association between a fungus, the mycobiont and one or more green algae or cyanobacteria, the photobiont, lichens are one of the most successful examples of symbiosis. Their biology has long fascinated scientists, as they have features that are unique, or occur rarely in other organisms. For instance, many lichens survive extreme desiccation, and can display tolerance of very high and very low temperatures. Lichens live in and are adapted to a great variety of habitats, from coastal fog zones of deserts to intertidal zones, from plant leaves in tropical rain forests to the glacial moraines of the Himalayas at altitudes as high as 7800 m above sea level, and they are dominant components of communities in circumpolar ecosystems. Possibly because of their tendency to grow in nutrient poor habitats, lichens are extremely efficient accumulators of atmospherically deposited pollutants, and have widely been used to monitor environmental pollution.

Lichens contain a wide range of secondary products that have not been found in other organisms. Pharmaceutical companies are interested in using some of these compounds for their fungicidal, antibacterial and antiviral properties. Studying these outstanding organisms may greatly improve our understanding of life. We still have a limited knowledge and may be surprised what they will offer.

Most of the methods used to study lichens are not unique to lichenology. However, in addition to the above, lichens differ from most higher plants because of, among other reasons, the lack of a cuticle, their "dual nature", small size, and cell walls that are frequently unusually thick. Lichens are extremely difficult to culture, grow very slowly, and the presence of co-extracted secondary metabolites very often complicates the analyses of other compounds. As a result, techniques used to study them often require modification before they can be applied to lichens. The major aim of this manual is to provide a good range of protocols which are well-tested for lichens, so that this will encourage more researchers to study these unusual organisms, and improve these methods or develop new ones.

We have tried to cover most of the commonly used techniques, that are appropriate to lichenology wherever we could find authors who had time and expertise to write a chapter, but are aware that there are still several topics which could have been included. In some cases we allowed overlapping to occur between chapters so that each chapter can stand on its own; for instance the same culture media are given in several chapters that deal with culturing lichens.

The 32 chapters of this manual are arranged in seven sections. The first section contains protocols for isolating and culturing lichen symbionts, and how to re-combine them, as well as techniques for growing thalli in tissue culture, and how to cultivate them in the field. A short section describes methods for studying lichen ultrastructure, and the following comprises techniques for investigating lichen physiology and ecophysiology, including stress physiology. The next part includes protocols for determination of compounds that are often uniquely present in lichens, as well as identification and quantification of compounds which also occur in other organisms, but present particular problems when analysing lichens. The following section on nucleic acids shows the application of some recent developments in molecular biology to lichens. Thereafter protocols for using lichens to monitor environmental pollution are outlined. We did not include a chapter on lichen identification, because information on identification techniques and excellent keys are already available, and page restrictions prevented us from including chapters on identification techniques. However, the last section of the book focuses on the documentation of lichen biodiversity, including very recent protocols based on computer aided information systems which make use of the internet to identify lichens.

We thank the many reviewers who took the time to make helpful suggestions. Special thanks are extended to Dr. Ute Scardelli for helping to check the literature citations, and to Dr. Martin Grube for stimulating discussions, and for always making himself available to help us in many matters. We very much appreciate the advice and encouragement given by Dr. Dieter Czeschlik and Dr. Jutta Lindenborn from Springer-Verlag throughout the editing process. We acknowledge that work on this book was made possible only by grants from the Austrian Academy of Sciences (APART 428) and the Austrian Science Foundation (P12690-BIO) to Ilse Kranner, and in part by a grant from the South African National Research Foundation (GUN 2039625) to Richard Beckett. Finally, we would like to express sincere thank to all the authors who contributed to this book for sharing their expertise with the reader.

The editors very much hope that the present volume conveys something of their own and the authors' enthusiasm for lichens, and will encourage more scientists to work on these fascinating organisms.

ILSE KRANNER, RICHARD BECKETT, and AJIT VARMA

List of Contents

Physiology and Ecophysiology

Lichen Compounds

Culture and Cultivation

Isolation and Culture of Lichen Photobionts and Mycobionts

ISAO YOSHIMURA, YOSHIKAZU YAMAMOTO, TAKETO NAKANO, and JEFFREY FINNIE

Introduction

Lichens are dual organisms formed from a symbiotic association of a fungus, the mycobiont, and an alga and/or cyanobacterium, the photobiont, in which numerous photosynthetic cells are intertwined in a matrix of fungal hyphae. Such definitions raise the question as to whether lichens are technically individual organisms. Many aspects of lichen biology are concerned with the interactions of these different organisms. The separation, isolation and culture of the symbionts offers the scientist a fascinating opportunity to study the components and contribute to the understanding of the nature of the symbiosis in lichens. The culture of mycobionts, photobionts and lichen thalli is central for the establishment of experimental systems for lichens, needed to solve questions associated with symbiosis biology. In addition, they are essential to solving the many fundamental problems of lichen physiology, morphogenesis and molecular biology.

Cultures of lichen symbionts were thought to be too difficult to study mainly because of the time consuming, long-term techniques necessary for successful culture of the symbionts. However, Ahmadjian (1967b) in his ground-breaking research stimulated interest in the culture of mycobionts and photobionts for many lichenologists. In the past two or three

✉ Isao Yoshimura, Kochi Gakuen College, Biology Laboratory, 292 Asahitenjincho, Kochi, 780-0955, Japan (*phone* +81-88-840-1121, ext. 432; *fax* +81-88-840-1124; *e-mail* isaoyosh@inforyoma.or.jp)
Yoshikazu Yamamoto, Akita Prefectural University, Department of Biological Production, Faculty of Bioresource Sciences, Laboratory of Advanced Bio-production Science, 241-7, Kaidobata-nishi, Shimoshinjyo-nakano, Akita, 010-0195, Japan
Taketo Nakano, Hiroshima Institute of Technology, Department of Environment and Information, 2-1-1, Miyake, Saiki-ku, Hiroshima, 731-5193, Japan
Jeffrey Finnie, University of Natal, Department of Botany and Zoology, Pietermaritzburg, Private Bag X01, Scottville, 3209, Republic of South Africa

decades, research on mycobionts, photobionts, and on resynthesis of lichens from isolated symbionts has made considerable progress. Cultures of lichen mycobionts and photobionts can be obtained by different methods as outlined in Fig. 1.

Lichen symbionts (mycobionts and photobionts) are usually very difficult to culture using nutritionally rich media. This is because contaminating bacteria, foreign algae, and fungi may grow more rapidly than the slowly growing lichen symbionts. Contaminating organisms such as moulds, yeasts, and bacteria are located not only on the surface of the lichen thallus, but also in the medulla, and the gelatinous sheath of the photobionts, especially of cyanobacteria. To obtain sterile lichen symbionts several aseptic steps are necessary. After obtaining both symbionts in axenic condition, one may make an association of the two symbionts to reform a lichen thallus. Unfortunately, only a few examples of the reformation of new lichen thalli have been reported (see Chapter 3).

The aim of this chapter is to describe protocols for the isolation and culture of mycobionts and photobionts from lichen thalli. Subprotocol 1 describes alternative methods for culturing mycobionts, and Subprotocol 2 provides several alternative techniques for culturing photobionts. These culture techniques have been reported in the literature, but we have added additional notes based on our own observations. Comprehensive reviews of the techniques that can be used for the isolation of symbionts have been described by Ahmadjian (1967a,b, 1973) and Galun (1988).

Outline

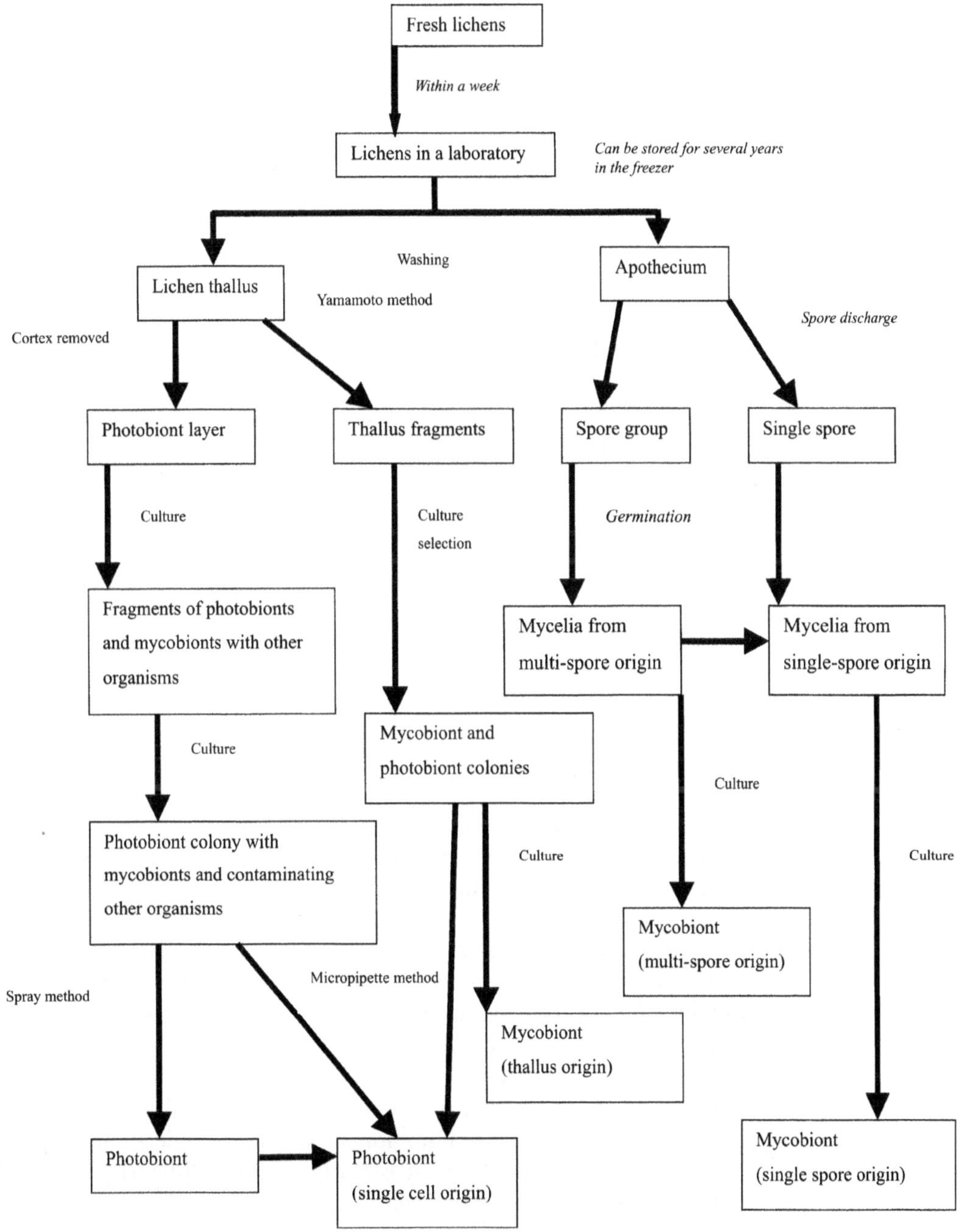

Fig. 1. Flow chart outlining the origin of mycobiont and photobiont cultures from lichen thalli.

Subprotocol 1
Mycobiont Culture

Note: You must carry out all treatments on a clean laminar flow bench or under sterile conditions, except for the pre-washing stage. All equipment should be autoclaved (15 - 20 min, 121°C, 1 atm) or oven-dried (30 min, 180°C) before use.

Materials

Equipment

- Compound microscope
- Dissecting microscope
- Inverted microscope
- Autoclave
- Incubator
- Sonicator
- Centrifuge
- Laminar flow bench or Clean box

Sources of mycobionts

We recommend that lichens should be freshly collected from the field and used within one week. However, lichens may be stored in the desiccated state for a few weeks, or in a freezer for several years (Yoshimura et al. 1990).

Mycobionts can be isolated from ascospores, conidia, isidia, soredia, and thallus fragments (Ahmadjian 1993). For laboratory cultures the most usual method of isolating mycobionts is to start from discharged spores, primarily ascospores. Another useful method for obtaining mycobionts and photobionts is to dissect out thallus fragments. This may result in large quantities of purified mycobiont. The use of thallus fragments for isolation of mycobionts or photobionts is described in detail in Chapter 2.

Mycobiont cultures stored at Akita Prefectural University and Kochi Gakuen College are listed in Table 1.

Table 1. Mycobiont cultures stored at Akita Prefectural University and Kochi Gakuen College

Acarospora fuscata (Nyl.) Arnold
Alectoria lata Tayl.
Alectoria ochloreuca (Hoffm.) Massal.
Amygdalia panaeola (Ach.) Hertel & Brodo
Anaptychia palmurata (Michux) Vain.
Anzia colpota Vain.
Anzia gregoriana Muell. Arg.
Anzia hypoleucoides Muell. Arg.
Anzia japonica (Tuck.) Muell. Arg.
Anzia leucobatoides Zahlbr.
Anzia opuntiella Muell. Arg.
Arthonia tumidula (Ach.) Ach.
Asahinea chrysantha (Tuck.) Culb. & Culb.
Asahinea kurodakensis (Assah.) Culb. & Culb.
Baeomyces absolutes Tuck.
Baeomyces placophyllus (Lamb.) Ach.
Bryocaulon divergens (Ach.) Kaernef.
Bryoria furcellota (Fr.) Brodo & Hawksworth
Calicium japonicum Asah.
Caloplaca leptopisma Zahlbr.
Caloplaca scopularis (Nyl.) Lettau
Candelariella vitellina (Hoffm.) Muell. Arg.
Cetraria delisei (Bory) Th. Fr.
Cetraria islandica (L.) Hoffm.
Cetraria islandica subsp. orientalis (Asah.) Koernef.
Cetraria nivalis (L.) Ach.
Cetraria ondontella (Ach.) Ach.
Cetrelia japonica (Zahlbr.) Culb. & Culb.
Cetrelia nuda (Hue) Culb. & Culb.
Chaenotheca brunnaola (Ach.) Muell. Arg.
Cladia aggregata (Sw.) Nyl.

Table 1. Continuous

Cladina arbuscula (Wallr.) Hale & Culb.
Cladina mitis (Sandst.) Hustich
Cladina portensoa (Dufour.) Follm.
Cladina rangiferina (L.) Nyl.
Cladina stellaris (Opiz.) Brodo
Cladonia bacilioformis (Nyl.) Glueck
Cladonia bellidiflora (Ach.) Schaerer
Cladonia boryi Tuck.
Cladonia cristatella Tuck.
Cladonia cyathomorpha Wats.
Cladonia graciliformis Zahlbr.
Cladonia humilis (Wirth) Laundon
Cladonia merochlorophaea Asah.
Cladonia pityrea (Floerke) Fr.
Cladonia poccillum (Ach.) O. Rich
Cladonia ramulosa (Wirth) Laundon
Cladonia subpityrea Sandst.
Cladonia vulcani Savicz.
Cornicularia aculeatum (Schreber) Link.
Cyphellium tigirare Ach.
Dactylina ramulosa (Hook.) Tayl.
Dermatocarpon miniatum (L.) Mann.
Dermatocarpon reticulatum Magnusson
Diploschistes scuruposus (Schreber) Norm.
Durietzia crenulata (Hook.) Yoshim.
Eriooderma velligerum Tuck.
Evernia divaricata (L.) Ach.
Evernia esosrediosa (Muell. Arg.) Du Rietz
Evernia prunastri (L.) Ach.
Everniastrum cirrhatum (Fr.) Sipman
Faraminella ambigua (Wulf. In Jaeg.) Fricke Meyer
Flavoparmelia caperata (L.) Hale

Table 1. Continuous

Graphis cervina Muell. Arg.
Graphis connectans Zahlbr.
Graphis proserpens Vain.
Graphis scripta (L.) Ach.
Graphis tenella Ach.
Gymnoderma lineare Yoshim. & Sharp
Haematomma ochrophaeum (Tuck.) Massal.
Haematomma ventosum (L.) Massal.
Heterodermia diademata (Tayl.) Awasthi
Heterodermia obscurata (Nyl.) Trev.
Heterodermia pandurata (Kurok.)
Heterodermia pseudospeciosa (Kurok.) Culb.
Hypogymnia physodes (L.) Nyl.
Icmadophila ericetorum (L.) Zahlbr.
Lasalia papulosa (Ach.) Llano
Lasallia asiae-orientalis Asah.
Lasallia papulosa (Ach.) Llano
Lasallia pensylvanica (Hoffm.) Llano
Lecanora argopholis (Ach.) Ach.
Lecanora expectans Darb.
Lecanora fuscata Nyl.
Lecanora muralis (Schreber) Rabenh.
Lecanora pulverulenta Muell. Arg.
Lecanora stenospora Stiz.
Lecidea confluens (Weber) Ach.
Lecidea inopsis Th. Fr.
Leprocaulon arbuscula (Nyl.) Nyl.
Letharia columbiana (Nutt.) Thomson
Letharia vulpina (L.) Hue
Lobaria adscripturiens (Nyl.) Hue
Lobaria linita (Ach.) Rabenh.
Lobaria spathulata (Inum.) Yoshim.

Table 1. Continuous

Lopadium ferrugineum Muell. Arg.
Megalospora sulphurea Meyen
Melanelia stygia (L.) Esslinger
Menegazia terebrata (Hoffm.) Massal.
Nephroma arcticum (L.) Torss.
Nephroma helveticum Ach.
Nephromopsis endocrocea Asah.
Nephromopsis ornata (Muell. Arg.) Hue
Niebla homalea (Ach.) Rundel & Bowler
Normandina pulchella (Borr.) Nyl.
Ochrolechia parellula Muell. Arg.
Ochrolechia trochophora (Vain.) Oshio
Ochrolechia yasudae Vain.
Parmotrema austrosinense (Zahlbr.) Hale
Parmotrema tinctorum (Delise ex Nyl.) Hale
Peltigera aphthosa (L.) Willd.
Peltigera canina (L.) Willd.
Peltigera neckelli Hepp ex Muell. Arg.
Peltigera polydactyla (Necker) Hoffm.
Peltigera ponijensis Gyelnik
Pertusaria corallina (L.) Arn.
Pertusaria laeviganda Nyl.
Pertusaria ophthamaliza Nyl.
Phaeographina pseudomontagnei Nakanishi
Phaeophyscia endococcina (Koerber) Moberg
Physcia adscendens (Fr.) H. Olivier
Physcia phaea (Tuck.) Thomson
Pilophorus clavatus Th. Fr.
Platismatia interrupta Culb. et Culb.
Platismatia lacunosa (Ach.) Culb. & Culb.
Porpidia albocaerulescens (Wulfen) Hertel & Knoph
Porpidia macrocarpa (DC. In Lam. & DC.) Hertel & Schweb..

Table 1. Continuous

Pseudevernia intensa (Nyl.) Hale & Culb.
Pseudocyphellaria aurata (Ach.) Vain.
Pseudoephebe pubescens (L.) M.
Pseudoevernia furfurcea (L.) Zopf
Pseudoevernia olivetorum Zopf.
Puncteria rudecta (Ach.) Krog.
Pyrenula japonica Kurok.
Ramalina exilis Asah.
Ramalina leiodea (Nyl.) Nyl.
Ramalina litoralis Asah.
Ramalina menziesii Tayl.
Ramalina pacifica Asah.
Ramalina roesleri (Hochst.) Nyl.
Ramalina subbreviuscula Asah.
Ramalina subfraxinea var. *leiodea*
Ramalina subgeniculata Nyl.
Rhizocarpon flavum Dodge & Baker
Rhizocarpon geographicum (L.) DC.
Rimelea reticulata (Tayl.) Hale & Fletcher
Roccella fusiformis DC.
Solorina crocea (L.) Ach.
Solorina saccata (L.) Ach.
Sphaerophorus meiophorus (Nyl.) Vain.
Stereocaulon alpinum Laurer ex Funck
Stereocaulon azureum Yoshim. & Weber
Stereocaulon curtatum Nyl.
Stereocaulon dactylophyllum Floerke
Stereocaulon grande (Magn.) Magn.
Stereocaulon paschale (L.) Hoffm.
Stereocaulon sorediiferum Hue
Stereocaulon subcoralloides (Nyl.) Nyl.
Stereocaulon tomentosum Fr.

Table 1. Continuous

Stereocaulon vesuvianum Pers.
Sulcaria sulcata (Lev.) Bystr.
Teloschistes flavicans Norm.
Thamnolia subuliformis (Ehrh.) Lamb.
Thamnolia vermicularis (Swartz) Ach. Ex Schaerer
Thelotrema lepadium (Ach.) Ach.
Thelotrema subtile Tuck.
Trapeliopsis granulosa (Hoffm.) Lumbsch
Tremolechia atrata (Ach.) Ach.
Trypetheliopsis boninensis Asah.
Tuckermannopsis sepincola (Ehrh.) Hale
Umbilicaria aprina Nyl.
Umbilicaria caroliniana Tuck.
Umbilicaria cylindrica (L.) Delise ex Duby
Umbilicaria decusata (Vill.) Zahlbr.
Umbilicaria deusta (L.) Baumg.
Umbilicaria esculenta (Miyoshi) Mink.
Umbilicaria hyperborean (Ach.) Hoffm.
Umbilicaria kisovana Kurok.
Umbilicaria mammulata (Ach.) Tuck.
Umbilicaria muhlenbergii (Ach.) Tuck.
Umbilicaria polyphylla (L.) Baumg.
Umbilicaria proboscidea (L.) Schrader
Umbilicaria torrefacta (Lightf.) Schrader
Umbilicaria vellea (L.) Ach.
Umbilicaria virginis Schaerer
Usnea arizonica Mot.
Usnea bismolliuscula Zahlbr.
Usnea diffracta Vain.
Usnea flexillis Stirt.
Usnea hirta (L.) Weber ex Wigg.
Usnea longissima Ach.

Table 1. Continuous

Usnea misamiensis (Vain.) Mont.
Usnea montis-fuji Mot.
Usnea roseola Vain.
Usnea rubescens Stirt.
Usnea strigosa (Ach.) A. Eaton
Usnea sulphurea (Koenig) Th. Fr.
Vermilacinia combeoides (Nyl.) Spjut & Hale
Vulpicida juniperinus (L.) J. E. Mattsson & M. J. Lai
Vulpicida pinastri (Scop.) J. E. Mattsson & M. J. Lai
Xanthoparmelia subpolyphyloides (Geyln.) Kurok.
Xanthoria elegans (Link.) Th. Fr.
Xanthoria mandsculica (Zahlbr.) Asah.
Xanthoria mawsonii Dodge

Culture media for mycobionts

Ahmadjan (1993) and Pyatt (1973) recommend that the culture media used for spore collection and germination should have a low nutrient content (i.e. should be plain or mineral agar). We have obtained good results using 4 % water agar media at 15°C in the dark. Media for culture of mycobionts include the following:

WA4 medium

4% distilled water agar medium

Agar	4 g
Distilled water	make up to 100 ml

MY medium

Malt /Yeast extract medium (Ahmadjian 1967a)

Malt extract	20 g
Yeast extract	2 g
Agar	20 g
Distilled water	make up to 1000 ml

LB Medium Lilly and Barnett's medium (Lilly and Barnett 1951)

Glucose	10.0 g
Asparagine	2.0 g
KH_2PO_4	1.0 g
$MgSO_4 \cdot 7H_2O$	0.5 g
$Fe(NO_3)_3 \cdot 9H_2O$	0.2 mg
$ZnSO_4 \cdot 7H_2O$	0.2 mg
$MnSO_4 \cdot 4H_2O$	0.1 mg
Thiamine	0.1 mg
Biotin	5 µg
Distilled water	make up to 1 l

For a solid medium, add 15-20 g of agar to the above ingredients and make up to 1 l

LBG medium **Lilly and Barnett's Gelrite medium** (Yamamoto et al. 1998)
Lilly and Barnett's Medium containing 1% w/v Gelrite instead of agar.

Note: Autoclave all media before use and pour in Petri dishes (5 mm thick) or test tubes (5 ml) in the laminar flow bench.

Procedure

Isolation of mycobionts from spores

Spore discharge

1. Clean thalli collected from the field, then leave for a few days after collection to equilibrate with the environment. Alternatively, clean and freeze material, and before use allow a few hours for equilibration.
2. Remove the apothecia or perithecia from the thallus and place into dishes containing distilled water and allow them to soak for about 4 h, or alternatively wash them in running water. Blot dry the spore bearing structures to remove excess water.
3. Fix these structures to the bottom of a plastic petri dish using petroleum jelly and place 4 % water agar medium on the top cover of the Petri dish (Fig. 2). Placing the media in the upper lid limits contamination of the agar.

4. Ensure that the agar is within the discharge range of the spores (ca. 5 - 10 mm). Discharged spores attach to the agar surface either singly or in groups. If you want to carry out single spore isolation, it may be necessary to reduce the discharge time or increase the distance between the ascocarp and the water agar, because multiple spore discharge may occur. Replace the top cover of the Petri dish with new covers containing fresh medium several times at the appropriate interval (according to the discharge time, normally one day).
   Alternatively, spores can be discharged onto glass slides or onto sterilised Parafilm in a damp environment, and the spores washed off with distilled water. Then, transfer the spores to a medium immediately.
5. Seal Petri dishes containing apothecia with Parafilm and store in an incubator at 15°C in the dark.
6. To monitor germination, observe spores discharged onto the water agar medium under an inverted microscope. Remove discharged spores onto a glass microscope "well slide" containing agar medium. Keep the slides in a Petri dish in humid conditions and observe continuously. Monitor germination of spores and mycelial growth either in water or stain using lactic-glycerol-cotton blue.

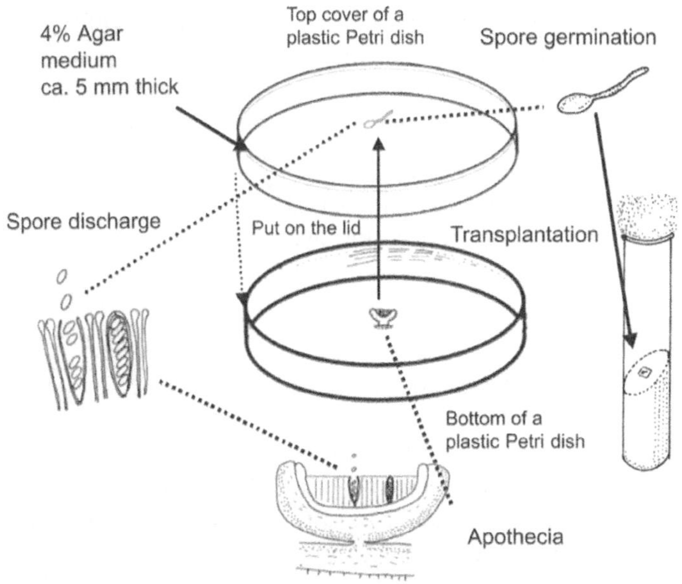

Fig. 2. Isolation of spores after discharge from an apothecium.

Spore germination and mycelial growth

In some lichens, spores germinate within one day after dispersal.

- After germination, excise and transfer to culture tubes or Petri dishes (containing nutrient culture media) small blocks of agar containing the spores.
- Malt-Yeast Extract Medium (Ahmadjian 1967a) and Lilly and Barnett's Medium (Lilly and Barnett 1951) are the culture media most frequently used.

Isolation of mycobionts from the thallus

Where mycobionts cannot be obtained from the spores because of a lack of apothecia, mature spores or poor spore germination, other mycobiont sources may be applicable, e.g. conidia (Vobis 1977) and soredia (Honegger and Bartnicki-Garcia 1991; Honegger et al. 1993). In our experience, isidia are usually too heavily contaminated by epiphytic micro-organisms to be used for isolating lichen mycobionts. Yamamoto et al. (1985) have outlined a method using thallus fragments, see Chapter 2.

- Cut pieces from a fresh thallus using a sterilised razor blade and store in small test tubes containing water (without nutrients) or on wet filter paper in humid conditions at about 15°C. New medullary hyphae usually elongate after a couple of weeks. Excise a portion of the newly elongated hyphae using aseptic technique and transfer to fresh culture media in test tubes. We have found this method useful for obtaining mycobionts of *Cladia rangiferina* and *Parmotrema tinctorum.*
- Sufficient numbers of replicates should be prepared to ensure that the fungal growth obtained is likely to be that of the mycobiont and not a foreign fungus growing on or in the thallus.

Maintaining mycobiont cultures

Mycobionts can be stored for long periods (about one year). However, we recommend that you subculture every 2 to 3 months as follows.

Cut cultured mycobiont colonies into several segments (usually about 5 mg) with a scalpel. Place segments on either MY or LB medium in petri dishes and culture for 2 to 3 months at 15°C in the dark. Repeat this procedure every 2 to 3 months.

Mycobionts in general show maximum growth between 15 and 20°C. The pH of the medium has a significant effect on the development of mycobiont cultures. Each species has an optimum, usually in the range of pH 5-6; significantly higher or lower pHs will retard growth. Light is not necessary for maintaining mycobiont cultures.

Lichen mycobiont cultures can be cryopreserved under liquid nitrogen where they remain viable for extended periods of time.

Comments

Spore Discharge

In most lichens, spore discharge can be observed within one day of placing an apothecium in the culture dish. However, in some lichens, spore discharge may be observed only after 2-3 weeks (Table 2). The time after which the first spores discharge after positioning apothecia in the petri dish varies widely and depends on the species, the developmental and metabolic condition of the apothecia at the time of collection, treatment of the thalli after collection and the age of the individual ascocarps. Similarly, the amount and duration of spore discharge also varies widely depending on the above conditions. Maximum spore discharge occurs from ascocarps that are soaked in water for 15 min to 24 h, blotted dry, and then placed in a humid atmosphere (90% RH) where they will dry slowly (Ahmadjian 1993).

In some lichens, e.g. *Porpidia albocaulescens*, *Graphis cervina* (Table 2), the spores separate rather easily after their discharge from the ascocarp, and fall onto the agar surface. In such cases, single-spore culture can be easily carried out. However, in *Umbilicaria vellea* and other species (Table 2), spores remain together in packets of eight (or less). In these species, single-spore isolations are rather difficult. Ahmadjian (1993) recommends further techniques for isolation (e.g. the micropipette method described below).

Yamamoto et al. (1998) found that spore discharge from the apothecia of many lichens was influenced by collecting seasons, storage temperatures and storage periods. Tested species had more or less an endogenous rhythm of sporulation. Winter and spring were good seasons for spore discharge in temperate lichens, although we have found that *Umbilicaria* spores collected in Canada in summer germinated well at this time.

Spore germination

Pyatt (1968) reported that while some spores may germinate as quickly as 2-4 h after discharge, others might take up to 4 or 5 days. In most lichens, spores germinate a few days after their dispersal. In our experience, the germination period varies from 1 to 21 days after dispersal (Table 2). Some spores germinate at the same time as spore dispersal. Lawrey (1984) re-

Table 2. Time required after setting apothecia for spore discharge and germination to occur in a range of lichen species. Culture: Lilly and Barnett's Gelrite Medium at 15 °C under dark conditions.

Lichen names	Number of days required for first spore discharge after setting apothecia	Number of days for spore germination	Number / condition of discharged spores
Anzia hypoleucoides	1	6	single spores
Anzia opuntiella	1	7	single spores
Caloplaca sp.	1	8	single spores
Graphis cervina	1	1	single spores
Graphis cicatricosa	5	14	single or together
Heterodermia pandurata	1	7	single spores
Lasallia papulosa	1 or 2	6	single spores, but muriform
Ochrolechia parerulla	18	18	single spores
Peltigera praetextata	1	4	single spores
Porpidia albocaulescens	1	1	single spores
Ramalina boninensis	1	21	single spores
Sarcographa melanocarpa	4	ca. 30?	single spores
Trypethelium boninensis	3	7	8 spores together
Umbilicaria proboscidea	2	not germinated	8 spores together
Umbilicaria torefacta	2	ca. 4?	single or a few spores together
Umbilicaria vellea	2	6	8 spores together

ported that spores of *Cetraria ciliaris* only germinate six weeks after their dispersal. Rousssard (1969) and Mathey and Hoder (1978) reported that some spores germinate in their ascus. In general, single-cell spores of crustose lichens can germinate faster than two- or many-cells spores. Spores of foliose and fruticose lichens require more time for germination than those of crustose lichens. Muriform spores or single spores with many nuclei may require more time before they germinate.

Spore germination is influenced by medium composition, initial medium pH and culture temperature. Most spores of the species tested germinated on plain agar medium at pH 6 at 15°C. Spores of *Letharia* species could not germinate on agar-medium, but did on Gelrite-medium. Malt-yeast extract medium prevented spore germination of a few species.

Spores of some *Peltigera* species germinated on media supplemented with resins that can absorb phenols that inhibit spore germination.

Modifications of culture media

Studies on the utilisation of nitrogen, either as amino acids or other nitrogenous substances, have yielded such a diversity of results that no general conclusions can be drawn. Addition of most amino acids permit good growth. Only cysteine, cystine, phenylalanine, and tryptophane fail consistently to sustain good growth. Most of the hexoses used as a carbohydrate source permit satisfactory growth. Mannitol, maltose and lactose allow good growth, while citrate, acetate, erythritol and trisaccharides are poor carbohydrate sources (Ahmadjian 1967b; Hale 1983). Both nitrogen and carbohydrate sources can change or modify the morphological and physiological characteristics of the mycobiont. Mycobionts from a variety of species have a requirement for both thiamine and biotin, while some have a requirement for biotin or thiamine alone.

Use of filters

The mycobiont may be cultured directly on agar or supported on filters. Filters can be made from cellulose, esters of cellulose acetate and nitrate, or glass fibre (Oliver et al. 1989, Honegger and Kutasi 1990). The advantage of using filters is that they can be transferred to new media.

Liquid media

Submerged culturing in liquid media necessitates regular replacement of the culture fluid by fresh medium (Honegger and Kutasi 1990, Honegger et al. 1993). Because most taxa tend to form hard, cartilaginous colonies with only marginal growth, it is advantageous to grind the material at regular intervals with a sterile homogeniser (Honegger and Kutasi 1990, Armaleo 1991). It is difficult to generalise about suitable growth media for culture of mycobionts, as different species appear somewhat individualistic in their nutrient requirements (Bubrick 1988).

Effects of pH and light

The effects of the pH of the medium and of light on the morphology or physiology of the mycobionts have received little attention but are probably important. As heterotrophic organisms, lichen mycobionts would not be expected to respond to differences in irradiance or duration of light. However, in non-lichenized fungi light has a greater effect on reproduction than on vegetative growth.

Subprotocol 2
Photobiont Culture

Early investigators placed thin slices of a lichen into an illuminated damp chamber and waited for the moisture and light conditions to cause outgrowths of the algal symbiont and disintegration of the fungal tissue (Ahmadjian 1967b).

The simplest method, in terms of equipment and time, for obtaining photobiont culture was described by Ahmadjian (1967a,b). Nakano modified Ahmadjian's method and has described it in detail (Nakano 1987). Nakano and his co-workers have obtained many photobionts from Japanese lichens (Nakano 1988, Takeshita et al. 1989).

Note: You must carry out all treatments on a clean laminar flow bench or under sterile conditions, except for the pre-washing stage. All equipment should be autoclaved or oven-dried before use.

Materials

Equipment

- Compound microscope
- Dissecting microscope
- Autoclave
- Incubator
- Sonicator
- Centrifuge
- Laminar flow bench or Clean box
- Capillary tube
- Micropipettes

Preparation of a special micropipette (Fig. 3)

1. Heat a glass tube (4-5 mm inside diameter, 20 cm long) at the centre and stretch from both ends, and divide at the centre (a Pasteur pipette can be used).
2. Place a cotton stopper at the wider end of the pipette, and connect a long rubber tube.

3. Wrap the micropipettes in aluminium foil and sterilise. Before use, in the laminar flow bench, heat the small end of pipette and stretch to make a capillary tube by grasping with forceps. Make the diameter of the capillary several times wider than a typical algal cell (about 50 to 75 μm).

Culture media for photobionts

Most photobionts grow easily in culture. While few photobionts have absolute requirements for organic carbon or nitrogen sources, some green algal photobionts grow much faster after the addition of glucose and / or proteose peptone to the culture medium. Media for culturing photobionts include the following.

BBM

BBM Bold's Basal Medium (Deason and Bold 1960; Bischoff and Bold 1963)

$NaNO_3$	250 mg
KH_2PO_4	175 mg
K_2HPO_4	75 mg
$MgSO_4 \cdot 7H_2O$	75 mg
$CaCl_2 \cdot 2H_2O$	25 mg
NaCl	25 mg
EDTA	50 mg
KOH	31 mg
$FeSO_4 \cdot 7H_2O$	4.98 mg
H_3BO_3	11.42 mg
$ZnSO_4 \cdot 7H_2O$	8.82 mg
$MnCl_2 \cdot 7H_2O$	1.44 mg
MoO_3	0.71 mg
$CuSO_4 \cdot 5H_2O$	1.57 mg
$Co(NO_3)_2 \cdot 6H_2O$	0.49 mg
Make up to 1 l with distilled water	

For a solid medium, add 15-20 g of agar to the above ingredients and make up to 1 l with distilled water.

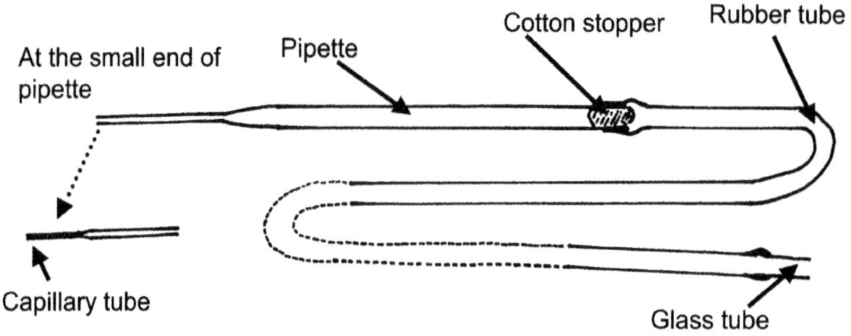

Fig. 3. Micropipette used for a single cell separation of photobionts (after Nakano 1987).

3xN BBM

3xN BBM Modified Bold's Basal Medium with three times more nitrogen (Brown and Bold 1964)	
As above except	
$NaNO_3$	750 mg

For a solid medium add 15-20 g of agar to the above ingredients and make up to 1 l with distilled water.

***Trebouxia* Organic Nutrient Medium**

Trebouxia Organic Nutrient Medium (Ahmadjian 1967a)	
1 x N BBM	970 ml
Proteose peptone	10 g
Glucose	20 g

For a solid medium add 15-20 g of agar to the above ingredients and make up to 1 l with distilled water.

MDM medium

MDM medium for cyanobacteria instead of BBM (Watanabe 1960)	
KNO_3	1 g
$MgSO_4 \cdot 7H_2O$	250 mg
K_2HPO_4	250 mg
NaCl	100 mg
$CaCl_2 \cdot 2H_2O$	10 mg
Fe solution	1 ml
A5 solution	1 ml
Make up to 1 l with distilled water	

For a solid medium add 15-20 g of agar to the above ingredients and make up to 1 litre with distilled water.

Fe Solution

$FeSO_4$	1 g
Distilled water	500 ml
Concentrated H_2SO_4	2 drops

A5 solution

H_3BO_3	286 mg
$MnSO_4 \cdot 7H_2O$	250 mg
$ZnSO_4 \cdot 7H_2O$	22.2 mg
$CuSO_4 \cdot 5H_2O$	7.9 mg
Na_2MoO_4	2.1 mg
Distilled water	100 ml

Other media or media constituents have been recommended by Kratz and Myers (1955), Stanier et al. (1971), Starr (1980), Nichols (1973), Allen (1968; 1973), Archibald (1975; 1977) and Carr et al. (1973).

Note: Autoclave all media before use and pour into Petri dishes (5 mm thick) or test tubes (5 ml) in the laminar flow bench.

Procedure

Pre-treatments

- For macrolichens (foliose and fruticose):
  Cut off about 1 cm^2 from the apex of the thallus, then place into tap water for about 5 to 10 min. Brush the surface of the thallus in running tap water using a paintbrush, and then wash in sterilised water.
- For microlichens
  If the lichen thallus is small (most crustose lichen thalli), place it into a small test tube with 1-2 ml sterilised water and one drop of Tween 20 and ultrasonicate for about 3 min. Centrifuge (2000 rpm) to separate the epiphytes detached from the surface of lichen thalli.

Producing a thallus homogenate

Note: You must carry out all treatments on a clean laminar flow bench or under sterile conditions, except for the pre-washing stage. All equipment should be autoclaved (15 -20 min, 121°C, 1 atm) or oven-dried (30 min, 180°C) before use. Sterilise all glassware and implements before use, clean slides (if dirty) with detergent or acid and wash well in distilled water.

Macrolichens (foliose and fruticose species)

1. After the washing procedure mount the lichen thallus on a sterilised glass microscope slide.
2. Under a dissecting-microscope, carefully scrape or shave away the surface of the thallus (cortex) using a small knife made by filing a small needle.
3. Under the microscope, remove the photobiont layer and transfer it onto a new, sterilised glass slide.
4. Add one drop of sterilised water to the slide, cover the excised portions of photobiont layers with another glass slide and grind into smaller fragments using light pressure. Photobionts are mechanically separated from mycobionts, although some mycobiont hyphae may remain attached. Both symbionts are suspended in the liquid.

Microlichens

After washing, place small portions of the thallus on a sterilised microscope slide and grind between two microscope slides by applying light pressure. Unlike the above protocol for macrolichens, the cortex is still present, so apply more pressure. The resulting suspension contains both photobionts and mycobionts. Break up larger pieces using a blender, or mortar and pestle (Yamamoto 1987). However, fragile cells may be destroyed in blenders, and thus it is best to use less "violent" techniques such as grinding between two glass slides (Nakano 1987).

Variations in the method how to wash thalli, and recommendations for best equipment have been reported by several researchers e.g. a wooden board by Ahmadjian (1967a) and a small washing chamber, cup and filter by Yoshimura et al. (1993). With cyanolichens, use thin slices (up to 40 μm thick), obtained with a freezing microtome for inoculation.

Isolation of photobionts

1. Add a few drops of solution containing a suspension of photobionts and mycobionts onto solid agar media in petri dishes. For green algae use 1 x N BBM culture medium. For cyanobacteria use MDM culture media. Alternatively, thinly spray (see spray method) a solution containing a suspension of symbionts over the surface of an agar medium in a petri dish.
2. Culture at 15°C in an incubator. Generally the cultures should be kept cool at 15-20°C and a light intensity of 10-27 μmol $m^{-2}s^{-1}$ (PPFD, photosynthesis photon flux density). We recommend that initially you expose the cultures to lower light intensities.
3. After about 1 month (depending on the photobiont species) small photobiont colonies appear on the surface of the agar medium. If you have done enough washing of thalli previously, almost all of the algal colonies that appear are true photobionts.
4. It is important to confirm that the isolated algae are the true photobiont. For filamentous algae, trace the algae back to a cell that is affiliated with fungal tissue within the fragment. For single-celled algae, make a search for cells that retain fragments of the fungal hyphae on their walls.

Obtaining axenic photobiont cultures

Amongst the photobiont colonies developing on an agar medium, many of the colonies may be contaminated. In order to obtain axenic photobiont cultures, the following alternatives are given.

Direct method

1. Under the stereo-microscope, photobiont colonies without contamination are selected and are transplanted onto suitable solid agar (either petri dish or test tube) media. *Trebouxia* Organic Nutrient Medium (Ahmadjian 1967a) is used for green algae, MDM medium containing 1% glucose for cyanobacteria (Watanabe 1960).
2. If uncontaminated, a true axenic photobiont culture is obtained. However, photobiont colonies are usually contaminated with bacteria or mycobionts, or some other organisms. Additional spray regimes or micropipette methods are necessary for obtaining true axenic photobionts cultures.

Spray method The spray method (Fig. 4) can be useful for single cell green algae isolation. This technique was developed by Wiedeman et al. (1964) and can be used to both isolate single cell green algae and produce axenic cultures of algae (without contamination of bacteria, yeast etc. and without mycobiont hyphae attached to the photobiont surface).

1. To obtain cultures derived from a single cell, select colonies containing low levels of contamination (or no contamination), from the photobiont colonies growing on the agar plate, and transplant onto 1 x N BBM slant agar media in test tubes.
2. Culture the colonies for several weeks.
3. Transfer the colonies into 10 ml centrifugation tubes with 1 ml of sterilized water and one drop of Tween 20.
4. Ultrasonicate. This results in the photobiont colonies dissociating from contaminating bacteria and mycobionts attached to the surface of their cell walls.
5. Centrifuge (1000 rpm, 5-10 min) the mixture of dissociated photobiont cells and other organisms.
6. Remove the supernatant and add 1 ml of sterilised water and a drop of Tween 20 to the photobiont cells remaining in the centrifuged tube. Repeat this treatment about 10 times.
7. Insert a capillary tube in the bottom of the centrifuge tube and hold in place.
8. Direct compressed air through a small opening across the top of the capillary tube extending from the centrifuge tube.
9. The algal suspension is drawn up the microtube and atomised into a fine spray.
10. Quickly pass a Petri dish containing medium (usually *Trebouxia* Organic Nutrient Medium) through the spray. The dish will become coated with a suspension of algal cells.
11. After 1 or 2 weeks, remove non-contaminated algal colonies and transplant them onto suitable culture media.

Micropipette method The micropipette method is a useful and reliable way to isolate green algae, although you cannot use it for long filamentous algae.

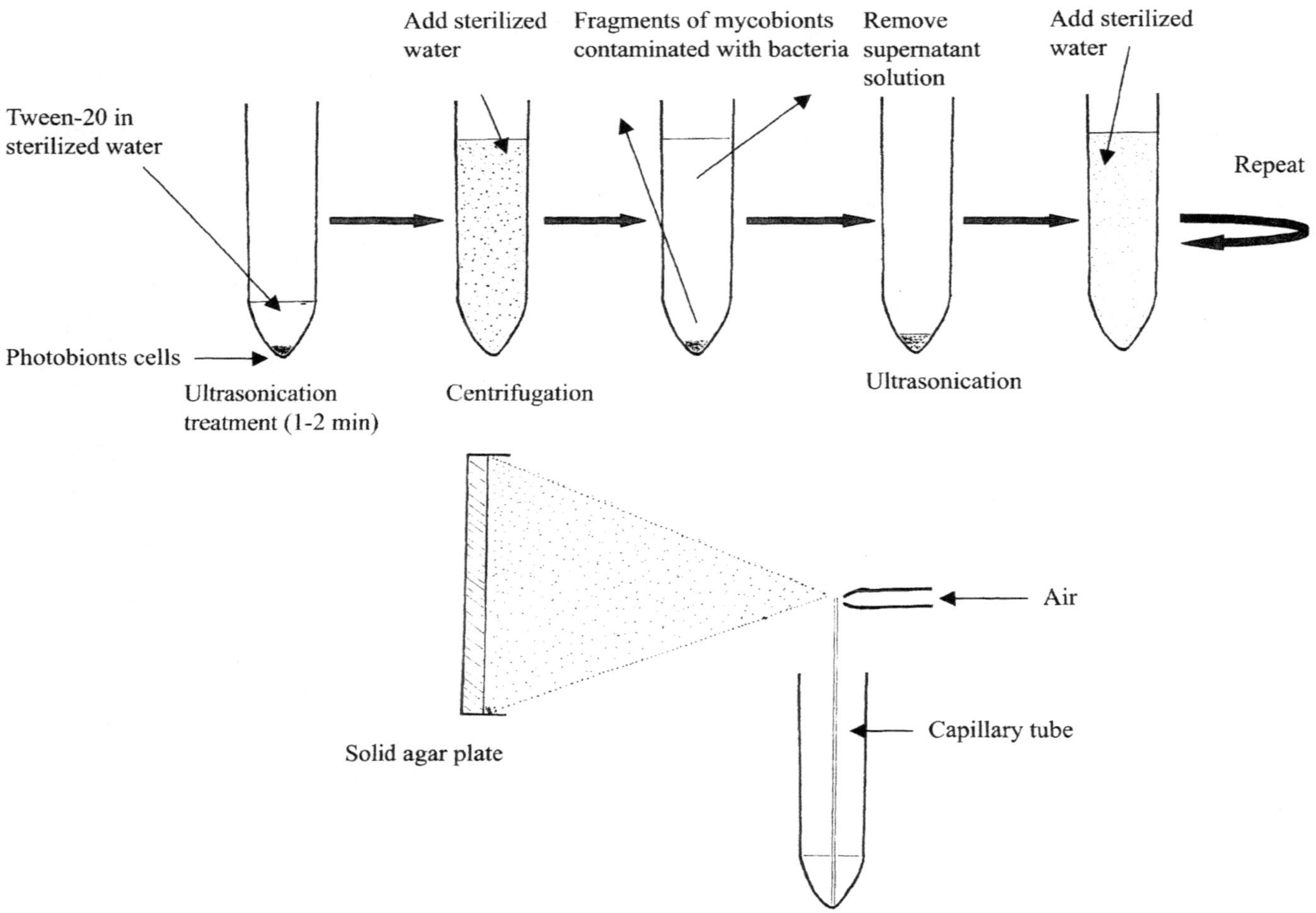

Fig. 4. Isolation protocol of photobionts using the spray method (after Nakano 1987).

1. Add a drop of sterilised water or sterilised culture medium to each well of a glass microscope slide with 5 wells. Remove a developed algal colony from the nutrient medium and place into the first well. If algal colonies are compact then they can be diffused by ultrasonication.
2. Suck up about 5 to 10 algal cells using the micropipette under the stereo-microscope.
3. Expel the cells into the next well, by blowing through the glass tubing attached to the rubber end of the micropipette.
4. Repeat the technique more than 5 times. The micropipette tip can be steamed to reduce contamination. Sometimes, when the algal cell is introduced into a new drop of water, it floats on the surface. It is impossible to suck up the cell in this position with a micropipette. Generally, most cells sink after a few minutes (Ahmadjian 1967b).
5. If contamination is serious, the procedure must be repeated again using more sterilised water or sterilised culture medium.
6. After the final wash, pick up a single algal cell, and transplant it to a new culture medium (transplant about 10 to 50 algal cells).
7. The following culture media should be used to test that your culture is successful: *Trebouxia* Organic Nutrient Medium for green algae, and MDM media containing glucose for cyanobacteria. When contaminating bacteria or yeast are present, they rapidly grow on rich nutrient culture medium. It is very difficult to remove some bacteria in the gelatinous sheath of cyanobacteria; however, they grow in harmony with cyanobacteria. Probably the complete purification of cyanobacteria can be made by using micro-manipulation methods, although we have never successfully achieved this in our laboratory.
   Make an average of fifteen single-cell isolates for each lichen species.

Cutting method

It is difficult to obtain axenic cultures of filamentous green algae, e.g. *Trentepohlia*, using only the methods described above. The cutting method may help you to obtain axenic cultures of these algae.

1. Cut the newly grown algal filaments off at their apices using sterilised forceps, and transplant them onto new culture media.
2. Apply Yamamoto's method (see Chapter 2), and after the second filtration select greenish pieces under a stereo-microscope, then transfer them to test tubes with solid nutrient or plain agar.

3. Maintain many (ca. 50) test tubes with agar slants for a few weeks, and remove any contaminated test tubes. Contamination of culture tubes depends on the part and condition of the thallus, and the species.

4. In the uncontaminated tubes, new growth of the symbionts (photobionts and mycobionts) occurs after about 4 weeks of incubation. The photobiont can be further isolated using additional treatments (the spray method or micropipette method) to get a genetically pure photobiont colony derived from a single cell.

Centrifugation methods

Protocols and guidelines for centrifugation have been published by Richardson (1971).

1. Use suspension derived from the procedure "Producing a thallus homogenate" (Subprotocol 2), and centrifuge (100 - 400 g, about 10 min). In general, most small photobionts remain in the supernatant after low speed centrifugation (100 - 200 g, 10 min). For larger photobionts you must test various combinations of speed and time. Use low-speed centrifugation (100 g) to remove large fragments of thalli or clumps of tissue, and higher speeds (400 g) to separate intact cells from cell fragments and debris.

2. Filter your sample through a nylon sieve mesh 10 - 30 μm, or through other large-pored filters to remove larger size debris before centrifugation (Bubrick 1988).

Maintaining photobiont cultures

Under low illumination conditions, photobionts can be cultured for long periods (about one year). However, we recommend that you subculture every 2 to 3 months. Most lichen algae have an optimum temperature range between 15-20°C. However, the temperature in which the lichen naturally grows should be considered when selecting culture conditions. The optimum pH range is 4.0 - 7.0

The optimum light intensity range lies between 16 - 27 μmol m^{-2} s^{-1} (PPFD) (Ahmadjian 1967a; b). *Trebouxia* from pigmented lichens are more sensitive to high irradiances whereas those from lichens with a non-pigmented cortex are more light tolerant.

1. Cut cultured photobiont colonies with agar media into several segments (about 5 mm^2) with a scalpel. Place colonies on a plate with appropriate medium in Petri dishes and culture for 2 to 3 months.

2. Every 2 to 3 months, transfer growing colonies to fresh medium of the same composition and culture under the same conditions.

Lichen photobiont cultures can also be cryopreserved under liquid nitrogen where they remain viable for extended periods of time.

Comments

Some *Trebouxia* strains lose their colour when cultured at light intensities above 11 µmol m^{-2} s^{-1} (Ahmadjian 1967a). We recommend culturing *Trebouxia* at about 5 µmol m^{-2} s^{-1} illumination to preserve strains for a long time. However, for taxonomical observations, *Trebouxia* strains must be cultured on 3 x N BBM under 22°C, 33 µmol m^{-2} s^{-1}, and a light-dark cycle of 12 h light and 12 h dark (Archibald 1975).

Troubleshooting

- Ascospores do not germinate
  In the absence of compatible photobiont cells, the ascospores of some taxa (e.g. the Peltigerales) may germinate, but fail to grow (Lallemant and Bernard 1977; Ahmadjian 1989). In these cases, germination can be stimulated by including photobiont extracts and growth may be stimulated by adding whole photobiont cells to the agar (Bubrick 1988). See also Chapter 3.
  Various conditions that influence lichen spore germination have been well summarised by Pyatt (1973), Ahmadjian (1993) and Yamamoto et al. (1998). Spore germination may be affected by various environmental factors. These factors include collection season (Pyatt 1969, Ostrofsky and Denison 1980), culture humidity (Garrett 1971), medium pH (Pyatt 1968, Chrismas 1980, Ostrofsky and Denison 1980), culture temperature (Ostrofsky and Denison 1980), alternation of light and dark (Pyatt 1968), salinity (Ramkær 1978), natural extracts (Ostrofsky and Denison 1980), the presence of glucose (Belandria et al. 1989), secondary lichen compounds (Whiton and Lawrey 1982, 1984), air pollutants (Pyatt 1969, Belandria et al. 1989), heavy metals (Pyatt 1976), low oxygen concentration (Kofler 1970), and influence of bark extract (Ostrofsky and Denison 1980).

- Contamination
  To rid cyanobacterial photobionts from bacteria that are normally present in the gelatinous sheaths enveloping the algal cells, irradiate with ultraviolet light (sufficient to destroy bacteria but not the algal cells) or treat them with antibiotics (Ahmadjian 1967b). During the culture, mites often eat cultured algae or mycobionts. Vinyl tape shields are often effective against mites.

Acknowledgements. We thank Prof. V. Ahmadjian of Clark University for his encouragement and advice during our studies on lichen culture. Thanks are extended to Prof. M. Nakanishi of Hiroshima University for kindly helping us to establish our spore-derived cultures of lichen mycobionts.

References

Ahmadjian V (1967a) The Lichen Symbiosis. Blaisdell Publishing Company, Massachusetts

Ahmadjian V (1967b) A guide to the algae occurring as lichen symbionts: isolation, culture, cultural physiology and identification. Phycologia 6: 127-160

Ahmadjian V (1973) Methods of isolation and culturing lichen symbionts and thalli. In: Ahmadjian V, Hale ME (eds) The Lichens. Academic Press, New York, pp 653-660

Ahmadjian V (1989) Studies on the isolation and synthesis of bionts of the cyanolichen *Peltigeria canina* (Peltigeraceae). Pl Syst Evol 165:29-38

Ahmadjian V (1993) The Lichen symbiosis. John Wiley and Sons, New York

Allen MM (1968) Simple conditions for growth of unicellular blue-green algae. J Phycol 4: 1

Allen MM (1973) Methods for Cyanophyceae. In: Stein JR (ed) Handbook of Phycological Methods. Culture Methods and Growth Measurements. Cambridge University Press London, pp 127-138.

Archibald PA (1975) *Trebouxia* de Puymalay (Chlorophyceae, Chlorococales) and *Pseudotrebouxia* gen. nov. (Chlorophycee, Chlorosarcinales). Phycologia 14: 125-137

Archibald PA (1977) Physiological characteristics of *Trebouxia* (Chlorophyceae, Chlorococcales) and *Pseudotrebouxia* (Chlorophyceae, Chlorosarcinales), Phycologia 16: 295

Armaleo D (1991) Experimental microbiology of lichens: mycelia fragmentation, a noval grow chamber, and the origins of thallus differentiation. Symbiosis 11: 163-177

Belandria G, Asta J, Nurit F (1989) Effects of sulfur dioxide and fluoride on ascospore germination of several lichens. Lichenologist 21: 79-86

Bischoff HW, Bold HC (1963) Some soil algae from enhanced rock and related algal species. Phycological Studies IV. Univ. Texas Publ. No. 6318, Texas, 95 pp

Bubrick P (1988) Methods for cultivating lichens and isolated bionts. In: Galun M (ed) Handbook of Lichenology, vol 3. CRC Press, Boca Raton, pp 127-138.

Brown RM Jr, Bold HC (1964) Comparative studies of the algal genera *Tetracystis* and *Chlorococcum*. Phcological Studies V. Univ. Texas Publ. No. 6417, Texas, 213 pp

Carr NG, Komarek J, Whitton BA (1973) Notes on isolation and laboratory culture, In: Carr NG, Whitton, BA, (eds) The Biology of Blue-Green Algae, University of California Press, Berkeley, pp 525-530.

Chrismas M (1980) Ascospore discharge and germination in *Xanthoria parietina*. Lichenologist 12: 403-406

Deason TR, Bold HC (1960) Phycological studies I. Exploratory studies of Texas soil algae. Univ. of Texas Publication Nr. 60022

Galun M (1988) Handbook of Lichenology Vol 3. CRC Press, Boca Raton.

Garrett RM (1971) Studies on some aspects of ascospore liberation and dispersal in lichens. Lichenologist 5: 33-44

Hale ME (1983) The Biology of Lichens. Edward Arnold, London

Honegger R , Bartnicki-Garcia S (1991) Cell wall structure and composition of cultured mycobionts from the lichen *Cladonia macrophylla, Cladonia caespiticia, and Physcia stellaris* (Lecanorales, Ascomycetes). Mycol Res 95: 905-914

Honegger R, Kutasi V (1990) Anthraquinone production in three aposymbiotically cultured teloschistalean lichen mycobionts: the role of the carbon source. In: Nardon P, Gianinazzi-Pearson V, Grenier MA, Smith DC , Margulis L (eds) Endocytobiology IV. 4th International Symposium on Endocytobiology and Symbiosis, Lyon, July 4-8, 1989175-8. Paris: Institute National de la Recherche Agronomique

Honegger R, Kutasi V, Rufflier HP (1993) Polyol patterns in eleven species of aposymbiotically cultured lichen mycobionts. Mycol Res 97: 359

Kofler L (1970) A method to use spores in quantitative studies on germination. Bryologist 73: 602-606

Kratz WA, Myers J (1955) Nutrition and growth of several blue-green algae. Am J Bot 42: 282

Lallemant R , Bernard T (1977) Obtention de cultures pures des mycosymbiontes du *Lobaria laetevirens* (Lightf.) Zahlbr. et du *Lobaria pulmonaria* (L) Hoffm.: le roles des gonidies. Rev Bry Lichen 43: 303-308

Lawrey JD (1984) Biology of Lichenized fungi. New York, Praeger Publishers

Lilly VG, Barnett HL (1951) In Physiology of Fungi. McGraw-Hill, New York

Mathey A, Holder A (1978) Fluorescence, Iuminescence et pouvoir germinatif des spores dans la famille des Trypethéliacées (lichens pyrénomzcètes). Nova Hedwigia 30:79-110

Nakano T (1987) Isolation and culture methods of lichen photobionts. [In Japanese] Lichen 6: 1-3

Nakano T (1988) Phycobionts of some Japanese species of Graphidaceae. Lichenologist 20: 353-360

Nichols HW (1973) Growth media-fresh water. In: Stein JR (ed) Handbook of Phycological Methods. Culture Methods and Growth Measurements. Cambridge University Press, London, p 7

Oliver E, Crittenden PD, Beckett A, Brown DH (1989) Growth of lichen-forming fungi on membrane filters. Lichenologist 21:387-391

Ostrofsky A, Denison WC (1980) Ascospore discharge and germination in *Xanthoria polycarpa*. Mycologia 72: 1171-1179

Pyatt FB (1968) An investigation into conditions influencing ascospore discharge and germination in lichens. Rev Bryol Lichen 36:323-329

Pyatt FB (1969) Studies of the periodicity of spore discharge and germination in lichens. Bryologist 72: 48-53

Pyatt FB (1973) Lichen propagules. In: Ahmadjian V, Hale ME (eds) The Lichens. Academic Press, New York pp 117-145
Pyatt, FB (1976) Lichen ecology of metal spoil tips: Effects of metal ions on ascospore viability. Bryologist. 79: 172-179
Ramkær K (1978) The influence of salinity on the establishing phase of rocky shore lichens. Bot. Tidsskrift 72: 119-123
Richardson DHS (1971) Lichens. In: Booth C (ed) Methods in microbiology, vol 4. Academic Press, New York, pp 267-293
Roussard M (1969) Contribution a l'étude des ascospores de lichens et de leur germination. Bull Soc Mycol France 85: 355-366
Stanier RY, Kunisawa R, Mandel M,Cohen-Bazire G (1971) Purification and properties of unicellular blue-green algae (order Chroococcales). Bact Rev 35:171
Starr RC (1980) The culture collection of algae at the University of Texas at Austin. J Phycol 14: 483
Takeshita S, Nakano T, Iwatsuki Z. (1989) Phycobionts of some Japanese species of *Pertusaria* (Pertusariaceae). Pl. syst. Evol. 165: 49-54
Vobis G. 1977. Studies on the germination of lichen conidia. Lichenologist 9:131-136
Watanabe A (1960) List of algal strains in collection at the Institute of Applied Microbiology, University of Tokyo. J. Gen. Appl. Microbiol. 6: 283-292
Whiton JC, Lawrey JD (1982) Inhibition of *Cladonia cristatella* and *Sordaria fimicola* ascospore germination by lichen acids. Bryologist 85: 222-226
Whiton JC, Lawrey JD (1984) Inhibition of crustose lichen spore germination by lichen acids. Bryologist 87: 42-43
Wiedeman VE, Walne PL, Trainor FR (1964) A new technique for obtaining axenic cultures of algae. Can J Bot 42: 958
Yamamoto Y (1987) Tissue cultures of lichens. Proceedings of symposium on Tissue Culture of Lichen and Bryophytes. Osaka, Japan: Nippon Paint company, 14-25
Yamamoto Y, Mizuguchi R, Yamada Y (1985) Tissue cultures of *Usnea rubescens* and *Ramalina yasudae* and production of usnic acid in their cultures. Agric. Biol. Chem. 49: 3347-3348
Yamamoto Y, Kinoshita Y, Takahagi T, Kroken S, Kurokawa T, Yoshimura I (1998) Factors affecting discharge and germination of lichen ascospores. J. Hattori Bot. Lab. 85: 267-278
Yoshimura I, Kurokawa T, Kanda H (1990) Tissue culture of some Antarctic lichens preserved in the refrigerator. Proc. NIPR symp. Polar Biol. 3: 224-228
Yoshimura I, Kurokawa T, Yamamoto Y, Kinoshita Y (1993) Development of lichen thalli *in vitro*. Bryologist 96: 412-421

Culture of Thallus Fragments and Redifferentiation of Lichens

Yoshikazu Yamamoto, Yasuhiro Kinoshita, and Isao Yoshimura

Introduction

Lichens have been used as medicines, dyes, perfumes, and food and drink stuffs since ancient times all over the world. Within the last two decades many pharmacologically active compounds have been isolated from lichens. However, mass harvesting of lichens as an industrial resource may lead to extinction of species. Therefore, if lichens are to be used in industrial applications, they must be cultured in vitro.

A standard method for obtaining cultures of lichen mycobionts is to initiate them from their spores. However, this method has several disadvantages (reviewed by Ahmadjian, 1993; see also Chapter 1). For example, apothecia may not discharge spores, spores do not always germinate in vitro, and not all species regularly produce apothecia. In our laboratories, we have succeeded in isolating, culturing and maintaining mycobionts as well as photobionts of about 400 species of lichens.

Since Schwendener (1868) proposed that lichens are composed of fungal and algal symbionts, thallus resynthesis in vitro from isolated symbionts has been a challenge for lichenologists. Very few lichenologists have achieved this, and usually not under aseptic conditions (for review see Ahmadjian, 1973a).

If large amounts of cultured lichens need to be produced for physiological experiments to study the metabolism of secondary compounds, or even more so for industrial use, it is not advantageous to use pure my-

✉ Yoshikazu Yamamoto, Akita Prefectural University, Department of Biological Production, 241-7, Kaidobata-nishi, Shimoshinjyo-nakano, Akita, 010-0195, Japan (*phone* +81-18-872-1646; *fax* +81-18-872-1678;
e-mail yyamamoto@akita-pu.ac.jp)
Yasuhiro Kinoshita, Nippon Paint Co., Ltd., Basic Research Department, 1-15-4, Minamishinagawa, Shinagawa, Tokyo, 140-8675, Japan
Isao Yoshimura, Kochi Gakuen College, Biology Laboratory, 292 Asahitenjincho, Kochi, 780-0955, Japan

cobiont cultures in most cases. This is because the secondary compounds produced by the isolated mycobiont almost always differ considerably from those present in the intact thallus. Another problem when using mycobiont cultures for industrial purposes is that they grow slowly. Furthermore, it is very difficult and extremely time consuming to achieve resynthesis of a lichen thallus using isolated photobionts and mycobionts grown in axenic culture.

In 1985 we described a method for using thallus fragments, the "lichen tissue culture method" ("Yamamoto method") (Yamamoto et al., 1985). When we started to develop this method, we were concerned that our cultured thallus fragments would be contaminated. According to Ahmadjian's comment in the book "The Lichens", thallus-derived cultures are always contaminated by micro-organisms present in the thallus (Ahmadjian, 1973b). We found that we could reduce this problem by using carefully selected, very small thallus fragments (several hundred micrometers in size).

In 1990, Kon et al. and Yoshimura et al. independently reported the redifferentiation of microthalli in vitro. They succeeded in aseptically redifferentiating several *Usnea* species by culturing thallus fragments on agar-medium. Later, Yoshimura and Yamamoto (1991) succeeded in redifferentiating cyanobacterial lichens in vitro, e.g. *Peltigera* spp. In total, we have achieved thallus redifferentiation from thallus fragments of five genera (*Peltigera*, *Lobaria*, *Pannaria*, *Cladonia*, and *Usnea*) (Yamamoto, unpublished data).

In this chapter we describe the above method that uses thallus fragments to produce lichen thalli in vitro. In addition, this method can be used as a basis for deriving axenic mycobionts and photobiont cultures. Other chapters of this book refer to this method (Chapters 1 and 3). While the method of using thallus fragments for redifferentiation of lichens does not overcome the problem of slow growth rates, it is much faster than the method using isolated symbionts to resynthesize a lichen thallus.

Outline

The protocol of the culture method of thallus fragments is outlined in Fig. 1.

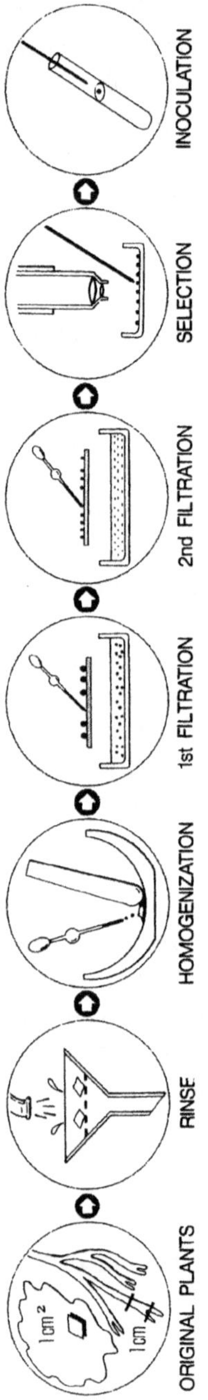

Fig. 1. The culture method of thallus fragments ("lichen tissue culture").

Materials

Note: You must carry out all treatments on a clean laminar flow bench or under sterile conditions, except for the pre-washing stage. All equipment should be autoclaved (15 - 20 min, 121°C, 1 atm) or oven-dried (30 min, 180°C) before use.

Equipment

- Autoclave
- Bamboo sticks (15 cm in length for "Yakitori" sold in Japanese stores).
- Laminar flow bench
- Incubator
- Nylon sieves (150 μm and 500 μm meshes, e.g. Kyoshin Riko Co, Tokyo)
- Dissecting microscope

Lichen material

Specimens can be either used within one week after collection from the field, or alternatively, put in a paper bag and stored in a freezer at -25°C within one week after collecting. Frozen specimens can be used for culture purposes following storage for more than one year.

Culture media

MY medium, Malt/Yeast extract medium (Ahmadjian, 1961)

Malt extract (Difco Laboratories, Detroit)	20 g
Yeast extract (Difco Laboratories, Detroit)	2 g
Agar	20 g
Deionised water or distilled water	make up to 1000 ml

LB medium, Lilly-Barnett medium (Lilly and Barnett, 1951)

Glucose	10.0 g
L-Asparagine	2.0 g
KH_2PO_4	1.0 g
$MgSO_4 \cdot 7H_2O$	0.5 g
$Fe(NO_3)_3 \cdot 9H_2O$	0.2 mg
$ZnSO_4 \cdot 7H_2O$	0.2 mg
$MnSO_4 \cdot 4H_2O$	0.1 mg
Thiamin hydrochloride	100 μg
Biotin	5 μg
Deionized water or distilled water	make up to 1000 ml

WA medium, Water-agar medium

Agar	20 g
Deionised water or distilled water	make up to 1000 ml

Autoclave all media before use, and in the laminar flow bench, pour 15 ml into Petri dishes or 5 ml into test tubes.

Procedure

"Culture of thallus fragments" and "Redifferentiation from thallus fragments or tissue cultures" comprise similar steps. The light conditions and the media used are important differences between these two methods. The former is carried out in the dark on MY or LB medium, but the latter is done in the light or dark/light regime on nutrient poor or nutrient free media. While the first method leads to formation of undifferentiated cell-aggregates, the second produces microthalli.

a) Culture of thallus fragments ("Lichen tissue culture method")

1. Separate a segment (c. 1 cm in length of a fruticose lichen or 1 cm^2 of a foliose or crustose lichen) from a thallus by scissors or knife and wash it in tap water for 30 min to 1 h.
2. In the laminar flow bench, transfer the segment to a mortar and homogenise with c. 3 ml sterile water (use a sterilised pipette).
3. Filter the homogenate through a nylon sieve mesh 500 μm. Discard what remains on the filter. Filter the solution through a nylon sieve mesh 150 μm. This double filtration removes small fragments of damaged cells and large thallus pieces. This produces thallus fragments with sizes between 150 and 500 μm.
4. From the nylon mesh pick up thallus fragments with a sterilised bamboo stick under the dissecting microscope. Inoculate onto the surface of slant MY agar-medium (5 ml) in 25 to 100 test tubes (10.5 cm in length). Place an (autoclaved) aluminium cap on top.
5. Maintain the tubes at 15°C in the dark in an incubator. Within one to two weeks rapidly growing contaminants such as fungi or yeasts will appear in some or many tubes and rapidly cover the thallus fragments. Rigorously remove contaminated test tubes from the incubator before their spores are spread.

6. Two weeks after inoculation, mycobiont hyphae and/or algae will grow out of the thallus fragments.
7. Six months later, transfer growing cell-aggregates composed of the mycobiont and photobiont onto fresh MY agar-medium in plastic Petri dishes (60 or 90 mm diameter).
8. Transfer cell-aggregates to fresh medium every 3 to 6 months and maintain at 15°C in the dark.

b) Redifferentiation of microthalli from thallus fragments

Repeat Steps 1 to 4 as described above, then:

1. Maintain the tubes at 15°C in the light (2.6 W/m^2) or under a light/dark regime in an incubator. Within 1 week rapidly growing contaminants such as fungi or yeasts will appear in some or many tubes. Discard all contaminated tubes.
2. Six months later, transfer growing microthalli that developed from undifferentiated cell-aggregates composed of the mycobiont and photobiont onto fresh MY or WA agar-medium in plastic Petri dishes (60 or 90 mm diameter).

c) Separation of symbionts

After Step 7 in procedure a), the cell-aggregates comprising mycobiont and photobiont can be used to isolate the symbionts. Afterwards they can be grown axenically, and if required, used in resynthesis experiments.

1. Transfer small aggregates to a mortar and homogenise with sterile water (1 to 3 ml).
2. Dilute 1 ml of homogenised solution with sterile water ten times.
3. Spread 1 ml of solution onto a MY agar-plate in a Petri dish (90 mm in diameter) and incubate at 15 to 20°C.
4. After three to six months colonies appear on the surface of the medium. Pick up greenish colonies of photobionts and filamentous ones of mycobionts separately and transfer onto slant test tubes. After one to three months, remove test tubes which are contaminated with the other symbiont and if all test tubes are contaminated, go back to Step 1.
5. Maintain uncontaminated test tubes as pure cultures of symbionts.

Results

Culture of thallus fragments

Contamination

Table 1 shows the extent of contamination of cultures of thallus fragments of 16 lichen species. Lichens that have a smooth surface or contain antibiotics such as *Cetraria islandica*, *Evernia prunastri*, *Usnea diffracta* and *U. longissima* show lower contamination rates. However, the contamination rate is higher in lichens with soredia and soralia, or growing on soil such as *Cladonia coccifera*, *C. pleurota*, *Ramalina yasudae* and *Usnea rubescens*. This indicates that foreign micro-organisms occur in the thalli of these species. Lichen mycobionts form compact colonies, and may produce thick and short hyphae or club-shaped aerial hyphae. However, it is difficult to distinguish between lichen mycobionts and foreign fungi appearing one month after inoculation.

Seasonal variation

Table 2 gives evidence that collecting season does not influence the contamination and growth of thallus fragments of *Usnea bismolliuscula* and *U. rubescens*.

Variation in collection sites

Table 3 demonstrates that the contamination and growth of thallus fragments were only slightly affected by collection locality in species collected at various localities in Japan. We have obtained similar results with other lichen genera, e.g. *Cladonia* sp., *Cetraria* sp. and *Umbilicaria* sp. (Yamamoto, unpublished results).

Culture conditions: light and temperature

Yamamoto et al. (1987) report influences of light and temperature on the growth of undifferentiated cell-aggregates. The effect of light on the initial growth of undifferentiated cell-aggregates depended on the tested species. The growth of undifferentiated cell-aggregates of *Alectoria ochroleuca*, an alpine lichen, was induced only at 15°C. However, growth of undifferentiated cell-aggregates of subtropical forest lichens such as *Ramalina boninensis* and *R. pacifica* could be induced at 25°C as well as at 15°C (Yamamoto, unpublished data).

Culture conditions: media

Since Ahmadjian originally used MY medium to culture lichen mycobionts of *Cladonia* species, we often used this medium. However, MY may not be suitable for mycobiont and thallus fragment culture of some lichen genera; these included *Anzia*, *Gymnoderma* and many cyanolichens (*Lobaria*, *Nephroma*, *Peltigera*, *Solorina*, *Sticta* etc.) (Yamamo-

Table 1. Induction and contamination of thallus fragment cultures derived from various lichens on MY medium at 15°C in the dark. Rate of growing colonies (RGC) = Number of test tubes with growing colonies x 100 / number of inoculated tubes (IT) - number of contaminated tubes (CT). Rate of contaminated colonies (RCC) = CT x 100 / IT

Lichen species	RGC (%)	RCC (%)
Acarospora fuscata	67	74
Alectoria lata	11	26
Alectoria ochroleuca	67	31
Cetraria islandica	100	0
Cladonia coccifera	100	80
Cladonia pleurota	100	76
Evernia prunastri	96	16
Menegazzia terebrata	28	22
Ramalina boninensis	100	32-41
Ramalina yasudae	86	86
Sphaerophorus melanocarpus	86	2
Usnea bismolliuscula	94-100	2-32
Usnea diffracta	20-100	1-7
Usnea longissima	72	2-6
Usnea rubescens	100	30-80
Xanthoparmelia taractica	100	46

Table 2. Effects of collection month on contamination and induction of thallus fragment cultures incubated on MY medium at 15°C in the dark. Rate of growing colonies (RGC) = Number of test tubes with growing colonies x 100 / number of inoculated tubes (IT) - number of contaminated tubes (CT). Rate of contaminated colonies (RCC) = CT x 100 / IT

Lichen species	Month	RGC (%)	RCC (%)
Usnea bismolliuscula	Feb.	100	2
	Aug.	100	20
	Oct.	100	25
Usnea rubescens	Feb.	90	71
	Apr.	100	74
	Oct.	100	80

Table 3. Effects of locality on contamination and induction of thallus fragment culture incubated on MY medium at 15°C in the dark. Rate of growing colonies (RGC) = Number of test tubes with growing colonies x 100 / number of inoculated tubes (IT) - number of contaminated tubes (CT). Rate of contaminated colonies (RCC) = CT x 100 / IT

Lichen species	Locality	RGC (%)	RCC (%)
Evernia esorediosa	Hokkaido	27-60	36-53
	Nagano	80	12
	Tochigi	85-87	31-70
Usnea longissima	Hokkaido	72	2
	Tochigi	100	10
	Nagano	85	6
Usnea rubescens	Ehime	100	78
	Kyoto	90	71
	Nagano	97	30
	Wakayama	96	48

to, unpublished data). We observed that agar inhibited germination of lichen ascospores of *Letharia* species (Yamamoto et al. 1998).

Culture conditions: hormones

Phytohormones such as auxins, cytokinins and gibberellins regulate the growth and differentiation of plant cells. However, supplementing cultures of thallus fragments with phytohormones, vitamins and other organic compounds do not significantly promote growth (Yamamoto et al. 1987, and Yamamoto, unpublished data).

Storage of collected lichens

It is important to know how long lichens remain viable in the laboratory after collection. Table 4 shows effects of storage period at various temperatures on fragment growth in an incubator or refrigerator. When lichen thalli were stored at 25°C in the incubator, *Usnea* species died within one month, while other species, e.g., *Rimelia* and *Parmotrema* were still alive. When stored at -25°C in the refrigerator many lichens remained alive for 1 to 3 years, and for 5 years when stored at -80°C (Yamamoto, unpublished data).

Table 4. Effects of storage at various temperatures on induction of thallus fragment cultures incubated on MY medium in the dark. Rate of growing colonies (RGC) = Number of test tubes with growing colonies x 100 / number of inoculated tubes (IT) - number of contaminated tubes (CT)

Lichen species	Storage Temp.	Storage Period (months)	RGC (%)
Usnea rubescens	25°C	1	0
	5°C	2	100
	5°C	4	0
Usnea bismolliuscula	25°C	1	0
	5°C	2	100
	5°C	4	0
Rimelia clavulifera	25°C	1	95
	25°C	4	3
Parmotrema austrosinesis	25°C	1	86
	25°C	4	52

Table 5. Effects of temperature on redifferentiation of a lichen thallus from thallus fragments of *Usnea confusa* spp. *kitamiensis* incubated on MY medium in the light (2.6 W m^{-2}) for 150 days.

Temperature	Microthallus Formation Rate (%)
21°C	0
18°C	100
15°C	70
10°C	10

Redifferentiation of microthalli from thallus fragments or cell-aggregates

Effect of media

Yoshimura et al. (1990) reported the effects of nutrients in the medium on redifferentiation of *Usnea rubescens* and *Peltigera praetextata* from thallus fragments. *U. rubescens* preferred modified medium with less glucose and asparagine than in standard LB medium, and *P. praetextata* grew better on WA medium than MDM and MY medium. Kon et al. (1997) studied the effect of agar concentration in the MY medium on fragment culture of *U. confusa* ssp. *kitamiensis*. Lower water content (i.e. higher agar concentration) remarkably affected redifferentiation of microthalli.

Effects of temperature

Thallus development in nature is presumably optimal between 15 to 20°C. Table 5 shows the effect of culture temperature on redifferentiation of thalli from fragments of *U. confusa* ssp. *kitamiensis* (Kon et al. 1990). They indicated that optimum temperature for redifferentiation of this species was 18°C.

Troubleshooting

- Microbial contamination
  Contamination of the lichen thallus depends on their habitat. Thalli growing on soil are most difficult to culture. When only a few test tubes remain one month after inoculation, the specimen or species may be too difficult to culture. If fragments show high contamination rates, a filter with smaller mesh size may be useful. You may also try to vary the media used, the temperature or light/dark regime, or try to work with another specimen or species.
- Contamination by mites
  Usually lichen mycobionts as well as fungi attract mites that introduce contaminants into lichen cultures. Freezing Petri dishes where mites live on the agar plate at -25°C overnight can kill mites easily without damage to lichen symbionts.
- Fragments fail to grow
  If growth of thallus fragments from microlichens and cyanolichens cannot be induced using LB and MY medium, alternative media for culturing these lichens must be investigated. See also Chapter 3.

Comments

Growth factors

- Lichen cultures including fragments grow very slowly *in vitro*. These growth rates do not satisfy the requirements for industrial mass production of lichens. Therefore, we encourage other researchers to study mechanisms regulating lichen growth.

Effect of the alga on morphogenesis

- Kinoshita et al. (1993a) studied the capacity of the *Usnea hirta* mycobiont to form a thallus. The mycobiont formed small and branched microthallus-like filaments on MY agar-medium. However, they did not enlarge even after several months. This suggests that the algal partner plays an important role in thallus differentiation. Another approach to studying effects of the algae on lichen morphology are resynthesis experiments, where the original algal symbiont of a lichen is exchanged for that of a different species (Ahmadjian et al. 1980, Kon et al. 1993).

Studying photosymbiodemes

- Resynthesis of lichens from isolated mycobionts and photobionts can also be applied to study photosymbiodemes. For instance Yoshimura et al. (1994b) reported redifferentiation of *Peltigera aphthosa*, a cyanolichen with green algae. For more details see Chapter 3.

Applications

Production of secondary metabolites

Cultured thallus fragments and mycobiont cultures do not necessarily produce the same substances as intact lichens. Yoshimura et al. (1994a) investigated 100 samples of mycobiont and thallus fragment cultures. Twenty samples of these produced the same compounds as those found in natural thalli. Sixty of these produced unidentified substances that were different from compounds present in thalli collected from the field, while the last twenty failed to produce lichen substances.

Of course, culture conditions remarkably influence the production of secondary metabolites in cultures (see Chapter 18). It is well known that the content of some secondary metabolites depends on the extent of differentiation in tissue culture of higher plants. Kinoshita et al. (1993b) reported that the content of usnic acid (a dibenzofuran), a well-known lichen secondary metabolite, increased in proportion to the degree of morphological complexity in cultured thallus fragments of *Usnea hirta*.

This method may be a first step to commercially utilising lichens, although considerably more research is needed.

References

Ahmadjian V (1961) Studies on lichenized fungi. Bryologist 64: 168-179

Ahmadjian V (1973a) Resynthesis of lichens. In: Ahmadjian V, Hale M (eds) The Lichens. Academic Press, New York, pp 565-579

Ahmadjian V (1973b) Methods of isolating and culturing lichen symbionts and thalli. In: Ahmadjian V, Hale M (eds) The Lichens. Academic Press, New York, pp 653-659

Ahmadjian V (1993) The mycobiont (fungal symbiont). In: The Lichen Symbiosis. John Wiley & Sons, New York, pp 8-15

Ahmadjian V, Russel LA, Hildreth KC (1980) Artificial reestablishment of lichens. I. Morphological interactions between the phycobionts of different lichens and the mycobionts *Cladonia cristatella* and *Lecanora chrysoleuca*. Mycologia 72: 73-89

Kinoshita Y, Hayase S, Yamamoto Y, Yoshimura I, Kurokawa T, Ahti T, Yamada Y (1993a) Morphogenetic capacity of the mycobiont in *Usnea* (lichenized ascomycete). Proc Japan Acad 69B: 18-21

Kinoshita Y, Yamamoto Y, Yoshimura I, Kurokawa T, Yamada Y (1993b) Production of usnic acid in cultured *Usnea hirta*. Bibliotheca Lichenol 53: 137-146

Kon Y, Iwashina T, Kashiwadani H (1997) The effects of water content on the resynthesis, growth of mycobiont and photobiont and lichen substances production. Bull Natn Sci Mus, Tokyo, Ser B 23:137-142

Kon Y, Kashiwadani H, Kurokawa S (1990) Induction of lichen thalli of *Usnea confusa* Asah. ssp. *kitamiensis* (Asah.) Asah. *in vitro*. J Jap Bot 65: 26-32

Kon Y, Kashiwadani H, Masada M, Tamura G (1993) Artificial syntheses of mycobionts of *Usnea confusa* ssp. *kitamiensis* and *Usnea orientalis* with their natural and non-natural phycobionts. J Jap Bot 68:129-137

Lilly VG, Barnett HL (1951) Physiology of fungi. McGrow-Hill, New York

Schwendener S (1868) Über die Beziehungen zwischen Algen und Flechtengonidien. Bot Ztg 26: 289-292

Yamamoto Y, Kinoshita Y, Takahagi T, Kroken S, Kurokawa T, Yoshimura I (1998) Factors affecting discharge and germination of lichen ascospores. J Hattori Bot Lab 85: 267-278

Yamamoto Y, Mizuguchi R, Yamada Y (1985) Tissue cultures of *Usnea rubescens* and *Ramalina yasudae* and production of usnic acid in their cultures. Agric Biol Chem. 49: 3347-3348

Yamamoto Y, Mizuguchi R, Yamada Y (1987) Effects of culture conditions on growth of Usneaceae lichen tissue cultures. Plant Cell Physiol 28: 1421-1426

Yoshimura I, Kurokawa T, Kinoshita Y, Yamamoto Y, Miyawaki H (1994a) Lichen substances in cultured lichens. J Hattori Bot Lab 76: 249-261

Yoshimura I, Kurokawa T, Yamamoto Y, Kinoshita Y (1990) Thallus-formation of *Usnea rubescens* and *Peltigera praetextata* in vitro. Bull Kochi Gakuen Coll 21: 565-576 (in Japanese)

Yoshimura I, Kurokawa T, Yamamoto Y, Kinoshita Y (1994b) In vitro development of the lichen thallus of some species of *Peltigera*. Crypt Bot 4: 314-319

Yoshimura I, Yamamoto Y (1991) Developent of *Peltigera praet*extata lichen thalli on culture. Symbiosis 11: 109-117

Suppliers

Kyoshin Riko Co. - 5-13-1, Hongo, Bunkyo, Tokyo, 113 Japan.
Phone +81-3-3813-1073.

Abbreviations

LB	Lilly-Barnett medium
MY	malt-yeast extract medium
WA	water-agar medium

Resynthesis of Photosymbiodemes

ELFIE STOCKER-WÖRGÖTTER

Introduction

In foliose lichens, sometimes two phototypes are formed by probably the same mycobiont which lives in symbiosis with either green or cyanobacterial photobionts. Moreover, the same mycobiont can also form two different morphotypes, which are present as cyanobacterial fruticose thalli growing on green lobate thalli. This occurs particularly in the genera *Peltigera*, *Lobaria*, *Nephroma*, *Sticta* and *Pseudocyphellaria*. The "one fungus-two photomorphs" hypothesis (Goffinet and Bayer 1997) implies that one lichen fungus can live in a symbiotic relationship with either (prokaryotic) cyanobacteria and/or (eukaryotic) green algae. The often-used term "photosymbiodeme", derived from the former phycosymbiodeme (Renner and Galloway 1982), includes lichens forming dimorphic thalli, thalli with different green algal or cyanobacterial photobionts ("cyanobionts") and cephalodiate thalli. The origin of photosymbiodemes is far from understood, as is their presence or absence in natural ecosystems. Perhaps symbiotic ascomycetous fungi evolved a high phenotypic plasticity that allowed them to respond to changing conditions in their environment. Field studies of *Sticta filix* revealed that light intensity determines whether the green phototype or the cyanotype differentiates (Poelt 1986). Between the green and cyanobacterial morphotypes, mixed phototypes (photosymbiodemes) are found.

Mycobionts of triple symbiotic systems show obligate "cyanotrophy" (Poelt and Mayrhofer 1988). Cyanotrophy may be defined as a dependence of some lichen fungi on nutrients provided by cyanobacteria. Many of these lichens have large thalli and grow on nutrient-poor substrates. They depend on an effective nitrogenase system occurring in the heterocysts of the cyanobionts. This system has been analysed in cephalodia and

Elfie Stocker-Wörgötter, University of Salzburg, Institute of Plant Physiology, Hellbrunnerstr. 34, Salzburg, 5020, Austria (*phone* +43-662-8044-5528 or 5594; *fax* +43-662-645450; *e-mail* elfriede.stocker@sbg.ac.at)

various cyanobacterial thalli by Feige (1976a,b), Rai et al. (1980) and Rai (1988). The isolation of "cyanotroph" mycobionts has always been a challenge, as they need contact with cyanobacteria to start growth. One method involves growing them in contact with isolated cyanobacterial colonies. Alternatively, it may be possible to find conditions that compensate for the presence of cyanobacteria by growing them on media enriched with nitrogenous compounds.

During the past twenty years, many new photosymbiodemes have been described and investigated (e.g. James 1975; James and Henssen 1976; Brodo and Richardson 1978; Tønsberg and Holtan-Hartwig 1983, Ott 1988; Armaleo and Clerc 1991; Holtan-Hartwig 1993; Vitikainen 1994; Stocker-Wörgötter and Türk 1994; Stocker Wörgötter 1995, 1998; Goffinet and Bayer 1997, Goffinet and Goward 1998). Recently, studies in photosymbiodemes have entered a new era. Modern laboratory manipulations allow the dissociation and resynthesis of lichens in an artificial environment under controlled laboratory conditions. These investigations are of great interest for answering questions on symbiont specificity, formation of secondary compounds (lichen substances) in resynthesised thalli and also for an understanding of the evolution of symbiotic systems in general. In the future, molecular biology techniques may be used to create new symbioses with predetermined characters, but this will require a detailed knowledge about factors that influence the establishment of natural and artificial symbiotic associations.

Early resynthesis experiments (e.g. Ahmadjian 1989; Ahmadjian and Jacobs 1982, 1983, 1985, 1987) mainly dealt with lichens where one mycobiont lives together with one type of photobiont. However, in a few studies triple and multiple symbioses have been examined (Brodo and Richardson 1978, Honegger 1982, Ott 1988), and most of these investigations

Fig. 1. Resynthesis of *Peltigera britannica* and *Lobaria fendleri*. **1** *Peltigera britannica* photosymbiodeme in nature. Left: phototype with subfoliose cephalodia; right: cyanobacterial photomorph with green cups; bar = 5 mm; **2** *Nostoc* isolate (*Peltigera britannica*) with gelatinous sheath and heterocysts (arrow); bar = 10 µm; **3** *Coccomyxa* isolate (*Peltigera britannica*) in liquid BBM; bar = 11 µm; **4** Mycobiont (*Peltigera britannica*) growing on cyanobacterial colonies; bar = 600 µm; **5** Mycobiont (*Peltigera britannica*) on MIX medium; bar = 800 µm; **6** Green photomorph (*Peltigera britannica*) protruding from cyanobacterial photomorph in nature; bar = 4 mm; **7** Resynthesised green photomorph (14 months old) growing on the margin of a cyanobacterial photomorph (*Peltigera britannica*; cephalodia are starting to develop, arrow); bar = 600 µm; **8** *Lobaria fendleri* photosymbiodeme, natural thallus bearing coralloid cyanobacterial photomorph (arrow); bar = 600 µm; **9** Resynthesized cyanobacterial photomorph (*Lobaria fendleri*) on soil substrate; bar = 400 µm.

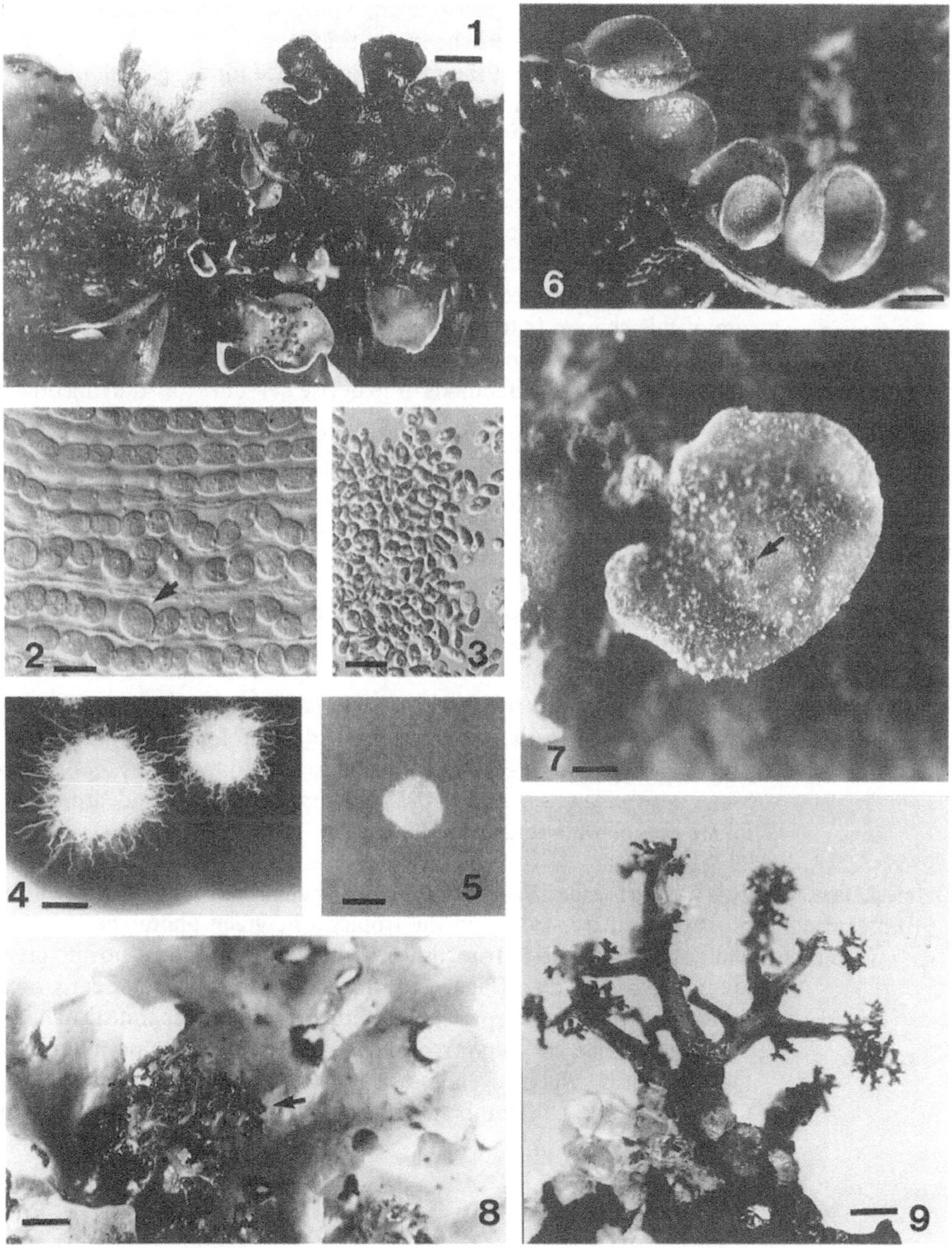
1
2
3
4
5
6
7
8
9

refer to observations of developmental processes, ultrastructure and secondary compounds of lichens in the field.

Our knowledge about the conditions needed for the growth of photosymbiodemes is very limited, and no general method exists for photosymbiodeme resynthesis. It is necessary to modify traditional techniques for isolating and culturing the individual symbionts, and the intact lichens. The methods presented in this chapter refer to two photosymbiodemes (*Peltigera britannica* and *Lobaria fendleri*) that have been repeatedly resynthesised during the past five years. Voucher specimens and resynthesised thalli have been compared by chemical methods (TLC, HPLC). The aim of this chapter is to review the information that is available on the resynthesis of these two particular phytosymbiodemes, and to encourage workers to adapt these methods or to devise new ones for studying others.

Materials

Plant material

Voucher specimen and cultures of the following species are shown in Figure 1 (1.1 - 1.9).

***Peltigera britannica* (Gyeln.) Holt. Hartw.** This species forms green and cyanobacterial phototypes (Fig. 1.1) in arctic habitats. Both photo- and cyanomorphs develop lobes. The adult green phototypes are sprinkled with cephalodia, formed by tiny cyanobacterial lobules. Juvenile green, cup-shaped phototypes often protrude from adult lobate cyanobionts (Fig. 1.6).

***Lobaria fendleri* (Tuck.ex.Mont.) Lindau** This species is especially common in moist and shady habitats, growing on moss-covered tree-trunks in the tropics. The green phototype is lobate, while the cyanotype is fruticose/shrubby and tightly fixed to the green lobes. *Lobaria fendleri* (Fig. 1.8) forms dimorphic thalli, and early taxonomists mistakenly interpreted the fruticose part as a separate lichen species that they named "*Dendriscocaulon*". This tropical photosymbiodeme, that comprises shrubby cyanomorphs growing from a green *Lobaria* thallus, is to some extent comparable to the *Lobaria amplissima* photosymbiodeme.

Culture media

Sabouraud 4% glucose agar

- 10 g polypeptone (casaminoacids)
- 40 g glucose
- 18 g agar

Make up to 1000 ml with double distilled water.

Malt Yeast (MY) extract medium

- 20 g malt extract
- 2 g yeast extract
- 20 g agar (Merck)

Make up to 1000 ml with double distilled water.

MIX medium

- 8 g peptone (from meat)
- 8 g peptone (from casein)
- 20 g malt extract
- 3 g yeast extract
- 5 g NaCl
- 40 g glucose
- 15 g Agar (Difco)

Make up to 1000 ml with double distilled water.

Murashige and Skoog Medium (modified) (MS)

- 2 g malt extract
- 2 g caseine hydrolysate
- 20 g mannitol
- 40 g sucrose
- 18 g agar
- 1 mg Murashige mineral salts

Make up to 1000 ml with double distilled water.

A useful medium for the isolation of the photobiont of *Peltigera britannica* (*Coccomyxa sp.*) is MY agar (Honegger and Kutasi 1990; Yamamoto 1990). For culture, grow isolated *Coccomyxa* and *Dictyochloropsis* colonies in BBM (Bold's Basal Medium, Bischoff and Bold 1963; Deason and Bold 1960).

Bold's Basal medium

6 macro-element solutions:

- 1 g $NaNO_3$
- 1 g $CaCl_2$
- 3 g $MgSO_4 \cdot 7H_2O$
- 3 g K_2HPO_4
- 7 g KH_2PO_4
- 1 g NaCl

Make up each solution to 400 ml double distilled water.

4 trace-element solutions:

- 11.42 g H_3BO_3
- 4.98 g $FeSO_4 \cdot 7H_2O$, 8.82 g $ZnSO_4 \cdot 7H_2O$, 1.44 g $MnCl_2 \cdot 4H_2O$
- 0.71 g MoO_3, 1.57 g $CuSO_4 \cdot 5H_2O$, 0.49 g $Co(NO_3)_2$
- 50 g EDTA (Titriplex II), 31.0 g KOH

Make up each solution to 1000 ml double distilled water.

Take 10 ml of each macro-element solution and 1 ml of each trace-element solution and make up to 1000 ml with double distilled water.

Subprotocol 1
Isolation of Symbionts

Procedure

Mycobionts

The following procedure is based on the method of Yamamoto (1990), see also Chapter 2.

1. Wash single thallus lobes from all different photo- and cyanomorphs (*Peltigera britannica, Lobaria fendleri*) in sterile double distilled water and add a drop of Tween 80 (a surfactant). Change the water several times during a washing procedure lasting for about 4 h.
2. Select clean fragments under a dissecting microscope and grind gently using a homogeniser at low speed (5000-10000 rpm for 15 sec) or a sterile mortar and pestle.
3. The resulting suspension should contain minute fragments of algae and fungi. Filter using sieves of two different meshes (sizes: 500 μm and then 150 μm).
4. Pick up approx. 150 μm pieces using inoculation needles or bamboo sticks under a dissecting microscope. Inoculate agar slants with one piece of every sample in each tube. The original Yamamoto method (Yamamoto 1990) used MY agar for all mycobiont isolations. My recent experiments have shown that many mycobionts grow better on other media.
5. *Peltigera* and *Lobaria* mycobionts are well known for their cyanotrophy (Stocker-Wörgötter and Türk 1994) and do not show any growth reactions on MY medium. Rather, first grow these mycobionts together with cultured colonies of their natural photobionts (Fig. 1.4). Later, introduce a very rich nutrient medium (MIX or MS) to compensate for the presence of the photobionts by providing the mycobiont with nitrogenous compounds.
6. Keep the mycobiont cultures for 2 - 3 months covered by aluminium foil to suppress growth of the algae or cyanobacteria present in the fragments. Keep the temperature low (approx. 10 - 15°C) for mycobiont isolates of temperate species like *Peltigera britannica*. For tropical lichens e.g. *Lobaria fendleri* raise the temperature to 23°C.

7. After 3 months, select mycelia, free of contamination and transfer them using an inoculation needle to new medium (MIX medium).

Green photobionts

8. I also recommend the Yamamoto method (Yamamoto 1990 and Chapter 2) for isolating the algae. In this case, keep agar slants containing isolations of algae under a light regime of 50 -100 µmol photons m^{-2}. Within 2 to 3 weeks algal colonies develop next to the fungus.

9. If the isolate is free of contaminants, the algae/cyanobacteria can be easily removed by an inoculation needle and transferred to a new nutrient medium, e.g. BBM or MY medium. *Coccomyxa*, the green photobiont of *Peltigera britannica* has been isolated by this method (Fig. 1.3). The green photobiont of the main thallus of *Lobaria fendleri*, probably a species of the algal genus *Dictyochloropsis* can be grown in liquid BBM (with soil extract). To prepare soil extract, dilute 500 g soil in 2000 ml double distilled water and autoclave two times for 3 h. Adjust the pH of the supernatant to pH 6, filter under sterile conditions and keep this stock solution in the refrigerator.

Cyanobacteria

10. Isolate the cyanobionts, e.g. *Nostoc* from *Peltigera britannica* and *Lobaria fendleri*, by a different method (modified after Ahmadjian 1973). Crush a carefully washed thallus fragment in a drop of sterile water between two microscope slides.

11. Examine the resulting cyanobacterial suspension under a dissecting microscope. Pick up the cyanobionts using a micropipette and transfer them to a petri dish containing BBM or MY agar.

12. The first isolates of cyanobacteria are very often highly contaminated by bacteria. The solid substrate allows to localise the contaminants and to get rid of them through successive transfers and subcultures.

13. Alternatively, another useful isolation method for cyanobacteria involves using vegetative reproductive units of the filamentous cyanobacteria (e.g. *Nostoc, Stigonema*), termed hormogonia. Hormogonia usually have no gelatinous sheath, and can move on agar surfaces if they are attracted by a light source. Movement on or through the agar substrate often loosens attached bacteria (Boissière et al. 1987). Isolating hormogonia leads more rapidly to axenic cultures of the required *Nostoc* colonies than following Steps 10 - 12.

14. Within two weeks, isolated hormogonia of *Nostoc* can develop new filaments (Fig. 1.2). For *Peltigera britannica*, grow the cyanobionts and also the green photobionts at 15°C during a 14 h day and at 10°C during a 10 h night. Grow cyanobacteria and green algae isolated from *Lobaria fendleri* at 28°C during a 12 h day, and at 23°C during a 12 h night. Keep the light intensity at 50-100 μmol photons $m^{-2}s^{-1}$.

Subprotocol 2
Resynthesis

Procedure

1. Collect soil from the habitat of the lichen you want to resynthesise (e.g. *Peltigera aphthosa*).
2. Sieve 500 g of soil (mesh 1 - 2 mm) and then add 100 ml of double distilled water.
3. Add moist soil to 100 x 15 mm glass Petri dishes to a height of 8 mm, autoclave and leave for 24 hours, then autoclave again.
4. Inoculate the soil with axenically cultured photobionts and cyanobionts, and maintain the cultures for 3 - 4 weeks under the above described culture conditions (Step 14 in Subprotocol 1).
5. To get the fungal isolates ready for resynthesis, transfer them to a sugar deficient liquid medium (e.g. BBM) and maintain them there for about one month.
6. Spread homogenised, segmented hyphal filaments over the algal/cyanobacterial colonies.
7. Maintain the resynthesis cultures in a culture chamber adjusted to the conditions described for their photobionts.

Results

Peltigera britannica

Axenic *Nostoc* (isolated from hormogonia) and *Coccomyxa* colonies were obtained after 3 or 4 subcultures on agar plates (Fig. 1.2, 1.3). Further culturing was conducted in liquid BBM, where both photobionts were characterised by high growth rates that allowed them to be subcultured every

3 - 4 weeks. Hyphal growth was first observed on Sabouraud 2% glucose agar plates inoculated with the green or cyanobacterial bionts (Fig. 1.4). *Nostoc* colonies promoted the development of the mycobiont into a mycelium. After 3 months, washed and fragmented mycelia were transferred to MIX medium. On this very nutrient-rich medium, mycobiont growth occurred without contact with the cyanobacteria. After 6 months of culture, fungal colonies of an average size of 1 cm diameter (from photo- and cyanomorphs) were available for initiating a resynthesis experiment.

Resynthesis and relichenization only took place on the soil substrate. The very early developmental stages were tiny, cyanobacterial/fungal globules, obviously a "hormocyste" infected by mycobiont hyphae. After approximately 6 months, these early associations slowly differentiated into juvenile, lobate cyanotypes. After 8 to 12 months small stages of the green phototype developed on a layer of *Nostoc* colonies. In one case, a cup-shaped green thallus differentiated at the margin of a cyanobacterial thallus lobe (Fig. 1.7). After 14 months the green cup was colonised by fungus-infected *Nostoc* colonies that grew into tiny lobules. Comparable green cups, with subfoliose cephalodia protruding from the cyanotypes, are common in field material (Fig. 1.6).

Lobaria fendleri

The mycobiont of *Lobaria fendleri* grew very slowly on MIX and Sabouraud media. On MS agar the growth rates (3 - 5 mm increase of radial growth per month) were comparable with those of the *Peltigera britannica* mycobiont on MIX medium. For resynthesis, the fragmented mycobiont isolates were transferred to the soil substrate and inoculated on a mixture of *Dictyochloriopsis* and *Nostoc* cells. After 4 months the substrate was covered by minute globose primordia of the cyanotype. The green colonies had become invisible and were completely overgrown by *Nostoc*. After a further period of two months, the cyanotype-primordia had differentiated into branched structures of 2 - 3 mm height (Fig. 1.9). They were comparable to the juvenile stages of the cyanomorph, normally growing on well-developed green lobes of *Lobaria fendleri* (Fig. 1.9). After one year, the substrate was colonized by well-developed shrubby cyanomorphs. Resynthesis of the green photomorph, which normally grows as an epiphyte on tree-trunks under tropical conditions, has not yet been achieved in an in vitro experiment.

Analysing secondary compounds

It is very interesting to test if resynthesised thalli form the whole spectrum of secondary compounds as the voucher specimens did, especially if cultures of the isolated symbionts do not. Methods for analysing for secondary metabolites are outlined in Chapters 17 and 18. Results for the two photosymbiodemes studied here can be summarised as follows:

Peltigera britannica

- Voucher specimen
  Tenuiorin, methyl-gyrophorate, trace of gyrophoric acid; triterpenoides: trace of zeorin, 2 unidentified, phlebic acid A and B.
- Green algal photomorph
  Tenuiorin, methyl-gyrophorate, trace of zeorin, other triterpenoides: phlebic acid A and B.
- Cyanobacterial photomorph
  Tenuiorin, methyl-gyrophorate, trace of zeorin, trace of gyrophoric acid.
- Lobate cephalodium
  Tenuiorin, trace of of zeorin.
- Cultured mycobiont
  Methyl-lecanorate, methyl-gyrophorate.
- Green algal morphotype (resynthesised)
  Tenuiorin, methyl-gyrophorate, trace of zeorin, but no other triterpenoids.
- Cyanobacterial photomorph (resynthesised)
  Tenuiorin, methyl-gyrophorate, methyl-lecanorate.
- Green cup fixed to cyanobacterial photomorph (resynthesised)
  Tenuiorin, methyl-gyrophorate, trace of zeorin, all other triterpenoids.

Lobaria fendleri

- Green algal photomorph
  Gyrophoric acid, 4-O-methylgyrophoric acid, first isolated from *Lobaria dissecta* (Sw.) Räusch from Central America (Culberson 1970).
- Coralloid cyanobacterial photomorph
  No secondary compounds.

- Cultured mycobiont
  Gyrophoric acid (found for the first time in a mycobiont culture); formed on aerial hyphae (crystals: fine curved needles) after 6 months.
- Cultured cyanobacterial photomorph
  No secondary compounds.

The voucher specimen of *P. britannica* contains mainly tridepsides and triterpenoides. Tenuiorin, the most prominent tridepside in species of the genus *Peltigera* was present in the different photomorphs and also in the lobate cephalodia. However, the cultured mycobiont did not form tenuiorin, but methyl-lecanorate and methyl-gyrophorate. In all resynthesised stages tenuiorin was found, but the pattern of triterpenoides except zeorine clearly deviated from the voucher photomorphs. These results indicate that the resynthesised stages in lab culture always lack compounds; the whole variety of secondary compounds in *P. britannica* seems to be present only in a fully developed and differentiated symbiotic system. In the case of *Lobaria fendleri*, the voucher specimen has gyrophoric acid and 4-O-methyl-gyrophoric acid (a substance found in a few species of *Lobaria* from North America, Latin America, Madeira, East Africa and Japan). Surprisingly, the cultured mycobiont formed the major compound of *L. fendleri*, gyrophoric acid, but did not produce the 4-O-methylated form, found in the voucher specimen. Both, the resynthesized cyanobacterial photomorph and the natural cyanomorph did not produce any secondary substances. It remains uncertain why the mycobiont from *Peltigera* can produce a tridepside when it lives together with cyanobacteria and green algae, whereas the mycobiont from *Lobaria* obviously can form tridepsides only in symbiosis with green algae. This is particularly surprising, because the chemical structures of tenuiorin and gyrophoric acid are very similar.

Comments

My experiments have shown that mixing of the symbionts in BBM before inoculation is less effective. Resynthesis and formation of pre-thallus stages are promoted if the hyphal isolates come into contact with division stages of the algae/cyanobacteria. In the case of *Nostoc*, hormocysts (persistent stages) are often formed on soil. Perhaps surprisingly, hyphae seem to preferentially envelop these cell types, possibly explaining why the primordial stages of cyanobacterial lichens are almost globose. In contact with hyphae, the hormocysts slowly transform and release new vegetative filaments thus initiating the next, more advanced stage of development.

Acknowledgements. I am very grateful to the "Fonds zur Förderung der wissenschaftlichen Forschung" for supporting this investigation by the grant P12789-BIO. Furthermore I would like to thank Chicita F. Culberson and Anita M. Johnson (Duke University, USA) for identifying the compounds of *Lobaria fendleri*.

References

Ahmadjian V (1989) Studies on the isolation and bionts of the lichen *Peltigera canina* (Peltigeraceae). Pl Syst Evol 165: 29-39

Ahmadjian V, Jacobs JB (1982) Artificial reestablishment of lichens. III. Synthetic development of *Usnea strigosa*. J Hattori Bot Lab 52: 393-399

Ahmadjian V, Jacobs JB (1983) Algal fungal relationships in lichens: recognition, synthesis and development. In: Goff, JL (ed) Algal symbiosis. Cambridge University Press, Cambridge, pp 147-172

Ahmadjian V, Jacobs JB (1985) Artificial reestablishment of lichens. Comparison between natural and synthetic *Usnea strigosa*. Lichenologist 17: 149-165

Ahmadjian V, Jacobs JB (1987) Studies on the development of synthetic lichens. Bibliotheca lichenologica 25: 47-58

Armaleo D, Clerc P (1991) Lichen Chimeras: DNA Analysis suggests that one fungus forms two morphotypes. Exp Mycol 15: 1-10

Bischoff HW, Bold HC (1963) Some soil algae from enchanted rocks and related species. Phycol Studies 4: 1-95

Boissière JC, Boissière MC, Champion-Arnaud P, Lallemant R, Wagner J (1987) Le cycle de *Nostoc* des genres *Peltigera* et *Collema* en cultures in vitro et dans le thalle lichénique. Can J Bot 65: 1468-1477.

Brodo IM, Richardson DH (1978) Chimeroid associations in the genus *Peltigera*. Lichenologist 10: 157-170

Culberson CF (1970) Supplement to "Chemical and Botanical guide to lichen products". Bryologist 73: 177-377

Deason TR, Bold HC (1960) Phycological studies. I. Exploratory studies of Texas soil algae. Univ Texas Public 6022: 70.

Feige B (1976a) Untersuchungen zur Physiologie der Cephalodien der Flechte *Peltigera aphthosa* (L) Willd. I. Die photosynthetische ^{14}C-markierung der Lipidfraktion. Zeit Pflanzenphysiol 80: 377-385

Feige B (1976b) Untersuchungen zur Physiologie der Flechte *Peltigera aphthosa* (L) Willd. II. Das photosynthetische ^{14}C-Markierungsmuster und der Kohlenhydrattransfer zwischen Phycobiot und Mycobiot. Zeit Pflanzenphysiol 80: 386-394.

Goffinet B, Bayer RJ (1997) Characterization of mycobionts of photomorph pairs in the Peltigerinae (lichenized Ascomycetes) based on ITS sequences of specifically amplified fungal ribosomal DNA. Fungal Gen Biol 21: 228-237

Goffinet B, Goward T (1998) Is *Nephroma silvae-veteris* the cyanomorph of *Lobaria oregana*? Insights from molecular, chemical and morphological characters. In: Glenn MG, Harris RC, Dirig R, Cole MS (eds) *Lichenographia Thomsoniana*: North American lichenology in honour of JW Thomson. Mycotaxon Ltd, Ithaca, New York pp 41-52

Holtan-Hartwig J (1993) The lichen genus *Peltigera*, exclusive of the *P. canina* group, in Norway. Sommerfeltia 15: 1-77

Honegger R (1982) Cytological aspects of the triple symbiosis in *Peltigera aphthosa.* J Hattori Bot Lab 52: 379-391

Honegger R, Kutasi V (1990) Anthraquinone production in three aposymbiotically cultured teloschistalean mycobionts: the role of the carbon source. In: Gianninazzi Pearson V, Nardon P, Margulis L, Smith DC (eds) Endocytobiology IV. Proceedings of the IVth International Colloquium on endocytobiology and symbiosis, Lyon. pp 175-178

James P (1975) Lichen chimeras. Reports of the British Museum (Natural History) 1972-74: 37-42.

James P, Henssen A (1976) The morphological and taxonomic significance of cephalodia. In: Brown DH, Hawksworth DL, Bailey RH (eds) Lichenology, Progress and Problems. Academic Press, London pp 27-77

Ott S (1988) Photosymbiodemes and their development in *Peltigera venosa.* Lichenologist 20: 361-368

Poelt J (1986) Morphologie der Flechten. Forschritte und Probleme. Ber Dtsch Bot Ges 99: 3-29

Poelt J, Mayrhofer, H (1988) Über Cyanotrophie bei Flechten. Pl Syst Evol 158: 265-281

Rai AN, Rowel P, Steward WDP (1980) NH_4^+assimilation and nitrogenase regulation in the lichen *Peltigera aphthosa* (L) Willd. New Phyt 85: 545-555

Rai AN (1988) Nitrogen metabolism. In: Galun, M (ed) Handbook of Lichenology. Vol I. CRC Press, Boca Raton pp 201-237

Renner B, Galloway DJ (1982) Phycosymbiodemes in *Pseudocyphellaria* in New Zealand. Mycotaxon 16 :197-231

Stocker-Wörgötter, E, Türk R (1994) Artificial resynthesis of the photosymbiodeme *Peltigera leucophlebia* under laboratory conditions. Crypt Bot 4: 300-308

Stocker-Wörgötter, E (1995) Experimental cultivation of lichens and lichen symbionts. Can J Bot (Suppl. 1): 579-589

Stocker-Wörgötter, E (1998) Culture methods and culture of selected tropical mycobionts and photobionts as exemplified by South American lichens. In: Marcelli, MP, Seaward MRD (eds) Lichenology in Latin America: history, current knowledge and applications. CETESB. Sao Paulo pp 145-154

Tønsberg T, Holtan-Hartwig J (1983) Phycotype pairs in *Nephroma*, *Peltigera* and *Lobaria* in Norway. Nordic J Bot 3: 681-688.

Vitikainen O (1994) Taxonomic revision of *Peltigera* (lichenized Ascomycotina) in Europe. Acta Bot Fen 152: 1-96

Yamamoto Y (1990) Studies of cell aggregates and the production of natural pigments in plant cell culture. Nippon Paint Publications

Abbreviations

BBM	Bold's Basal medium
MIX	Mixture of compounds derived from other media
MS	Murashige and Skoog medium
MY	Malt yeast medium

Protoplast Isolation from Lichen Mycobionts

YASUHIRO KINOSHITA

Introduction

Molecular biology has advanced remarkably, and gene transformation is now possible in many bacteria, yeasts, plants and animals. Biotechnological methods such as cell fusion have been so improved that it is now easy to establish cell hybrids. It is expected that such techniques will be applied to lichens in the near future. However, there remain barriers for applying these techniques, and one of them is protoplast isolation. There are two reports of protoplast isolation from lichen mycobionts (Ahmadjian et al., 1987; Kinoshita et al., 1991). The procedure outlined here is derived from these results.

Protoplast isolation is not an end in itself, but a stage in preparation for subsequent experiments. Because it is very difficult to make protoplasts aseptic after isolation, it is necessary to obtain them from axenically cultured mycobionts. The condition of the cultured mycobiont will of course affect the quality of protoplasts and the isolation efficiency, which in turn will affect subsequent experiments. Good quality cultured mycobiont will yield more protoplasts and less cell debris during protoplast isolation. Chapters 1 to 3 provide information on mycobiont culture.

Materials

Equipment

- Laminar flow bench
- Centrifuge
- Microscope
- Shaker

Yasuhiro Kinoshita, Nippon Paint Co., Ltd, Basic Research Department, 1-15, 4-Chome, Minamishinagawa, Shinagawa-Ku, Tokyo, 140-8675, Japan
(*phone* +81-3-3740-1142; *fax* +81-3-3740-1399; *e-mail* ponkichi@mtj.biglobe.ne.jp)

Chemicals

- Citrate-phosphate (CP) buffer (50 mM, pH 5.8)
- CP buffer containing 10 mM mercaptoethanol
- Novozym 234 (Novo Nordisk A/S). Alternatively use a mixture of cellulase (e.g. Meicelase, Meiji Seika Kaisha Ltd.), chitinase (e.g. from Sigma) and protease (e.g. Ficin, Wako Pure Chemical Industries Ltd).
- $MgSO_4$

Procedure

Note: All steps must be done under aseptic conditions.

1. Harvest cultured mycobiont cells in the logarithmic phase by filtration with nylon mesh of pore size 150 μm. In our studies involving *Cladonia cristatella* mycobiont subcultured every 3 weeks, an 8-day old culture was used.
2. Homogenate harvested cells gently with a pestle and mortar in CP-buffer. The purpose of this step is to break up cell aggregates. Take care not to damage mycobiont cells.
3. Filter mycelial homogenate with a nylon mesh of pore size 10 μm to remove cell debris.
4. Suspend the mycelia remaining on the filter in CP buffer containing 10 mM mercaptoethanol for 1 h at 25°C using a shaker (80 rpm).
5. Collect mycelia with a 10 μm nylon mesh.
6. Suspend mycelia in CP buffer and collect them with a nylon mesh. Repeat the washing at least twice to remove mercaptoethanol.
7. Suspend 0.8 g wet weight of mycelia in 80 ml of CP buffer containing 0.1% w/v Novozym 234 and 0.6 M $MgSO_4$ and incubate for 12 h at 25°C, 60 rpm. Novozym 234 can be substituted by the following mixture of enzymes: 1% w/v cellulase, 0.02% w/v chitinase and 0.1% w/v protease.
8. Filter mycelial suspension with nylon meshes of pore size 20, 10, and 5μm.
9. Centrifuge the filtrate (5000 rpm, 10 min) to collect protoplasts.

10. Suspend protoplasts with 80 ml of 0.6 M $MgCl_2$ gently and then collect protoplasts by centrifugation (5000 rpm, 10 min). Repeat this step (washing protoplasts) at least twice.

Results

The method described above was successfully applied to five species of mycobionts in the Cladoniaceae. More than ten thousand protoplasts were isolated from each gram fresh weight of cultured mycobiont. The size of the protoplasts ranged from 2 to 8 μm in diameter.

Troubleshooting

If protoplast isolation is not successful, e.g. if isolation efficiency is low or much contamination by cell debris results, I recommend that you change the experimental conditions. For example, try varying the times between sub-culturing the mycobiont, the timing of harvest following sub-culture, the number of washings with buffer before enzyme treatment, the components of enzyme solution, or the length of time you treat lichens with enzyme. It may be helpful to refer to methods described for the isolation of other symbiotic fungi. For instance, a protocol for isolation of mycorrhizal fungi is given by Hampp et al. 1998.

Comments

- The procedure outlined above produced the best results in the studies of Ahmadjian et al. (1987) and Kinoshita et al. (1991). However, both these studies were carried out on lichen mycobionts from the Cladoniaceae. Modification of several conditions in the experiment might be necessary for mycobionts from other lichen families.
- It is necessary to purify protoplasts before using them for gene transformation or other experiments because undigested cells and cell debris are present as contaminants. Filtration by paper filters, cheese cloths or nylon meshes or centrifugation with density gradient using Percol (Pharmacia, with which a density gradient can be made without changing osmotic pressure) may be helpful here. Optimum conditions have not yet been determined. Enzyme treatment will cause some da-

mage to mycobiont cells. To obtain viable protoplasts, isolation procedures should be carried out as quickly as possible.

- Culture of protoplasts is also still waiting to be studied. Adjustment of the osmotic pressure with sugar or sugar alcohol will be necessary in the medium. Information on sugar preference (Yamamoto et al. 1993) will be helpful for selecting an osmoticum.

References

Ahmadjian V, Chadeganipour M, Korien AM, Paracer S (1987) DNA and protoplast isolations from lichens and lichen symbionts. Lichen Physiol. Biochem. 2: 1-11

Hampp R, Stülten C, Nehls U (1998) Isolation and regeneration of protoplasts from ectomycorrhizal fungi. In Varma A (ed) Mycorrhiza Manual, Springer, Berlin, pp 115-126

Kinoshita Y, Hayase S, Yamamoto Y, Yamada Y (1991) Improvement of protoplast isolation from lichen mycobints. Agric Biol Chem 55: 1891-1892

Yamamoto Y, Miura Y, Higuchi M, Kinoshita Y, Yoshimura I (1993) Using lichen tissue cultures in modern biology. Bryologist 96: 384-393

Differentiation Processes in Lichens - *in Vivo* Cultivation and Transplantation Methods

SIEGLINDE OTT and HANS MARTIN JAHNS

Introduction

When the symbiotic nature of lichens was discovered, scientists attempted to culture these organisms. As the lichens consist of a mycobiont and a photobiont, an obvious aim was to separate the two partners and then re-synthesise intact lichens. Later experiments included the culturing of isolated symbionts (Ahmadjian 1973, 1989, Bubrick & Galun 1988) and the growth and transplantation of entire thalli (Brodo 1961, Armstrong 1993). All these experiments were less successful than expected, as the complexity of the interactions between the partners, and thus the sensitivity of the internal equilibrium of the symbiosis were underestimated. Although no culture method can solve all the scientific problems that exist in lichenology, several methods are now available that can be used successfully for different purposes. The most important areas of research where techniques for culturing lichens have been useful include:

1. The isolation and culturing of the separated symbionts in tissue culture (see Chapters 1 and 2).
2. Experiments on re-synthesis (see Chapter 3).
3. The growth of thalli from fragments and vegetative diaspores (see Chapter 2).
4. Transplantation of thalli.

The appropriate methods depend upon the questions being asked. Studies on lichen physiology, morphology, ecology and pollution all require dif-

✉ Sieglinde Ott, Universität Düsseldorf, Botanisches Institut, Universitätsstr. 1, Düsseldorf, 40225, Germany (*phone* +49-211-81-13537; *fax* +49-211-81-1288; *e-mail* otts@uni-duesseldorf.de)
Hans Martin Jahns

ferent experimental approaches. In this chapter we describe techniques for conducting transplantation experiments and in vivo cultivation of lichens.

Materials

- Silicon adhesive: silicone rubber (e.g. Aquaria sealant; N.V. Dow Corning S.A. Seneffe, Belgium) or other products without fungicide!
- Clay from local pottery. Look for material without lime.

Subprotocol 1
In Vivo Cultivation of Fragments and Diaspores in the Field

For in vivo cultivation of fragments and diaspores in the field, use soredia, isidia and small or larger fragments of thalli (Schuster et al. 1985). Larger fragments have the advantage of a sufficiently large biomass for uninterrupted growth. Smaller fragments frequently degenerate into an undifferentiated lump of tissue that consists of fungal hyphae and algal cells from which new thalli differentiate. Apart from size, the age of the tissues is also important. Comparisons of the growth of older parts of the thallus with younger marginal parts with meristematic zones can give insights into growth processes and the degree of determination. Isidia often germinate slowly. Possibly, in the intact lichen the thallus inhibits isidial growth, and residual inhibition is present in detached isidia. In some lichen species, globose isidia may become spatulate as a first step towards the differentiation of secondary lobes (e.g. in *Parmelia saxatilis*). The growth of these spatulate isidia is not inhibited in cultures (Klöckner 1998).

Procedure

1. Obtain small fragments by cutting lichen thalli. Alternatively, grind air-dried thalli in a mortar. Isidia can be cut individually or scraped from thalli in larger numbers. Soredia can be brushed from thalli and collected in a Petri dish.
2. As new thalli never develop from a single diaspore in nature, use a large number to ensure close contacts.

3. In the field, fragments and diaspores may be sown directly onto a natural substrate. Sow many propagules, as many diaspores will be lost through rain and wind action.

4. The propagules easily adhere to soil, but if you use stone as a substrate you must attach them with glue, or cover them with a net. Silicone adhesive (without fungicide!) is suitable, as it remains pliable even during extreme climatic conditions. Many cultivated lichens cover the adhesive and use it as a substrate. Sterile gauze or squares from nylon stockings may be used as net (Fig. 1).

5. On tree bark, nets and glue may also be used and the nets can be fixed with staples. The diaspores may also be placed in horizontal wedge-shaped cuts in the bark. In all cases, shading by the net or sowing into fissures provides optimal conditions for development, as most primordia prefer low light intensities and a good water supply.

6. Blocks of burned clay are also suitable for fieldwork. In the field, large numbers of these blocks should be placed adjacent to one another and surrounded with a metal fence to prevent displacement by rain or by small animals. If diaspores are not glued to the surface of the clay they may be washed off by rain. Blocks may be protected for a few weeks by a net placed about 30-50 cm above the cultures.

7. Initial developmental stages are best examined by SEM. The normal specimen holders for SEM with pieces of bark or rock glued to their surface may be used as a substrate. In the field, holders can be placed on plastic discs or artificially drilled holes in the rock (Figs. 2 and 3). Note however, that these methods may prevent close contact between the culture and the surrounding natural substrate and therefore change the microclimate.

8. Remove all nets when you see that the diaspores have attached themselves by outgrowing hyphae, usually after 6-12 months. This process depends on microclimate, and is species specific.

9. Experiments on growth and development in the field usually take several years, and photographic recording of growth is necessary. Take your first photograph directly after inoculation.

10. To examine intermediate developmental stages you will need many parallel experiments. Complete samples must be removed with the substrate at intervals for SEM viewing, and clearly such harvesting is destructive. The intervals depend on the rate of development in a

1 2 3 4

certain environment and no general rules can be given. Definite changes may occur in 2-3 weeks or may take months. Monitor the sites regularly! Our observations have shown that most samples reach a similar stage of development if the propagules have been sown at the same time.

Subprotocol 2
In Vivo Cultivation of Fragments and Diaspores in the Laboratory

The methods for in vivo cultivation of fragments and diaspores in the laboratory are virtually the same as those in the field. A natural substrate can be used and, particularly for terricolous lichens, no substrate seems to be an adequate substitute for soil. Terricolous lichens are cultivated best in small clay pots (Stocker-Wörgötter 1991). If defined media are used in preference to natural substrates, special experimental conditions are necessary which will be discussed below. Axenic conditions are superfluous as the diaspores and fragments are never sterile and cultures are only rarely overgrown by contaminating fungi. Initially the fragments often seem to be overgrown by fungi (Dibben 1971), but in many instances this is only the mycobiont leaving the symbiosis and forming cotton-like mycelia (Jahns 1993). A major problem in the laboratory is the regulation of the microclimate. A change between dry and wet phases is necessary as continuous high humidity favours the growth of aerial hyphae instead of a closed cortex.

Procedure

1. Lichens growing on a natural substrate in a growth chamber should be sprayed regularly (every second day) with deionised water. The optima for temperature and light are species specific, but low temperatures (15°C) and low irradiation are usually best.

Fig. 1. Culture of fragments of *Placynthium nigrum* covered by a net made from nylon stockings. **Fig. 2.** Cultures on carriers for SEM placed on a disc in the Antarctic. **Fig. 3.** Cultures on carriers for SEM in artificially drilled holes in rock. **Fig. 4.** Cultures on blocks of fired clay on a metal grid in a glass container.

2. For fragments growing on artificial substrates, a major problem is the requirement for alternating cycles of thallus wetting and drying. Therefore, we do not recommend using agar media, but have obtained satisfactory results with small blocks of burned clay. With a rolling pin spread the clay on foil (about 0.5 cm thick) and cut into squares of 1x1 cm. With a scalpel cut several grooves on the upper surface (Jahns 1993).

3. Fire the air-dried blocks at 900°C.

4. It is important to select clay that does not change the pH of solutions that come into contact with it. Filter deionised water through a piece of burned clay, and if the pH changes use a different type of clay.

5. If you wish to view your material under the SEM, include a thin layer of gauze 1-2 mm below the upper surface of the clay (Honegger 1993). This gauze layer is carbonised during the firing of the clay. The upper part can be separated at the end of the cultivation experiment, resulting in a smaller block suitable for SEM viewing.

6. Place the blocks on a metal grid with 1 cm high supports in a small glass container with a cover (Fig. 4).

7. Pour the nutrient solution into the container until it just reaches the base of the blocks where it is absorbed by the clay (Klöckner 1998). This is much better than spraying the cultures. The mineral media for algal cultures (see Chapters 1 and 3) may be used, but dilute it to 5% or 10% to prevent excessive algal growth. You can also add soil extract, hormones and other nutrients if you want to test the effect of these on lichen development.

8. Keep the containers in a growth chamber with regulated light and temperature conditions (10°C during the night, 15°C during day). Light intensities of 10 to 90 µmol photons m^{-2} s^{-1} are suitable. The best light intensity to use will depend on habitat of the lichen. Always use low light intensities for germinating spores and diaspores.

9. At regular intervals remove the grids and place in dry open containers. After this dry phase return the cultures to the nutrient solution. Dry and wet phases should last for a minimum of two days; a longer dry phase usually improves differentiation.

10. For recording of experiments and examination of samples see Subprotocol 1.

Subprotocol 3
Transplantation of Young and Adult Thalli

Transplanted lichens can be used to evaluate air pollution (Werner 1993), but we do not recommend this method for rigorous scientific experiments. Thalli from unpolluted habitats are attached to a board under standardised conditions and exposed under field conditions. The discoloration and decomposition of the thalli is recorded at intervals. Even after an adaptation time that is obligatory in these experiments, major problems remain, because the physiological state, the age, and the degree of adaptation to the original location of the lichens used are unknown. Older thalli suffer in particular from a change of microclimatic conditions and may be unable to adapt to the new location. Growth of transplanted young stages, or lobes from old thalli are the best indicators of environmental conditions. Degeneration of older thalli can be induced either by an unfavourable environment or by changes in the conditions that are not necessarily unfavourable. In ecological research both young and adult thalli should be transplanted to examine the problem of pre-adaptation and their physiological properties. The microclimate of the old and the new site should always be measured and compared. Light and water availability are of special importance. The best transplantation method will depend on the substratum and growth form of the lichen.

Procedure

1. Select appropriate material for transplantation.
   - For terricolous macro-lichens, transplant with parts of the original soil, or detach if the thalli are only loosely attached to substratum. Try to use similar ecological conditions at the new site, e.g. between moss cushions.
   - For lichens that are attached with their complete lower surface to bark or rock, separate from the substrate with a sharp knife and glue them to the new surface with silicone adhesive (Armstrong 1993). This method works well for species of *Parmelia*. This is recommended for very young stages with a size of only a few mm.
   - For foliose or fruticose lichens that are only attached to the substratum with small parts of their lower surface or with a holdfast, detach the whole thallus.

Note: Species of *Cladonia* and *Peltigera* usually do well at new sites. The effects of transplantation are not as crucial for these lichens as the influences of the microclimate at the new site. Water supply, light intensity and the possible existence of mosses and phanerogams appear decisive. For example, *Peltigera aphthosa* requires low irradiation and high air humidity, and grows best between mosses. However, it may be overgrown and outcompeted by large moss cushions. The narrow ecological amplitude of this species implies that it may be more difficult to transplant than, for example *Peltigera canina.*

2. Glue the thalli to the new substratum, or staple large thalli (e.g. of *Lobaria*) to bark (Scheidegger et al. 1995). Unfortunately, these areas often tear and the thalli fall off. As an alternative, cover whole thalli with a net. For large foliose lichens, nets with thin filaments and wide meshes are suitable (e.g. nets used for packing oranges). Tiny thalli on bark or stone can be covered with pieces of nylon stockings (Fig. 1). For close contact remove pieces of bark with an attached thallus and a corresponding part from a tree at the new location. Glue the transplantation sample into the aperture (Brodo 1961).
   If transplanted pieces of substratum are loosely attached to another tree or rock the transplanted lichen will be exposed to an altered microclimate. For example, water running down a tree trunk may fail to reach the samples. The growth of the lichen from the transplanted substratum to the new habitat is virtually impossible if the contact between old and new substratum is not very close.

3. In long term experiments remove the nets after 6-12 months when, under favourable conditions, the thalli have become attached by outgrowing hyphae (check regularly for this).

   Note: The number of observations, and the method used to make observations, will depend on the problems being investigated and on the growth rate of individuals at a certain site. Photographing, sketching, external morphology and anatomical observations are all possibilities. No general rule can be given. When you start experiments for the first time, check cultures in the field as often as possible to observe any change in your transplants, and then determine the necessary interval between observations.

Comments

Transplantation experiments and the cultivation of vegetative diaspores or fragments will result in lichen thalli with normal anatomy. Thalli with nearly normal developmental stages and even ascocarps can develop in the laboratory on a natural substratum. Experiments in the field and under nearly natural conditions in the laboratory are useful for ecological investigations and for the observation of morphogenesis. To date, attempts to cultivate lichens under defined conditions have only been partly successful. Nevertheless, defined culture conditions (media and microclimate) must be used if you want to determine the influence of single factors on growth and differentiation. The effects of external influences are best observed as small variations in growth patterns. If the development of a complete thallus were the required criterion for a successful experiment it would be impossible to show the influence of a single factor on the different steps. This is because a single correlation will be lost in the complex interactions in a lichen thallus. Therefore, even results from cultures that do not develop into mature thalli will help us to understand morphogenesis.

References

Ahmadjian V (1973) Methods of isolating and culturing lichen symbionts and thalli. In: Ahmadjian V, Hale M E (eds) The lichens. Academic Press, London, pp 653-658

Ahmadjian V (1989) Studies on the isolation and synthesis of bionts of the cyanolichen *Peltigera canina* (Peltigeraceae). Plant Syst Evol 165: 29-38

Armstrong R A (1993) The growth of six saxicolous lichens transplanted to lime-rich and lime-poor substrates in South Gwynedd, Wales. Symbiosis 15:257-267

Brodo I M (1961) Transplant experiments with corticolous lichens using a new technique. Ecology 42: 838-841

Bubrick P, Galun M (1988) Methods for cultivating lichens and isolated bionts. In: Galun M (ed) Handbook of Lichenology, CRC Press, Boca Raton, Vol. 3, pp. 127-138

Dibben M J (1971) Whole-lichen culture in a phytotron. Lichenologist 5: 1-10

Honegger R (1993) A simple outdoor culturing system for the foliose macrolichens *Xanthoria parietina* (L.) Th.Fr. and *Parmelia sulcata* Taylor. Bot Helv 103:223-229

Jahns H M (1993) Culture experiments with lichens. Plant Syst Evol 187:145-174

Klöckner P (1998) Wachstums- und Differenzierungsprozesse vegetativer Flechtenstrukturen unter kontrollierten Bedingungen. Inaugural-Dissertation, Düsseldorf

Scheidegger C, Frey B, Zoller S (1995) Transplantation of symbiotic propagules and thallus fragments: methods for the conservation of threatened epiphytic lichen populations. In: Scheidegger C, Wolseley P A, Thor G (eds) Conservation Biology of lichenized fungi. Mittl. Eidgen. Forschungsanstalt Wald Schnee Landschaft, Birmensdorf, pp. 1-62

Schuster G, Ott S, Jahns H M (1985) Artificial cultures of lichens in the natural environment. Lichenologist 17:247-253

Stocker-Wörgötter E (1991) Thallus formation of two cyanobacterial lichens: *Peltigera didactyla* and *Peltigera praetextata*, under laboratory conditions. Bul Soc Bot France 138, Lettres botaniques: 179-187

Werner A (1993) Aktives Biomonitoring mit der Flechte *Hypogymnia physodes* zur Ermittlung der Luftqualität in Hannover. Bibl Lichenol 49: 1-113

Isolation and Culture of Lichenicolous Fungi

JAMES D. LAWREY

Introduction

The lichenicolous fungi are characterised by their habit of growing on lichens. Hawksworth (1982) has estimated that as many as 1000 fungal species within 300 genera can be assigned to this group, and it is clear that they form a myriad of associations with lichens. Many attack living lichens and are obviously parasitic (mycoparasites) or pathogenic, but with levels of virulence that vary considerably. There are parasites that cause massive destruction of lichen tissues, but many cause little or no damage. Of these, some (called parasymbionts) appear to be lichen-forming fungi that share the photosynthate produced by another lichen's captured photobiont cells. Still others are saprophytic and colonise only dead lichen tissues. Because lichens are now known to harbour many opportunistic fungi that are not restricted to lichens (e.g., Petrini et al. 1990), some investigators (Rambold and Triebel 1992) prefer to exclude saprophytes from the lichenicolous fungi. I will restrict my discussion here to those lichenicolous fungi that are obviously parasitic or pathogenic.

Thorough laboratory study of lichen fungal parasites requires that they be isolated and brought into culture. As a general rule, standard mycological techniques can be employed to isolate these fungi. However, there are some special considerations to be taken. It is my objective in this chapter to summarise techniques for the isolation and culture of lichen mycoparasites. To encourage investigators to bring more of these fungi into culture, I will also briefly list some of the research questions that can be addressed using these cultured fungi as experimental organisms.

James D. Lawrey, George Mason University, Department of Biology, Fairfax, Virginia, 22030, USA (*phone* +01-703-993-1059; *fax* +01-703-993-1046; *e-mail* jlawrey@gmu.edu)

Subprotocol 1
Isolation of Lichenicolous Fungi From Living Collections

Materials

Equipment

- Laminar flow bench
- Incubator
- Autoclave

Chemicals

- 70% Ethanol
- 10% - 20% sodium hypochlorite

Culture media

- Sabouraud's medium with dextrose (SDA) or maltose (SMA)
- Potato dextrose agar (PDA)
- Cornmeal agar (CMA)
- Malt-yeast extract (MYA)

For more details see also Chapters 1-3 and 18.

Procedure

1. Lichenicolous fungi are usually collected attached to lichens from normal lichen substrates. It is sometimes not immediately evident that a fungal parasite is included with a lichen collection. What is frequently seen in the field are discoloration of lichen thalli or oddly coloured spots where parasites have developed fruiting structures. When discoloured lesions are observed microscopically in the laboratory, fungal parasites can frequently be seen and identified. Some general comments about collections:
   - Identification generally requires fruiting material (ascomata, basidiomata, conidia-forming structures, etc.).
   - Accurately labelled voucher specimens should always be kept, and specimens should be sent to specialists for verification of identifications.
2. In general, isolation of pure cultures of lichen mycoparasites follows steps similar to those for the isolation of fungi from other substrata (Stevens 1974):

- Wash infected lichen tissue in sterile water.
- In the laminar flow bench, using flamed forceps or needles, remove fruiting structures (ascomata, basidiomata, sporodochia, perithecia, sclerotia, etc.) and plate directly onto solidified agar media (various listed below).
- After 1-2 days incubation at room temperature, examine the dishes for growth of fungal mycelium from the fruiting structure.
- For badly contaminated lichen tissue, surface sterilise with solutions of ethanol (70%) or sodium hypochlorite (10-20%) for 30 sec, followed by several rinses in sterile water. Then proceed with tissue isolation onto agar.

3. Isolation of single-spore or single-conidium cultures provides the greatest certainty of the identity of the isolated fungus. It is also desirable for population genetic studies and subsequent extraction of DNA for sequencing.
   - Mature fruiting structures can be macerated with a flamed needle in cool sterile water on sterile glass microscope slides. This releases spores or conidia into the water.
   - Pipette solutions containing spores or conidia onto solidified agar and incubate at room temperature for 1-2 days.
   - Examine spores or conidia through glass Petri dish covers and remove those that germinate to a fresh agar surface with a flamed needle.
   - Keep accurate records concerning the origin of each isolate (single-spore, mycelial culture, sclerotial culture, etc.), when it was isolated and from what voucher specimen.
   - Give each isolated culture a unique number. Refer to all cultures afterwards with the same unique number.

Lichenicolous fungi grow on a wide variety of media in the laboratory. The choice of media to use will ultimately depend on information obtained during the isolation and culturing of these fungi. Standard liquid and solid media used for fungi (Stevens 1974, see also Chapters 1-3 and 18), including Sabouraud's medium with either dextrose (SDA) or maltose (SMA), potato dextrose (PDA), cornmeal (CMA), malt-yeast extract (MYA) agars or broth have all been used successfully with nearly all of our isolates. Occasionally, a low-N medium enhances the growth of these fungi, and sometimes growth depends on the presence of lichen tissues in the medium, but this is rare. It is likely that interesting nutritional requirements will be discovered for these fungi as more investigators work with them. Other culture requirements, for example variations in light and temperature conditions, may also be tested. Very little is presently known about the cultural requirements of these fungi.

Depositing cultures in culture collections

Of the many fungal culture collections available for deposit, I have used these listed here most frequently. Prior to deposit of cultures in a culture collection, one should contact a mycologist at the institution to ensure that proper procedures are followed and sufficient information is sent along with the culture.

- ATCC - American Type Culture Collection, 10801 University Boulevard, Manassas, Virginia 20110-2209, email: http://www.atcc.org
- CBS-Centraalbureau voor Schimmelcultures, P.O. Box 273, 3740 AG Baarn, Netherlands, http://www.cbs.knaw.nl
- IMI-International Mycological Institute, Egham, Surrey TW20 9TY, U.K., http://www.cabi.org/institut/imi/imi.htm
- NRRL-The National Center for Agricultural Utilization Research (previously called the Northern Regional Research Laboratory), 1815 N. University Street, Peoria, IL 61604, http://nrrl.ncaur.usda.gov

Subprotocol 2
Growing Cultured Lichenicolous Fungi on Lichen Tissues in the Laboratory

Materials

Equipment

- Laminar flow bench
- Incubator
- Waring Blender
- Homogenizer (e.g. Wiley mill)
- Laboratory steriliser

Chemicals

- Anhydrous calcium sulphate
- Acetone

Procedure

Once they are isolated, cultures of lichenicolous fungi can be used in various laboratory experiments. In our laboratory, lichen tissues are inoculated with these fungal isolates to determine how growth of the lichenicolous fungus is affected by lichen secondary products.

1. Collect lichens from the field, clean debris from thalli, wash briefly in distilled water, and air dry. Grind dried tissues in a Wiley mill to produce small (1 mm diameter) fragments. These should be stored in a desiccator over anhydrous calcium sulphate or silica gel prior to use.
2. If the experiment is designed to test the effect of lichen secondary compounds on lichenicolous fungal growth, lichen tissues can be washed in acetone to remove these compounds.
   - Washing several times in cold acetone usually results in removal of most lichen compounds.
   - The effectiveness of the washing (especially the first time it is done) should be tested by thin-layer (TLC) or high-pressure liquid chromatography (HPLC), see also Chapter 17.
3. Small samples (100 mg) of sterilised lichen tissues (washed and unwashed) are weighed and placed in glass Petri dishes.
   - Sterile technique is not necessary at this point, but accurate weights must be obtained.
   - Produce replicate plates of all treatments.
4. Since lichen thalli harbour numerous micro-organisms and cannot be maintained long under laboratory conditions, it is necessary to sterilise tissues before inoculating with fungal cultures.
   - Generally, heat treatment in a laboratory steriliser is used, but this will inevitably affect the secondary products of the lichen tissues. It is best to monitor these changes using TLC or HPLC.
   - Alternative methods include the use of various sterilising chemicals. For example, ethylene oxide vapours are effective in sterilising lichen tissues, but they may also alter lichen chemistry. Radiation can be used to sterilise tissues, but the equipment necessary to do this is not commonly available.

5. In the laminar flow bench, inoculate all sterilised plates with fungal inoculum. Inoculum is made from active (< 1 week old) cultures blended in sterile water to make a slurry with known concentration of fungal mycelium.
   - Using sterilised instruments, scrape fungal mycelium from a culture plate into sterile water in a sterilised Waring blender and blend for 15 seconds.
   - Take a small volume of the slurry produced, filter and weigh to obtain the concentration of fungal mycelium per unit volume; this will indicate how much of the slurry should be used in the inoculum.
   - Inoculate each sterilised substratum with the same amount of fungal mycelium (usually between 0.2 and 0.5 mg dry weight). Inoculate control plates to make certain the fungus grows and is not contaminated.

6. Wrap all plates in plastic film to prevent drying and place in a controlled temperature chamber for the experimental period. We routinely use a 12 h day/12 h night cycle at 18°C. We have generally used a 30-day period to determine degradative activity of fungi.

7. After 30 days, dry each plate at 105°C and store in a desiccator until it can be weighed.

8. Scrape out the dried substratum from each of the dishes.
   - If lichen tissues are to be used later, freeze in plastic containers. We have analysed this tissue using TLC and HPLC to identify changes in secondary products caused by fungal degradation. Control tissues are those maintained for the same length of time in sterile water without fungal inoculum, dried and frozen.
   - Clean plates thoroughly and dry them at 105°C. The dry weight of tissues is determined by subtracting the weight of the dried glass petri dishes from the weight of the plates plus the tissues.
   - Note that the weight of the substratum at the end of the experiment includes the weight of fungus tissue mixed with it; weight losses caused by tissue degradation by lichenicolous fungi are therefore net losses.
   - Some fungi fail to degrade lichen tissues after 30 days; in these cases, the dry weight of the substratum at the end of the experiment should be the initial weight plus the dry weight of the inoculum.

9. Express weight losses after 30 days as a percentage of the initial sample weight. Compare percent weight losses of tissues containing lichen

compounds with those of tissues washed with acetone to remove these compounds.

Results

- As an example of results that can be obtained from the growth of isolated lichenicolous fungi on lichen tissues in the laboratory, consider the following experiment. An isolate of the common basidiomycetous lichen pathogen *Marchandiomyces corallinus* (ATCC 200796) was obtained from sclerotia removed from thalli of the lichen *Flavoparmelia baltimorensis* collected in Maryland, U.S.A. In the laboratory, the fungus grows luxuriantly on both SDA and PDA and forms masses of coral-coloured mycelium typical of this species. To assess the growth of this fungus on various lichens, the fungus was inoculated onto four lichen substrata, *Flavoparmelia baltimorensis* (a typical host for this species in the mid-Atlantic region of the U.S.A.), *Punctelia rudecta*, *Lasallia papulosa*, and *Peltigera canina*. Also inoculated were lichen tissues that had been washed in acetone to remove lichen compounds. After 30 days, growth of the fungus is estimated by determining the net loss of dry weight of lichen tissues inoculated with the fungus.
- Results of this experiment (Fig. 1) demonstrate that *M. corallinus* degrades *F. baltimorensis* without regard to the presence of acetone-soluble lichen products. It is significantly inhibited by compounds of the other lichens used in the experiment, especially those of *L. papulosa* and *P. rudecta*. It is interesting that removal of compounds from tis-

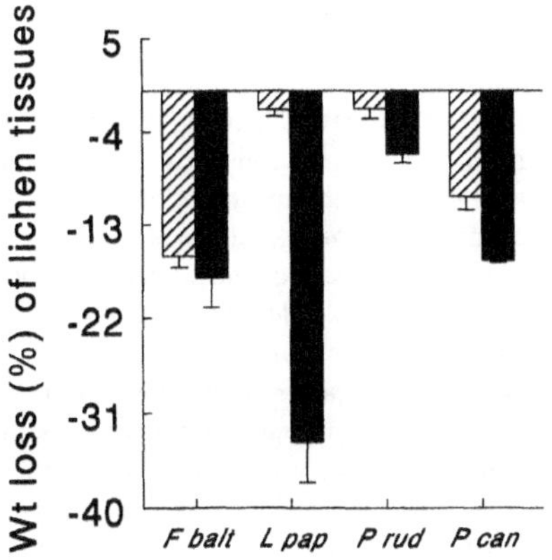

Fig. 1. Degradative activity (measured as mean percentage weight loss of lichen tissues (± SE of the mean) of the lichenicolous fungus *Marchandiomyces corallinus* inoculated onto four lichen substrata (*Flavoparmelia baltimorensis, Lasallia papulosa, Punctelia rudecta* and *Peltigera canina*) that have either been left untreated (lichen compounds present, hatched bars) or washed with acetone to remove lichen substances (compounds absent, solid bars).

sues of *L. papulosa* produces the best substratum for growth of *M. corallinus*; this is true also for most of the lichenicolous fungi maintained in our laboratory. Washed tissues (without lichen compounds) of *L. papulosa* are therefore used to maintain fungal cultures that do not grow well on other media.

- If these results are compared with those obtained in the same way using a nonlichenicolous relative of *M. corallinus*, an as-yet-undetermined *Marchandiomyces* sp. (ATCC MYA-299, Fig. 2), obvious differences in the degradative behaviour of the two species are observed. The nonlichenicolous species degrades *F. baltimorensis*, *L. papulosa* and *P. canina*, but only very slightly; it is completely inhibited by *P. rudecta*. Removal of lichen compounds significantly enhances growth on *L. papulosa* (as it does on *M. corallinus*), but not on the other lichens. These results suggest that certain cell wall-degrading enzymes are common to the two close relatives, but that those of *M. corallinus* are more effective against certain lichens (notably *F. baltimorensis*) and their secondary products.
- Another recent comparison of the degradative abilities of nonlichenicolous relatives of lichenicolous fungi (Lawrey et al. 1994) yielded similar results. In this study, nonlichenicolous relatives of the common lichenicolous fungus *Nectria parmeliae* proved to be generally unable to degrade lichens regardless of the presence of lichen compounds. Although preliminary, these results indicate that the evolution of a lichenicolous habit in fungi depends on the elaboration of cell wall-degrading enzymes which are either not present or not active in their nonlichenicolous ancestors.

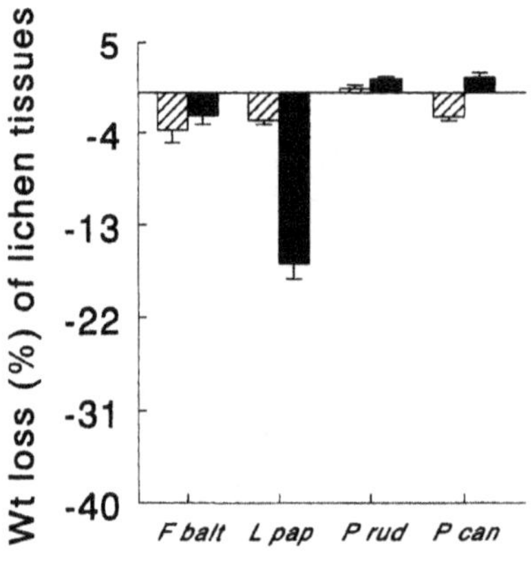

Fig. 2. Degradative activity (measured as mean percentage weight loss of lichen tissues (± SE of the mean) of the as-yet-unidentified nonlichenicolous fungus *Marchandiomyces* sp. inoculated onto four lichen substrata (*Flavoparmelia baltimorensis, Lasallia papulosa, Punctelia rudecta* and *Peltigera canina*) that have either been left untreated (lichen compounds present, hatched bars) or washed with acetone to remove lichen substances (compounds absent, solid bars).

Research questions to consider

A number of interesting research questions can be addressed using lichenicolous fungi as subjects, a few of which are listed here. Since so little is known about these fungi, much work must be done to understand even their basic life cycles and culture requirements. Given the uniqueness of lichenicolous fungi, however, it is hoped that as more investigators collect and isolate them and study them in the laboratory, discoveries will open entirely new avenues for research. Some of the questions we have raised in our work include the following:

1. How has the lichenicolous habit evolved? How do lichenicolous fungi differ from their near relatives? Is parasitism an ancestral or derived condition? Is there evidence for increased host-specificity or reduced virulence in lichen parasites?
2. Why are so many lichenicolous fungi host-specific? How are host-specific fungi different from those with broad host ecologies?
3. To what extent is host specificity determined by lichen secondary products? Which lichen compounds are most effective as defensive agents? What other factors are involved in host specificity?
4. To what extent has the evolution of host-specificity led to ecotypic differentiation in lichenicolous fungi?
5. To what extent do lichenicolous parasites degrade mycobiont tissues exclusively? Photobiont tissues exclusively? Both tissues? To what extent do lichenicolous fungi share photobiont production with the mycobiont (parasymbiotic fungi)?
6. Can one identify unique degradative enzymes necessary for a lichenicolous habit? To what extent are enzymes uniquely designed to degrade lichen cell walls (mycobiont and photobiont)? To what extent are enzymes uniquely tolerant of lichen secondary compounds?
7. To what extent are the culture conditions of lichenicolous fungi different from those of nonlichenicolous relatives? Has the exploitation of lichens as substrates altered the nutrient requirements (especially N) of lichenicolous fungi? Do parasites of lichens with cyanobacterial and chlorophyte photobionts have different requirements for nutrients (especially N)?

References

Hawksworth DL (1982) Secondary fungi in lichen symbioses: parasites, saprophytes and parasymbionts. Journal of the Hattori Botanical Laboratory 52: 357-366

Lawrey JD, Rossman AY, Lowen R (1994) Inhibition of selected hypocrealean fungi by lichen secondary metabolites. Mycologia 86: 502-506

Petrini O, Hake U, Dreyfuss MM (1990) An analysis of fungal communities isolated from fruticose lichens. Mycologia 82: 444-451

Rambold G, Triebel D (1992) The inter-lecanoralean associations. Bibliotheca Lichenologica 48: 1-201

Stevens RB (1974) Mycology guidebook. University of Washington Press, Seattle and London

Ultrastructure

Preparative Techniques for Transmission Electron Microscopy and Confocal Laser Scanning Microscopy of Lichens

ASUNCIÓN DE LOS RÍOS and CARMEN ASCASO

Introduction

Transmission electron microscopy

Ultrastructural study of lichen symbionts can provide valuable information about symbiotic performance complementary to that obtained using other techniques (Lallemant et al. 1986). Transmission Electron Microscopy (TEM) was first used to describe the cellular organelles of both symbionts (Jacobs and Ahmadjian 1969; Boissière 1972; Galun et al. 1970, 1974; Peveling 1973, 1974, 1976; Ascaso and Galvan 1975, 1976). Later, different aspects of lichen symbiosis were studied, for example cellular membranes and cell wall with the freeze-etching electron microscopy technique (Ellis and Brown 1972; Peveling and Robenek 1980; Ascaso et al. 1985; Honegger 1986a; Rapsch et al. 1986). TEM has contributed to the understanding of different types of mycobiont-photobiont relationships in lichens, e.g. by observing the physical contacts between symbionts (for reviews see Honegger 1984, 1985, 1986b). The study of storage bodies in both symbionts provides indirect information on biotrophic relationships (Valladares and Ascaso 1994). Some authors have described the variability of lichen ultrastructure in relation to season or environment (Holopainen 1982; Scott and Larson 1986; Fiechter and Honegger 1988; Balaguer et al. 1999). In some investigations, TEM techniques have revealed structural changes due to different experimental conditions, ranging from desiccation to environmental pollution (Eversman and Sigal 1984, 1987; Ascaso et al. 1986, 1988; Brown et al. 1987, 1988; Balaguer et al. 1996, 1997; Tarhanen et al. 1997).

✉ Asunción de los Ríos, Centro de Ciencias Medioambientales (CSIC), Serrano 115 dpdo, Madrid, 28006, Spain (*phone + fax* 34-915640800; *e-mail* arios@ccma.csic.es)
Carmen Ascaso, Centro de Ciencias Medioambientales (CSIC), Serrano 115 dpdo, Madrid, 28006, Spain

As poikilohydric organisms, lichens have no means of controlling their water relations, and they can spend long periods in a desiccated state. Therefore it is important not only to know the ultrastructure of hydrated thalli but also the ultrastructure of desiccated thalli. However, conventional techniques for TEM involve immediate rehydration of the thalli, which implies that it is impossible to study all of the morphological and ultrastructural details of lichens in the desiccated state (De los Ríos et al. 1999). Cryotechniques have been used to overcome this problem (Honegger and Peter 1994; Honegger et al. 1996) but their application in the preparation of lichen thalli for TEM presents many methodological problems. Cryotechniques rely on low temperatures to fix and stabilize specimen ultrastructure, which should then reflect the living state accurately. Optimising TEM cryotechniques in lichens is an important goal to study lichens with high resolution in different states of hydration and to discern artefacts originating in conventional procedures.

The capacity to use electron microscopy in tandem with molecular probe techniques (e.g. immunolabelling of molecules with antigenic capacity and in situ hybridisation of nucleic acids) has permitted the development of methods that allow in situ localisation of cellular components and processes. Physiological and taxonomic data can be obtained, integrating ultrastructural data with biochemical, physiological and molecular work. Immunolabelling techniques based on the reaction of labelled antibodies with cellular antigens exposed at the cut surface of a section, has already been applied successfully in lichen thalli (Janson et al. 1993; Ascaso et al. 1995; Balaguer et al. 1996, 1999; Palmqvist et al. 1997). In situ hybridisation of nucleic acids, the aim of which is to localize nucleic acids sequences (DNA or RNA) in the cytoplasm, organelles, chromosomes or nuclei of biological material is currently being optimised for lichens (De los Ríos et al. 2001; Grube and De los Ríos 2001). Labelled probes (oligonucleotides, DNA or RNA) that hybridise specifically to nucleic acids with complementary sequences, are used for the in situ localization of these sequences.

Confocal laser scanning microscopy (CLSM)

Confocal microscopy was invented in 1957 by Minsky (Minsky 1961, 1988) but has only recently become a practical tool for general use. With the advent of relatively economical personal computers, PC-based imaging software and small air-cooled lasers, commercial instruments became available in 1987 (Amos et al. 1987). Light and electron microscopy have played vital roles in lichen studies as important tools for analysing cellular structure, physiology and function. However, confocal microscopy can provide new kinds of information because it constitutes a means to observe living cells and tissues in three dimensions without fixation or physical sectioning artefacts (Wright et al. 1993). This kind of microscopy is beginning to be used in the conventional study of lichens (De los Ríos and Ascaso 1996; Grube and Matzer 1997; Ascaso et al. 1998; De los Ríos et al. 1999).

A confocal microscope is a microscope where the optical geometry of a light microscope has been modified so that all of the laser incident illumination (excitation) is focused to a diffraction-limited spot. Light emanating from the spot is simultaneously focused on an aperture in front of a detector (Known et al. 1993). This special configuration of sequential illumination makes it possible to avoid out-of-focus information that often obscures structures of interest, especially in thick specimens or if overlapping structures exist (Wright et al. 1993). Confocal microscopy has the unique ability to create images of a section through a sample. Three-dimensional reconstructions can be made by collecting images at different planes of focus (optical sections) followed by running image reconstruction algorithms (Known et al. 1993).

The confocal principle is especially valuable in fluorescence microscopy and this is the predominant optical mode used for biological confocal microscopy. Combination of molecular and biochemical techniques with confocal microscopy can allow the design of specific fluorescent probes to detect and visualise biological molecules including proteins, nucleic acids, carbohydrates, lipids and ions in cells or in cell free extracts. Also, different organelles or macromolecular assemblies can be observed simultaneously within cells by using more than one fluorescent probe at the same time.

The possibility of easily obtaining three-dimensional reconstructions will enable scientists to clearly and conveniently visualise various cellular organelles or/and structures in living or fixed cells, and provide a better understanding of their distribution in the cell. In this way CLSM can become a fundamental tool in molecular, taxonomic and physiological studies of lichens.

Outline

Figure 1 illustrates the steps that must be followed in the protocols for preparation of lichen material for TEM described in this chapter. The three basic kinds of preparation, i.e. conventional, for immunocytochemistry and with use of cryomethods are described in Subprotocols 1 - 3.

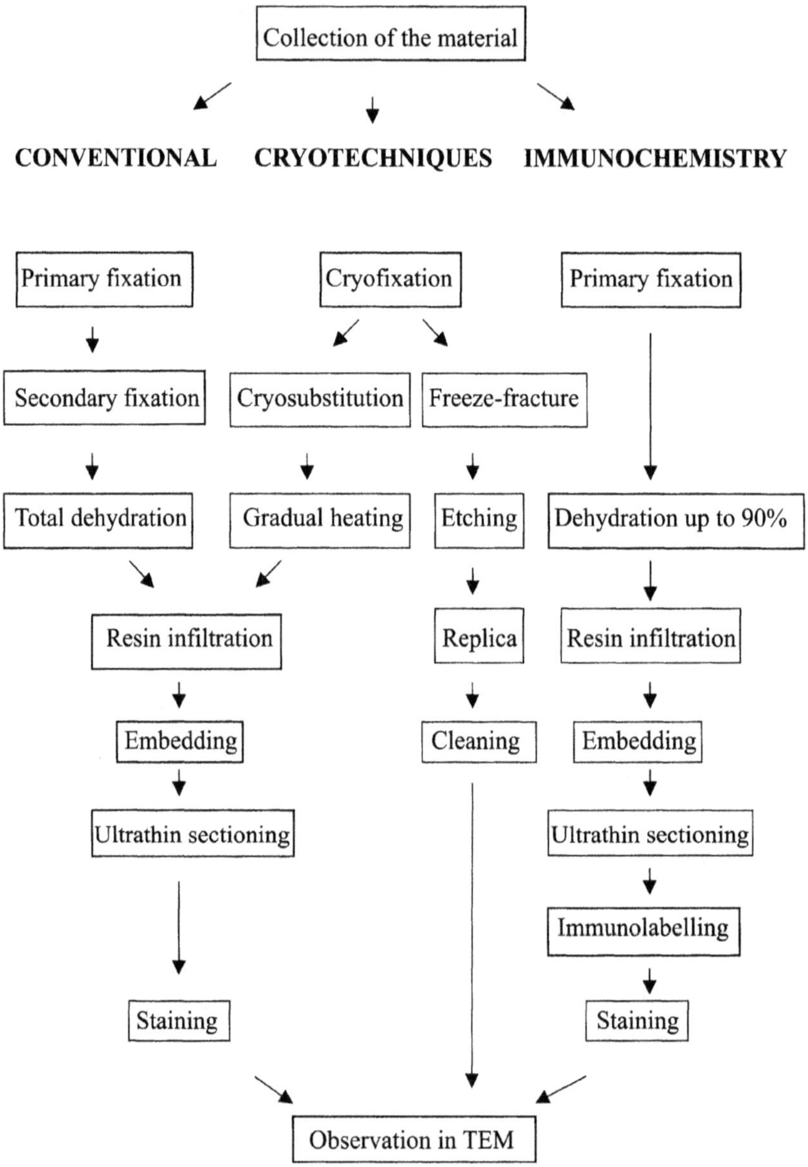

Fig. 1. Schematic diagram illustrating the steps in preparation of lichen samples for TEM.

Materials

Equipment

Conventional preparation and immunocytochemistry

- Vacuum pump (glass desiccator and rotatory pump)
- Fume hood
- Ultramicrotome with effective anti-vibration mounting
- Polymerisation oven

Cryotechniques

- Cryovacuumblock e.g. Reichert-Jung (Leica), for cryofixation by slamming
- Liquid Propane plunging system (Halpern and Quintana, 1989), for cryofixation by plunging
- Freezer Cryosystem (RUA) and conventional freezers, for cryosubstitution
- Polaron E7500 (shadowing unit)/ P650A (vacuum unit), for Freeze-Fracture techniques

Chemicals

Conventional Preparation

- 3% Glutaraldehyde (EM grade) in 100 mM sodium phosphate buffer (pH 7.1)
- 1% Osmium tetroxide in 50 mM sodium phosphate buffer (pH 7.1)
- 100 mM sodium phosphate buffer (pH 7.1)
- Ethanol
- Propylene oxide
- Lead citrate solution (Reynolds, 1963)
- Copper grids covered by pioloform (a formvar-like membrane)
- Spurr's resin: Low viscosity ("Spurr") embedding kit

Immunocytochemistry

- 50 mM sodium cacodylate buffer (pH 7.4)

- 2.5% glutaraldehyde (EM grade) in 50 mM sodium cacodylate buffer (pH 7.4)
- Ethanol
- LR-White acrylic resin
- 2% aqueous uranyl acetate solution
- Lead citrate solution
- Nickel grids covered by pioloform
- 10 mM Tris-HCl buffer (pH 7.4) (TBS)
- 2% BSA (Bovine Serum Albumin) in 10 mM Tris-HCl buffer (pH 7.4)
- 0.2% BSA in 10 mM Tris-HCl buffer (pH 7.4)
- Rabbit primary antibody
- Goat anti-rabbit antibody coupled to 15 nm colloidal gold particles (secondary antibody)
- 0.1% (v/v) Triton X-100 and 0.2 % BSA in 10 mM Tris-HCl buffer (pH 7.4)
- Whatman no. 1 filter

Cryotechniques

- Liquid propane
- Liquid nitrogen
- Liquid helium
- Rubber foam
- Double sided tape
- Mica sheets
- Methanol
- Acetone
- Spurr's resin: Low viscosity ("Spurr") embedding kit
- 2% Osmium tetroxide in methanol
- 0.1% Uranyl acetate in methanol

- 2% Osmium tetroxide in acetone
- 0.1% Uranyl acetate in acetone
- Carbon and platine for sputter coat
- Chromic mixture (5% $Cr_2O_7K_2$ and 10% H_2SO_4 in water)

Subprotocol 1
Conventional Preparation for TEM

Procedure

There are many ways to prepare material, and including all possibilities is impossible. We describe the protocols that are used in our laboratories that have been modified for lichens.

To preserve the structure of the lichens for ultrastructural examination, the lichen material must be freshly collected. Herbarium and material collected previously can also be prepared but generally the results achieved are inferior.

1. Prepare a 3% glutaraldehyde solution in 100 mM sodium phosphate buffer (pH 7.1) (primary fixative) and maintain this at 4-5°C.
2. Cut the lichen thalli (preferentially hydrated) into small pieces not larger than 1-2 mm^2. This will allow even fixation and minimise fixative penetration time.
3. Place the fragments of the specimen in small glass tubes (5 cm high, 1 cm in diameter) where all the following preparation steps except final polymerisation are carried out.
4. Add the primary fixative to the tubes with the fragments of the specimen and leave with the fixative for 3 h at 4-5°C. Vacuum infiltrate inside a desiccator connected to a rotary pump (for two periods of 10 min approximately) at start of fixation. This facilitates the penetration of the fixative into the structure of the lichen thalli that would otherwise simply float on the surface of the fixative because of the associated air spaces.
5. Rinse fragments of the specimen three times, for 30 min each at room temperature in 100 mM sodium phosphate buffer (pH 7.1).

6. Place fragments of the specimen for 5 h, at room temperature and in darkness, in a secondary fixative composed of 1% osmium tetroxide in 50 mM sodium phosphate buffer (pH 7.1).
7. Rinse fragments of the specimen three times, for 30 min each, at room temperature, in 100 mM sodium phosphate buffer (pH 7.1).
8. At room temperature, dehydrate the fragments of the specimen by taking them through a series of ethanol solutions as follows:
   - 30% ethanol for 15 min,
   - 50% ethanol for 15 min,
   - 70% ethanol for 15 min,
   - 90% ethanol for 30 min,
   - 100% ethanol three times, for 1 h each.
9. Place fragments of the specimen in propylene oxide twice for 20 min each, at room temperature.
10. Place fragments of the specimen in (1:1) propylene oxide:Spurr's resin overnight, at room temperature.
11. Place fragments of the specimen in (1:2) propylene oxide:Spurr's resin, for 4-6 h at room temperature.
12. Change to pure Spurr's resin and leave for two days at 4-5°C, in tightly sealed tubes, in a refrigerator.
13. Place specimen fragments in fresh Spurr's resin in flat embedding moulds.
14. Place the embedding moulds in the oven and polymerise at 70°C for 24 h, then turn off the oven.
15. Remove the blocks of the resin-embedded specimen from the mould, once the oven has reached room temperature.
16. Trim (remove excess resin around the specimen) and prepare the resin embedded blocks for sectioning in an ultramicrotome. For this purpose, first trim the specimen by hand with a razor blade on the ultramicrotome block trimming holder to produce a pyramid-shaped block. Thereafter, cut semi-thin sections until the face of the block is smooth.
17. Cut semi-thin (0.35-0.5 μm) sections of resin-embedded specimens on an ultramicrotome using glass or diamond knives with a water-filled boat. Stain the semi-thin sections with toluidine or methylene blue to determine the orientation of the specimen.

18. Cut ultra-thin (70-90 nm) sections on an ultramicrotome using a diamond knife and deposit the sections on copper grids covered previously with pioloform.

19. Stain the ultra thin sections with lead citrate solution (Reynolds 1963) inside a Petri dish trying to avoid exposure of the stain to air. Place a fresh sheet of Parafilm in the bottom of a Petri dish and quickly place drops of lead citrate onto the film. Put the section side of grids on the drops. After 10-12 min, remove the grid from the drop of stain and wash it in a series of three water drops by placing the grid in each of the drops.

Comments

Precautions

As all fixation solutions are designed to react with biological material, it is necessary to be extremely careful when handling them and to work in a fume hood.

Spurr's resin

Spurr's resin is prepared following the manufacturer's instructions and stored at -20°C until use. Keep it for a maximum of two months. Warm the resin to room temperature before opening to prevent condensation of water. The components of Spurr's resin are harmful so it should be handled with extreme caution. We advise that you wear protective latex gloves and always use a fume hood when making up or handling the uncured resin or its components. In spite of the high toxicity of Spurr's resin, we continue to use this resin because, in our experience, it provides the best results with lichens.

Chemicai fixation

We recommend hydration of the specimen before fixation, as this will improve subsequent fixation. This step is very important in lichens as their anatomy makes the penetration of the fixatives difficult. The time necessary for optimal hydration depends on the species. Also, the time required for vacuum infiltration during primary fixation can differ among species. Double fixation gives a better resolution. The aldehyde component (primary fixative) preserves mainly proteinaceous structures. Osmium tetroxide complements the aldehydes by fixing or immobilising mainly lipids.

Dehydration

For dehydration you can use acetone instead of ethanol. We use ethanol because it causes less extraction of cellular contents and lichen substances than acetone. Acetone rinsing of lichens is a standard method for extracting secondary lichen products (Culberson and Kristinsson 1970).

Staining Reynold's lead staining has given good results in lichen samples embedded in Spurr's resin. Aqueous uranyl acetate does not stain samples embedded in Spurr's resin as effectively (Dykstra 1992). Sometimes, globose lead stain deposits are observed in lichen sections stained with lead citrate. Reynolds' lead citrate has a pH of 12-13 and can precipitate in the presence of acidic compounds such as lichen substances.

Subprotocol 2
Preparation for Immunocytochemistry

Procedure

1. Prepare a 2.5% glutaraldehyde solution in 50 mM sodium cacodylate buffer (pH 7.4) (primary fixative) and maintain this at 4-5°C. If glutaraldehyde affects the antigenicity of the component you wish to localize, solutions of 2-4% paraformaldehyde must be used.
2. Cut the lichen thalli into pieces not larger than 1-2 mm^2.
3. Place the fragments of the specimen in small glass tubes (5 cm high, 1 cm diameter) where all the following preparation steps except final polymerisation will be carried out.
4. Add the primary fixative to the tubes with the fragments of the specimen for 2.5 h at 4-5°C. Vacuum infiltration must be carried out for two periods of 10 min, at the start of fixation (see Subprotocol 1, Step 4).
5. Rinse fragments of the specimen twice, for 20 min each, in 50 mM sodium cacodylate buffer (pH 7.4), at 4°C.
6. Dehydrate the fragments of the specimen at 4°C using a series of ethanol solutions:
   - 30% ethanol for 15 min,
   - 50% ethanol for 15 min,
   - 70% ethanol for 15 min,
   - 90% ethanol for 30 min,
   - 90% ethanol twice, for 1 h each.
7. Infiltrate for 4 h at 4°C in (2:1) 90% ethanol:LR-White resin.
8. Infiltrate for 4 h, at 4°C, in (1:2) ethanol absolute:LR-White resin.

9. Change to pure LR-White resin and keep the tubes with fragments of the specimen in LR-White resin, well closed, in a refrigerator for three days. Change the resin twice during this period.

10. Place fragments of the specimen in fresh LR-White in capped gelatine capsules to reduce the presence of oxygen that can inhibit resin polymerization.

11. Place the gelatine capsules in the polymerization oven and polymerise the LR-White at 60°C for 48 h. Turn off the oven and leave the capsules inside until they reach room temperature.

12. Trim and prepare the resin embedded blocks for sectioning in an ultramicrotome (see Subprotocol 1, Step 16).

13. Cut semi-thin (0.35-0.5 μm) sections of resin-embedded specimens on an ultramicrotome using glass or diamond knives with a water-filled boat. Stain the semi-thin sections with toluidine or methylene blue to determine the orientation of the specimen.

14. Cut ultrathin (70-90 nm) sections on an ultramicrotome using a diamond knife and deposit the sections on nickel grids covered previously with the formvar-like membrane "pioloform". The ultra thin sections have to be cut just before the immunolabelling processes.

Gold immunolabelling

15. Protocols can vary depending on the component you wish to localise. A protocol used for Rubisco localisation within photobiont will be described here. Ultra thin sections on nickel grids are put, section side down, on drops of the following solutions. The drops are placed on a fresh sheet of Parafilm in the bottom of a Petri dish.
    - 2% Bovine serum albumin (BSA) in 10 mM Tris-HCl buffer pH 7.4 (TBS), for 1 h, at 37°C
    - Rabbit anti-*Euglena* Rubisco (primary antibody), diluted 1:1000 (v/v) in 2% BSA in TBS, for 1 h, at 37°C
    - 0.2% BSA in TBS, five times, for 3 min each, at room temperature
    - Goat anti-rabbit antibody coupled to 15 nm colloidal gold particles (secondary antibody), diluted (1:25 v/v) in 0.2 % BSA in TBS, for 1 h, at 37°C
    - 0.1 % (v/v) Triton X-100, 0.2 % BSA in TBS, three times, for 1 min each, at room temperature
    - Deionised water, five times, for 1 min each, at room temperature

Process the following set of controls in parallel: (i) treat some grids with 2% BSA in TBS instead of anti-Rubisco antibodies; (ii) treat some grids with 0.2% BSA in TBS instead of secondary antibody.

16. Stain with uranyl acetate in the dark. Put the section sides of the grids on drops of 2% aqueous uranyl acetate. After 10-12 min, remove the grid from the drop of stain and wash in a series of three drops of water, then blot the grid completely dry on Whatman no. 1 filter paper before proceeding.
17. Post-stain with lead citrate solution (9 min; Reynolds 1963) to improve the visualisation of ultrastructural details (see Subprotocol 1, Step 19).

Comments

Fixation In the procedure of preparation of the samples for immunocytochemistry there are several common steps with conventional procedures (Fig. 1). However, to preserve antigenicity some modifications are necessary. A specimen preparation technique is required which preserves both the antigenicity of the sample and retains the morphology of the specimen without movement of the antigen. This usually involves a compromise (Robards and Wilson 1993), but is helpful to carry out all processes at 4°C. In addition, osmium tetroxide is omitted from the immunocytochemical fixation protocol because it may mask antigenic sites, causing a substantial reduction in antibody recognition.

The best fixative to use in inmunocytochemistry processes depends on the antigen that you wish to localise. Glutaraldehyde fixation results in better ultrastructural preservation, but can diminish or completely prevent antibody recognition of protein (Mariac et al. 1992; Dykstra 1992). If this happens, we recommend using paraformaldehyde instead of glutaraldehyde.

Resin In conventional specimen preparation you use a different resin from that required in immunocytochemistry. In conventional protocols, epoxy resins are used as they give good structural preservation and stability in the electron beam. However, the hydrophobic nature of these resins reduces the antigenic capacity. As a result, for immunocytochemical purposes, polar acrylic resins are preferred due to their hydrophilic nature. The use of these hydrophilic resins prevents total dehydration. Ultrastructural preservation with acrylic resins gives poorer results, and therefore it is always advisable to prepare samples for conventional observation in parallel. In samples embedded in acrylic resins, double staining (uranyl acetate followed by lead staining) is recommended, as uranyl acetate is a mordant for lead.

Cryotechniques

In the preparation of the samples for immunocytochemistry, a good alternative procedure to the conventional protocol described above is the use of cryotechniques.

Subprotocol 3
Preparation Using Cryotechniques

As mentioned in the introduction, using cryotechniques to study lichens is difficult. We describe three protocols that we have used with lichens in our laboratories. These protocols include two different methods for cryofixing the specimen, freezing by slamming onto cold surfaces (slamming) and freezing by high-speed plunging into liquid cryogens (immersion). Biological systems usually contain more than 70% water, and as a result the formation of ice crystals becomes the main source of artefacts in cryofixation (Quintana 1994). The use of a higher cooling rate and/or higher pressure during cryofixation reduces ice crystal size. Chemical fixatives are also frequently added to stabilize the ultrastructure and to prevent the loss of cellular compounds.

In the two first protocols, cryofixation is followed by a cryosubstitution in which substitution agents replace water in the sample ensuring rapid cessation of biological activity as a result of cryofixation. Conventional embedding then follows this. However, the third (freeze-etching) consists of obtaining a replica of a fractured surface from a cryofixed sample. In this technique, we observe in the microscope the replica, and not ultra thin sections of embedded material, as in the other protocols.

If you want to cut ultra thin sections from your specimen and immunolabel them, the protocols below have to be modified: omission of osmium tetroxide fixation and the use of hydrophilic resins are recommended (Fig. 1).

Procedure

Cryofixation by slamming, cryosubstitution and embedding

1. Cut lichen thalli into small fragments.
2. Fasten the specimen with a drop of 2% agar to a small piece of paper and place it with the paper side on a small fragment of mica sheet. The paper is useful to identify the sample during the different steps and to determine the orientation for ultra thin sectioning.
3. Place the specimen with the mica side on the stub. This stub has rubber foam with double-sided tape where the mica sheet is stuck.
4. Mount the stub with the specimen at the end of the plunger from the Cryovacuum block.
5. Slam the sample onto a highly polished copper surface cooled with liquid helium at -265°C.
6. Transfer the specimen quickly to liquid nitrogen for storage until subsequent processing steps.
7. Cryosubstitution: Transfer the specimen to a container inside the RUA cryosystem containing 100% acetone (substitution agent) with 2% osmium tetroxide and 0.1% uranyl acetate. Leave sample in this solution for 72 h at -90°C, to replace all the water with the substitution agent. The RUA cryosystem is a commercial freezer, with a platform to work and that supports the cold organic medium containers where the samples are cryosubstituted.
8. Slowly heat the samples:
   Increase the temperature from -90°C to -40°C at 5°C h^{-1}. Maintain at -20°C for 1 h then at least for 1 h at 4°C.
   The cryosystem RUA is used for the first step, and other freezers set at fixed temperatures are used for the other steps.
9. Clean the samples at room temperature with 100% acetone for 2 h, changing to fresh acetone every 5 min.
10. Infiltrate for 8 h in (1:3) Spurr's resin:acetone, at room temperature.
11. Infiltrate in (1:1) Spurr's resin:acetone, overnight, at room temperature.

12. Infiltrate in (3:1) Spurr's resin:acetone for 8 h, at room temperature.
13. Infiltrate for 24 h in Spurr's resin with at least two changes of fresh resin, at room temperature.
14. Place the specimens in fresh Spurr's resin in flat embedding moulds.
15. Transfer the embedding moulds to the oven and polymerise the Spurr's resin at 60°C for 48 h, then turn off the oven.
16. Remove the blocks of the resin-embedded specimen from the mould, once the oven has reached room temperature.
17. Trim and prepare the resin embedded blocks for sectioning in an ultramicrotome (see Subprotocol 1, Step 16).
18. Cut semi thin (0.35-0.5 μm) sections of resin-embedded specimens with an ultramicrotome using glass or diamond knives with a water-filled boat. Stain the semi-thin sections with toluidine or methylene blue to determine the orientation of the specimen.
19. Cut ultrathin (70-90 nm) sections with an ultramicrotome using a diamond knife and deposit on copper grids covered previously with pioloform.
20. Stain with lead citrate solution (10-12 min; Reynolds 1963) (see Subprotocol 1, Step 19).

Notes

Equipment

Cryofixation and cryosubstitution instruments used in our work have been described by Quintana (1991a,b, 1992, 1994). More information about other equipment can be obtained from Robards and Sleytr (1985) and Sitte et al. (1977). Also, new commercial equipment for cryofixation, cryosubstitution, freeze drying and low temperature embedding has recently been developed. Examples include the Leica Metal Mirror Cryofixation system EM MM 80E, the Leica EM CPC Universal Cryoworkstation and the Leica EM AFS Automatic Freeze Substitution System.

Advantages of slamming

Cryofixation by slamming has the advantage that the specimen is frozen using a solid cryogenic medium. Solid metals like copper are better refrigerants than fluids because they may be cooled to a lower temperature than liquids (limited by their melting point) and their thermal conductivities are higher than those of liquid cryogens. These involve very high cooling velocities. As a result, they are very useful for studying a time course of any cellular activity, for which specimens must be prepared at fixed times, or for investigations into ultrastructural changes produced during hydration or dehydration processes in lichens. However, this method gives only 10-20 µm of good ultrastructure from the surface of the sample. For this reason, we recommend that you prepare a small flat sample section of the lichen material. Cryo-ultamicrotomy can be used for the preparation of these flat sections from hydrated samples but it is very difficult with dehydrated ones. Another disadvantage is that although the sample is mounted on a small piece of rubber foam on the stub, slamming can induce some degree of surface deformation.

Substitution agents Different substitution agents can be used for cryosubstitution e.g. acetone, methanol and diethylether. Depending on the agent used, the step lengths and temperatures used in specimen preparation can differ. During cryosubstitution, it is important to avoid the production of artefacts by the growth of secondary ice crystals from sample water. For this reason, it is important not to allow samples to reach temperatures above -85°C during this period. Chemical fixatives dissolved in the substitution agent, diffuse throughout the sample with the agent in the cryosubstitution process. Fixatives interact simultaneously with all the parts of the sample when the sample is slowly warmed and when they reach their particular reactive temperature. Specific recommendations for the manipulation of Spurr's resin are included in Subprotocol 1.

Cryofixation by plunging, cryosubstitution and embedding

1. Cut the lichen thalli into small fragments.
2. Fasten the specimen with a drop of 2% agar in water to a small piece of paper and place it, with the paper side on a small fragment of mica sheet. The paper is useful to identify the sample during the different steps and to determine the orientation for ultra thin sectioning.
3. Cryofixation: Immerse the specimen into liquid propane by injection with a mechanical device. The damping system used plunges the sample at a velocity of 6 m s^{-1}. The liquid propane is cooled at -188°C using liquid nitrogen.

4. Keep the sample immersed for approximately 10 s and then transfer into liquid nitrogen for storage until subsequent processing steps.
5. Cryosubstitution: Replace water of the sample with 2% osmium tetroxide and 0.1% uranyl acetate in 100% methanol, for 20 h at -90°C, in a RUA cryosystem.
6. Slowly heat the sample. Increase the temperature from -90°C to -60°C at 5°C h^{-1}, maintain at -40°C for 1 h, at -20°C for 1 h, then at least 1 h at 4°C. The cryosystem RUA is used for the first step, and other freezers set at a fixed temperature are used for the other steps.
7. Place the samples in 100% methanol for 2 h, changing to fresh methanol every 5 min.
8. Pre-infiltration: Place the samples in propylene oxide, three times for 10 min each, at room temperature.
9. Infiltrate in (1:3) Spurr's resin:propylene oxide, overnight, at room temperature.
10. Infiltrate in (1:1) Spurr's resin:propylene oxide, for 7 h, at room temperature.
11. Infiltrate in (3:1) Spurr's resin:propylene oxide, overnight, at room temperature.
12. Infiltrate for 48 h in Spurr's resin, changing to fresh resin after 24 h, at room temperature.
13. Place fragments of the specimen in fresh Spurr's resin in flat embedding moulds.
14. Place the embedding moulds in the oven and polymerise the Spurr's resin at 60°C for 48 h.
15. Remove the blocks of the resin-embedded specimen from the mould, once the inside of the oven has reached room temperature.
16. Trim and prepare the resin embedded blocks for sectioning in an ultramicrotome (see Subprotocol 1, Step 16).
17. Cut semi-thin (0.35-0.5 µm) sections of resin-embedded specimens on an ultramicrotome using glass or diamond knives with a water-filled boat. Stain the semi-thin sections with toluidine or methylene blue to determine the orientation of the specimen.

18. Cut ultra-thin (70-90 nm) sections on an ultramicrotome using a diamond knife and deposit on pioloform coated copper grids.

19. Stain with lead citrate solution (10-12 min; Reynolds 1963) (see Subprotocol 1, Step 19).

Notes

Precautions This method consists of quick plunging of samples into liquid propane. Due to the flammability of propane, which forms a potentially explosive mixture with liquid oxygen condensed from the air, the sample cannot be handled too roughly.

Advantages This method is the best for starting to freeze samples. It can be established in any laboratory using simple and inexpensive apparatus [see Robards and Sleytr (1985) and Quintana (1991a)]. However, unless plunge conditions are optimised, poor cooling rates may be obtained. The shape and size of the specimen are factors that must be considered. The sample has to be as small and flat as possible to facilitate a homogenous freezing. Hydrated samples must be handled fast enough to avoid desiccation. Do not keep your sample too close to the cool liquid bath, as it could become precooled as a result of cold gas evaporations. As with most freezing techniques, optimal sample freezing occurs no more than 10-20 µm from the surface of the sample surface. Lichen thalli are irregularly cryofixed in transverse section due to their anatomy and the presence of air spaces inside the thalli.

Freeze drying Instead of using cryosubstitution in these protocols, cryofixed cells can be cryodehydrated by freeze-drying. The advantage of this technique is that a native frozen specimen can be infiltrated with resin without having been exposed to aqueous fixatives, aqueous buffers or dehydrating agents. The advantage of this is that it will give better retention of diffusible elements (Dykstra, 1992). However, cryosubstitution has the advantage of being simpler, giving clearer and more reproducible preservation of fine structure and easier resin infiltration (Quintana, 1994). Therefore, unless you want to localise water-soluble components, we recommend cryosubstitution.

Freeze-etching

1. Cryofix the lichen thallus. For this purpose, it is possible to use the plunging method as previously described (immersion of the sample in liquid propane at -188°C).
2. Mount the stub with the sample in the Polaron E7500/P650A or similar equipment.
3. Fracture the sample at a temperature between -150°C and -110°C, generally with the help of a razor blade which is used as a knife, in a chamber with a vacuum of 10^{-6} mbar.
4. Immediately etch the ice from the fractured surface; this could be superficial (freeze-etching) or profound (deep-etching). In our experience, good etching can be obtained by raising the temperature to -100°C for 90 s.
5. Coat the sample with Platinum-Carbon, thus producing a mould or "replica" of the fractured surface.
6. Before observing the replica under the TEM, you must clean it of organic and inorganic matter. For this purpose, place the replica in chromic mixture (5% $Cr_2O_7K_2$ and 10% H_2SO_4 in water), for 1-3 days at room temperature.
7. After cleaning, carefully rinse the replicas several times in distilled water.
8. Finally, pick up the replica from the surface of the distilled water by pressing a grid against it using forceps and lifting it away.

Notes

Studying lichen symbionts

For studying lichen symbionts, we recommend making a paste of photobionts or mycobionts under a dissecting microscope rather than attempting to use thallus pieces. If you are interested in examining the membranes of photobionts, make a paste enriched in photobiont cells under a magnifier (although obviously some mycobiont cells will also be present). Cryofix a drop of this mixture for later fracture. A razor blade is the best tool for fracturing lichen samples. For cleaning the replica, sodium hypochlorite is used traditionally. However, we have found that bleach inflates the walls of the mycobiont cells producing strong tensions that break the replica, and we thus recommend a chromic acid mixture.

LTSEM Low Temperature Scanning Electron Microscopy (LTSEM) is an alternative to freeze-etching and TEM observations in the study of the plasma membranes. This technique is based on the same principles of cryofixation, etching and shadowing with a metal (in this case gold). With this technique details of the cellular membranes (e.g. invaginations) can be observed. However, due to the lower resolving power of the SEM in comparison to the TEM, intramembranous particles cannot be observed. High resolution scanning electron microscopy (HRSEM) connected to a high resolution Cryotrans camera solves this problem. For details on LTSEM see Chapter 8.

Results

Ultrastructure of both symbionts can be studied in ultra thin sections prepared as described in Subprotocol 1. In Figure 2, a lichenized *Trebouxia* cell is in the centre of the figure. The cell contains a large chloroplast that occupies the greater part of the cell volume and contains a pyrenoid at its centre. The pyrenoid is a structure that has electron-dense globules called pyrenoglobuli immersed in the pyrenoidal matrix. Thylakoidal membranes identical to those that constitute the chloroplast are observed at the centre of the pyrenoid. Mitochondria, vesicular complexes and marginal lipid bodies are observed in the cytoplasm of the *Trebouxia* cell. In this figure, some mycobiont cells occur around the *Trebouxia* cell. These fungal cells have a thinner cell wall than the algal cells and concentric bodies, lipid bodies and different kind of vacuoles in the cytoplasm.

In lichenized *Trebouxia* cells of lichen thalli, the immunocytochemistry protocol described above (Subprotocol 2.), has been used to study the enzyme Rubisco. Figure 3 shows gold particles concentrated mainly in the pyrenoid matrix, demonstrating the pyrenoidal localisation of this enzyme. The quality of ultrastructural detail preserved in samples prepared for immunocytochemistry (Fig. 3) is lower than those prepared with the conventional procedure (Fig. 2) (e.g. compare the chloroplast thylakoids of the two images). The suppression of osmium tetroxide fixation also changes the appearance of the specimens. For instance, pyrenoglobuli appear black when the sample is fixed with osmium tetroxide (Fig. 2) but white when the samples are prepared for immunocytochemistry (without osmium tetroxide fixation) (Fig. 3).

Observation of ultra thin sections of the samples treated using cryofixation and cryosubstitution (Fig. 4) reveals some ultrastructural differences when compared with samples prepared by conventional methods

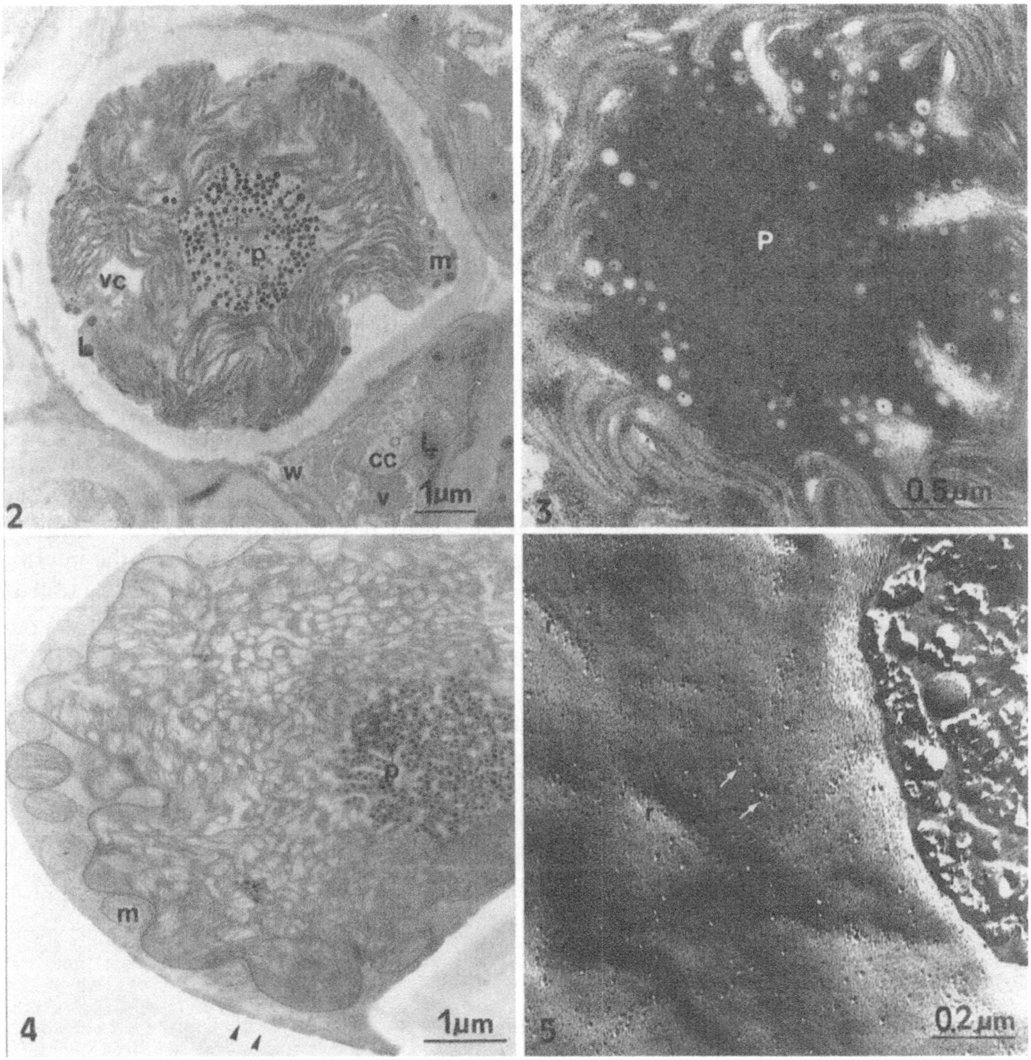

Fig. 2. Electron micrograph of *Trebouxia* photobiont and mycobiont cells from *Lasallia hispanica* processed with conventional protocols. cc, concentric bodies; L, lipid bodies; m, mitochondria; p, pyrenoid; v vacuoles; vc, vesicular complex; w, cell wall. **Fig. 3.** Immunogold-TEM localization of Rubisco in the *Trebouxia* photobiont of *Parmelia omphalodes*. p, pyrenoid. **Fig. 4.** Electron micrograph of a cryoprocessed (slamming and freeze-substitution) *Trebouxia* cell from *Neofuscelia pokonyi*. m, mitochondria, p, pyrenoid. Arrows indicate the depressions of the plasma membrane. **Fig. 5.** EF face (exoplasmic fracture face) of the *Trebouxia* plasmalemma from a freeze-fracture preparation of *Evernia prunastri*. r, ridges. Arrows indicate intramembranous particles.

(Fig. 2). Cryoprocessed cells and their organelles appear more turgid. This can be clearly seen in mitochondria that appear more rounded in cryofixed cells (Fig. 4). More details of the plasma membrane can be observed in cryoprocessed cells. The membrane looks smoother and is in close contact with the cell wall. Also, in cryofixed cells, it is possible to distinguish at irregular intervals tiny depressions in the plasma membrane. Plasmolysis is sometimes observed in samples treated with conventional techniques, but not observed in samples treated with cryomethods. Ultrastructural studies using cryotechniques reveal details not preserved by conventional chemical fixation and produce ultrastructural images more representative of the living state. The study of the relationships between lichen symbionts is facilitated by more precise information than this technique gives. We particularly recommend these techniques in studies where chemical fixation is too slow to reflect the precise cellular location of ions and molecules, and those where it is necessary to avoid the hydration produced in the conventional chemical fixation.

Freeze-etching is a good alternative to the transverse sections of membranes, because the preferential plane of the fracture exposes the interior of the lipid bilayer, thus exposing plasmalemma invaginations, as well as the proteins that constitute the intra-membranous particles. It allows determination of the size, distribution and density (number per unit area of the membrane) of intra-membranous particles. Fig. 5 shows the exoplasmic fracture (EF) face of the plasmalemma, where ridge and particle structures are visible (i.e. intra-membranous particles). Ridges correspond to the tiny depressions observed in ultra thin sections of cryo-processed samples (Fig. 4). Varying density and size of intra-membranous particles are connected with physiological changes of the lichen thalli (Rapsch et al. 1986). This way, these investigations can be very useful for physiological studies.

Troubleshooting

- Artefacts produced by fixation
  Aldehyde and osmium tetroxide fixation and dehydration of lichen thalli can occasionally produce artefacts. The fixative artefacts could include extraction of material, distortion of organelles, displacement of chemical components and anomalous deposits (Robards and Wilson 1993). Whenever possible it is better to interpret results in the light of information gathered as a result of other, complementary types of microscopy such as LTSEM that do not generate these artefacts.

- Incomplete dehydration
  If the dehydration is incomplete (the level is determined by the nature of the resin), resin infiltration will not be complete and it will be impossible to cut good ultrathin sections. Sections appear with holes or soft parts inside the resin. Incorrect dehydration can occur, if the ethanol is no longer anhydrous, the sample size is too large, if the dehydration process was too rapid or it can be caused by specific features of the sample. This last aspect is especially important in some species of lichens as their anatomy or presence of some components in the thallus, makes the process of dehydration difficult. Holes in a specimen can result in the breakage of ultra thin sections when the electron beam penetrates the sample. The presence of holes or irregularities is especially problematic when the section is going to be subjected to immunolabelling processes. Antibodies can join to these holes and produce high background.

- Embedding
  If the embedding is made at a lower temperature than required, the resin will be soft and the sections obtained irregular in thickness.

- Secondary lichen products
  The presence of secondary lichen products complicates the preparation of samples for TEM. Modifications of the pH of the fixative solutions by lichen substances are common and may cause problems during the fixation process. Lichen substances can also produce precipitates of the staining solutions on the ultra thin sections.

- Cryofixation
  As has already been discussed, so far, it is not possible to study totally dehydrated thalli with conventional protocols. The use of cryotechniques in TEM permits their study. However, these techniques are not fully optimised and often cryofixation is irregular and only occurs in a thin layer from the surface of the sample. Better results are obtained with hydrated samples than with dehydrated ones although both are irregularly cryofixed. Small specimen size (< 0.1 mm in all dimensions) improves the chance of getting fast cooling. The use of cryoprotectants can improve the cryofixation by immersion of hydrated samples, but a comparison with ultra-fast freezing results is required to show if artefacts are produced by the presence of the cryoprotectant. Cryotechniques are not easy to use with lichen thalli, but we must try to

improve them because they can be a useful tool to understanding dynamic processes during dehydration and hydration of fungal and photobionts cells. Chemical fixation takes longer and some rapid processes and ultrastructural changes cannot be fixed quickly enough.

Subprotocol 4
Confocal Laser Scanning Microscopy (CLSM)

Procedure

Fresh material

1. Make sections of the lichen thalli. Different kinds of sections can be used e.g. hand, cryostat or vibratome sections (lichen thalli mounted in agar blocks).
2. Optional: Staining of the lichen thalli with specific dyes that can reveal the structure that you want to observe (see Butt et al. 1989; Rost, 1995). For instance, lichen thalli can be stained with acridine orange (50 mg ml^{-1} aqueous solution).
3. Observation in the microscope. Combination of different wavelengths excitation lights (argon laser, 476/488 nm; krypton laser, 568 nm and He/Ne laser, 623 nm), excitation filters and emission filters permits the detection of different fluorophores and fluorescent components of the lichen thalli.

Material for immunolabelling or in situ hybridization

In specimens of lichen thalli that are going to be subjected to immunolabelling or in situ hybridization processes, fixation is required for facilitating the immobilisation of the cell components and probe accessibility. The vast majority of biological confocal imaging is accomplished with fixed material (Bacallao et al. 1995). Fixatives used in this type of microscopy can retain the three dimensional structure of the specimen. If the sample is going to be studied with fluorescence microscopy, careful selection of the fixative is required because some of them modify the fluorescent features of the material. Fixation with 4% paraformaldehyde in PBS

(130 mM ClNa and 10 mM sodium phosphate buffer at pH 7.4) have provided good results with lichens.

Saxicolous lichens

The ability to observe lichens while still attached to their substrate allows the analysis of the interface, and the relationships with other lithobiontic micro-organisms present in the same zone. Fixing and embedding the specimen facilitates the study of saxicolous lichens. Specimen preparation of this material is similar to that shown for TEM (Subprotocol 1). After embedding, the block of resin which includes lichen with its substrate is treated to produce a surface that permits studying it by microscopy. The method used is the same as that in the technique developed by Wierzchos and Ascaso (1994) for the study of lichen-substrate interface by scanning electron microscopy operating in Back Scattered electron mode and called "technique SEM-BSE".

Results

The application of Confocal Laser Scanning Microscopy to lichens has allowed the study of the spatial organization of the lichen symbionts in the lichen thallus. In thalli stained with acridine orange both symbionts can be visualised using argon laser excitation and filtering the emission with a long pass filter (> 515 nm) or a band pass (515-545 nm)(Fig. 6). A series of sectional images may be used to create three-dimensional images, where the spatial relationship of the photobiont and mycobiont in the thallus can be analysed (De los Ríos et al. 1999). This can be applied to physiological studies in which changes in the organisation of algal and fungus cells are of interest. For example, three-dimensional images of hydrated and dehydrated thalli have been compared and it has been shown that they have different spatial organisation. In hydrated thalli cells appear more independent and widely spaced in comparison with dehydrated thalli (De los Ríos et al. 1999).

Although the resolution of confocal laser microscope cannot be compared with TEM, it is also possible to study in detail some organelles and/or structures in the cells of mycobiont and photobiont (Fig. 6). Samples without staining can be studied for autofluorescence of some components such as chlorophylls in the photobiont. The fact that CLSM is not an invasive technique allows recognition of artefacts produced in the conventional preparation of the sample for TEM (De los Ríos et al. 1999). CLSM

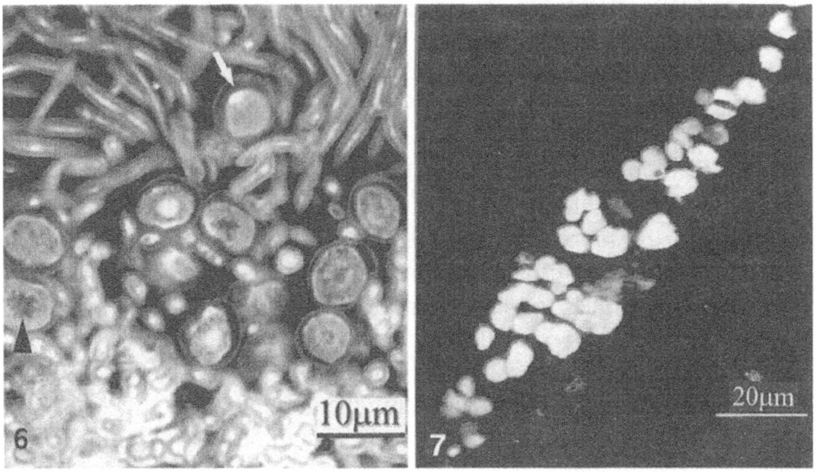

Fig. 6. CLSM image of a *Parmelia omphalodes* thallus stained with acridine orange (argon laser: 488 nm; emission band pass: 515-545 nm). The *black arrow* indicates a pyrenoid and the *white* the nucleus of the photobiont. **Fig. 7.** CLSM image of a fissure occupied by endolithic micro-organisms and situated close to a *Lasallia hispanica* thallus.

can profit from all advantages that fluorescence microscopy has. The use of different fluorochromes and filters makes it possible to distinguish different organelles or specific residues inside or on the cell but with the possibility of a three dimensional observation.

New aspects of the interface saxicolous lichen, lithic substrate and lithobionthic micro-organisms can be studied by confocal microscopy. This technique is unique for visualising endolithic micro-organisms and organisms in their natural undisturbed condition. Although the spatial resolution of CLSM images is not the same as that with the technique of scanning electron microscopy SEM-BSE (see Wierzchos and Ascaso, 1994), its application is an important complement to cytological information obtained with the cited method. The use of CLSM and "technique SEM-BSE" on the same sample zone allows, what Ascaso et al. (1998), have termed the "correlative microscopy" strategy, to be carried out. When auto-fluorescence is detected, the living material attached to the rock and inside of the mineral substrate can be visualised (Fig. 7). In addition, the possibility of producing 3D reconstruction by means of CLSM permits the visualisation of the spatial organisation of lithobionthic micro-organisms, and saxicolous lichen thalli, which occupy a determined volume of a fissure or cavity.

Troubleshooting

- Autofluorescence
  Before studying samples stained with fluorescent dyes it is essential to check the autofluorescence of lichen thalli at different wavelengths in order to avoid misinterpretations. Also, as discussed above, if fixed samples are used, the choice of fixative is important because fixatives affect tissue fluorescence.
- Photobleaching of fluorescence labelling agents
  Fluorescent preparations nearly always photobleach during irradiation. Photobleaching or progressive losts of fluorescent intensity during irradiation, is due mainly to photochemical reactions induced by the light used for excitation (Rost, 1995). In confocal microscopy this irreversible photochemical reaction can be particularly annoying because the entire three-dimensional sample volume is illuminated during a single scan cycle with the signal only collected in the plane of the focus (Known et al. 1993). We recommend that you illuminate fluorescent labels for as short a time as is needed to obtain a useful signal, and always use optimal collection conditions. The use of an anti-quenching mounting medium can help alleviate photobleaching, although, as it is highly toxic to living cells, you should only use it with fixed material.

Acknowledgements. The authors gratefully acknowledge Virginia Souza (Centro de Ciencias Medioambientales) for her assistance in the cryotechniques procedure and for producing Figure 4, and Dr Jacek Wierzchos (Universidad de Lerida) for his help on the confocal procedure. Thanks also to M Jose Villoslada and Fernando Pinto for their skillful technical assistance (Centro de Ciencias Medioambientales) and Beryl Ligus-Walker for the English revision. DGICYT PB95-0067 and the Austrian Science Foundation (FWF) P11998-Gen provided financial support.

References

Amos WB, White JG, Fordham M (1987) Use of confocal imaging in the study of biological structure. Appl Optics 26: 3239-3234

Ascaso C, Galvan J (1975) Concentric bodies in three lichen species. Arch Microbiol 105: 129-130

Ascaso C, Galvan J (1976) The Ultrastructure of the symbionts of *Rhizocarpon geographicum, Parmelia conspersa* and *Umbilicaria pustulata* growing under dryness conditions. Protoplasma 87: 409-418

Ascaso C, Brown DH, Rapsch S (1985) Ultrastructural studies of desiccated lichens. In: Brown DH (ed) Lichen physiology and cell biology. Plenum Press, New York, pp 259-274

Ascaso C, Brown DH, Rapsch S (1986) The ultrastructure of the phycobiont of desiccated and hydrated lichens. Lichenologist 18: 37-46

Ascaso C, Brown DH, Rapsch S (1988) The effect of desiccation on pyrenoid structure in the oceanic lichen *Parmelia laevigata*. Lichenologist 20: 31-39

Ascaso C, Valladares F, De los Ríos A (1995) New ultrastructural aspects of pyrenoids on the lichen photobiont *Trebouxia* (Microthamniales, Chlorophyta). J Phycol 31: 114-119

Ascaso C, Wierzchos J, De los Ríos A (1998) *In situ* investigations of lichens invading rocks at cellular and enzymatic level. Symbiosis 24: 221-234

Bacallao R, Kiai K, Jesaitis L (1995) Guiding principles of specimen preservation for confocal fluorescence microscopy. In: Pawley J B (ed) Handbook of biological confocal microscopy. Plenum Press, New York, pp 311-323

Balaguer L, Valladares F, Ascaso C, Barnes JD, De los Ríos A, Manrrique E, Smith EC (1996) Potential effects of rising trophospheric concentrations of CO_2 and O_3 on green algal lichens. New Phytol 132: 641-652

Balaguer L, Manrique E, Ascaso C (1997) Predictability of the combined effects of sulphur dioxide and nitrate on the green-algal lichen *Ramalina farinacea*. Can J Bot 75: 1836-1842

Balaguer L, Manrique E, De los Ríos A, Ascaso C, Barnes JD, Palmqvist K, Forham M (1999) Long term response of the green-algal lichen *Parmelia caperata* to a naturally CO_2 enriched atmosphere. Oecologia 119: 166-174

Boissière MC (1972) Cytologie du *Peltigera canina* (L.) Willd. en microscopie électronique: I. Premières observations. Rev Gen Bot 79: 167-185

Brown DH, Ascaso C, Rapsch S (1987) Ultrastructural changes in the pyrenoid of the lichen *Parmelia sulcata* stored under controlled conditions. Protoplasma 136: 136-144

Brown DH, Ascaso C, Rapsch S (1988) Effects of light and dark on the ultrastructure of lichen algae. Ann Bot 62: 625-632

Butt TQ, Hoch HC, Staples RC, Leger J St (1989) Use of fluorochromes in the study of fungal cytology and differentiation. Exp Mycol 13: 303-320

Culberson CC, Kristinsson H (1970) A standard method for the identification of lichen products. J Chom 46: 85-93

De los Ríos A, Ascaso C (1996). Changes in the photobiont related to the hydrated status of the lichen thallus. IAL 3 Progress and problems in lichenology in the nineties, Abstract book. pp. 72.

De los Ríos A, Ascaso C, Wierzchos J (1999). New prospects in the study of lichens with different state of hydration motivated by means of the combination of low temperature scanning electron microscopy and confocal laser scanning microscopy. Int Microbiol 2: 251-257

De los Ríos A., Ascaso C, Grube M (2001) Infection mechanisms of lichenicolous fungi studied by various microscopic techniques. Bibliotheca lichenologica (in press)

Dykstra MJ (1992) Biological electron micoscopy. Theory, techniques and trouble-shooting. Plenum Press, New York.

Ellis EA, Brown RM Jr (1972) Freeze-etch ultrastructure of *Parmelia caperata* (L.) Ach. Trans. Amer. Microsc. Soc. 91: 411-421

Eversman S, Sigal LL (1984) Ultrastructural effects of peroxyacetyl nitrate (PAN) on two lichen species. Bryologist 87: 112-118

Eversman S, Sigal LL (1987) Effects of SO_2, O_3 and SO_2 and O_3 in combination on photosynthesis and ultrastructure of two lichen species. Can J Bot 65: 1806-1818

Fiechter E, Honegger R (1988) Seasonal variations in the fine structure of *Hypogymnia physodes* (lichenized Ascomycetes) and its *Trebouxia* photobiont. Plant Syst Evol 158: 249-263

Galun M, Paran N, Ben-Shaul Y (1970) Electron microscopic study of the lichen *Dermatocarpon hepaticum* (ach.) Th. Fr. Protoplasma 73: 457-468

Galun M, Behr L, Ben-Shaul Y (1974) Evidence for protein content in concentric bodies of lichenized fungi. J Microscopie 19: 193-196

Grube M, Matzer M (1997) Taxonomic concepts of lichenicolous *Arthonia* species. In: Turk R and Zorer R (eds) . Progress and Problems in Lichenology in the nineties-IAL3, Bibliotheca lichenologica 68. J Cramer in der Gebr Bornträger Verlagsbuchhandlung, Berlin, Stuttgart, 1-17.

Grube M, de los Ríos A (2001) Observations in *Biatoropsis usnearum* (lichenicolous heterobasidiomycetes) and other gall-forming lichenicolous fungi using different microscopic techniques. Mycological Research 105 (in press)

Holopainen TH (1982) Summer versus winter condition of the ultrastructure of the epiphytic lichens, *Bryoria capillaris* and *Hypogymnia physodes* in central Finland. Annales Botanici Fenici 19: 39-52

Halpern S, Quintana C (1989) Two simple devices for quick-freezing and low temperature embedding. Biol Cell 67: 10a.

Honegger R (1984) Cytological aspects of the mycobiont-phycobiont relationship in lichens. Haustorial types, phycobiont cell walls types, and the ultrastructure of cell wall surface layers in some cultured and symbiotic myco- and phycobionts. Lichenologist 16: 111-127

Honegger R (1985) Fine structure of different types of symbiotic relationships in lichens. In: Brown DH (ed.) Lichen physiology and cell biology. Plenum Press, New York, pp 287-305

Honegger R (1986a) Ultrastructural studies in lichens I. Haustorial types and their frequencies in a range of lichens with trebouxoid phycobionts. New Phytol 103: 785-789

Honegger R (1986b) Ultrastructural studies in lichens II. Mycobiont and photobiont cell wall surface layers and adhering crystalline lichen products in four Parmeliaceae. New Phytol 103: 797-808

Honegger R, Peter M (1994) Routes of solute translocation and the location of water in heteromerous lichens visualized with cryotechniques in Light and Electron Microscopy. Symbiosis 16: 167-186.

Honegger R, Peter M, Scherrer S (1996) Drought-induced structural alterations at the mycobiont-photobiont interface in a range of foliose macrolichens. Protoplasma 190: 221-232

Jacobs JB, Ahmadjian V (1969) The ultrastructure of lichens. I. A general survey. J Phycol 5: 227-240

Janson S, Rai AR, Bergman B (1993) The marine lichen *Lichina confinis* (O.F. Müll.) C. Ag.: ultrastructure and localization of nitrogenase, glutamine synthetase, phycoerythrin and ribulose 1,5-biphosphate carboxylase/oxygenase in the cyanobiont. New Phytol 124: 149-160

Known YH, Wells KS, Hoch HC (1993) Fluorescence confocal microscopy: applications in fungal cytology. Mycologia 85: 721-733

Lallemant R, Boissiere JC, Boissiere MC, Leclerc JC, Velly P, Wagner J (1986) Le symbiose lichenique: approches nouvelles. Bull Soc Mycol Fr 133: 41-79

Mariac C, Rougier M, Gaude T, Dumas C (1992) Effects of fixatives on the antigenicity of *Brassica* S-locus specific glycoproteins and rapid immunolocalization by using the tissue print technique. Protoplasma 166°: 223-227

Minsky M (1961) Microscopy apparatus. U.S. Patent 3013467.

Minsky M (1988) Memoir on inventing the confocal scanning microscope. Scanning 10: 128-138

Palmqvist K, De los Ríos A, Ascaso C, Samuelsson G (1997) Photosynthetic carbon acquisition in the lichen photobiont *Coccomyxa* and *Trebouxia* (Chlorophyta). Physiol Plant 101: 67-76

Peveling E (1973) Fine structure. In: Ahmadjian V, Hale M (eds) The lichens. Academic Press, New York, pp147-182

Peveling E (1974) Biogenesis of cell organelles during the differentiation of the lichen thallus. Port Acta Biol 14: 357-368

Peveling E (1976) Investigations into the ultrastructure of lichens. In: Brown DH, Hawksworth DL, Bailey RH (eds) Lichenology: progress and problems. Academic Press, London, pp 17-26

Peveling E, Robenek H (1980) The plasmalemma structure in the phycobiont *Trebouxia* at different stages of humidity of a lichen thallus. New Phytol 84: 371-374

Quintana C (1991a) X-ray microanalysis of cell nuclei. J. Electron Microsc. Tchn. 18: 411-423

Quintana C (1991b) The use of low temperature embedding in elemental microanalysis, immunocytochemistry and ultrastrutural studies. Abs Colloque Franco-Iberique Microscopie Electronique, Université de Barcelona, (ed.). Barcelona (Spain), pp 58-59.

Quintana C (1992) Development of a new cryosystem of freeze-substitution and cryoembedding for processing biological samples. In: Rios A, Arias JM, Megias L (eds) Electron microscopy 92, vol. 1. Universidad de Granada, Granada (Spain), pp 363-364

Quintana C (1994) Cryofixation, cryosubstitution, cryoembedding for ultrastructural, immunocytochemical and microanalytical studies. Micron 25: 63-99

Rapsch S, Ascaso C, Cifuentes B (1986) Effects of the lipoxygenase enzyme on morphological changes of the EF face in the plasmalemma of the *Evernia prunastri* phycobiont. Cryptogamie Bryol Lichénol 7: 257-266

Reynolds S (1963) The use of lead citrate at high pH as electron-opaque stain in electron microscopy. J Cell Biol 17: 200-211

Robards AW, Sleytr UB (1985) Low temperature methods in biological electron microscopy. In: Glauert AM (ed) Practical methods in electron microscopy Vol 10. Elsevier Science Publishers B, Amsterdam, pp 1-551.

Robards AW, Wilson AJ (1993) Procedures in electron microscopy. Wiley, Chicester.

Rost FWD (1995) Fluorescence microscopy. Vol II. Cambridge University Press, Cambridge.

Scott MG; Larson DW (1986) The effect of winter field conditions on the distribution of two species of *Umbilicaria*. II. Fine structure and storage bodies distribution. New Phytol 102: 313-326

Sitte H, Fell H, Hobli W, Kleber H, Neumann K (1977) Fast freezing device. J Microscopy 111: 35

Tarhanen S, Holopainen T, Oskanen J (1997) Ultrastructural changes and electrolyte leakage from ozone fumigated epiphytic lichens. Annals of Botany 80: 611-621

Valladares F, Ascaso C (1994) Quantitative approach to the fine structure of both symbionts on nine *Umbilicaria* species. Protoplasma 178: 168-178

Wierzchos J, Ascaso C (1994) Application of backscattered electron imaging to the study of the lichen rock interface. J Microscopy 175: 54-59

Wright SJ, Centonze VE, Stricker SA, DeVries PJ, Paddock SW, Schatten G (1993). Introduction to confocal microscopy and three-dimensional reconstruction. In: Methods in cell Biology vol. 38. Academic Press, Boston, Mass.

Suppliers

All reagents should be the highest quality available, e.g. from Merck (Darmstadt, Germany) with the following exceptions: Osmium tetroxide and Pioloform can be bought from Agar Scientific LTD (Essex, UK) and glutaraldehyde (EM grade) from Fluka. LR-White acrylic resin is available from London Resin Company (Hampschire, UK) or Agar Scientific LTD (Essex, UK), and Low viscosity (Spurr) embedding Kit from Energy Beam Sciences (Massachutsetts, USA).

Preparative Techniques for Low Temperature Scanning Electron Microscopy of Lichens

BEAT FREY and CHRISTOPH SCHEIDEGGER

Introduction

Scanning electron microscopy (SEM) is an important tool for studying the cellular and subcellular structures of lichens and SEM has been particularly valuable for examining the external and internal morphology of cells and plectenchyma (Hale 1973; Garty and Delarea 1987; Jahns 1987; Lumbsch and Kothe 1992; Valladares et al., 1994; Honegger and Peter 1994; Scheidegger 1995) and for studying lichen taxonomy (Tibell 1991; Scheidegger 1993). SEM alone or combined with micro-analytical techniques has been used in environmental studies of lichens to examine calcium oxalate on the thallus surface of *Parmotrema reticulatum* (Modenesi 1993), to evaluate the impact of air pollution on the integrity of cell membranes of *Ramalina duriaei* (Garty et al. 1993) and to investigate the lichen-rock interface (Ascaso and Wierzchos 1994; Williamson et al. 1998). For most biological SEM applications it is necessary to fix specimens, in order to stabilise structures and functional states. The main problem encountered during preparation of biological specimens for electron microscopy (EM) arises from the necessity to transform the aqueous biological sample into a solid in which it can resist the physical impact of the electron microscope (high vacuum, electron beam irradiation). Fixation is achieved either by chemical (chemical fixation) or by physical treatments e.g. freezing (cryofixation). Chemical fixatives react relatively slowly and cannot preserve all cellular components. Subsequently, chemically fixed specimens are normally dehydrated in a medium such as acetone or alcohol. Unfortunately, exposure to such organic solvents has a number of disadvantages, because they can be the source of artefacts (removal of

✉ Beat Frey, WSL Swiss Federal Research Institute, Zürcherstraße 111, Birmensdorf, 8903, Switzerland (*phone* +41-1-7392541; *e-mail* beat.frey@wsl.ch)
Christoph Scheidegger

crystalline secondary lichen products, shrinkage, disruption and increased permeability of membrane-bound cell compartments, relocalisation of mobile ions and solutes within specimens). Cryofixation, however, offers the only way of stabilising liquid and semi-liquid aqueous systems in the native state for subsequent detailed microscopical study. It is possible to capture dynamic events in many cell physiological processes much more readily by rapid freezing techniques than by chemical preparation. Cryofixation is fast enough to retain the original distributions of inorganic elements of tissue electrolytes sufficiently for micro-analytical studies (Zierold and Steinbrecht 1987). The most significant recent advance in the SEM of fragile plant and fungal specimens is the availability of low temperature scanning electron microscopy (LTSEM). This technique permits the direct examination of hydrated specimens and precludes the artefacts associated with chemical fixation (Jeffree and Read 1991; Echlin 1992). Freeze-fracturing and subsequent LTSEM renders any kind of substitution medium unnecessary and is therefore a promising method in lichenological research. For a comparative evaluation of LTSEM with conventional SEM, the reader is referred to the work done by Beckett and Read (1986).

Since the first studies by Brown et al. (1987) LTSEM of freeze-fractured thalli has been used to investigate the ultrastructure of water-saturated and desiccated lichens. The authors showed that artefacts in the shape of algal cells are a result of chemical fixation (e.g. rapid rehydration of desiccated algal cells during fixation). In the past few years, various authors have found LTSEM to be a suitable method for studying turgor-controlled morphology and detecting free water (Honegger and Peter 1994; Honegger 1995; Scheidegger 1994; Scheidegger et al. 1995; Valladares et al. 1998). Recent LTSEM studies revealed that desiccation of water-saturated lichens led to strong cytorrhysis of the photobiont cells (Honegger and Peter 1994) and to the cavitation of thick walled fungal hyphae (Scheidegger et al. 1995). Furthermore, LTSEM is an important tool in the study of ion location in the lichen thallus using X-ray microanalysis where minimal ion redistribution is required. Thalli rehydrated with a solution containing tracers (La for apoplastic pathway and Rb for cell-to-cell transport) helps to identify pathways of solutes (Scheidegger et al. 1997). Attention in this chapter is directed to freezing techniques prior to LTSEM and energy dispersive X-ray microanalysis (EDX), because many research questions related to structural functionality of lichens cannot be answered with conventional SEM preparation techniques. Therefore, conventional wet chemical preparation steps will not be considered further here and the reader is referred to standard protocols of SEM laboratory techniques (Postek et al., 1980; Hall and Hawes 1991; Fowke 1995).

Outline

Modern LTSEM equipment consists of a high vacuum preparation chamber directly attached to the scanning microscope. In the preparation chamber, which is an evacuated chamber containing a cold stage cooled by liquid nitrogen, the specimen can be kept at a controlled temperature. It may be retained intact or fractured; and kept either fully frozen hydrated, partially freeze-dried ("etched"), or fully freeze-dried (Fig. 1). The samples may be coated for subsequent observation in the SEM. In non-dedicated systems the preparation chamber stands separate from the SEM, whereas in the dedicated system the preparation chamber is connected to the SEM by a high vacuum valve. The transfer of the sample is easily done by opening the gate valve between preparation chamber and SEM. The sample is then transferred onto a temperature controlled stage in the SEM where it is examined at very low temperature (e. g. -160°C). This cold stage located in the electron microscope specimen chamber is indirectly cooled (thermally insulated) from an external source.

To visualise an object by LTSEM, an electron beam is rastered across the surface of the specimen. In contrast to TEM, electrons do not penetrate the specimen, but are reflected as either secondary electrons (SE) or backscattered electrons (BSE). Secondary electrons originate from the specimen itself and are emitted after being pushed out from the electron shell by the electron beam. The signal produced is collected with an SE detector.

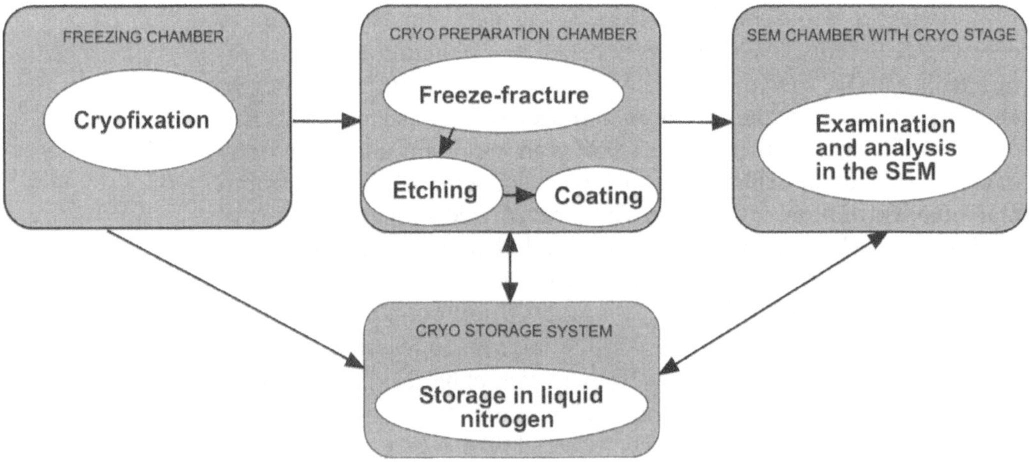

Fig. 1. Schematic flow sheet illustrating the steps for low temperature scanning microscopy

The BSE-signal is used to detect e.g. immunolabelled gold particles and is detected using a special BSE detector. For further details see Ascaso et al. (1998).

Materials

Equipment and Reagents

Equipment and materials can be obtained from electron microscope suppliers (see section Suppliers).

- Scanning electron microscope with a cold stage
- Secondary electron detector and / or backscattered electron detector
- An energy dispersive X-ray detector
- Software for the acquisition of spectra and spectra processing, and an image processing package for the display of images and digital X-ray maps
- Cryopreparation chamber attached to the SEM (dedicated system) or cryopreparation chamber isolated from the SEM (non-dedicated system)
- Sputter coating unit, fitted to the cryopreparation chamber
- Quartz thin film thickness monitor inside the cryopreparation chamber
- Vacuum pumps (rotary pump, turbomolecular pump) at the cryopreparation chamber
- Transfer freezing device (TFD)
- Mounting stubs
- SEM specimen support holder (copper support)
- Double-sided adhesive tape
- Liquid nitrogen dewar
- Cryotank for liquid nitrogen
- Cryo-glue (e.g. Tissue-Tek)
- Primary cryogen (liquid nitrogen)
- Secondary cryogen (e.g. propane)

- High purity argon gas
- Coating metals

Caution. Low temperature procedures carry a risk of cold burns, and care must be taken when handling liquid nitrogen and any cooled metal objects!

Procedure

Cryofixation

1. Cut the lichen sample in small pieces using a scalpel and subsequently blot the excised sample with filter paper to remove adhering surface water.
2. Place the sample on a sample support. Set the samples vertically on stubs using a cryo-glue (e.g. Tissue-Tek) to make freeze-fractures or mounted horizontally with double-sided adhesive tape for the analysis of surfaces.

   Note: Standard stubs provided with LTSEM systems have a flat surface. These stubs can be readily modified to suit particular specimen. Mounting the specimen in a hole is useful for a cylindrical object while mounting in a slit is suitable for transverse cryofractures of tissue slices. A wide range of custom-made SEM stubs should be now available, with holes, slots etc. of various dimensions which are designed to hold a small, excised piece of tissue. SEM stubs are commonly made of carbon or aluminium. The choice of material from which the stubs are made is only important in EDX analysis and depends on the elements of interest and their X-ray energies.
3. Using forceps, plunge-freeze the mounted specimen in a primary or secondary cryogen for at least 5 s. Then quickly transfer the specimen holder to the basket in the LN_2 and release it from the forceps.

   Note: If the water content of the samples should not be altered due to an aqueous cryoglue, then cryofixation without prior mounting in a cryoglue is essential. Specimens are then mounted afterwards onto the stubs. Mounting is either conducted mechanically or with a low-temperature glue such as a DMSO in water. For further details see Scheidegger (1994).

4. Store the frozen specimens in a liquid nitrogen dewar until required for observation. **Note:** If fresh tissue samples are plunge-frozen immediately after harvest without prior mounting on a stub, they should be kept at low temperature ($< -120°C$) until examination in the SEM. Frozen samples can then be mounted under low temperature in a transfer-freezing device (TFD).

Cryopreparation

5. Mount the stub with the specimen on a pre-cooled SEM specimen holder.

6. Transfer the specimen holder to the preparation chamber set at -80°C and partially freeze-dry in a high vacuum ($P < 2 \times 10^{-4}$ Pa) for 5-10 min. Sublimation of the specimen can be monitored visually in the preparation chamber.

   Note: Freezing rapidly transforms freezable cellular and extracellular water into its solid state (ice) and the specimen is considered to be fully frozen-hydrated (FH). Etching will remove superficial water droplets and films derived from environmental sources either naturally or as contaminants. If the sample is etched, the specimen may be then considered partially freeze-dried (PFD) and features of the surface can be exposed that would otherwise be obscured.

7. Fracture the sample with a microtome at -90°C. After fracturing, etch the fracture plane (if necessary) by keeping specimens for 30 s to 1 min at -85°C.

8. Purge the preparation chamber with argon gas raising the pressure to 2.2 Pa prior to coating and set the current at about 60 mA (for coating with platinum). Allow coating to proceed until a programmed thin film thickness monitor terminates at a specific coating thickness (5 - 15 nm).

   Note: For EDX analysis, leave the samples uncoated or use chromium or carbon (5 nm) as coating materials that normally do not interfere with the X-ray energies of elements of interest.

Examination in the SEM

9. After coating, transfer the specimen with a manipulator through the sliding vacuum valve onto the SEM cold stage with the temperature set to below -160°C (dedicated system) or transfer the specimen in a high vacuum cell onto the SEM cold stage (non-dedicated system).

10. Observe the frozen lichen samples (surface or fracture plane) by secondary emission (SE) electron mode of the SEM at accelerating voltages between 10 and 15 kV (higher kV for EDX).

11. Acquire images or EDX spectra of selected cell compartments and tissues.

Results

LTSEM shows an excellent structural preservation of the specimen and the resolution power of the SEM is not limited by the cryo-scanning unit (Fig. 2). The specimens retain their shape and surfaces are not distorted, ruptured, or collapsed. Preparation time for LTSEM is an order of magnitude faster than conventional preparation and is in most cases less than half an hour per specimen. Because neither chemical fixation nor organic solvents are needed in the preparation protocol for LTSEM, crystalline lichen compounds and their localization on the symbionts are well preserved (Fig. 2). Conventional preparation protocols including dehydration of the lichen thallus with an organic solvent and subsequent critical point drying usually dissolves crystalline lichen substances. Similarly, Honegger (1986) showed that freeze-drying of cryofixed lichen specimens proved to be

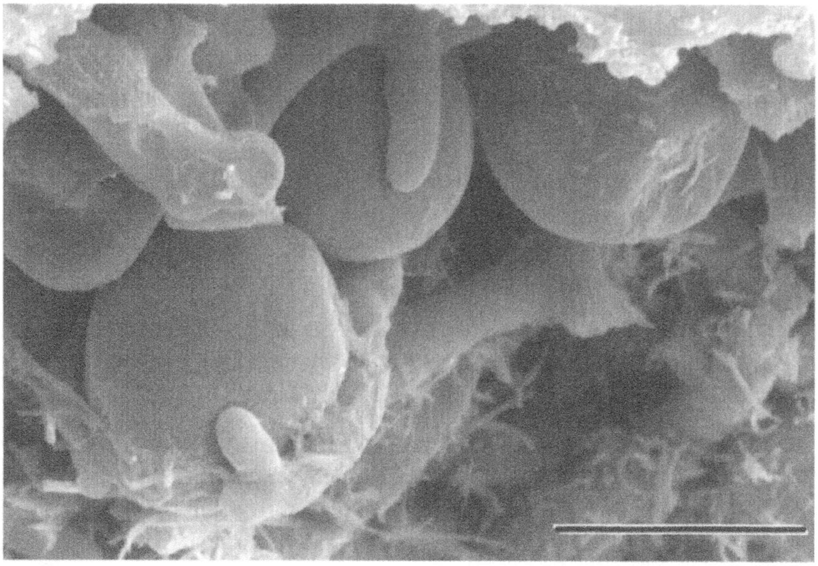

Fig. 2. LTSEM of freeze-fractured, partially freeze-dried thallus of *Umbilicaria aprina*. Appressoria-forming hyphae are short-celled and form contact zones with the photobiont. Medullary hyphae and the smooth surface of the turgid photobiont are covered with crystalline lichen substances. Bar = 10 μm

a useful method for the preservation of the cystalline secondary compounds on algal and fungal cell wall surfaces.

LTSEM is a suitable method where turgor-related functional states of lichens are investigated. Turgor-induced structures are preserved due to the physical fixation of the complete amount of water. Structural changes related to different turgor pressures at given water contents have previously been demonstrated (Scheidegger et al. 1995; Scheidegger et al. 1997). LTSEM of freeze-fractured thalli of air-dried *Lobaria pulmonaria* showed heavily collapsed cells of the photobiont *Dictyochloropsis reticulata* (Fig. 3a). Cross-fractured fungal hyphae revealed one cavity in the symplast of most cortical and medullary hyphae (Fig. 3a). LTSEM demonstrated that rehydration of a dry lichen thallus progressed gradually from the upper cortex to the medullary hyphae and the photobionts. Uptake of water led to a rapid refilling of the protoplast of cortical hyphae. Increasing water uptake led to a gradual unfolding of the green algal photobiont cells. After 100 s most photobiont cells were globular (Fig. 3b). During rehydration no free water was detected in the intercellular spaces in

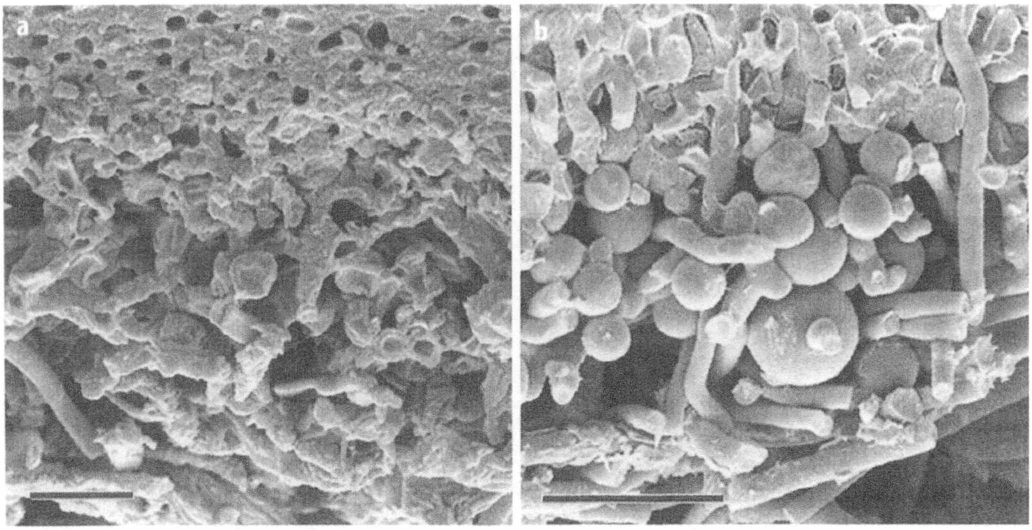

Fig. 3. a LTSEM-micrograph of a freeze-fractured thallus of *Lobaria pulmonaria*. Air-dried lichen thalli were mounted on aluminium stubs and plunged in liquid propane. Photobionts are heavily collapsed. Cortical and medullary hyphae are cavitated. Bar = 10 μm.
b LTSEM-micrograph of a freeze-fractured thallus of *Lobaria pulmonaria*. Air-dried lichen thalli were soaked in tap water for 100 s, mounted on aluminium stubs and plunged in liquid propane. Uppermost and lower photobiont cells are globular and fungal hyphae were refilled. Bar = 10 μm.

the medulla or in the photobiont layer (Fig. 3b). Water transport to the photobiont cells in *L. pulmonaria* was predominantly from the fungal apoplast through the apoplastic host-symbiont interface to the photobiont apoplast and symplast, as demonstrated by energy dispersive X-ray (EDX) microanalysis in the LTSEM of thalli rehydrated with a solution containing tracers for apoplastic (La) and cell-to cell-transport (Rb) (Scheidegger et al. 1997). This means that during the rapid rehydration of a desiccated lichen thallus, cell-to-cell water translocation played a negligible role and the rapid rehydration was predominantly supported through the apoplastic space.

LTSEM has not only proved to be a suitable method for the study of structural changes due to desiccation and rehydration at ambient temperatures but also for detecting extracellular water in the spongiostratum of *Anzia japonica* and for showing extracellular ice formation during slow freezing in *Umbilicaria aprina* (Fig. 4). When water saturated *U. aprina* were slowly cooled at sub-zero temperatures, extracellular freezing of

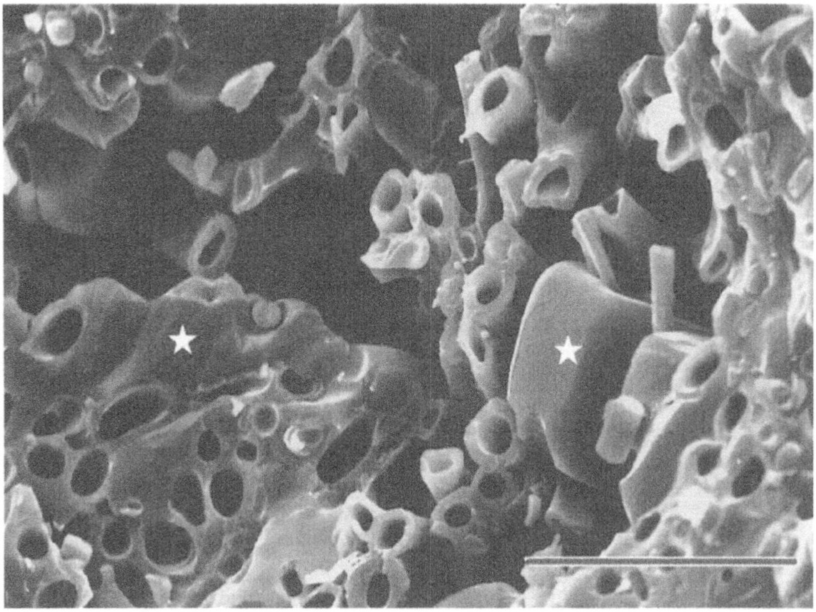

Fig. 4. LTSEM micrograph of a freeze-fractured water-saturated thallus of *Umbilicaria aprina*, frozen at slow cooling rates (1°C min^{-1}) to -20°C. Cortical and medullary hyphae are cavitated. Intercellular spaces are partly filled with extracellularly frozen hexagonal ice (asterisk). Bar = 10 µm

water was detected (Schroeter and Scheidegger, 1995). Ice crystals occupied a considerable amount of the intercellular air space, and often completely surrounded medullary hyphae and photobiont cells (Fig. 4). If water-saturated slowly frozen thalli were thawed after extracellular ice formation had occurred, all water that was previously frozen extracellularly on medullary hyphae and photobionts was absorbed through the apoplast and again no free extracellular water could be detected. Furthermore, the same study showed that the extracellular ice formation leads to cytorrhysis in the photobiont cells and to cavitation in the mycobiont cells. Both processes were reversible if the lichen thallus was thawed. Similar structural changes occurred during the processes of desiccation and rehydration at ambient temperatures (Honegger and Peter 1994; Scheidegger 1994).

LTSEM-EDX in combination with mercury intrusion porosimetry is a suitable tool to study the pore-size distribution in lichens. Fig. 5 is an X-ray map of mercury in a freeze-fracture across a thallus of *Lobaria pulmonaria* intruded with mercury. The elemental map shows the distribution of deposits containing mercury in a cross-fractured thallus (Fig. 5b) and clearly shows that mercury is located in the medullary intercellular spaces of the lower part of the algal layer (the gas filled interior) and is lacking in the

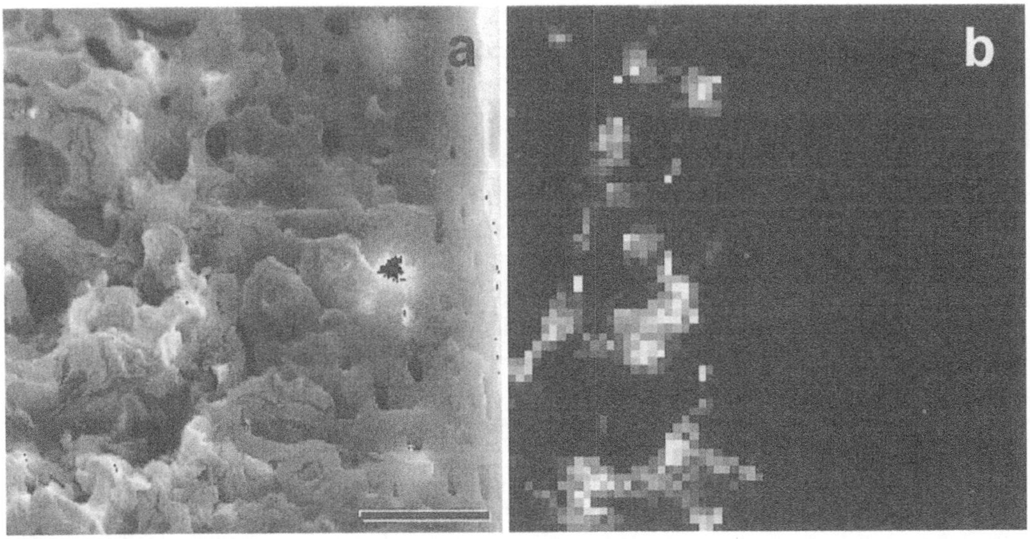

Fig. 5. **a** LTSEM micrograph of a freeze-fractured thallus of *Lobaria pulmonaria* intruded with mercury (resolution 512 x 512). Scale bar = 10 µm.
b Mercury mapping of the same freeze-fractured thallus intruded with mercury obtained with the energy dispersive X-ray microanalysis system (resolution 64 x 64).

upper cortex (Fig. 5a). This means that no fine pores exist in the upper cortex. Therefore, water uptake through the intercellular space (fine pores) can be excluded in *Lobaria pulmonaria*. Mercury porosimetry is therefore a valuable estimate for the intercellular medullary air space in *Lobaria pulmonaria* and other lichens with a hydrophobic surface of the medullary hyphae. However, mercury porosimetry is not usually a valid estimate for water storage capacity, except for lichens with easily wettable medullary hyphae (Scheidegger et al. 1997; Valladares et al. 1998).

Troubleshooting

Specific technical problems related to the electron microscope, the detectors, and the vacuum pumps are beyond the scope of this chapter and the reader is advised to contact the manufacturer's technicians. Difficulties with the preparation of the specimen may appear to varying degrees during a number of stages in specimen preparation and observation.

- Mounting the specimen
  The sample must be securely attached to the stub so that during manipulations inside the specimen chamber it will not fall off. Furthermore, the stub must be firmly attached to the specimen holder in order to guarantee high thermal conductivity and maximize the heat exchange rate at the cryogenic surface.
- Freezing
  One of the most critical parameters for successful cryofixation is the initial sample treatment. Specimens and their holders should be as small or as thin as possible. The first 20 µm will ultimately become the well cryofixed zone. Rapid freezing is important, because the higher the freezing rate, the less time is available for the formation of large ice crystals. This can be achieved by plunging the samples into liquid propane. Beware of cold gas layers above coolants that can freeze the specimen prematurely.
- Handling of the frozen specimen
  Once the sample is cryofixed, care must be taken to assure that rewarming does not occur during subsequent manipulations. Always use pre-cooled forceps when touching frozen specimen. The transfer of the specimen to the preparation chamber should be as fast as possible in order to prevent contamination due to frost.

- Freeze-fracturing
  The fracture plane is difficult to control. A small mound of cryo-adhesive in the hole or slit will freeze around the specimen to form a rigid sleeve, and the specimen often breaks close to the top of this, at or above the thinnest point. Unless the fracture plane of interest is exposed, repeat fracturing by transferring specimens again to the cold stage of the preparation chamber until the fracture plane of interest is exposed. In some cases it is advisable to fracture the specimen (very thick samples or very delicate samples) outside the preparation chamber in LN_2. Repeated fractures of the specimen can be made outside and should be done with the aid of a dissecting microscope. If the fracture knife is not sufficiently pre-cooled prior to fracturing or has insufficient clearance angle then smearing of the fracture face may occur.

- Condensation of water vapour
  Condensation of water vapour on the specimen surface is a common artefact in LTSEM and may have various sources (vacuum leakage, time of etching too short, etching temperature too low). If condensation of water vapour on the fracture surface occurs (which is normally not the case), then fracturing at a higher temperature may help to overcome this problem. Generally a longer etching time reduces the possibility of having condensation of water vapour on the surface. It should be taken into account that etching in the preparation chamber must be performed empirically. The extent of the etching process is unknown until the specimen is inspected on the cold stage of the microscope. If insufficient etching has been achieved, the process may be repeated only when the sample has not been coated before.

- Electric charging in the SEM
  Specimen charging is one of the most common but potent sources of image deterioration and artefact in LTSEM. Prolonged coating is useful to overcome charging on the specimen surface. Excessively thick coatings may, however, mask fine structures. The amount of coating may vary with specimen type. It is therefore necessary, that each user adopt his or her own routine for coating.

Comments

Conventional SEM at room temperature is no doubt a more widely used method for examining the morphology of lichens than the LTSEM technique. LTSEM needs special instruments (SEM with a cryo-stage) and the use of cryotechniques requires experienced users. In contrast, many aspects related to structural functionality in the lichen research can only be investigated with the LTSEM method. Frozen material is very useful for histochemical or X-ray microanalysis studies because the freezing process stabilizes some diffusible elements which could not be preserved by chemical fixation methods and frozen specimens may be freeze-fractured to reveal complex fracture planes through internal structures, which could not be visualised by other means (e.g. conventional SEM). Besides several advantages of cryofixation, e.g. the ease and speed of specimen preparation, there are several difficulties that should be taken into account when studying lichens using LTSEM. Cryofixation of water-containing samples is a very complex physical process. Many artefacts common in LTSEM arise from the properties and behaviour of water during cryofixation and the cryopreparation of specimens. As ice crystals grow during the freezing process, they sweep solutes and solid structures into a eutectic boundary between neighbouring ice crystals. Thus cryofixation can have a significant effect in altering the spatial relationship between structures and the distribution of solutes in a specimen. A comprehensive review of the field as a whole can be found by Echlin (1992).

References

Ascaso C, Wierzchos J, De los Rios A (1998) In situ cellular and enzymatic investigations of saxicolous lichens using correlative microscopical and microanalytical techniques. Symbiosis 24:221-233

Ascaso C, Wierzchos J (1994) Structural aspects of the lichen-rock interface using backscattered electron imaging. Bot Acta 107: 251-256

Beckett A, Read ND (1986) Low temperature scanning electron microscopy. In: Aldrich HC, Todd WJ (eds) Ultrastructure techniques for microorganisms. Plenum Press, New York London, pp 45-86

Brown DH, Rapsch S, Beckett A, Ascaso C (1987) The effect of desiccation on cell shape in the lichen *Parmelia sulcata* Taylor. New Phytol 105:295-299

Echlin P (1992) Low-temperature microscopy and analysis. Plenum Press, New York

Fowke LC (1995) Transmission and scanning electron microscopy for plant protoplasts, cultured cells and tissues. In: Gamborg OL, Phillips GC (eds) Plant cell, tissue and organ culture. Springer Lab Manual. Springer, Berlin Heidelberg New York, pp 229-238

Garty J, Delarea J (1987) Some initial stages in the formation of epilithic crustose lichens in nature: a SEM study. Symbiosis 3:49-56

Garty J, Karary Y, Harel J (1993) The impact of air pollution on the integrity of cell membranes and chlorphyll in the lichen *Ramalina duriaei* (De Not.) Bagl. transplanted to industrial sites in Israel. Arch Environ Contam Toxicol 24:455-460

Hale MJ (1973) Fine structure of the cortex in the lichen family Parmeliaceae viewed with the scanning-electron microscope. Smithsonian Contributions to Botany 10:1-92

Hall JL, Hawes C (1991) Electron microscopy of plant cells. Academic Press, London

Honegger R (1986) Ultrastructural studies in lichens. II. Mycobiont and photobiont cell wall surface layers and adhering crystalline lichen products in four parmeliaceae. New Phytol 103:797-808

Honegger R, Peter M (1994) Routes of solute translocation and the location of water in heteromerous lichens visualized with cryotechniques in light and electron microscopy. Symbiosis 16:167-186

Honegger R (1995) Experimental studies with foliose macrolichens: fungal responses to spatial disturbance at the organismic level and to spatial problems at the cellular level during drought stress events. Can J Bot 73 (Suppl 1): 569-578

Jahns HM (1987) New trends in developmental morphology of the thallus. In: E. Peveling (ed.) Progress and problems in lichenology in the Eighties. Bibl. Lichenol. 25: 17-33

Jeffree CE, Read ND (1991) Ambient- and low-temperature scanning electron microscopy. In: Hall JL, Hawes C (eds) Electron microscopy of plant cells. Academic Press, London, pp 313-413

Lumbsch HT, Kothe HW (1992) Thallus surfaces in Coccocarpiaceae and Pannariaceae (lichenized ascomycetes) viewed with scanning electron microscopy. Mycotaxon 43:277-282

Modenesi P (1993) An SEM study of injury symptoms in *Parmotrema reticulatum* treated with paraquat or growing in sulphur dioxide-polluted air. Lichenologist 25:423-433

Postek MT, Howard KS, Johnson AH, McMichael KL (1980) Scanning electron microscopy - a student's handbook. Ladd Res. Industries

Scheidegger C (1993) A revision of European saxicolous species of the genus Buellia De Not. and formerly included genera. Lichenologist 25: 315-364

Scheidegger C (1994) Low temperature scanning electron microscopy: the localization of free and perturbed water and its role in the morphology of the lichen symbionts. Crypt Botany 4: 290-299

Scheidegger C (1995) Reproductive strategies in *Vezdea* (Lecanorales, lichenized Ascomycetes): a low temperature scanning electron microscopy study of a ruderal species. Crypt Bot 5: 163-171

Scheidegger C, Schroeter B, Frey B (1995) Structural and functional processes during water vapour uptake and desiccation in selected lichens with green algal photobionts. Planta 197: 399-409

Scheidegger C, Frey B, Schroeter B (1997) Cellular water uptake, translocation and PSII activation during rehydration of desiccated *Lobaria pulmonaria* and *Nephroma bellum*. Bibl Lichenol 67:105-117

Schroeter B, Scheidegger C (1995) Water relations in lichens at subzero temperatures: structural changes and carbon dioxide exchange in the lichen *Umbilicaria aprina* from continental Antarctica. New Phytol 131: 273-285

Tibell L (1991) A new species of *Stenocybe* from Western North America. Bryologist 94: 413-415

Valladares F, Ascaso C, Sancho LG (1994) Intrathalline variability of some structural and physical parameters in the lichen genus *Lasallia*. Can J Bot 72:415-428

Valladares F, Sancho LG, Ascaso C (1998) Water storage in the lichen family Umbilicariaceae. Bot Acta 111:99-107

Williamson BJ, McLean J, Purvis OW (1998) Application of X-ray element mapping across the lichen-rock interface. J Microsc 191:91-96

Zierold K, Steinbrecht RA. (1987) Cryofixation of diffusible elements in cells and tissues for electron probe microanalysis. In: Steinbrecht RA, Zierold K (eds) Cryotechniques in biological electron microscopy. Springer, Berlin Heidelberg New York, pp 272-282

Suppliers

Principal Suppliers of microscope manufacturers, materials and equipment for EM specimen preparation and EM Accessories can be found in the internet under:
http://www.microscopy-analysis.com/linkscom.htm
http://www.microscopy-analysis.com/services.htm

Abbreviations

BSE	backscattered electrons
DMSO	dimethyl sulphoxide
EDX	energy dispersive X-ray
EM	electron microscopy
FH	frozen hydrated
LN_2	liquid nitrogen
LTSEM	low temperature scanning electron microscopy
PFD	partially freeze-dried
SE	secondary electrons
SEM	scanning electron microscope
TEM	transmission electron microscope
TFD	transfer freezing device

Physiology and Ecophysiology

Measurement of Chlorophyll Fluorescence in Lichens

MANFRED JENSEN

Introduction

Chlorophyll fluorescence principles

Chlorophyll fluorescence (CF) is the absorption of blue or red photons (= excitation) by chlorophyll molecules and the emission of far red photons (the energy difference is lost as heat). In contrast to chloroform solutions of chlorophyll or to fluorescent dyes, the emission from active photosynthetic systems does not remain constant, but changes in a complicated way after a constant excitation light is switched on. These changes (= variable fluorescence) are due to the onset of photosynthetic reactions that consume absorbed quanta. Accordingly, several derived fluorescence parameters (see below) provide diagnostic information about the state of the photosynthetic apparatus.

The Kautsky curve, i.e. the emission time course in the range of about 0 to 1500 ms, is very sensitive to changes at the acceptor side of photosystem II (e.g. PS II herbicides) and at the donor side of PS II (e.g. desiccation stress). The term "induction curve" is used with the same meaning, but may also comprise the period of dark/light induction of photosynthetic dark reactions, i.e. the time regime of a few minutes.

In conventional (non-modulated) fluorometers only one light type serves as actinic light (= light that is strong enough to drive photosynthesis) and as excitation light for fluorescence. The more sophisticated fluorometers use two separate light sources: a strong light source for driving the photosynthetic reactions and a very weak measuring beam of pulse modulated light for an extra-excitation of the chlorophylls. In principle, the modulation technique is somewhat analogous to a TV remote control

Manfred Jensen, Universität-GH Essen, FB 9 - Botanik, Universitätsstr. 5, Essen, 45117, Germany (*phone* +49 (0)201 1832450; *fax* +49 201 1834290; *e-mail* manfred.jensen@uni-essen.de)

where modulated infrared light switches the programs, but continuous infrared light from the sun is ignored (electronically filtered sensor). In the corresponding fluorometers the modulated light acts as a weak excitation beam; the instrument measures only that part of the fluorescence emission that results from excitation by the pulse-modulated light. Changes in ambient light will not disturb the signal: additional excitation by sunlight (which increases the emission intensity by several orders of magnitude) is simply ignored. Modulated systems measure proportional changes to the emission yield and not the absolute emission intensity. The variable fluorescence emission yield of photosynthetic systems is precisely monitored. In addition, modulated systems use a third type of light: the saturation pulse light, i.e. very bright excitation flashes (intensity > sunlight) of about 1 s in duration. This light saturates all photosystem II centres for a short time, resulting in a transient maximum of emission yield. After the saturation pulse the fluorescence yield returns to the F level (or to the F_0-level if dark adapted). For a more comprehensive overview see Schreiber and Bilger (1993).

It should be noted that all output from modulated fluorometers is an emission yield, but some manufacturers of fluorometers identify the term "yield" with the derived CF parameter $\Delta F/F_m$'. This inaccuracy unfortunately causes confusion. Note also the distinction between fluorescence yield and measures of quantum yield calculated from derived fluorescence parameters. A brief description of CF levels and CF derived parameters is given in the following:

Levels of CF emission

F_p	Peak level of emission in Kautsky curves (lower than F_m)
F_m	Transient maximum emission yield of dark adapted samples during a single saturation pulse
F_m'	Transient maximum emission yield of samples illuminated with continuous actinic light **and** a single saturation pulse
F_o	Emission yield before the onset of actinic light (modulated measuring light on)
F_o'	Emission yield after switching the actinic light off. Some machines use an additional far red light pulse after the actinic light for a better determination of the F_0' level (lower values), but caution is needed. In lichens, too strong a far red pulse may increase rather than decrease F_0'. Therefore, instruments with automatic far red pulse are not necessarily automatically better than those without far red pulse.

CF derived parameters (after Schreiber and Bilger 1993)

	Definition	Meaning	Restriction
F_v/F_m	$(F_m - F_o) / F_m$	maximum efficiency of photosystem II (dark adapted samples); monitors a sensitive part of the photosynthetic light reaction; it is independent of the ambient temperature	dark adaptation necessary
$\Delta F/F_m'$	$(F_m' - F) / F_m'$	efficiency of photosystem II of illuminated samples (during running photosynthesis)	$\Delta F/F_m'$ decreases when light intensity increases, therefore the concomitant measurement of the actinic light intensity is important. Some machines make use of stored light intensity values, that have been determined previously.
Φ_{PSII}	identical to $\Delta F/F_m'$		
ETR	(relative) electron transport rate given by the formula $ETR = \Delta F/F_m' * PPFD * 0.42$ [µmol electrons * m^{-2} * s^{-1}]; PPFD = photosynthetic photon flux density = incident light intensity in the wavelength range 380 - 710 nm (because of the critical proportionality factor 0.42: **relative** ETR); divide the ETR by 4 for estimation of the photosynthetic CO_2 fixation rate during steady state photosynthesis		includes alternative electron transport types that do not contribute to CO_2 fixation like photorespiration and electron transport to O_2 (Mehler reaction)
NPQ	$(F_m - F_m') / F_m'$	nonphotochemical quenching, largely related to energy dissipation by heat production. Indicates photoprotection or photoinhibition (more sensitively than F_v/F_m)	mechanistic relation to ΔpH and xanthophyll cycle pigments not exactly known (Li et al. 2000)
q_{NP}	$(F_m - F_m' + F_o' - F_o) / (F_m - F_o)$	like NPQ, but less well correlated to xanthophyll cycle pigment ratios than NPQ; for automatic calculation of q_N, some machines omit $F_o' - F_o$ in the numerator see comments p. 146	
q_P	$(F_m' - F) / (F_m' - F_o')$	photochemical quench parameter, indicates the oxidation state of the PS II acceptor side: $q_P = 0$ if the acceptor side is fully reduced, $q_P = 1$ if the acceptor side is oxidised; for automatic calculation of q_P, some instruments replace F_o' by F_o, see comments p. 146	

Chlorophyll fluorescence of lichens

In recent years, progress in commercial CF equipment has made possible measurements not only in the laboratory but also in the field. In principle, CF should replace conventional, costly gas exchange measurements and make (eco-) physiological work accessible to more people and applicable to the measurement of crustose lichens. However, gas exchange measurements cannot be abandoned altogether, because their relation to chlorophyll fluorescence-derived electron transport rates is often non-linear (Schroeter et al. 1995; Leisner et al. 1997; Green et al. 1998). Therefore the quantitative estimation of CO_2 fixation rates or primary production of the photobionts is not possible by simple chlorophyll fluorescence techniques. Rather, the value of fluorescence techniques is in the diagnosis and analysis of photoinhibition (Manrique et al. 1993; Leisner et al. 1995; Gauslaa and Solhaug 1996; Gauslaa et al. 1996), environmental stress (Scheidegger and Schroeter 1995; Calatayud et al. 1996) and indicating when a lichen is photosynthetically active. Recently, some approaches have been published that extend the application of fluorescence techniques. For example, estimation of the photosystem II efficiency of cyanobacterial photobionts now appears possible (Sundberg et al. 1997). In addition, it is possible to carry out 2-dimensional imaging of the fluorescence of lichen thalli (Jensen and Siebke 1997). This chapter outlines the uses of some fluorescence techniques and parameters derived from them, and the simultaneous measurement of O_2 evolution.

Subprotocol 1
Determination of the Chlorophyll Fluorescence Parameter F_v/F_m

Materials

It is possible to use any of the pulse modulated systems outlined in Subprotocol 2.

Here are some simpler instruments:

- Hansatech Instruments Ltd.
  PEA: illumination by LED's, only for rapid screening of fluorescence induction and F_v/F_m, integral data storage
- ADC Bioscientific
  OSI30: for rapid screening of F_v/F_m, integral data storage

Procedure

Measurement of green algal lichens

1. If the samples are dry, spray them with water. Remove excess water with absorbent paper, but take care to keep the sample wet during the experiment (unless you are investigating the effects of desiccation).
2. If the samples are not fresh, and you want to take measurements in the laboratory, activate lichens by storing them wet in the dark (1 - 3 days, depending on the collection/storage conditions) or at low light [30 µmol photons m^{-2} s^{-1}; cf. Palmqvist (1993)] at relatively low temperature (e.g. 4 - 7°C in a refrigerator).
3. In the field, mount the distance or dark adaptation clips over the samples (adhesive tape, nails or you may need to make some special device). Install the light guide of the fluorescence measuring system.
4. Set the saturation pulse width to 1 s (see Troubleshooting).
5. The necessary pre-darkening period should if possible be determined by preliminary trials. For this purpose, illuminate samples with a relatively high light intensity for 10 min (400 µmol photons m^{-2} s^{-1} or less: 200 for shade adapted samples) and observe the recovery of F_v/F_m following these periods in the dark: 10 s, 30 s, 60 s, 2 min, 4 min, 8 min, and 16 min. From the time course of F_v/F_m recovery estimate the shortest pre-darkening period needed to get stable maximum values.
6. Darken the samples by black foil (a large area of black foil is needed, available from photo supply shops) or black velvet for the time determined in Step 5. Alternatively, use 30 min if the optimal time is not known.
7. Fire the saturation pulses (= taking a measurement). Low light while mounting the light guide from one clip to another is not critical, but the respective sample should be darkened for one further minute.

Measurement of cyanolichens or cephalodia

In principle, the same protocol as for green algal lichens can be used. However, this does not yield the maximal F_v/F_m values. Instead of pre-darkening, the lichens should be kept in low light prior to the F_v/F_m determinations [(5 - 20 µmol photons m^{-2} s^{-1}; cf. [Sundberg et al. 1997)]. Estimate the appropriate low light intensity by pilot tests at 0, 5, 10, 20 and 30 µmol photons m^{-2} s^{-1}.

Troubleshooting

- When a miniature fibre-optic light guide (c. 2 mm active diameter, Walz) is used for fluorescence measurements, the applied light intensities may be extremely high. With this instrument care must be taken that the saturation pulse intensities do not exceed 6000 µmol photons m^{-2} s^{-1} (set saturation pulse intensities to the lowest values).
- For the proper determination of F_v/F_m (and $\Delta F/F_m'$, see Subprotocol 2) values, there are differences in the optimal duration of the saturation pulse that is used to get the maximal fluorescence F_m. For example, with the PAM 2000 apparatus, too long or too short saturation pulse widths have to be avoided. Otherwise, the transient maximum of fluorescence is not determined correctly. From the analysis of chlorophyll fluorescence induction it can be shown that for *Trebouxia* photobionts shorter pulse widths are needed than for *Coccomyxa* or *Dictyochloropsis* (Jensen et al. 1999). With most instruments a saturation pulse width of 1 s is a good choice (the lag phase for switching the pulse light is critical). However, at low temperatures (e.g. that may occur when carrying out field work), the F_m value may be reached later. The saturation pulse width must be carefully controlled.

Comments

- General vitality check
  As the determination of F_v/F_m is the quickest way (only seconds for a single measurement) to get data about a large number of specimens and as this parameter is the only one that is not temperature sensitive, it appears to be an ideal tool for vitality screening of lichens not only in the laboratory but also in the field. F_v/F_m is thought to monitor the maximal photochemical efficiency of photosystem II. It is remarkably constant (value ca. 0.83) among higher plants (Björkman and Demmig 1987) but is usually lower in algae and lichen photobionts (0.63 - 0.76). The main prerequisites for a proper determination are premoistening and dark adaptation of the samples. The dark adaptation time depends on the preillumination (time and intensity) of the samples. The shortest dark adaptation times required to get unbiased values should be determined for a given population of lichens specimen. For active, wet samples a dark adaptation time of 30 min is long enough, but often much shorter times (20 - 5 min) are sufficient. In the procedure a recommendation is given on how to estimate the optimal times.

- SO_2 effects and air pollution studies
  Whether the parameter F_v/F_m can monitor the health of lichens or not depends on the kind of stress you wish to study. Fumigation studies with a concentration of 1.0 ppm SO_2 or higher caused a large decline in F_v/F_m and other chlorophyll fluorescence parameters after 6 hours (Gries et al. 1995). These alterations could be detected before any other signs of injury. In nature, however, such severely stressful conditions are rare. At least during longer periods of exposure they will lead to the death of lichens. It is less clear how accurately F_v/F_m monitors milder damage. For *Parmelia quercina* populations measured in areas subjected to different levels of pollution no significant differences in F_v/F_m could be established (Calatayud et al. 1996). Field studies on *Hypogymnia physodes* growing in different areas revealed a remarkable constancy of F_v/F_m (Jensen 1994). Even thalli infected with *Lichenoconium erodens* did not show a decrease in F_v/F_m in the non-infected margin parts, while there was a negligible chlorophyll fluorescence level in the inner parts because of chlorophyll degradation. Evidently, the reaction of the photobionts was all or nothing. A recent air pollution study, on the other hand, clearly revealed reductions in F_v/F_m from the more polluted areas for the species *Hypogymnia physodes, Pseudevernia furfuracea,* and *Platismatia glauca* (Niewiadomska et al. 1998). An investigation on the influence of airborne acidic deposition on *Lobaria pulmonaria* also demonstrated that F_v/F_m reduction was correlated with chlorophyll degradation and decreasing pH of thallus water (Gauslaa et al. 1996). Thus this type of damage to *L. pulmonaria* thalli can be monitored by measurements of F_v/F_m. Senescing photobionts occur naturally in mat forming lichens. The cell material at the more basal regions is necrotic or dead while the material at the tip is healthy (Crittenden 1991). Accordingly, the senescence of photobionts in lower parts of *Cetraria cucullata* is indicated by low F_v/F_m values. Generally it seems to be evident that conditions that lead to thallus senescence or disintegration actually can be monitored by F_v/F_m measurements.

- Photodestruction studies
  F_v/F_m measurements have also been used to investigate the susceptibility of lichens to photodestruction. In such laboratory studies, lichens were artificially exposed to high light (1000 µmol photons m^{-2} s^{-1}) for several hours and the recovery of chlorophyll fluorescence parameters was measured after low light periods. In this case F_v/F_m sensitively indicated the extent of photodestruction. The response of lichen species to these treatments varied according to the microhabitat of their nor-

mal growing conditions (Manrique et al. 1993). For some *Lobarion* species, even irradiation in the dry state for 2-3 days could be deleterious and lead to irreversible reductions in F_v/F_m (Gauslaa and Solhaug 1996). Obviously, high light intensities can lead to permanent damage or slowly recovering F_v/F_m values. As discussed above, the normal dark adaptation times of lichens are in the range of minutes. Thus, low predawn F_v/F_m values indicate maintained inhibition of PS II. However, low values alone do not indicate whether permanent destruction or harmless photoinhibition that can be overcome after some days of low irradiation has occurred. For this reason, F_v/F_m studies with material collected or measured shortly after long periods of strong sunlight may give anomalous results.

- Cyanobacterial lichens
  F_v/F_m values of dark-adapted lichens with cyanobacterial photobionts (Demmig-Adams et al. 1990) appear to be relatively low (0.41 - 0.61). This mainly results from the well-known state transition of cyanobacteria, which redistributes the excitation energy in the dark in favour of photosystem I at the expense of photosystem II (Papageorgiou and Govindjee 1968) and by non-variable phycobilin fluorescence (Ghosh and Govindjee 1966). Low F_v/F_m values may therefore monitor environmental damage, photoinhibition, a marked state transition or a high phycobilin content. Recently, it has been proposed that F_v/F_m should be measured after low light treatment instead of darkness (Sundberg et al. 1997) to overcome state transition effects. The optimal intensity of the low light treatment varies between different species but values between 5 and 20 µmol photons m^{-2} s^{-1} should be used (Sundberg et al. 1997).
  At present, most information is available on *Peltigera rufescens*, as its chlorophyll fluorescence parameters have been measured continuously over a complete year. For this lichen, no maintained photoinhibition was detected by dawn measured F_v/F_m values in the field (Leisner et al. 1996), but nevertheless wet thalli can be protected against light stress by the formation of canthaxanthin (Lange et al. 1999). After a period of dry and hot weather artificial wetting in the afternoon (in darkness) did not restore any variable fluorescence for about 0.5 h, probably indicating transient photoinhibition.

Subprotocol 2
Measurement of Chlorophyll Fluorescence Parameters in Continuous Light

The parameters $\Delta F/F_m'\phi_{PSII}$, derived electron transport rate (ETR), and the quenching parameters q_P, q_N, NPQ are dependent on the ambient temperature and light intensity. Therefore it is important not only to measure chlorophyll fluorescence but also temperature and light. In addition, it is necessary to determine F_v/F_m within the same run of measurements, as the F_m value and (partly) the F_o and the F_o' value must be known for the correct calculation of the CF parameters. Users should keep in mind that the correct determination of F_o' and therefore of q_N and q_P is critical. Furthermore, the water content of the lichen samples plays a role.

Materials

- PAM 2000: all important parameters, infrared light for more correct determination of F_o'; connection to micro quantum / temperature sensor for non-leaf-like samples; miniature fibreoptic light guide (2 mm diameter) available; computer necessary; special fluorescence standard in order to assess absolute fluorescence intensities. **Walz GmbH**
- MiniPAM: all important parameters, burst mode for the reduction of the measuring beam intensity; integral data storage; accessories like PAM 2000 (data must be uploaded into a computer after using the instrument in the stand-alone mode).
- PAM-200 (= teaching PAM): all important parameters; all illumination by LED's (less light intensity); less accessories; computer necessary (new model PAM-210: computer not necessary); only small area measurements (ca. 2 - 3 mm diameter).
- A new system for simultaneous measurement of CO_2 gas exchange and chlorophyll fluorescence in the field ('Klapp'-cuvette, Lange et al. 1997) is available, but only on special order (Walz).
- FMS2: all important parameters; infrared light; integral data storage (data must be uploaded into a computer after using the instrument in the stand-alone mode); connection to a leaf clip with integrated micro quantum / temperature sensor. **Hansatech Instruments Ltd.**

ADC Bioscientific

- OSI 1 FL: all important parameters, integral data storage
- OSI 5 FL: all important parameters, integrated screen, data storage on diskettes
- OSI50: for determination of chlorophyll density via the ratio F690/F735, integral data storage

Procedure

1. Prior to the measurements, determine the F_v/F_m value of a wet sample (see Subprotocol 1).
2. Mount the light guide in a distance clip (or in a dark adaptation clip, if stored light intensity values can be used).
3. Use an accessory for the measurement of temperature and light intensity.
4. If internal (artificial) light sources are used for excitation of photosynthesis in the field, place a miniature light sensor directly at the margin of the lichen sample (requires special construction) or use stored light intensity values obtained in an identical geometry. Otherwise any light sensor in the neighbourhood of the sample can be used. If possible, determine the water content of the samples (Chapters 14 and 15).

Results

A typical measurement demonstrating fluorescence quenching of *Peltigera aphthosa* near the CO_2 compensation point is shown in Fig. 1. Typically, it is accompanied by a dramatic F_o quench (low F_o') which is larger then normally observed in other plants. At the same time, the nonphotochemical quenching parameter NPQ increases. Interestingly, a high NPQ and F_o quench can be observed already at moderate light intensities, i.e. at 360, 200 and to a lesser extent at 90 µmol photons m^{-2} s^{-1}. The high NPQ value found at limited CO_2 supply may partly indicate reversible photoinhibition of the zeaxanthin type (Gilmore 1997), but the slow biphasic recovery of F_o points to additional effects (cf. Li et al. 2000).

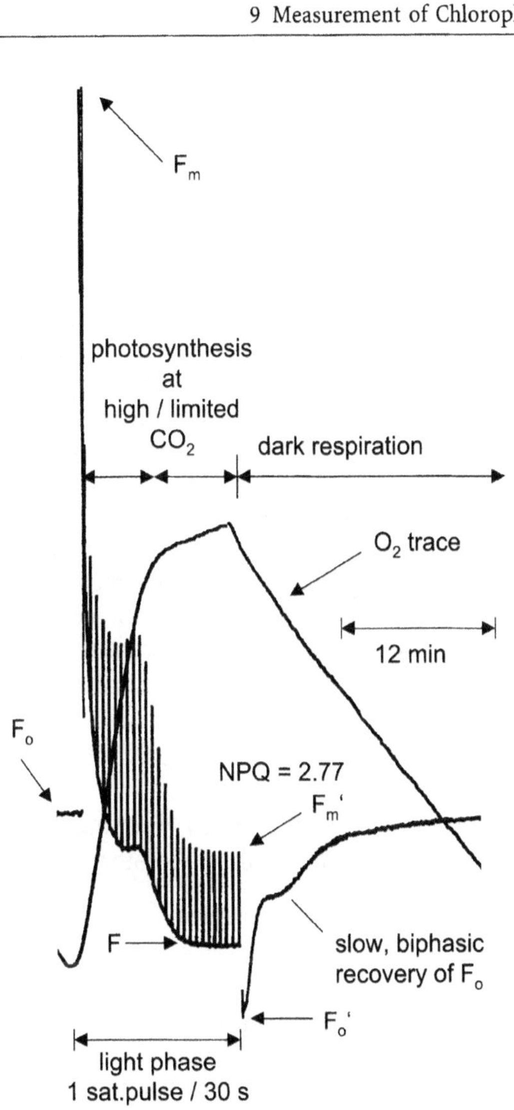

Fig. 1. Typical traces of chlorophyll fluorescence and O_2 gas exchange of *Peltigera aphthosa* in a closed cuvette (LD2, no addition of $KHCO_3$). After dark respiration and the onset of illumination (360 µmol photons m^{-2} s^{-1} white light), CO_2 is fixed at a high rate until there is approximation to the CO_2 compensation point (low CO_2). At limited CO_2, severe quenching of chlorophyll fluorescence is observed (F, F_m'). Finally, F_o' is more than 30 % below the F_o value. Note the slow recovery of F_o in the dark.

Comments

It is a good idea to start comprehensive measurements with a light effect curve (dependency of CF on light intensity), which can be performed automatically by some of the instruments. However, this is time consuming, as the illumination at each light intensity should not be shorter than 4 min in order to get near steady state values. Generally, an actinic light intensity of 100 μmol photons m^{-2} s^{-1} will provide a good balance between photochemical and non-photochemical quenching. The time period needed to get steady state values of F and F_m' at this light intensity is about 8 min, if not too many saturation pulses are fired (one at the end is sufficient). If an induction curve for the time-dependent analysis of quench parameters is desired (cf. Fig. 1), the interval between the saturation pulses should not be shorter than 30 s.

Quench analysis (parameters NPQ, q_P and q_{NP}) may be critical if a large decay in F_o occurs (cf. Fig. 1). This may result from the lichen or in some cases from the CF instrument. Since F_o and F_o' are not needed for the calculation of NPQ, the determination of this parameter is more reliable than that of q_P and q_{NP}.

Subprotocol 3
Simultanous Measurement of O_2 or CO_2 Exchange and Chlorophyll Fluorescence in the Laboratory

Materials

- LD2 leaf disc O_2 electrode system (Hansatech Instruments Ltd.) Measures O_2 concentration in a closed cuvette (no gas stream). The main illumination is from the top; there are 3 additional optical side ports. Adapters for the mounting of the light guides of all chlorophyll fluorescence instruments can be easily constructed (some are commercially available).
- Infrared gas analyser systems for CO_2 (e.g. PP systems, ADC Bioscientific, LiCOR, Walz)
- O_2 gas analyser AEI S3A/DOX Laboratory oxygen analyser (ADC Bioscientific), can be used in a gas stream (the resolution is $>$ 3ppm)

- Two channel strip chart recorder or data acquisition system (many manufacturers)

Procedure

Although the maintenance of an O_2 electrode system is not easy (see Troubleshooting) and there are many restrictions (only laboratory measurements, no crustose lichens, mainly measurement at high CO_2 concentrations, relatively low sensitivity), its use (possibly in combination with fluorescence measurements) is a low-priced alternative to CO_2 gas exchange measurements. If photosynthetic capacity has to be assessed, note the following points:

- Pay attention to the application book distributed by the supplier (Walker 1987). In particular, calibrate the electrode system and test it for leaks by increasing the pressure (use a gas tight syringe). Before and after sample measurements the O_2 signal must be checked for drift.
- Moisten the lichen samples (see previous section, measurement of green algal lichens, Step 2.)
- Soil, moss and other extraneous material must be carefully removed from the lower surface of the samples.
- The system is temperature controlled, but the temperature of the cuvette must not be set too low (more than 4°C below room temperature) because of condensation on the top window. If the samples are water saturated, the cuvette has to be opened every 30 min to control the formation of condensation water within the cuvette.
- Avoid contact of the lichen cortex with the cuvette top window, i.e. do not use the sponge material etc. provided by the supplier. Instead, put black velvet underneath the lichen (or nothing as in Fig. 1).
- If high concentrations of CO_2 within the cuvette are desired, moisturise the margin of the velvet with 3 drops of 0.5 M $KHCO_3$.
- Lichen material should be cut into pieces of 2 - 2.5 cm diameter before putting it into the centre of the cuvette. Cutting does not influence the signals.

Troubleshooting

- Maintenance of O_2 electrodes
  For the maintenance of O_2 electrodes, it is good practice to polish the silver anode thoroughly with fine diamond powder paste and to clean it for 30 min with a small volume of 1 M KCN (caution!). To prevent crystallisation of the electrolyte, a KCl concentration of 1 M instead of 3 M can be chosen. To assemble the membrane, the "membrane applicator" should be used. A second O-ring near the platinum cathode can be omitted as this may lead to mechanical problems. If the electrode is not new, it may be necessary to extend the equilibration period (polarisation voltage on) to 24 h or more. All electrical connections must be tight (some tend to loosen).

Comments

In C_4 plants, ETR values can be used to calculate the corresponding CO_2 fixation rates (µmol CO_2 fixed m^{-2} s^{-1}) (Krall and Edwards 1992). In *Umbilicaria aprina*, however, the relationship between fluorescence parameters and CO_2 is very complex because of strong fluorescence quenching effects (Green et al. 1998). Unfortunately, both under- and overestimation of photosynthetic gas exchange from fluorescence based ETR were observed. However, measuring O_2 evolution will of course enable you to determine photosynthetic rates. Note that at high thallus water content (suprasaturation) and increased CO_2 diffusion resistance in *Lecanora muralis* (Leisner et al. 1997) and *Peltigera rufescens* (Lange et al. 1999), CO_2 fixation rate is reduced but electron transport to O_2 may take place (Mehler reaction) indicated by a high ETR. This would explain at least one type of overestimation. Still, it is not clear, what kind of electron transport is occurring in *Peltigera aphthosa* in gas stream of pure N_2 (Jensen and Siebke 1997). This may reflect an unexpected CO_2 source.

References

Björkman O, Demmig B (1987) Photon yield of O_2 evolution and chlorophyll fluorescence characteristics at 77 K among vascular plants of diverse origins. Planta 170:489-504

Calatayud A, Sanz MJ, Calvo E, Barreno E, DelValle-Tascon S (1996) Chlorophyll a fluorescence and chlorophyll content in *Parmelia quercina* thalli from a polluted region of Northern Castellón (Spain). Lichenologist 28:49-65

Crittenden PD (1991) Ecological significance of necromass production in mat-forming lichens. Lichenologist 23:323-331

Demmig-Adams B, Maguas C, Adams III WW, Meyer A, Kilian E, Lange OL (1990) Effect of high light on the efficiency of photochemical energy conversion in a variety of lichen species with green and blue-green phycobionts. Planta 180:400-409

Gauslaa Y, Solhaug KA (1996) Differences in the susceptibility to light stress between epiphytic lichens of ancient and young boreal forest stands. Func Ecol 10:344-354

Gauslaa Y, Kopperud C, Solhaug KA (1996) Optimal quantum yield of photosystem II and chlorophyll degradation of *Lobaria pulmonaria* in relation to pH. Lichenologist 28:267-278

Ghosh AK, Govindjee (1966) Transfer of the excitation energy in *Anacystis nidulans* grown to obtain different pigment ratios. Biophys J 6:611-619

Gilmore, AM (1997) Mechanistic aspects of xanthophyll cycle-dependent photoprotection in higher plant chloroplasts and leaves. Physiol Plant 99:197-209

Green TGA, Schroeter B, Kappen L, Seppelt RD, Maseyk K (1998) An assessment of the relationship between chlorophyll a fluorescence and CO_2 gas exchange from field measurements on a moss and lichen. Planta 206:611-618

Gries C, Sanz MJ, Nash TH (1995) The effect of SO_2 fumigation on CO_2 gas exchange, chlorophyll fluorescence and chlorophyll degradation in different lichen species from western North America. Crypt Bot 5:239-246

Jensen M (1994) Assessment of lichen vitality by the chlorophyll fluorescence parameter F_v/F_m. Crypt Bot 4:187-192

Jensen M, Siebke K (1997) Fluorescence imaging of lichens in the macro scale. Symbiosis 23:183-196

Jensen M, Linke K, Dickhäuser A, Feige GB (1999) The effect of agronomic photosystem-II herbicides on lichens. Lichenologist 31:95-103

Krall JP, Edwards GE (1992) Relationship between photosystem-II activity and CO_2 fixation in leaves. Physiol Plant 86:180-187

Lange OL, Reichenberger H, Walz H (1997) Continuous monitoring of CO_2 exchange of lichens in the field: short-term enclosure with an automatically operating cuvette. Lichenologist 29:259-274

Lange OL, Leisner JMR, Bilger W (1999) Chlorophyll fluorescence characteristics of the cyanobacterial lichen *Peltigera rufescens* under field conditions II. Diel and annual distribution of metabolic activity and possible mechanisms to avoid photoinhibition. Flora 194:413-430

Leisner JMR, Bilger W, Lange OL (1995) Does photoinhibition occur under natural field conditions in a cyanobacterial lichen? In: Mathis, P (ed) Photosynthesis: from Light to Biosphere, vol 4. Kluwer Academic Press, Dordrecht, pp 259-262

Leisner JMR, Bilger W, Lange OL (1996) Chlorophyll fluorescence characteristics of the cyanobacterial lichen *Peltigera rufescens* under field conditions. I. Seasonal patterns of photochemical activity and the ocurrence of photosystem II inhibition. Flora 191:261-273

Leisner JMR, Green TGA, Lange OL (1997) Photobiont activity of a temperate crustose lichen: long-term chlorophyll fluorescence and CO_2 exchange measurements in the field. Symbiosis 23:165-182

Li X-P, Björkman O, Shih C, Grossman AR, Rosenquist M, Jansson S, Niyogi K (2000) A pigment-binding protein essential for regulation of photosynthetic light harvesting. Nature 403:391-395

Manrique E, Balaguer L, Barnes J, Davison AW (1993) Photoinhibition studies in lichens using chlorophyll fluorescence analysis. Bryologist 96:443-449

Niewiadomska E, Jarowiecka D, Czarnota P (1998) Effect of different levels of air pollution on photosynthetic activity of some lichens. Acta Soc Bot Pol 67:259-262

Palmqvist K (1993) Photosynthetic CO_2-use efficiency in lichens and their isolated photobionts - the possible role of a CO_2-concentrating mechanism. Planta 191:48-56

Papageorgiou G, Govindjee (1968) Light-induced changes in the fluorescence yield of chlorophyll a *in vivo* I. *Anacystis nidulans*. Biophys J 8:1299-1315

Scheidegger C, Schroeter B (1995) Effects of ozone fumigation on epiphytic macrolichens - ultrastructure, CO_2 gas exchange and chlorophyll fluorescence. Environ Poll 88:345-354

Schreiber U, Bilger W (1993) Progress in chlorophyll-fluorescence research: major developments during the past years in retrospect. Prog Bot 54:151-173

Schroeter B, Green TGA, Kappen L, Seppelt RD, Maseyk K (1995) The relationship between electron transport rate through PS II and CO_2 gas exchange in Antarctic cryptogams. In: Mathis P (ed) Photosynthesis: from Light to Biosphere, vol 5. Kluwer Academic Publishers, Dordrecht, pp 893-896

Sundberg B, Campbell D, Palmqvist K (1997) Predicting CO_2 gain and photosynthetic light acclimation from fluorescence yield and quenching in cyano-lichens. Planta 201:138-145

Walker DA (1987) The use of the oxygen electrode and fluorescence probes in simple measurements of photosynthesis. Oxygraphic Ltd., Sheffield

Suppliers

ADC BioScientific Ltd
Unit 12 Spurling Works
Pindar Road, Hoddesdon
Herts, EN11 0DB
England
phone: 44 (0)1992 445995
fax: 44 (0)1992 444567
e-mail: support@adc.co.uk
internet: http://www.adc.co.uk/

Hansatech Instruments Ltd.
Narborough Road, Pentney,
King's Lynn, Norfolk PE32 1JL,
England.
phone: +44 (0)1760 338877
fax: +44 (0)1760 337303
e-mail: info@hansatech-instruments.co.uk
internet: http://www.hansatech-instruments.com

LI-COR, Inc.
P.O. Box 4425
Lincoln, NE 68504
USA
phone: +1 402-467-3576
fax: +1 402-467-2819
e-mail: envsales@env.licor.com
internet: http://www.licor.com

PP Systems
In North America:
241 Winter Street, Haverhill, MA 01830, USA
phone: +1 (978) 374 1064, fax: +1 (978) 374 0972
email: support@ppsystems.com
internet: http://www.ppsystems.com/
In Europe:
Unit 2, Glovers Court, Bury Mead Road, Hitchin, Hertfordshire, SG5 1RT, UK
phone: +44 (0)1462 453411
fax: +44 (0)1462 431090

Heinz Walz GmbH, Eichenring 6, D-91090 Effeltrich, Germany
phone: +49-(0)9133/7765-0
fax: +49-(0)9133/5395
email: info@mail.walz.com
internet: http://www.walz.com

Characterising Photosynthesis and Respiration in Freshly Isolated or Cultured Lichen Photobionts

KRISTIN PALMQVIST and BODIL SUNDBERG

Introduction

The fixation of CO_2 in photobiont photosynthesis provides the basis for lichen survival and development, producing the carbohydrates needed for metabolism, structural components, osmotic regulation, cryo- and drought protection. The release of CO_2 in lichen respiration is subsequently related to the energy requirements of cell maintenance, nutrient acquisition, and growth. To understand basic metabolism and the potential for growth of a lichen, we need to know much more about how these processes are regulated and integrated in lichen photo- and mycobionts.

This chapter describes techniques that can be used to characterise some carbon acquisition traits of lichens and their photobionts. The first set of techniques describes how to isolate and/or culture large amounts of photobionts (algae or cyanobacteria) prior to physiological measurements, followed by a chlorophyll extraction protocol. The second set of protocols describes two applications of the liquid phase oxygen electrode; namely how to obtain light- and CO_2 response curves from cultured or recently isolated photobiont cells. Thereafter, a particular CO_2 gas exchange (IRGA) technique is described, providing one line of evidence for the presence of a CO_2 accumulating mechanism in lichens. Finally, a recently developed method for quantifying the amount of Rubisco in lichens and their photobionts is presented.

✉ Kristin Palmqvist, Umeå University, Department of Ecology and Environmental Science, Umeå, 901 87, Sweden (*phone* +46 90 786 91 84; *fax* +46 90 786 67 05; *e-mail* kristin.palmqvist@eg.umu.se)
Bodil Sundberg, Umeå University, Department of Plant Physiology, Umeå, 901 87, Sweden

Subprotocol 1
Rapid Isolation of Lichen Photobionts for Immediate Physiological Measurements

Photosynthesis in lichens is the result of the summed photosynthetic activity of their algal or cyanobacterial photobiont population. Therefore, the upper limit for gross photosynthetic rate (i.e. photosynthetic capacity) of a lichen will be determined both by the inherent photosynthetic properties of each photobiont cell, and the amount of photobiont cells the particular lichen contains (Palmqvist et al. 1998), see also Björkman (1981) for a comprehensive survey on factors determining photosynthetic capacity in general. It may sometimes be fruitful to study the characteristics of isolated lichen symbionts (see Chapter 1), an approach that has for example been used to characterise photobiont CO_2 acquisition strategies (cf. Palmqvist 1993, Palmqvist et al. 1994a, 1994b, 1995, 1997).

Experiments can be made either with cells freshly isolated from the lichen thallus (Subprotocol 1) or with cells that have been cultured under defined environmental conditions (Subprotocol 2). If one wishes to characterise the cells as they are in the lichen thallus, the former procedure may be preferred, even though it can be difficult to isolate a sufficient quantity of representative cells. Therefore, when large quantities of cells are required it may be necessary to grow cultures of the isolated photobiont. However, it is important to keep in mind that physiological and biochemical characteristics of these cultured photobionts may be different from their lichenized state. For example, net photosynthetic capacity may be altered (Palmqvist 1993, Palmqvist et al. 1997). In addition, photobiont carbon export and possibly overall carbon metabolism changes rapidly after isolation from the mycobiont (Smith 1974, 1992). Whether this presents a problem is dependent on the specific question of the study.

Large quantities of intact photobiont cells can be isolated from foliose and fruticose lichens, and most easily from species that have relatively high photobiont amounts, such as *Lobaria pulmonaria*, *Peltigera aphthosa* and *P. canina*.

The procedure described in Subprotocol 1 was originally developed by Drew and Smith (1967), whereby mild homogenisation of the thallus and differential centrifugation is adopted. The procedure is relatively simple, can be performed at room temperature and takes 30 - 40 min, but yields of intact photobiont cells may be relatively low (10-20%). Also, the method is only suitable for unicellular photobionts, because filamentous species may easily be disrupted. Yields and purity can be increased if sucrose, Percoll, $CsCl_2$ or KI gradient centrifugation is adopted (cf. Bubrick 1988 and references therein). However, such procedures generally take longer and require extensive washing of the isolated cells before they can be used for physiological measurements. Please also note that the following protocol should not be adopted for the isolation of axenic cultures. Such techniques are described in detail in Chapter 1.

Materials

Equipment

- Low speed centrifuge (clinical centrifuges are handy for this procedure) for tube volumes between 5-20 ml, preferably with swing-out rotor
- Centrifuge for Eppendorf vials
- Porcelain mortar and pestle
- Glass homogeniser (Potter-type; 10 or 20 ml) with a relatively wide piston clearance
- Light Microscope
- Teflon screens of varying pore diameter sizes (e.g. from 10 up to 50 μm)

Media

Any of the organic buffers suggested below can be used for the homogenisation medium. The buffer serves to neutralize pH changes due to liberation of lichen acids during preparation. Choice of buffer will be dependent on the final application. For instance if an assay medium of high pH is to be used in the final experiment, it is most convenient to use a high pH buffer throughout.

Organic homogenisation buffers

- 25 mM 4-morpholinoethane-sulfonic acid (MES) pH 5.5
- 25 mM bis-tris-propane (BTP) for pH 6.0 and/or 8.0
- 25 mM 4-(2-hydroxyethyl)-1-piperazinepropanesulfonic acid (Hepps) pH 8.0

Dissolve buffer in distilled water or, if a particular assay medium must be used in the final application, it is also possible to use the assay medium as a solvent. If present in the chosen medium, omit Ca^{2+} or replace with Mg^{2+} to avoid flocculation of cells during preparation (Bubrick, 1988). See also Comments for additional suggestions concerning the preparation of media.

Procedure

1. Spray the lichen with water and reactivate for 1-3 days [e.g. 10-15°C; 16 h Light (50-100 μmol photons m^{-2} s^{-1}): 8 h Dark; 97-100% RH]. Lichen thalli typically contain 50-500 mg Chl m^{-2} or 0.5-3 mg Chl g DW^{-1}, so 2 g DW or 200 cm^2 lichen material may then yield 0.1-1 mg Chl, assuming a yield of intact photobiont cells from the thallus of 10%. **Activate the lichen**

2. Brush away all debris and wash the upper thallus surface with distilled water to remove any epiphytic algae. We have found that photobiont yields are increased and purity of the preparation is better, i.e. there may be less interfering subcellular contaminants (see Comments), if marginal lobes are used rather than older parts of the thalli. If possible, remove rhizinae and the lower fungal medulla with forceps. Cut the thalli into 0.5-1 cm^2 pieces, place them in a porcelain mortar and add 5 ml of one of the above homogenisation media per 1 g DW of lichen. Homogenize the lichen by gently grinding until the liquid is darkly green. Yields may also be increased if a smaller quantity is ground each time and then pooled, and if an additional squeezing step using a glass homogeniser is adopted. **Photobiont liberation**

3. Filter the homogenate through a teflon screen, from 10 to 50 μm diameter pore size, depending on photobiont size, and collect the filtrate in centrifuge tubes. If the thallus homogenate still contains green fragments, the grinding process can be repeated once or twice. The filtrate contains intact photobiont cells, disrupted photobiont cells, smaller hyphal fragments, and extracellular and subcellular compounds.

4. Centrifuge at room temperature at 300-400 g for 3 min to precipitate hyphal fragments and photobionts. Higher speeds may be required to pellet smaller photobiont cells such as *Nostoc* and *Coccomyxa*. Discard the brown supernatant containing extracellular and subcellular compounds. The pellet consists of a dark green zone of algal cells below a thin top layer of fine white debris. Scrape off the green layer with a spatula and resuspend in homogenisation medium. **Differential centrifugation**

5. The resulting suspension is centrifuged step-wise at 100-150 g 7 times, first for 30 s, then for 40 s, 50 s, 60 s, 70 s, 80 s and finally 90 s. Centrifugation time is taken from switching on to switching off the centrifuge. To reduce centrifugation time, manual braking may be applied after switching off, but be careful, manual braking can present a hazard when centrifugation speeds are high. However, at the relatively low **Pool and repeat**

speeds used here manual braking should not present a major hazard. After each centrifugation save each pellet. Carry out the next centrifugation with the supernatant.

6. Examine a small sample from each of the 7 pellets in the microscope. The first pellets are generally rich in broken fungal hyphae. Check the final supernatant and discard it if there are only few intact photobiont cells left, otherwise centrifuge between 300 g and 1000 g for 2-5 min, examine the pellet and pool with the others.
7. Pool all photobiont-enriched pellets by dissolving one of the pellets in a small volume of homogenization medium yielding a dark green suspension. Transfer this solution to the next pellet; if necessary add more medium, dissolve pellet, and transfer suspension to the next pellet and so on until all are pooled. Add enough medium to fill a centrifuge tube and repeat Step 5.
8. The obtained intact photobiont cells can be used for many kinds of experiments where it is necessary to start from an enriched photobiont fraction, so depending on final application dissolve the resulting pellet in the assay medium required for the particular study. For photosynthesis measurements, add pH buffered CO_2-free medium (Subprotocol 4) until the photobiont suspension is pale green with a Chl concentration of 10 ± 5 μg ml^{-1}.
9. For Chl determinations transfer 1 ml of the cell suspension to an Eppendorf tube and pellet at the highest speed in an Eppendorf centrifuge (c. 20000 g for 10 min) and proceed from Step 2 in Subprotocol 3.

Comments

Because of differences in photobiont cell sizes, relative rigidity of fungal-photobiont interfaces and presence of secondary carbon metabolites, the above protocol may need to be adjusted to the particular lichen and photobiont species. Such adjustments include addition of substances such as PVP, PVPP, DTT or EDTA to the homogenisation buffer, and/or, as already indicated, changed centrifugation times and speeds. Excessive breakage of photobiont cells can occur when the photobiont is tightly bound to mycobiont hyphae, or if the cell wall of the photobiont is weak. In our experience, it is more difficult to obtain pure photobiont extracts from lichen species containing high concentrations of phenolics or other secondary metabolites, e.g. *Cetraria islandica* and *Stereocaulon* spp. In these cases we have obtained photobiont preparations that consume large amounts of oxygen, particularly in the light. Irrespective of the precise nature of these substances, they heavily interfere with photosynthesis measurements using an O_2 electrode. If this is the final application of the isolated photobionts, repeated washing and centrifugation of the sample could avoid the problem.

Subprotocol 2
Cultivation of Photobionts for Physiological Experiments

When large quantities of cells are required for an experiment, or when photobionts are particularly difficult to isolate in sufficient amount directly from the lichen thallus, it may be more convenient to grow the cells in isolation. Growing the cells at relatively low densities in liquid medium under strict environmental control will result in a culture of actively dividing vegetative cells, producing large amounts of material relatively rapidly. The following protocol describes how to obtain such cultures, while methods describing how to maintain axenic photobiont isolates are described in Chapter 1.

The vegetative "life cycle" of many algae and cyanobacteria may be very short, ranging from a few hours up to a few days when environmental conditions are favourable. We have experienced that this is also the case for isolated photobionts such as *Nostoc*, *Coccomyxa* and *Trebouxia* (Palmqvist 1993, Ögren 1993, Palmqvist et al. 1997, Sundberg et al. 1997). Also, many algae and cyanobacteria can acclimatise to changing environmental conditions within a few hours. Therefore, the physiological characteristics may be dramatically different between populations of the same algal or cyanobacterial strain, depending on light, CO_2, temperature and nutrient conditions during growth. All these factors, with the exception of temperature, may indeed vary significantly when the cells are grown in batch cultures for prolonged periods, as increased cellular density will reduce light quantity, alter light quality, reduce CO_2, change pH and reduce nutrient concentrations. Therefore, it is advisable that environmental conditions are both known and controlled during cell growth.

For physiological experiments the photobiont should preferably be grown in dilute liquid cultures. In this way, illumination will be more uniform and CO_2 and O_2 concentrations can be maintained constant in the growth medium by vigorously bubbling air through it. Depending on the type of study, cells can be grown in continuous light, which will create a mixture of cells representing all the phases of the vegetative "life-cycle".

The cells can also be synchronised to divide simultaneously by cultivation in repeated light-dark cycles. Depending on the species inherent capacity for growth and environmental conditions such as temperature, level of irradiance and CO_2 supply, cultures need to be diluted daily or up to weekly, to maintain cells in exponential or linear growth. Also, depending on species and experimental application cell densities can be allowed to vary between 1-2 up to 25-50 μg Chl ml^{-1}. *Trebouxia erici* grows better when densities are maintained at higher values (Palmqvist et al. 1997), while a strain of *Coccomyxa* isolated from *Peltigera aphthosa* (*Coccomyxa* PA) can be diluted to the lowest (Ögren 1993). However, at high cell densities, illumination will be less uniform and it may be difficult to maintain CO_2 concentrations in equilibrium between the source air and the liquid medium.

Materials

Equipment

- Laminar Flow Bench
- Temperature controlled growth cabinet (5-35°C) or transparent water bath
- Illumination (fluorescent tubes or halogen lamps) providing white light (400-700 nm) with an irradiance of 30-100 μmol m^{-2} s^{-1}
- Air-pump or alternatively compressed air free from oil vapour
- Magnetic stirrer and fleas and/or shaker
- Quantum sensor (e.g. Li-189, Li-Cor Inc., Lincoln, Nebraska, USA)
- Autoclave for sterilisation of media, glassware and tubing
- Polyetherurethane foam (mattresses and furniture filling, check local store for baby cot mattress for example)

Growth media

Chapter 1 outlines suitable growth media for *Trebouxia* photobionts (BBM) (Nichols and Bold 1965). However, take care to avoid media with organic carbon or nitrogen supplies if you want to study the autotrophic characteristics of the photobionts. Cyanobacterial *Nostoc* and green algal *Coccomyxa* grow well on the BG-11 medium (Stanier et al. 1971) described below.

BG-11

Prepare a bottle of each of the below nutrient stock solutions (N) and the trace element stock (T). Stock solutions do not need to be autoclaved if kept in a refrigerator. We routinely use chemicals of analytical grade for growth media.

Add to 100 ml distilled water the following:

N1. $NaNO_3$	15.0 g
N2. $MgSO_4 \cdot 7H_2O$	0.75 g
N3. $CaCl_2$	0.36 g
N4. K_2HPO_4	0.4 g
N5. Citric acid	0.06 g
N6. Fe ammonium citrate	0.06 g
N7. Na_2 EDTA	0.01 g
N8. Na_2CO_3	0.2 g
T. H_3BO_3	0.286 g
$MnCl_2 \cdot 4H_2O$	0.181 g
$ZnSO_4 \cdot 7H_2O$	0.0222 g

Prepare final medium by adding 10 ml of each nutrient solution (N1-N8) and 1 ml of the trace elements (T) to 919 ml distilled water. Seal the bottle, autoclave and adjust the pH to 7.1 after autoclaving and cooling.

pH buffering

Green algal photobionts, but not cyanobacteria (see Price and Badger 1985) can be grown in media buffered with 10 - 25 mM organic buffer, e.g. bis-tris-propane (BTP). In this way, the pH of the medium will remain constant even when nutrients have been consumed. Also, and most importantly, relative CO_2 and HCO_3^- concentrations remain constant in relation to each other in a pH buffered medium. Add the organic buffer to the medium prior to sterilisation and adjust the pH.

Procedure

Starting growth

1. Pour 10 - 25 ml of your chosen growth medium into a wide-bottomed 25 - 50 ml culture flask, add a magnetic flea, seal the flask with cotton wool and an aluminium cap, sterilise, and cool.

2. Transfer 2 - 3 loops of cells from an agar grown culture to the sterilised growth medium after cooling. See Chapter 1 on how to maintain an axenic, agar grown culture.

3. Place the culture on a magnetic stirrer in front of a weak light source providing c. 30 µmol m^{-2} s^{-1} of white light (400 - 700 nm). Many lichen photobionts grow better at temperatures below ambient room temperature, e.g. 12 - 15°C. However, once the culture is growing well, you can increase the temperature to 20 - 25°C. Stir the culture very gently until growth has started. When the suspension is visibly green, increase stirring. When the culture has become darker green, transfer the whole batch of cells to a larger flask containing 100-250 ml medium. If the culture does not become visibly green within 1 - 2 weeks, retry with new cells from agar and test another growth medium. Some cells may for example grow better on NH_4^+ instead of NO_3^- as nitrogen source. Suboptimal pH or lack of a specific nutrient or trace element may be other reasons for poor growth.

Increase culture volume

4. Increase batch culture volume step-wise, up to 0.5 - 1 litre. Depending on species and culturing conditions, batch cultures can be allowed to become rather dense. To obtain a homogenous culture density, put larger batch cultures on a shaker, or continue stirring with the magnetic flea.

Illumination

5. Illuminate the culture with 50 - 100 µmol m^{-2} s^{-1} of white light, depending on final applications. Higher irradiances, up to full daylight (250 - 2000 µmol m^{-2} s^{-1}), can be provided by halogen lamps. However, halogen lamps can cause considerable warming of the culture, so it may be necessary to place a transparent water filter or a fan in front of the cultures. Alternatively, the culture flasks can be placed in a temperature regulated water bath. Irradiance within the culture will be more uniform when:
   - illumination of the culture is provided from more than one direction and facing as large an area as possible;
   - cell densities are kept below 10-20 µg Chl ml^{-1};
   - culture flasks are cylindrical and not wider than 5 cm diameter;
   - the culture is vigorously stirred by air-bubbling.

CO_2 supply

6. CO_2 must be supplied when culture volumes are greater than 100 - 250 ml and visibly green (> 2-5 µg Chl ml^{-1}). Lichen photobionts such as *Trebouxia* and *Coccomyxa* are apparently inhibited by high CO_2 (Palmqvist et al., 1994a, 1997) so it is best to supply CO_2 by vigorously bubbling air through the culture. Avoid using gas supplies enriched

with CO_2. Use a membrane pump or compressed air and attach sterile tubing to the air outlet. If compressed air is used, remove contaminating oil vapour by passing the air stream through a filter. Attach a sterilised glass pipette to the tubing and fit the pipette through a cork sealing the culture flask. Corks can be made from polyetherurethane foam that is highly flexible, seals well and can be autoclaved. Placing a sterile filter between the air-source and the culture can reduce the risk of contamination from fungal spores and bacteria from the source air. Bubble air with a flow rate of 2 - 500 ml min^{-1}, depending on culture volume, cell densities, and other environmental conditions.

Subprotocol 3
Extraction and Quantification of Chlorophyll

Extraction of chlorophyll (Chl) from unicellular algae can often present a problem due to the relative rigidity of their cell wall. Chl extraction from these organisms therefore generally involves heating of the cells in an organic solvent. The following procedure is based on two recent publications (Barnes et al. 1992, Wellburn 1994), using DMSO as a solvent. Our experience is that this procedure is both reproducible and suitable for lichens, even though some lichen pigments can interfere with chlorophyll absorption. The whole procedure takes c. 60 min, with a 40 min long incubation step.

Materials

Equipment

- Heating block for Eppendorf tubes
- Centrifuge for Eppendorf tubes
- Tubes which are resistant to organic solvents such as DMSO and acetone
- Spectrophotometer
- Vortex

Chemicals

- Dimethyl-sulphoxide (DMSO)
- $MgCO_3$

Procedure

Sample preparation

1. The following protocol can be used for extraction of chlorophyll pigments from intact lichens, pulverised lichen material or pelleted photobiont cells. Intact thalli can be ground to smaller fragments directly in the test tube using a glass rod or spatula. Photobiont cells can be pelleted by centrifugation at c. 20000 g for 10 min in an Eppendorf centrifuge.
2. In a fume hood, add 2 ml DMSO and a spatula tip of $MgCO_3$ to 10 - 20 mg lichen material. Alternatively, add 1ml DMSO and a spatula tip of $MgCO_3$ to a photobiont pellet obtained from 1 - 5 ml culture, sample volume being dependent on cell density of the culture.

Incubation

3. Vortex the sample and incubate at 60°C for 40 min. Vortex a few times during incubation.
4. Centrifuge at room temperature in an Eppendorf centrifuge at maximum speed (c. 20000 g) for 5 min.
5. The resulting supernatant contains the Chl. If the pellet appears greenish, vortex thoroughly and incubate the sample for a few more minutes.

Measurement

6. Measure in a spectrophotometer (spectral resolution 1 - 4 nm) for absorption at 665 and 649 nm. Use pure DMSO as a blank or measure the baseline absorption at 750 nm. Absorption at 665 and 649 nm should preferentially be between 0.2 and 0.8, depending on the linearity of the spectrophotometer. Sample amount and DMSO volume should therefore be adjusted for the species being investigated.

Equations

7. Calculate chlorophyll *a* (C_a) and *b* (C_b) in mg l^{-1} according to equations 1 - 3 (Wellburn 1994):

$$C_a = 12.19A_{665} - 3.45A_{649} \qquad \text{Equation 1}$$

$$C_b = 21.99A_{649} - 5.32A_{665} \qquad \text{Equation 2}$$

Note that cyanobacteria lack Chl *b* so use equation 3:

$$C_a = 12.19A_{665} \qquad \text{Equation 3}$$

Some spectrophotometers have spectral resolutions of 0.1 - 0.5 nm in the visible spectrum. If these resolutions are used, other equations should be used as discussed by Wellburn (1994).

8. For estimation of chlorophyll degradation, the phaeophytinization quotient (OD_{435} : OD_{415}) can be used to express the ratio of chlorophyll *a* to phaeophytin *a*, as detailed by Barnes et al. (1992).

9. Apart from adjusting sample amount and DMSO volume when a new species is investigated it can also be useful to record an absorption spectrum from 750 to 400 nm, particularly if the lichen contains many other pigments which may interfere with the chlorophyll absorbency peaks. If additional pigments overlap with chlorophyll it may be necessary to adopt a method that can separate the pigments prior to their quantification, e.g. HPLC (see Chapter 22).

Subprotocol 4
Use of Oxygen Electrode to Characterize Photobionts

Liquid-phase O_2 -electrodes allow quantification of photosynthesis and respiration in algae, cyanobacteria and plant protoplasts as well as electron transport rates in isolated chloroplasts, thylakoids and mitochondria. The technique is relatively easy and the equipment is much less expensive compared to infra red CO_2 gas analyser (IRGA) equipment. However, in contrast to the IRGA technique, the O_2-electrode allows only indirect quantification of CO_2 and inorganic carbon (Ci) fluxes. Moreover, the oxygen electrode cuvette is a closed system where O_2 and CO_2 concentrations vary during the time course of the experiment. This section describes how to obtain light- and CO_2 response curves of photobiont photosynthesis using cells that have either been newly isolated from a lichen thallus or for cells that have been cultured.

Materials

Equipment

- Liquid-phase O_2 electrode (e.g. Hansatech, King's Lynn, Norfolk, UK) connected either to a chart-recorder or a computer
- O_2 electrode manual (e.g. Hansatech; Walker 1990)
- Thermostatically controlled water-bath (0 - 40°C) with circuit to electrode cuvette
- Light source
- Quantum sensor

- Centrifuges for tube volumes between 1 - 1000 ml
- Glass syringes with long and narrow needles for volumes between 1 up to 100 µl

Media

- pH buffered CO_2 free medium
  CO_2 free medium is obtained by bubbling an assay medium with CO_2 free air for 20-24 h. Use the photobiont growth medium as assay medium and buffer with 25-50 mM organic buffer, e.g. bis-tris-propane (BTP) to pH 6.0 or 8.0. Autoclave the buffered medium. Add 100 ml of this to a 200 ml bottle. Pass an air stream through a CO_2 absorbent (NaOH or soda lime) and bubble medium vigorously for a minimum of 20 h before use in your experiment. Semi-seal the bottle with parafilm to avoid evaporation of medium during bubbling. Please note that using pure N_2 (g) is not a good alternative, because in this case the medium will be devoid of O_2.
- HCO_3^- solutions
  Prepare 25 - 50 ml each of a 1, 5, 10, 50, 100 and 500 mM HCO_3^- stock solution (Table 1). These will be used to obtain the Ci response curve and must be prepared and used on the same day.

Table 1. Using HCO_3^- solutions, pH buffered medium and the oxygen electrode to estimate CO_2 and Ci response curves for photosynthetic cells. Please note that pKa values change with temperature and ionic strength of the medium (cf. Palmqvist 1993 or a textbook in Inorganic Chemistry).

			Approximate [CO_2] at 25° C if pKa_1 = 6.25 and sample volume is 1 ml		
HCO_3^- stock [mM]	Add to 1 ml [µl]	[Ci]tot [µM]	pH 6 [µM]	pH 7 [µM]	pH 8 [µM]
1	2	2	1.3	0.3	0.03
	5	5	3.2	0.8	0.09
5	2	10	6.4	1.5	0.2
10	2	20	12.8	3	0.4
	5	50	32	7.6	0.9
50	2	100	64	15	1.8
100	2	200	128	30	3.5
	5	500	320	75	8.7
500	2	1000	640	151	17.5

Procedure

Obtaining a light response curve with oxygen electrode

Preparation of the equipment

1. Install and calibrate the liquid-phase O_2 electrode according to the manufacturer's manual (e.g. Hansatech; Walker, 1990). The O_2 signal must be calibrated on an absolute scale to allow accurate quantification of photosynthetic rate.

2. Choose a light source and neutral density filters that can create up to 10 - 15 irradiance levels ranging from 1 - 5 up to 1000 - 2000 µmol photons m^{-2} s^{-1}. Measure the irradiance at the position of the sample.

Preparation of the sample

3. Prepare the sample in one of the following ways:
   - Transfer cells directly from the growth culture to the O_2 electrode cuvette, if the cell density is 10 ± 5 µg Chl ml^{-1}.
   - Mix a few ml of the growth culture with growth medium to obtain a more dilute sample.
   - If the culture is too dilute, pellet the amount of cells that will be needed for your experiments by centrifugation at 1000 - 3000 g for 5 - 10 min and dissolve the pellet in growth or assay medium to 10 ± 5 µg Chl ml^{-1}.
   - If measurements must be made in a specific medium, for example a particular buffer or CO_2 free medium, harvest 10 - 25 ml of cells and centrifuge at 1000 - 3000 g for 5 - 10 min depending on cell sizes. Wash the pellet in the particular assay medium and pellet again. Dissolve the pellet to 10 ± 5 µg Chl ml^{-1} in the assay medium.

Measurements

4. Add 1 ml sample to the O_2 electrode cuvette, seal with the lid and make sure that all air is evacuated.

5. Add HCO_3^- from one of the stock solutions with a glass syringe through the lid. Use the HCO_3^- concentration and volume that would yield c. 100 - 150 µM CO_2 at the particular pH of the sample medium (Table 1).

   Note: Please note that in Table 1 the sample is assumed to have a volume of 1 ml.

6. Start measurements in darkness and follow respiratory O_2 consumption until a steady-state rate is reached. Then switch on the light, start with the lowest irradiances and increase the irradiance step-wise until photosynthetic O_2 evolution becomes light saturated. Measure for at least 5 - 10 min at each irradiance to make sure that steady-state is reached. During prolonged measurements (> 30 min), and particularly

if high cell densities are used, the sample may suffer from CO_2 depletion in the cuvette. It is then better to use a new sample at each irradiance level, but always start with dark respiration. In general, the higher the irradiance, the longer the sample will take to reach steady-state.

Chl determination

7. Transfer each sample from the O_2 electrode cuvette to an Eppendorf tube for Chl determination. In some cases it can also be convenient to have a measure of cell number, which can be counted using a light microscope. For Chl determination, measure the sample volume before centrifugation at the highest speed in an Eppendorf centrifuge (c. 20000 g for 10 min) and proceed from Step 2 in Subprotocol 3.

8. Calculate dark respiration and photosynthesis rates from the data obtained, relating O_2 concentration changes to sample volume and Chl concentration and express O_2 exchange rates as µmol O_2 mg Chl^{-1} h^{-1} or nmol O_2 mg Chl^{-1} s^{-1}.

Data processing

9. The obtained data can be plotted as a function of irradiance to obtain a photosynthetic light response curve of the particular photobiont species or for cells representing a particular culturing condition. The data can further be fitted to a non-rectangular hyperbola equation (equation 4) giving values of maximum photosynthesis (P_{max}), dark respiration (R), the quantum yield (ϕ) and the convexity (θ) (Leverenz & Jarvis, 1979).

$$\theta P^2 - (\phi I + P_{max})P + \phi I P_{max} = 0 \qquad \text{Equation 4}$$

In this equation P is the rate of photosynthesis (y-variable), I is the irradiance (x-variable), ϕ is the maximum quantum yield, i.e. the slope of the line of the light limited part, P_{max} is the light-saturated rate of photosynthesis, and θ is the convexity, i.e. a measure of the curvature of the transition zone from the light limited to the light saturated part of the curve. This mathematical description is useful because it contains variables with a clear physiological meaning that can be derived from the light-response curve and used to model photosynthesis (Ögren 1993, Lambers et al. 1998). Note that it is not correct to fit the obtained data to a Michaelis-Menten equation because photosynthesis is composed of a combination of physical and enzymatic reactions. Also note that the quantum yield (ϕ) obtained from liquid phase O_2 electrode measurements has an arbitrary unit of mol O_2 (mg $Chl)^{-1}$ h^{-1} $(I)^{-1}$ where I represents incident irradiance in mol photons m^{-2} s^{-1} (see Sundberg et al. 1997). This is because photosynthetic rates are not obtained per sample area.

Obtaining Ci and CO_2 response curves with oxygen electrode

Preparation of CO_2 free medium

1. Start bubbling CO_2 free air through your pH buffered assay medium at least 20 h before experimental use. Longer times are required to get rid of all inorganic carbon (Ci) in high pH buffers compared to low pH buffers. Keep the medium at the same temperature that will be used for the experiments.

Preparation of equipment

2. Install and calibrate the liquid-phase O_2 electrode according to the manufacturer's manual (e.g. Hansatech; Walker 1990).
3. Rinse cuvette house chamber, lid and the tiny magnetic flea vigorously with distilled water. At all times, try to avoid uncontrolled contamination of O_2 electrode parts, pipettes and syringes with HCO_3^- or any other CO_2 source.

Determine light saturating irradiance

4. Use a cell sample from the growth culture to check the irradiance required for light saturation. If necessary, dilute the cells with growth medium to a Chl density of 10±5 μg Chl ml^{-1}. Put the cells in the O_2 electrode cuvette, and make sure that all air is evacuated when sealing the cuvette chamber with the lid as described in the manufacturer's manual. Then add HCO_3^- through the hole in the lid using a glass syringe. Final CO_2 concentrations should be between 100 –150 μM CO_2 (aq) (see Table 1). Switch on the light source, and illuminate the sample with an irradiance slightly below the growth light regime. Then increase the irradiance step-wise until photosynthetic O_2 evolution has reached light saturation, generally occurring at or slightly above the growth light regime.

Preparation of a CO_2 free sample

5. Harvest cells from their growth or isolation medium by centrifugation at 1000 - 3000 g for 5 - 10 min depending on cell sizes. Wash the cells in CO_2 free medium and pellet again twice to get rid of contaminating inorganic carbon. Resuspend the cells in CO_2 free medium to a final concentration of 10±5 μg Chl ml^{-1}. The washed cells can be maintained CO_2 free if bubbled with CO_2 free air and kept in darkness, and be used for measurements for up to 2 - 3 h after harvest. Please note that the sample should be kept at the same temperature as used for the experiments. Placing the sample on ice is not recommended.

Measurement

6. Add 1 ml of the cell suspension to the O_2 electrode cuvette chamber, seal with the lid and evacuate all air.
7. Switch on the light, which should be saturating for photosynthesis at CO_2 saturation (see Step 4 above). Allow the cells to deplete any re-

maining Ci before adding HCO_3^-. When there is photorespiratory O_2 consumption or no change in O_2 concentration in the cuvette, start adding HCO_3^-.

8. Add HCO_3^- from the stock solutions for stepwise increasing Ci and CO_2 concentrations (see Table 1). At least 8 - 10 steps will be required to obtain a response curve. Use an increasingly strong stock solution to minimise total volume additions. Make sure to rinse the glass syringes thoroughly with distilled water after each use.

9. At the lowest Ci and CO_2 concentrations there will generally be only a slight increase in O_2 evolution compared to dark respiratory consumption, and because CO_2 will soon be depleted, O_2 evolution will decrease again rather rapidly. Steady-state photosynthetic rates for longer times than 2 - 3 minutes will not occur until added concentrations are sufficient to significantly exceed the rate of consumption.

Chl determination

10. Transfer each sample from the O_2 electrode cuvette to an Eppendorf tube for Chl determination. In some cases it can also be convenient to have a measure of cell number, which can be counted using a light microscope. For Chl determination, measure the sample volume before centrifugation at the highest speed in an Eppendorf centrifuge (c. 20000 g for 10 min) and proceed from Step 2 in Subprotocol 3.

Data processing

11. Calculate net photosynthetic O_2 evolution from the obtained traces and express rates, for example as µmol O_2 mg Chl^{-1} h^{-1}.

12. Calculate total Ci concentration in the cuvette immediately after each HCO_3^- addition. At the lowest Ci concentrations it may be necessary to compensate for consumption of Ci during the previous measurement. This can be done by using the O_2 evolution rate obtained at the particular Ci concentration and subtracting the consumed Ci, assuming a 1:1 molar ratio of O_2 evolution to Ci consumption.

13. Use the data given in Table 1 to estimate the CO_2 concentration in the cuvette after each HCO_3^- addition. Please note that the values presented in Table 1 are based on a sample volume of 1 ml (see Step 6). However, because of the continuous consumption of Ci in the cuvette, it is unlikely that HCO_3^- and CO_2 will reach equilibrium, so this can only be an approximation of the CO_2 concentration. Also, due to its rapid consumption, the calculated CO_2 will be somewhat overestimated at the lowest Ci concentrations, an error that decreases with increasing CO_2. For this reason, cell samples should be diluted as much as possible.

14. Plot net photosynthesis as a function of Ci and/or CO_2 concentration. It is then possible to extract the parameters Ci or CO_2 compensation concentration and $K_{0.5}$ (Ci or CO_2) from the response curves (Fig. 1).

15. A low Ci or CO_2 compensation concentration of photosynthesis and relatively low $K_{0.5}$ (Ci or CO_2) values are indicative of the operation of a photosynthetic CO_2 concentrating mechanism (CCM) in the cells. However, additional measures are required to firmly establish that a CCM is present. This includes measurements of accumulated Ci pool sizes as outlined in Subprotocol 5, and a comparison of the cells' Rubisco characteristics with the cells' in vivo characteristics. Raven (1997) and Badger et al. (1998) discuss the matter in detail.

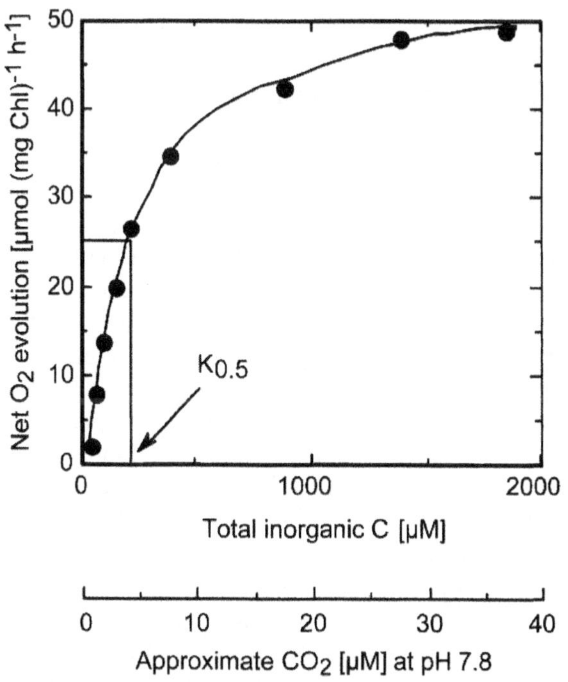

Fig. 1. A typical response of net photosynthesis to total inorganic carbon (Ci) concentration and to approximate CO_2 concentration at the particular pH of the assay medium. $K_{0.5}$ is equivalent to the Ci or CO_2 concentration that is required to reach half maximum rate of photosynthesis. Making measurements in assay media at different pH will yield information about the preferred Ci source (CO_2 or HCO_3^-) of the investigated species. See references in the text for additional applications.

Troubleshooting

- Problems with Ci contamination
  The major problem with this technique is Ci contamination of the assay medium or cuvette, typically evident as significant O_2 evolution immediately upon addition of the cells to the cuvette even before addition of HCO_3^-. If the harvested cells are left in darkness without bubbling CO_2 free air through the medium, respiration can be a significant source of Ci to the medium. More often, however, thorough rinsing with distilled water of O_2 cuvette chamber or lid has been neglected. If unexpected high rates of O_2 evolution occur after the addition of a low HCO_3^-concentration, this can be caused by contamination by a stronger HCO_3^- stock solution remaining in the glass syringe.

Subprotocol 5
Testing Intact Lichens or Isolated Photobionts for the Presence of Ci Accumulation

Absolute quantification of photosynthetic CO_2 fixation and respiratory CO_2 losses can be made with an infra red CO_2 gas exchange analyser (IRGA). This technique allows online measurements in flow-through gas exchange systems where CO_2 and O_2 concentrations can be controlled and held constant, using gas mixing and mass flow controllers. Generally, measurements are made in special cuvettes where other environmental parameters such as temperature, illumination and relative humidity are also controlled. The following protocol describes a particular IRGA technique whereby the ability of some lichen photobionts to accumulate an intracellular pool of inorganic carbon can be studied. The technique was originally developed by Badger et al. (1993), and later improved by decreasing the assay temperature whereby better resolution between Ci accumulation and CO_2 fixation could be obtained (Palmqvist et al. 1994b, 1997, Sundberg et al. 1997). The technique can be used for measurements of intact lichen thalli or for cultured or newly isolated photobiont cells. In the latter cases, the photobiont cells can be filtered on to a glass fibre or cellulose filter having narrow pore size diameters (< 10 µm). Such filters hold sufficient water to avoid desiccation effects during the measurement. Additional techniques that may reveal the presence of a CCM in a lichen or its photobiont include measurements of its carbon isotope discrimination characteristics (Máguas et al. 1993, Smith and

Griffiths 1996), or the CO_2/HCO_3^- disequilibrium technique developed by Badger et al. 1994. These latter techniques require a mass spectrometer that allows online measurements, and are not described further in this chapter.

Materials

Equipment

- Infra Red CO_2 gas analyser with a fast response (<5 s for 95% response to a change in the cuvette) (e.g. Heinz-Walz BINOS system, H-Walz, Effeltrich, Germany)
- Fast data acquisition system connected to the IRGA, e.g. a chart recorder or a computer logger
- Small cuvette (c. 5 ml) (can be hand made)
- Gas tubing between cuvette and IRGA (minimise the length)
- Glass fibre or cellulose filters for measurements of isolated photobionts (e.g. Micron Separations Inc., Westboro, Massachusetts, USA)

Buffers and inhibitors

- 50 mM 4-(2-hydroxyethyl)-1-piperazinepropanesulfonic acid (Hepps) pH 8.0
- 100 mM glycolaldehyde dissolved in 50 mM Hepps

Procedure

Activation of the lichen

1. Spray the lichen with water and reactivate for 1-3 days at for example 10-15°C; 16 h Light (50-100 µmol photons m^{-2} s^{-1}): 8 h Dark; 97-100% RH. Alternatively, use newly isolated or cultured photobiont cells obtained as described in Subprotocol 1.

Preparation of the IRGA

2. Install and calibrate the IRGA according to the manufacturer's manual.
3. Attach a small cuvette to the IRGA and reduce gas tubing to a minimal length. Reduced cuvette and tubing volumes will increase the resolution of CO_2 changes in the cuvette and decrease the response time between the sample and the IRGA.
4. Measurements can be performed at temperatures appropriate for the lichen, but keeping temperatures relatively low, e.g. at 5-10°C, will increase the resolution between CO_2 accumulation and fixation, because

the latter process has a slower induction at lower temperatures, while the accumulation process appears to be less sensitive to temperature changes.

Measurement

5. Put a sample in the cuvette. The sample can fill the cuvette, but make sure that the thallus surface is evenly illuminated. Alternatively, if isolated or cultured photobiont cells are used, filter these onto a glass fibre or cellulose filter to obtain a pale green film of cells on the upper surface of the filter. Press the filter gently between Kleenex tissue paper to reduce water volume held by the filter. High water concentrations in the filter will interfere with the IRGA measurements because CO_2 will dissolve in the water film. On the other hand, sufficient water must remain in the filter to avoid excessive desiccation of the photobiont sample.

Dark adaptation

6. Close the cuvette and let the sample equilibrate in darkness for a minimum of 10 min and wait until a steady-state dark respiration rate is reached. The signal can be recorded either on a chart recorder or the IRGA computer logger, provided data acquisition is fast enough. If present, CO_2 accumulation is initiated as soon as the light is switched on, and is revealed as a peak in CO_2 uptake that may appear within 10 s (Figure 2).

Light on

7. Switch on the light that should saturates photosynthesis, and leave the light on until photosynthetic CO_2 fixation has been fully induced and reached steady-state. This may require 15 - 20 minutes or sometimes longer.

Light off

8. Switch off the light, and continue recording until all transient CO_2 fluxes have relaxed and steady-state dark respiration is reached.

Inhibitor treatments

9. Remove the sample from the cuvette and incubate the sample using different inhibitors, depending on type of study. See Badger et al. (1993), Palmqvist et al. (1994b), Smith and Griffiths (1996), Sundberg et al. (1997) for different applications.

10. The same sample may be used for continuous measurements for up to 2-3 hours provided that the thallus water content can be controlled and held constant at optimal hydration for the particular specimen. In this way, the same sample can be used to make light or temperature response curves or be used as its own control prior to using it for inhibition treatments.

Quantification

11. When all measurements of the sample have been made, proceed with quantification of thallus area (if possible), thallus dry weight and Chl

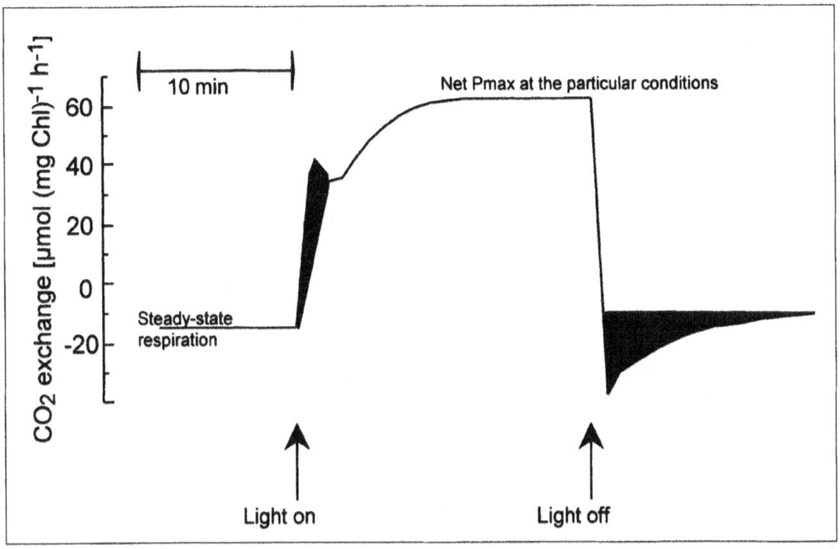

Fig. 2. Measurements of CO_2-uptake and release pools (shaded areas) in a lichen with cyanobacterial *Nostoc* photobionts. Lichens with green algal *Trebouxia* photobionts generally display the uptake pool only. The size of the pools can be calculated by integrating their respective area. See references given in the text for additional applications.

content as described in Subprotocol 3 (see also Chapter 21). In the case of a cellulose or glass fibre filter, it should be allowed to dry somewhat before placing the whole filter in a larger DMSO volume to extract the Chl.

12. Calculate CO_2 exchange rates from the acquired data and use either total Chl or Chl *a* content to compare between samples as in Figure 2.
13. Calculate initial CO_2 pool sizes and CO_2 efflux pools as outlined in Figure 2.

Subprotocol 6
Quantifying Rubisco in Lichens and Their Photobionts

Rubisco is the most abundant protein in cyanobacteria and in plant and algal chloroplasts, with up to 25% of all cellular nitrogen being invested in this enzyme. Such large investments into Rubisco are apparently the case also for lichenized algae and cyanobacteria, constituting about 4 - 5% of all thallus proteins (Palmqvist et al. 1998, Sundberg et al. 2001). Rubisco is deactivated in the dark, requiring re-activation in the light. Complete activation of all Rubisco in a cell may in fact require exposure to light saturating irradiances and relatively high CO_2 concentrations for up to 20-30 min. Measurements of only Rubisco amount is therefore insufficient for a complete understanding of how CO_2 fixation and Rubisco activity are regulated in a cell. So far no method has been developed whereby Rubisco activities can be obtained for lichens. This is primarily caused by a general difficulty in breaking algal and cyanobacterial cells. In addition, many lichens contain a range of secondary metabolites that may have inhibitory effects on enzyme preparations. As a first step to study Rubisco characteristics in lichens, we recently adopted the here presented procedure for the quantification of Rubisco amount. The procedure is still preliminary, and has only been tested for a few species (Palmqvist et al. 1998, Balaguer et al. 1999, Sundberg et al. 2001), but provides a basis for further development.

A crude protein fraction is prepared from pulverised lichen material and the extracted soluble proteins are separated on a SDS polyacrylamide gel. Rubisco is quantified by staining the gel with Commasie Brilliant Blue R-250 whereby the density of the large subunit band (LSU; 55kDa) can be compared with the density of known amounts of Rubisco LSU separated on the same gel.

Because algae and cyanobacteria are difficult to break, Rubiso extraction is often incomplete (cf. Palmqvist et al. 1998). To correct for this, we have assumed that Rubisco and Chl are equally extracted during the protein preparation procedure. Because Chl may be completely recovered in a DMSO extraction of a parallel sample (see Subprotocol 3), the Chl content in the Rubisco (protein) preparation in relation to Chl in DMSO can be used to estimate Rubisco yield as well. However, given their different sizes and solubility characteristics, it is possible that the yield of Rubisco is somewhat lower than for chlorophyll. It is also possible that the relative yields of Rubisco and Chl vary depending on the photobiont species. Yet, at present this method appears to be the best way to assess the amount of Rubisco in lichens. Moreover, the procedure has been compared with a method based on immuno-gold labelling of Rubisco, and both methods gave very similar results (Balaguer et al. 1999).

We have found that Chl yields are highly reproducible for a given lichen or photobiont species, with cultured photobionts having about the same yields as lichenized photobionts of the same species (Palmqvist et al. 1998). However, the yield may be as low as 40% and varies among species (Palmqvist et al. 1998, Balaguer et al. 1999), emphasising a need to improve the procedure for species where the photobiont is particularly difficult to disrupt. Such improvements may involve grinding of the lichen with liquid nitrogen before homogenisation (Step 1 of the procedure), adopting a French Press at Step 4, and increasing the number of freezing and thawing cycles (Steps 6-7).

The complete protocol requires a few days of laboratory work, starting with preparing buffers and gels, extracting lichen proteins, performing the gel electrophoresis and subsequent quantification of polypeptide densities. The lichen protein preparation (Steps 1-12) can be stored in a deep freezer awaiting gel analysis.

Materials

Equipment

- Glass homogeniser (Potter-type) (2 ml) with narrow piston clearance
- Heating block for Eppendorf tubes
- Centrifuge for Eppendorf vials
- Protein gel separation equipment (e.g. BIORAD Mini-Protean II, BIORAD, Hercules, California, USA)
- 10 % linear SDS Polyacrylamide gel with 20-30 μl wells
- Spectrophotometer
- Image analysis equipment and computer software - (e.g. Alpha Innotech Corporation, San Leandro, California, USA)

Chemicals

- Pefabloc (Boehringer Mannheim, Germany)
- DTT
- Purified Rubisco standard from plant, algal or cyanobacterial cells (Sigma)
- Molecular Weight standard
- 100% Ice cold acetone
- DMSO (see Subprotocol 3)

All chemicals should be of analytical grade.

Buffers

- Preparation buffer:
  Add to 80 ml distilled water

1 mM Na_2-EDTA	0.2 ml, 0.5 M
100 mM sucrose	3.42 g
100 mM NaOH	10 ml, 1.0 M
5 mM 2-mercaptoethanol	0.5 ml

Dissolve and mix all ingredients and add distilled water to a final volume of 100 ml

- Loading buffer

Distilled water	3.5 ml
0.5 M Tris-HCl, pH 8.8	1.0 ml
Glycerol	1.5 ml
SDS 10%	1.6 ml
Mercaptoethanol	0.4 ml
Bromo Phenol Blue	0.14 g

CBB Staining solution (0.10, 65:25:10)

- Add to 25 ml distilled water

Coomassie Brilliant Blue R-250	0.25 g
Acetic acid	7 ml
Methanol	50 ml

Dissolve and mix all ingredients and add distilled water to a final volume of 100 ml

Destaining solution (10:10:80)

- Add to 400 ml distilled water

Acetic acid	50 ml
Methanol	50 ml

Procedure

Preparation of sample

1. Weigh 25 mg pulverised lichen powder into a small (2 ml) glass homogeniser. Weigh an additional sub-sample of 10 mg lichen powder from the same source for Chl determination with DMSO extraction (see Step 22).
2. Add 1.5 ml Preparation buffer.
3. Add 1 mM Pefablock (10 µl, 100 mM per ml buffer) and 20 mM DTT (3.1 mg ml^{-1} buffer).

Cell disruption

4. Disrupt the photobiont cells by squeezing the homogeniser piston 20 - 25 times through the suspension.
5. Transfer the suspension to an Eppendorf tube (2 ml) and add 200 µl SDS (10%).

Freezing and thawing

6. To further disrupt the photobiont cells, place the tube in liquid nitrogen for 2 min.
7. Quickly transfer the tube to a heating block and heat for 5 min at 75°C.
8. Repeat Steps 6 and 7 two more times.
9. Centrifuge in an Eppendorf centrifuge at room temperature and at maximum speed (c. 20000 g) for 3 min.
10. The resulting supernatant contains soluble proteins. Transfer the supernatant to a new Eppendorf tube and add 200 µl SDS (10%).
11. Heat the sample at 75°C for 5 min.
12. The sample is now ready for protein separation by gel electrophoresis. The sample can be stored in a freezer at -18°C, or in a deep freezer, if separation is not performed immediately.

Preparation of Rubisco standard

13. Dissolve purified Rubisco in the preparation buffer. At least two different concentrations of Rubisco standards must be loaded on each gel and these concentrations should be within the span of the Rubisco concentrations of the sample. We have found that Rubisco contents can range from 0.7 to 5 mg g^{-1} DW of lichen. If the procedure above is followed, and if the gel allows 20 - 30 µl of sample per well, the Rubisco standards should range between 0.08 and 0.4 µg per lane. However, this may need to be adjusted depending on species.

Gel electrophoresis

14. Prepare a 10% SDS-PAGE gel, or use a prefabricated gel, and install the gel electrophoresis equipment.

15. Mix 2/3 sample or Rubisco standard with 1/3 of loading buffer. Prepare molecular weight standards according to the supplier's instructions.

16. Heat the obtained mixture at 100°C for 5 min and centrifuge briefly (1 s) at maximum speed in an Eppendorf centrifuge to remove droplets from the tube walls.

17. With a micropipette, load 20 µl of each lichen sample mixture, the two Rubsico standards, and the molecular weight standard, in one well each. Run the gel for 1 - 1.5 h according to the particular instructions for the electrophoresis equipment.

18. In a fume hood, stain the gel by bathing it for 30 min in the CBB Staining solution.

19. Destain the gel in destaining solution for at least 24 h. Change the destaining solution 5 - 6 times during this period.

20. The large subunit of Rubisco (LSU) will appear as a 55 kDa band and the small subunit (SSU) as a 15 kDa band on the gel.

Quantification

21. Quantify the density of the 55 kDa band (OD 55) of the lichen sample with OD 55 of at least two Rubisco standards, using image analysis and software. The amount of Rubisco present in the lichen can then be quantified in relation to the known concentration of the Rubisco standard on the gel. It is advisable to calibrate the quantification procedure by loading a full range of Rubisco concentrations on a separate gel to make a standard curve.

22. Calculate the amount of broken photobionts by comparing total Chl of a parallel sub-sample as determined after DMSO extraction (Subprotocol 3) with the Chl content of the lichen protein sample as follows. Add 200 µl of the protein sample to 800 µl pure ice cold acetone and freeze the mixture at - 18°C for 20 min. Centrifuge in an Eppendorf centrifuge at maximum speed (c. 20000 g) at 5°C for 5 min.

23. Measure the absorption spectrophotometrically at the wavelengths 663 and 646 nm and calculate chlorophyll *a* and *b* concentrations in mg l^{-1} as follows (Wellburn 1994):

$$C_a = 12.21A_{663} - 2.81A_{646} \qquad \text{Equation 5}$$

$$C_b = 20.13A_{646} - 5.03A_{663} \qquad \text{Equation 6}$$

For cyanobacterial lichens, calculate Chl *a* using the equation 7

$$C_a = 12.21A_{663} \qquad \text{Equation 7}$$

Troubleshooting

- Problems with photobiont cell disruption
  If the Chl yield of the lichen protein preparation is lower than 50% we recommend that you try to improve photobiont cell disruption as suggested above.

- Protein concentration
  As for gel electrophoresis experiments in general, total protein concentration in the loaded sample is critical. Too little protein will result in very weak bands, whereas higher concentrations may increase the background noise. Because Rubisco is apparently one of the most abundant proteins also in lichens (Palmqvist et al. 1998) low protein concentrations in the sample may be the best solution.

Acknowledgements. Numerous colleagues have contributed to the development of the above procedures, particularly Murray Badger, Luis Balaguer, Douglas Campbell, Asuncion de los Rios, Thomas Hiltonen, Stefan Jansson, Henric Johansson, Erling Ögren and Göran Samuelsson, as well as those specifically referred to in the text. The above protocols have also been developed, tested and put together thanks to grants from the Swedish Research Council for Environment, Agricultural Sciences and Spatial Planning and the Swedish Natural Science Research Council to Kristin Palmqvist, and the Kempe foundation to Bodil Sundberg.

References

Badger MR, Andrews TJ, Whitney SM, Ludwig M, Yellowlees DC, Leggat W, Price GD (1998) The diversity and coevolution of Rubisco, plastids, pyrenoids, and chloroplast-based CO_2-concentrating mechanisms in algae. Can J Bot 76: 1052-1071

Badger MR, Palmqvist K, Yu J-W (1994) Measurement of CO_2 and HCO_3^- fluxes in cyanobacteria and microalgae during steady-state photosynthesis. Physiol Plant 90: 529-536

Badger MR, Pfanz H, Büdel B, Heber U, Lange OL (1993) Evidence for the functioning of photosynthetic CO_2 concentrating mechanisms in lichens containing green algal and cyanobacterial photobionts. Planta 191: 57-70

Balaguer L, Manrique E, de los Rios A, Ascaso C, Palmqvist K, Fordham M, Barnes JD (1999) Long-term responses of the green-algal lichen *Parmelia caperata* to natural CO_2 enrichment. Oecologia 119: 166-174.

Barnes JD, Balaguer L, Manrique E, Elvira S & Davison AW (1992) A reappraisal of the use of DMSO for the extraction and determination of chlorophylls *a* and *b* in lichens and higher plants. Env Exp Bot 32: 85-100

Björkman O (1981) Responses to different quantum flux densities. In: Lange OL, Nobel PS, Osmond CB, Ziegler H. eds. Physiological Plant Ecology I. Responses to the physical environment. Encyclopedia of Plant Physiology 12 A. Springer, Berlin, pp 57-108

Bubrick P (1988) Methods for cultivating lichens and isolated bionts. In: Galun M, ed. CRC handbook of lichenology, vol III. CRC Press, Boca Raton, pp. 127-140

Drew EA, Smith DC (1967) Studies in the physiology of lichens VII. The physiology of the *Nostoc* symbiont of *Peltigera polydactyla* compared with cultured and free-living forms. New Phyt 66: 379-388.

Lambers H, Pons, TL, Chapin III FS (1998) Plant Physiological Ecology. Springer, Berlin

Leverenz J, Jarvis PG (1979) Photosynthesis in Sitka spruce VIII. The effects of light flux density and direction on the rate of net photosynthesis and the stomatal conductance of needles. J App Ecol 16: 919-932

Máguas C, Griffiths H, Ehleringer J, Serôdio J (1993) Characterization of photobiont associations in lichens using carbon isotope discrimination techniques. In: Ehleringer J, Hall A, Farquhar G, eds. Stable isotopes and plant carbon-water relations. Academic Press, New York, pp 423-458

Nichols HW, Bold HC (1965) *Trichosarcina polymorpha* gen. Et sp. nov., J Phyc 1: 34-38.

Ögren E (1993) Convexity of the photosynthetic light-response curve in relation to intensity and direction of light during growth. Plant Phys 101: 1013-1019

Palmqvist K (1993) Photosynthetic CO_2 use efficiency in lichens and their isolated photobionts: The possible role of a CO_2 concentrating mechanism in cyanobacterial lichens. Planta 191: 48-56

Palmqvist K, Ögren E, Lernmark U (1994a) The CO_2 concentrating mechanism is absent in the green alga *Coccomyxa*: a comparative study of photosynthetic CO_2 and light responses of *Coccomyxa, Chlamydomonas reinhardtii* and barley protoplasts. Plant Cell Env 17: 65-76

Palmqvist K, Samuelsson G, Badger MR (1994b) Photobiont-related differences in carbon acquisition among green-algal lichens. Planta 195: 70-79

Palmqvist K, Sültemeyer D, Baldet P, Andrews TJ, Badger MR (1995) Characterisation of inorganic carbon fluxes, carbonic anhydrase(s) and ribulose-1,5-bisphosphate carboxylase-oxygenase in the green unicellular alga *Coccomyxa.* Planta 197: 352-361

Palmqvist K, de los Rios A, Ascaso C, Samuelsson G (1997) Photosynthetic carbon acquisition in the lichen photobionts *Coccomyxa* and *Trebouxia* (Chlorophyta). Phys Plant 101: 67-76

Palmqvist K, Campbell D, Ekblad A, Johansson H (1998) Photosynthetic capacity in relation to nitrogen content and its partitioning in lichens with different photobionts. Plant Cell Env 21: 361-372

Price D, Badger MR (1985) Inhibition by proton buffers of photosynthetic utilization of bicarbonate in *Chara corallina.* Aus J Plant Phys 12: 257-267

Raven JA (1997) Putting the C in phycology. Eur J Phyc 32: 319-333

Smith DC (1974) Transport from symbiotic algae and symbiotic chloroplasts to host cells. Symp Soc Exp Biol 28: 437-508

Smith DC (1992) The symbiotic condition (review). Symbiosis 14:3-15

Smith EC, Griffiths H (1996) The occurrence of the chloroplast pyrenoid is correlated with the activity of a CO_2-concentrating mechanism and carbon isotope discrimination in lichens and bryophytes. Planta 198: 6-16

Stanier RY, Kunisawa R, Mandel M, Cohen-Bazire G (1971) Purification and properties of unicellular blue-green algae (order Chroococcales). Bac Rev 35: 171-205.

Sundberg B, Campbell D, Palmqvist K (1997) Predicting CO_2 gain and photosynthetic light acclimation from fluorescence yield and quenching in cyano-lichens. Planta 201: 138-145

Sundberg B, Näsholm T, Palmqvist K (2001) The effect of nitrogen on growth and key thallus components in the two tripartite lichens, *Nephroma arcticum* and *Peltigera aphthosa*. Plant Cell Env 24: 517 - 527

Walker D (1990) The use of oxygen electrode and fluorescence probes in simple measurements of photosynthesis. Oxygraphics Limited. Robert Hill Institute, University of Sheffield.

Wellburn AR (1994) The spectral determination of chlorophylls *a* and *b*, as well as total carotenoids, using various solvents with spectrophotometers of different resolutions. J Plant Phys 144: 307-313

Suppliers

Alpha Innotech Corporation
14743 Catalina St., San Leandro, CA 94577, USA, Internet: http://www.alphainnotech.com, Email: info@aicemail.com

Bio-Rad Laboratories Headquarters
1000 Alfred Nobel Drive, Hercules, California 94547, Internet: http://www.bio-rad.com

Hansatech Instruments Ltd.
Narborough Road, Pentney, King's Lynn, Norfolk PE321JL, England, UK, Email: info@hansatech-instruments.co.uk, Internet: http://www.demon.co.uk/hansinst/

Heinz Walz GmbH
Eichenring 6, D-91090 Effeltrich, Germany, Email: info@mail.walz.com, Internet: http://www.walz.com

Li-Cor Inc.
PO Box 4425, Lincoln, NE 68504, USA, Email: envsales@env.licor.com, Internet: http://www.licor.com

Micron Separations Inc.
135 Flanders Road, Box 1046, Westboro MA 01581-6046, Internet: http://www.msifilters.com/

Analysis of Ethylene and ACC in Lichens

Sieglinde Ott

Introduction

Much research has been done on plant hormones in higher plants, where they play a fundamental role in regulating plant growth, development and reproduction (Davies 1995). In addition, in lichens interactions between the mycobiont and the photobiont require regulation (Jahns and Ott 1990). Plant hormones are a group of natural organic substances that affect physiological processes at low concentrations. Auxins, cytokinins, gibberellins, abscisic acid and ethylene, as well as polyamines, jasmonates, salicylic acid and brassinosteroids are biologically effective substances widespread in higher plants. However, information on the role of these substances in algae and fungi is very limited. Plant hormones have been identified in a range of different lichens (Ott and Zwoch 1992). Indole acetic acid (IAA), abscisic acid (ABA) and ethylene have been discovered in several lichen species. For example, IAA and ethylene have been detected in *Ramalina duriaei* (Epstein et al. 1986, Lurie and Garty 1991), and ABA in *Ramalina maciformis* (Hartung and Gimmler 1994). Of these plant hormones, only the gaseous substance ethylene has been examined in detail (Garty et al. 1995, Ott and Schieleit 1994, Ott 1993).

Ethylene can be synthesized by different biosynthetic pathways using different substrates and precursors. The biosynthetic pathway of ethylene has been well characterized in higher plants. In the first step, methionine (MET) is transformed to S-adenosylmethionine (SAM). SAM is then broken down to 1-aminocyclopropane-1-carboxylic acid (ACC) and 5'-methylthioadenosin (MTA) by the enzyme ACC synthase. ACC is the direct precursor of ethylene in higher plants (Adams and Yang 1979, Yang and Hoffmann 1984). It is transformed to ethylene in the last step of the

Sieglinde Ott, Heinrich-Heine-Universität, Botanisches Institut, Universitätsstraße 1, Düsseldorf, 40225, Germany (*phone* +49-211-81-13537; *fax* +49-211-81-12881; *e-mail* otts@uni-duesseldorf.de)

ethylene producing reaction by the enzyme ACC oxidase (Ververidis and John 1991, Smith et al. 1992). ACC synthase and ACC oxidase are specific enzymes in the ethylene metabolism of higher plants (Kende 1989, 1993).

In bacteria and fungi the precursor of ethylene appears to be α-oxo-glutarate or glutamate (Nagahama et al. 1991, 1992). In some bacteria the substrate for ethylene is α-keto-γ-methylthio-butyric acid (KMBA) formed by the deamination of methionine (Thomas and Spencer 1977, Billington et al. 1979, Mansouri and Bunch 1989). In lichens, both ACC and ethylene have been detected (Ott and Zwoch 1992). The methods used for measuring ethylene and ACC are standard methods used in research with higher plants. Therefore, this chapter deals less with standard methods but more with problems concerning the application of these methods in lichenology, and particularities caused by the nature of lichens.

Subprotocol 1
Determination of Ethylene Emissions From Lichens and Their Isolated Symbionts

Materials

Equipment

- Gas chromatograph (GC), e.g. Carlo Erba GC Vega Series, with flame ionisation detector (FID)
- Column with stationary phase suitable for separation of permanent gases and small hydrocarbons, e.g. Carboxen 1000 (45 - 60 mesh, 2 ft x 1/8", Supelco)
- Integrator, e.g. Spectra Physics SP4270
- Gas tight glass vials (2.5 ml, 15 x 28 mm) with screw caps and septum (silicon, 12 mm in diameter)
- Gas tight syringe, e.g. Hamilton syringe, model 1005 TLL, 5 ml, and Hamilton needles, side port w/KEL-F®Hub
- Shaking water bath
- Centrifuge, e.g. Sorvall with rotor SS 34

For more details on detectors, column packing material, carrier and combustion gas see Abeles et al. (1992).

Chemicals

- Gases: He, H_2 and air
- Ethylene calibration gas (Messer Griesheim, Germany or respective companies in other countries)
- MOPS (Morpholinopropane sulphonic acid) buffer, 50 mM, pH 6
- Soil extract: collect 250 - 500 g soil from the habitat of the lichen, boil in 1000 ml distilled water for 1 h, filter and refill to 1000 ml
- Malt yeast extract (for details see Chapter 1)
- *Trebouxia* organic medium (TOM, for details see Chapter 1)

Procedure

Preparation of lichen material

Intact lichens

1. Store the lichen material in a growth chamber under constant conditions (14 h/10 h day/night cycle, 60 - 70 µmol photons $m^{-2} s^{-1}$, relative humidity 70-80 %, air temperature 10°C).
2. Wet the lichens by spraying with deionized water every second day for not longer than 6 weeks.
3. Use thallus fragments or thalli of a definite size to increase the precision of the estimate of ethylene production. As an alternative method, use entire thalli. Carefully remove adhering particles of soil and organic materials from the samples.
4. When dry, mix the pieces together, and take samples of a definite dry mass.
5. Wet the thallus pieces with a pipette by adding a definite amount of water (depending on size and the morphological/anatomical structure of the lichen species). Allow the water to fully hydrate the thallus pieces.
6. Ensure that moisture content is the same in each sample. Check this by determining the mass of wet thalli before GC analysis, and the dry mass afterwards. Control of the water content is necessary to exclude differences in ethylene production caused by unequal water content among the samples (Ott and Schieleit 1994).
7. Put each sample (c. 0.05 g FW) in a 2.5 ml glass vial.

8. Flush the vials with compressed air to ensure that they contain no ethylene at the beginning of the experiment.
9. Seal the vial by a screw cap with a silicon septum.

Mycobiont

Chapters 1-3 describe procedures for culturing mycobionts in detail. For physiological and biochemical experiments, mycobionts cultured on agar are not very suitable. Much time is needed to get sufficient experimental material, and many agar plates are required. Transferring the mycelia in liquid medium significantly accelerates the growth rate of the mycobiont, so less time is needed to obtain enough mycobiont. Using liquid medium instead of agar also helps to prevent residues from the medium from interfering with the assay. Soil extract (from the respective collection site) can be added to the liquid medium if the mycobiont is isolated from terricolous lichens obviously supporting the growth of mycelia.

1. Inoculate a fixed number (e.g. 12) of the mycelial colonies (c. 1 - 2 mm^2) into a 300 ml Erlenmeyer flask containing 50 ml of a liquid medium of malt-yeast extract (pH 5.2) and 40 ml l^{-1} soil extract.

   Note: To avoid contamination the whole procedure must be performed under sterile conditions.
2. Gently shake the liquid cultures at 25°C in the dark on a shaker.
3. Replacing the liquid medium every five weeks promotes the growth rate of the mycobiont.
4. Growth for 3 - 6 months, depending on species, should yield 2 - 3 g fresh weight of mycobiont culture.
5. Harvest the mycelia by vacuum filtration through 2 layers of filter paper, then wash several times (e.g. 4 times) with deionised water.
6. Remove the remaining water carefully with paper towels.
7. Put c. 0.05 g of mycelia in 2.5 ml vials and continue as described for intact lichens (Steps 8-9).

Photobiont

For preparation of culture media, see Chapters 1-3. Culture conditions depend on the algal species. For *Coccomyxa* spp. I recommend 25°C and 30 µmol photons m^{-2} s^{-1}. For species of *Trebouxia* use 11°C and 20-25 µmol photons $m^{-2}s^{-1}$. The time needed before cultures can be harvested depends on the species and its growth curve; ensure that you know this before starting the experiments.

1. Grow photobionts (green algae) on *Trebouxia* organic medium (TOM) with 1.5 % glucose, 1 % proteosepeptone and 1.5 % agar (Ahmadjian 1967; Friedl 1989).
2. Transfer the algae to liquid medium (Erlenmeyer flasks 250 ml, 50 ml TOM) and shake gently.
3. Harvest the algae when they have reached the stationary growth phase.
4. Centrifuge 50 ml of the harvested algal suspension for 10 min at 200 g and room temperature.
5. Decant 10 ml of the supernatant with a pipette connected to a water pump.
6. Wash the algal cells with 20 ml MOPS buffer three times and resuspend in 6 ml of the same buffer.
7. Check that the algal cells are still intact after this treatment with a compound microscope.
8. Pipette 200 µl of the algal suspension into 2.5 ml glass vials and add 150 µl MOPS-buffer resulting in a final volume of 350 µl and continue as described for intact lichens (Steps 8-9).

Incubation of samples

1. As a general rule, carry out the incubation in the dark, as standardized for investigations on higher plants. Place the glass vials containing the samples of the intact lichens, mycobionts or photobionts into a shaking water bath and incubate them at 30°C for 24 h.

   Note: Each sample of the mycelium and the intact lichens should have a definite weight of e.g. 0.05 g DW. For samples of the photobiont 200 µl algal suspension should be used. These sample sizes are appropriate for 2.5 ml glass vials.

2. Some photobionts show ethylene production under light (200 µmol photons m^{-2} s^{-1}) as well as under dark conditions (e.g. *Trebouxia irregularis*) while other photobionts produce ethylene only under light conditions (e.g. *Coccomyxa* spp.). When investigating a photobiont for the first time, prepare many replicates, and incubate using both conditions.

3. When studying material for the first time, we recommend taking measurements over a time course. Ethylene production is non-linear with time, but displays saturation kinetics. Estimate the rate of ethylene production from the initial slope of a graph of ethylene production as a function of time.

GC analysis

Use the following operating conditions for the GC: a pressure of 60 kPa for He (carrier gas). For the combustion gases use 75 kPa H_2 and 110 kPa synthetic air. Adjust temperatures to 160°C (oven), 225°C (injector) and 250°C (detector).

1. Calibrate the GC before running the experiments with the calibration gas ethylene.
2. Take a 1 ml gas sample of each vial after incubation.
3. Inject the gas sample into the GC.
4. Calculate the ethylene production following Equation 1:

$$E\,[\mathrm{nl\ g^{-1}\ h^{-1}}] = [(A - b)/a]\ [(V - FW)/(DW \times V_s t)] \qquad \text{Equation 1}$$

E	Rate of ethylene production
A	Peak area
a	Slope of the calibration curve
b	y-Axis intercept of the calibration curve
V	Volume of the vial
V_s	Sample volume (1 ml)
t	Incubation time (h)
DW	Dry weight (g)
FW	Fresh weight (g)

Results

Ethylene production can be influenced by temperature and water content (Ott and Schieleit 1994) and by light conditions (Ott 1993). There is also a clear seasonal dependency (Ott unpublished). Ethylene clearly influences the development of lichen thalli and parts of the thallus (Ott et al., in press). Investigations with photobionts clearly show a correlation of ethylene production with stage of growth (Ott unpublished), although this has not been observed in experiments with mycobionts.

Troubleshooting

- Results show high standard deviation
  Rates of ethylene production may well be highly variable. Ethylene production rates are particularly variable in experiments with entire thalli or thallus pieces. To accurately measure ethylene production, sample number has to be large (15-20).

Subprotocol 2
Determination of the ACC Content in Lichens and Their Isolated Symbionts

Two methods exist for the determination of ACC. The first method is based on the oxidation of 1-aminocyclopropane-1-carboxylic acid to ethylene, which is then determined by gas chromatography. This method was developed by Lizada and Yang (1979) and later modified by Rower and Schierle (1982), and is described in detail. The second method involves direct identification of ACC using HPLC-MS (Chauvaux et al. 1993). Because the equipment for this method is very expensive and not available in every laboratory the first method is most commonly used for ACC determination. However, to conclusively prove that an organism can produce ACC, determination by HPLC-MS is necessary.

Materials

Equipment

- GC and integrator (for details see Subprotocol 1)
- Gas tight vials (2 ml, 15 x 28 mm) with screw caps and septum (silicon/Teflon, diameter 12 mm)
- Gas tight syringe, e.g. Hamilton syringe, model 1005 TLL, 5 ml, and Hamilton needles, side port w/KEL-F®Hub
- Shaking water bath
- Minishaker
- Magnetic stirrer
- Pipetman®P (Gilson), or equivalent
- Filter paper circles (4.5 cm), Miracloth, pore width 22-25 µm (Calbiochem, Novabiochem Corporation)
- Filter funnels
- Water jet pump
- Mortar and pestle
- Centrifuge e.g. Sorvall with rotor SS 34
- Centrifuge tubes, P.P., 50 ml
- Round bottom flasks with ground joints, 100 ml
- Pear shape flasks with ground joints, 100 ml
- Rotary evaporator (e.g. IKA Typ RV05)
- Combitherm HCB (IKA)
- Vacuum controller (Büchi B-720)
- Membrane vacuum pump (e.g. KNF)
- Ultrasonicator (e.g. Sonorex super RK 102 H, Bandelin)
- Glass beads (0.17 - 0.18 mm) (B. Braun Biotech International)
- Separating funnels (250 ml)
- Columns, e.g. Econo 1 x 20 cm (Bio-Rad)

Chemicals

- Ethanol
- Diethylether
- Ammonia solution, 25% NH_3
- $HgCl_2$ 80 mM
- Oxidation mix: saturated NaOH solution: NaOCl solution (containing 4% active Chlorine, Aldrich) = 1:2
- MOPS buffer (Morpholinopropane sulphonic acid), 50 mM, pH 6
- ACC (1-aminocyclopropane-1-carboxylic acid, 99%) 1μM (Fa. Aldrich)
- Liquid nitrogen
- Cation exchange resin (e.g. AG® 50W-X8 Resin, 100 - 200 mesh, hydrogen form)

Procedure

Extraction of ACC from intact lichens and mycobionts

1. Prepare lichen material as described in Subprotocol 1.
2. Pulverize the lichen material or mycelium in liquid nitrogen with mortar and pestle.
3. Add 20 ml ethanol (80%) to the powder.
4. Stir the mixture for 60 min with a magnetic stirrer.
5. Filter the raw extract under a vacuum.
6. Evaporate the filtrate to dryness with a rotary evaporator. Set the water bath to 50°C and apply a pressure of 120 - 48 mbar.
7. Dissolve the residue in 20 ml of double distilled H_2O in the ultrasonic bath, filter again, and adjust the pH to 7 ± 0.3.
8. Shake the filtrate twice in diethyl ether.
9. When the aqueous phase is obtained, evaporate to dryness with the rotary evaporator (water bath: 50°C, pressure: 500 - 48 mbar).
10. Dissolve the residue in 1 - 2 ml of doubled distilled H_2O in the ultrasonicating bath and adjust the pH to 5 - 6.

11. Purify 1-2 ml extract using a cation-exchange resin, in a 20 x 1 (i.d.) cm column.

12. Load the extract onto the column, then wash with 50 ml of double distilled water and discard the elute.

13. Elute the ACC-fraction by using 50 ml 4 N NH_3 and to dryness (50°C, 500 - 48 mbar).

14. Store the dried elute at -20°C.

Extraction of ACC from photobionts

The volumes of algal suspension needed to obtain detectible levels of ACC depend on the algal species. When using larger volumes, remove the nutrient medium in several steps.

1. Decant the algal suspension into centrifuge tubes.

2. Centrifuge at 10°C and 3000 g for 10 min.

3. Carefully discard the supernatant.

   Note: All following steps are performed on ice.

4. Wash the algal pellets with MOPS-buffer. Centrifuge at 10°C and 3000 g, discard the supernatant, then repeat this process three times. The volumes of MOPS buffer added should be 16 - 20 ml, depending on the size of the pellet.

5. Resuspend and stir the algal pellet with MOPS-buffer, then transfer the algal suspension into two 40 ml centrifuge tubes, the weight of which has previously been determined.

6. Determine the total volume of the algal suspension. This is necessary for final calculation of the dry weight of the algal pellets, so that ethylene production can be expressed on a dry weight basis.

7. Vortex each centrifuge tube for 5 s, and pipette 6 aliquots of 200 μl of algal suspension from each tube into 2.5 ml glass vials to determine the dry weight. Dry the samples at 70°C for 48 h or until they reach a constant weight.

8. Centrifuge both centrifuge tubes from Step 7 (10 min, 3000 g), discard the supernatant and determine the fresh weight of the algal pellets.

9. Extract the algal cells with 10 - 15 ml 100% ethanol (4°C). To achieve efficient extraction, add glass beads (0.17 - 0.18 mm). Vary the amount of the glass beads according to the volume of the solvent.
10. Incubate the suspensions on ice for 1 h. During this period shake them 8 times for 40 s and sonicate 6 times for 15 s.
11. Centrifuge the suspensions for 10 min at 10°C and 12000 g.
12. Filtrate the supernatant through Miracloth filters.
13. Evaporate the extract to dryness in the rotary evaporator (50°C, 120 - 48 mbar).
14. Dissolve the residue completely in 20 ml of double distilled H_2O by sonication, and adjust the pH to 7.0 ± 0.3.
15. Shake twice with 20 ml diethyl ether.
16. Evaporate again the aqueous phase after partitioning (50°C, 500 - 48 mbar).
17. Store the dried extract at -20°C until you are ready to process it further.

GC analysis of ACC

1. Cool all solutions and reaction tubes (2.5 ml vials) on ice.
2. Add 5.5 ml of double distilled H_2O to the dried extract and dissolve in the sonicating bath.
3. Pipette 0.5 ml of the dissolved extract into the vials, two sets of 5 for each sample.
4. In the first set, add 0.1 ml double distilled H_2O and in the second, add 0.1 ml ACC solution.
5. Add 0.2 ml 80 mM $HgCl_2$ solution, close with screw caps with a silicon septum, and inject 0.2 ml of the oxidation mix with a disposable syringe.
6. Vortex the samples for 5 s, incubate on ice for 2.5 min, vortex again for 5 s and immediately remove a 1 ml gas sample with a gas-proof glass syringe (e.g. Hamilton 1005).
7. Inject the gas sample into the GC to quantify the ethylene produced by the assay.

8. Finally calculate the ACC content following Equation 2:

$$\text{ACC} - \text{content (nmol x g}^{-1}) =$$

$$\text{A x (B} - \text{A)}^{-1} \text{ x C x D}_{\text{total}} \text{ x D}_{\text{test}}^{-1} \text{ x DW}^{-1} \text{ x E}$$

A	Mean value of peak areas per gas sample
B	Mean value of peak areas of the extract with added ACC standard
C	ACC standard [0.1 nmol] (constant)
D_{total}	Volume of the extract (µl) (see Step 2)
D_{test}	Volume of the extract per test (µl) (see Step 3)
DW	Dry weight of the lichen, mycobiont or photobiont sample (g)
E	Dilution factor for samples with A > 10 000

Troubleshooting

- Intact thallus and mycobiont
  - Crushing material in liquid nitrogen may be difficult because the fungi can be very hard. Check that mycobiont cells have in fact been broken using a compound microscope.
  - Be careful when using the rotary evaporator, as samples may unexpectedly boil.
  - The ion exchange column is necessary for removing phenolic substances from the extract that may inhibit or influence the further experimental process. Phenolics and amides can disrupt the Lizada and Yang test because they can complex Hg^{2+}.
- Photobiont
  - During treatment with diethyl ether the phases often only separate after 20 - 40 min. Discard any greasy phase between the aqueous and the ether phase.
  - When using 80 % ethanol you may get delayed boiling in the rotary evaporator, so always carry out the extraction in 100% ethanol. If delayed boiling nonetheless occurs, add more 100% ethanol. *Trebouxia* forms cell clusters that can be difficult to break, reducing extraction efficiency.

- ACC
  - When preparing ACC as internal standard you should use sterile water only. Although ACC is a chemically stable substance (200°C), it may be used as a nitrogen source by some micro-organisms.
  - ACC solution may be kept in the freezer, but it should not be frozen twice.

References

Abeles FB, Morgan PW, Saltveit ME (1992) Ethylene in Plant Biology. Academic Press, New York 2nd.

Adams DO, Yang SF (1979) Ethylene biosynthesis: Identification of 1-aminocyclopropane-1-carboxylic acid as an intermediate in the conversion of methioninee to ethylene. Proc Natl Acad Sci USA 76:170-174

Ahmadjian V (1967) A guide to the algae occuring as lichen symbionts: Isolation, culture, cultural physiology and identification. Phycologia 6:127-160

Billington C, Golding BT, Primrose SB (1979) Biosynthesis of ethylene from Methionine - Isolation of the putative intermediate 4-Methylthio-2-Oxobutanoate from culture fluids of bacteria and fungi. Biochem J 182:827-836

Chauvaux N, van Dongen W, Esmans EL, van Onckelen HA (1993) Liquid chromatographic-mass spectrometric determination of 1-aminocyclopropane-1-carboxylic acid in tobacco. J Chromat 657:337-343

Davies PJ (1995) Plant Hormones. Physiology, Biochemistry and Molecular Biology. Kluwer Academic Publishers, Dordrecht London

Epstein E, Sagee O, Cohen JD, Garty J (1986) Endogenous auxin and ethylene in the lichen *Ramalina duriaei*. Plant Physiol 82:1122-1125

Friedl T (1989) Systematik und Biologie von *Trebouxia* (Microthamniales, Chlorophyta) als Phycobiont der Parmeliaceae (lichenisierte Ascomyceten). Inaugural Dissertation, Universität Bayreuth

Garty J, Kauppi M, Kauppi A (1995) Differential responses of certain lichen species to sulfur-containing solutions under acidic conditions as expressed by the production of stress-ethylene. Environ Res 69: 132 – 143

Hartung W, Gimmler H (1994) A stress physiological role for abscisic acid (ABA) in lower plants. Progress in Botany 55:157-173

Jahns HM, Ott S (1990) Regulation of regenerative processes in lichens. Bibl Lichenol 38:243-252

Kende H (1993) Ethylene biosynthesis. Annu Rev Plant Physiol Plant Mol Biol 44: 283 – 307

Lizada C, Yang SF (1979) A simple and sensitive assay for 1-aminocyclopropane-1-carboxylic acid. Anal Biochem 100:140-145

Lurie S, Garty J (1991) Ethylene production in the lichen *Ramalina duriaei*. Ann Bot 68:317-319

Mansouri S, Bunch AW (1989) Bacterial ethylene synthesis from 2-oxo-4-thiobutyric acid and from methionine. J Gen Microbiol 135: 2819-2827

Nagahama K, Ogawa T, Fujii T, Tazaki M, Goto M, Fukuda H (1991) L-Arginine is essential for the formation in vitro of ethylene by an extract of *Pseudomonas syringae*. J Gen Microbiol 137:1641-1646

Nagahama K, Ogawa T, Fujii T, Fukuda H (1992) Classification of ethylene-producing bacteria in terms of biosynthetic pathways to ethylene. J Ferm Bioengin 73:1-5

Ott S (1993) The influence of light on ethylene production by lichens. In: Feige GB and Lumbsch HT (eds) Festschrift S Huneck. Bibl. Lichenol 53:185-190

Ott S, Schieleit P (1994) Influence of exogenous factors on ethylene production by lichens. I. Influence of water content and water status conditions on ethylene production. Symbiosis 16:187-201

Ott S, Zwoch I (1992) Ethylene production by lichens. Lichenologist 24:73-80

Ott S, Krieg T, Spanier U, Schieleit P (2000) Phytohormones in lichens with emphasis on ethylene biosynthesis and functional aspects on lichen symbiosis. Phyton 40; 3: 83-94

Remmer SB, Ahmadjian V, Livdahl TP (1986) Effects of IAA (indole-3-acetic acid) and kinetin (6-furfuryl-amino-purine) on the synthetic lichen *Cladonia cristatella* and its isolated symbionts. Lichen Physiol Biochem 1:1-25

Rohwer F, Schierle J (1982) Effect of light on ethylene production: Red light enhancement of 1-aminocyclopropane-1-carboxylic acid concentration in etiolated pea shoots. Z. Pflanzen-physiol 107:295-300

Smith JJ, Ververidis P, John P (1992) Characterisation of the ethylene-forming enzyme partially purified from melon. Biochemistry 31: 1485-1494

Thomas KC, Spencer M (1977) L-methionine as an ethylene precursor in *Saccaromyces cerevisiae*. Can J Microbiol 23:1669-1674

Ververidis P, John P (1991) Complete recovery in vitro of ethylene-forming enzyme activity. Phytochemistry 30(3):725-727

Yang SF (1985) Biosynthesis and action of ethylene. Hort Sci 20:41-45

Yang SF, Hoffman NE (1984) Ethylene biosynthesis and its regulation in higher plants. Ann Rev Plant Physiol 35:155-189

Measuring Activities of the Enzymes Superoxide Dismutase and Glutathione Reductase in Lichens

Michael A. Thomas

Introduction

Superoxide dismutase (SOD, EC 1.15.1.1) and glutathione reductase (GR, EC 1.6.4.2) are ubiquitous enzymes, found in all organisms, and often as different isozymes in different compartments of the same cell. They are important defensive compounds, each having a primary function involving detoxification of reactive oxygen species. Superoxide dismutase is positioned at the beginning of the glutathione-ascorbate pathway (Fig. 1) (Foyer and Halliwell 1976, Nakano and Asada 1981, Kunert and Foyer 1993, Hess 1994), where it scavenges superoxide radicals, catalysing their conversion to hydrogen peroxide. Glutathione reductase, a flavoprotein, occupies a position at the opposite end of the pathway, where it catalyses the NADPH-dependent reduction of glutathione disulphide (GSSG) to glutathione (GSH):

$$\text{GSSG} + \text{NADPH} + \text{H}^+ \qquad\qquad 2\text{GSH} + \text{NADP}^+$$

In this way, GSH continues to be available for the reduction of dehydroascorbate, effectively recycling ascorbate.

SOD is actually the name for a general category of metallo-enzymes that can be subdivided into forms based on the nature of the reaction centres: the Cu,Zn-SOD form, and the Mn-SOD and Fe-SOD forms (Scandalios 1994, Asada et al. 1980). There is a very strong phylogenetic relationship between the existence of these forms and the nature of, and compartmentalisation within, the organism. In general, multicellular eukaryotes, including plants and phragmoplastic algae, have Cu,Zn-SOD present in the cytosol and in chloroplasts. The Mn-SOD form is also present in

Michael A. Thomas, University of Otago, Department of Biochemistry, PO Box 56, Dunedin, New Zealand (*phone* +64-3-476-7109; *fax* +64-3-479-7866; *e-mail* mike.t@sanger.otago.ac.nz)

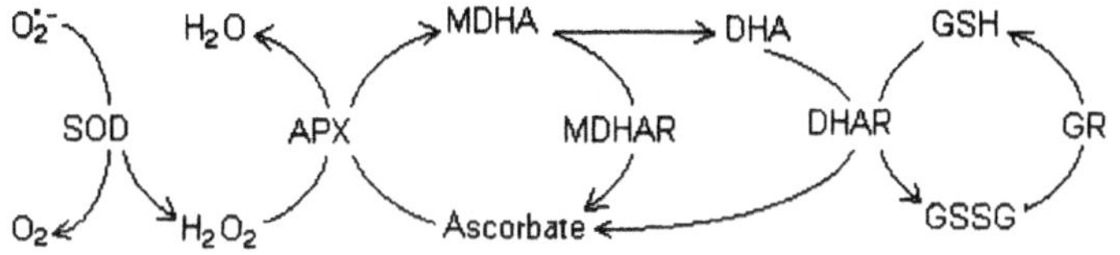

Fig. 1. The glutathione-ascorbate pathway for detoxification of reactive oxygen species. SOD = superoxide dismutase, APX = ascorbate peroxidase, MDHA = monodehydroascorbate, MDHAR = monodehydroascorbate reductase, DHA = dehydroascorbate, DHAR = dehydroascorbate reductase, GSH = glutathione (reduced), GSSG = glutathione disulphide, GR = glutathione reductase.

the chloroplasts and in the mitochondrial matrix of these organisms, and of the non-phragmoplastic algae (Asada et al. 1980). The latter group contains the Fe-SOD form in the stroma and in the mitochondrial matrix instead of the Cu, Zn-SOD form. Prokaryotes, in general, contain only the Mn-SOD and/or Fe-SOD forms (Scandalios 1994, Asada et al. 1980, Foyer and Hall 1980), with both forms present in cyanobacteria. Depending on the lichen, two or all three forms may be present. It is possible to distinguish between the three forms in assays. The Cu, Zn-SODs are inhibited by CN^{-1}, while both Cu, Zn-SOD and Fe-SOD are inhibited by hydrogen peroxide. By comparing the results of concurrent assays with and without inhibitors, the ratio of the various forms present can be determined. The assay described below allows Cu, Zn-SOD to be distinguished from the Mn-/Fe-SODs.

As a group, lichens are considered sensitive to air pollution (Gries 1996). The differential distribution of lichen species caused by their differing sensitivity to air pollutants has been recognised for at least 130 years (Nylander 1866). They are often used as passive indicators of air quality. One effective cause of damage by a number of air pollutants (e.g. ozone, sulphur dioxide, NO_x) is through reactive oxygen species (Pell 1987). Relating the differential sensitivity of lichen species to effective detoxification strategies through the potential activities of GR and SOD could have broad implications for environmental research, and reaffirm the utility of lichens as biomonitors. This chapter provides methods for assaying the activities of these enzymes.

Outline

Figure 2 gives an overview of the experimental procedure.

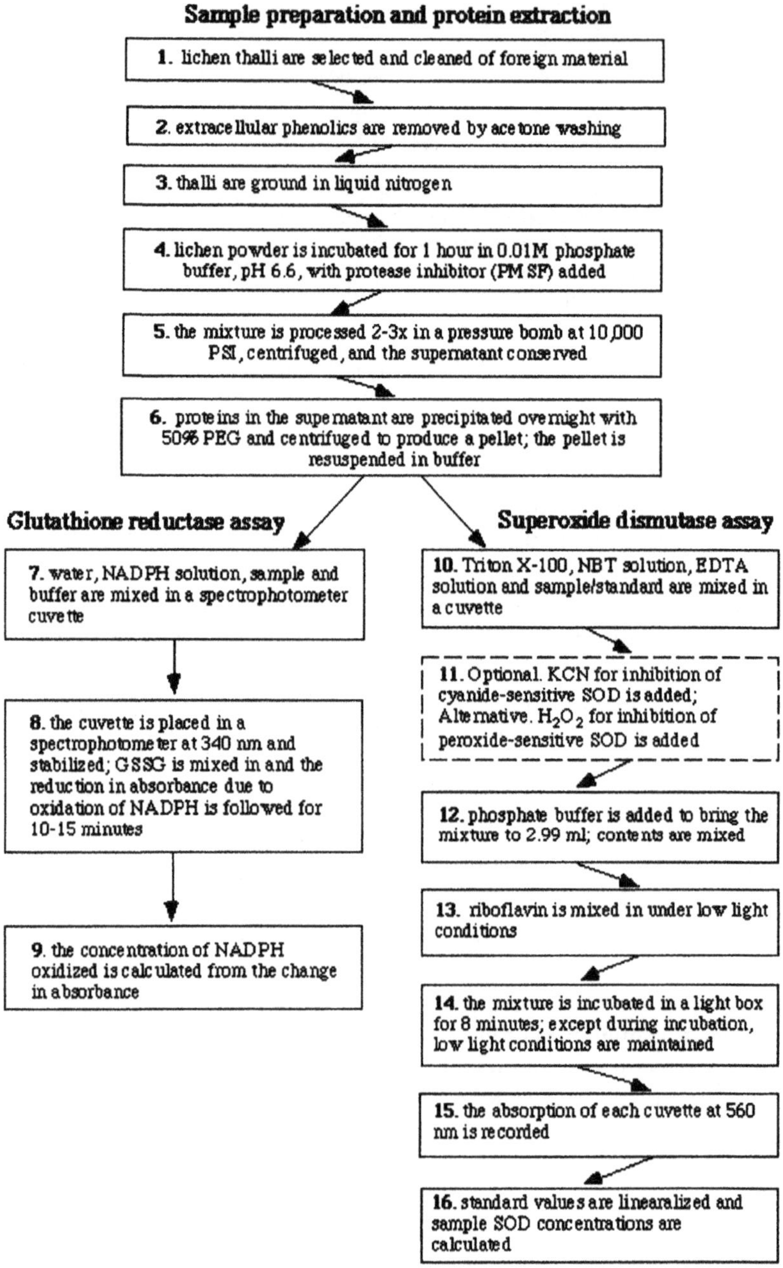

Fig. 2. Procedure flowchart. Steps 1-6 show sample preparation and protein isolation. Steps 7-9 are the assay for GR activity. Steps 10-16 demonstrate the SOD activity assay.

- **Sample preparation and protein extraction (Steps 1-6)**
  Precleaned lichen thalli are washed with pure acetone to remove extracellular secondary compounds. Cell disruption and release of intracellular contents from the dried lichens is achieved by grinding the thallus in liquid nitrogen, resuspending the powdered material in a slightly acidic, low salt buffer containing a protease inhibitor and processing this suspension through a French pressure cell at high pressures. The sample suspension containing the isolated soluble proteins is assayed for protein concentration. This mixture of known protein concentration is used for both the GR and SOD assays.

- **GR activity assay (Steps 7-9)**
  NADPH-dependent GR activity is determined by modifying the spectrophotometric assay of Carlberg and Mannervik (1985). The assay uses GR present in the sample to catalyse the reduction of GSSG to GSH using NADPH as a proton donor. A decrease in absorption of the solution at 340 nm occurs due to the oxidation of NADPH.

- **SOD activity (Steps 10-16)**
  Superoxide dismutase activity is determined using a modification by Beyer and Fridovich (1987) of the Beauchamp and Fridovich (1971) assay. In the presence of an electron donor, and light mediated, nitroblue tetrazolium (NBT) and riboflavin react to form an insoluble product, formazin, which has a peak absorption at 560 nm. This photometric assay takes advantage of the inhibition of formazin production by SOD, which competes for the electron. Because it is incomplete inhibition, the activity of SOD is relative to 50% inhibition of formazin production (Paoletti and Mocali 1990). To calibrate the activities, a standard for 100% formazin production (no SOD or sample addition) must be included in the assay. It is an exposure dependent, rather than time dependent assay, and all samples and standards must be run concurrently. Under the pH conditions for this assay there should be no difference in activity rates between the various forms of SOD (Asada et al. 1980). This assay is modified by the addition of the detergent Triton X-100, which keeps formazin soluble, to allow measurement in solution. Because it is an incomplete inhibition assay, a blank (no SOD addition) **must** be included with each run. The Cu,Zn-SOD isozymes are inhibited by cyanide and both this form and the Fe-SOD form are inhibited by the addition of H_2O_2, whereas the Mn-SOD isozyme is not affected by either (Asada et al. 1974, Giannopolitis and Ries 1977). Therefore, it is possible to distinguish between the three forms by including sample replicates containing these inhibitory compounds. This chapter describes the use of KCN to dis-

tinguish the Cu,Zn-SOD activity from the combined SOD activity and H_2O_2 to distinguish Mn-SOD activity.

Materials

Sample preparation, protein extraction and quantification

Equipment

- Refrigerated centrifuge, capable of 21,000 x G
- French press apparatus
- Sonicating water bath
- Spectrophotometer capable of UV and visible spectra

Other supplies

- Source of liquid nitrogen
- Dewar flask for liquid nitrogen
- Mortar and pestle (recommended size '0')
- Screw cap vials (for storage of powdered lichen)
- 15 ml screwtop centrifuge tubes
- 50 ml round bottom centrifuge tubes
- 1 cm spectrophotometer cuvettes, 3 ml capacity

Chemicals

All chemicals should be at least reagent grade.

- Pure acetone
- Liquid nitrogen
- Bicinchoninic acid (BCA) protein assay kit
- Distilled water

Buffers

- 10 mM potassium phosphate (K_2HPO_4/KH_2PO_4), pH 6.6

Note: For the potassium phosphate buffers, acidic potassium phosphate monobasic (KH_2PO_4) and basic potassium phosphate dibasic (K_2HPO_4) solutions are first made separately to the desired molar concentrations, and then titrated together to achieve the desired buffer pH. Store buffers at 4°C.

Stock solutions

- 50% w/v Polyethylene glycol (PEG) solution (25% w/v 8000 MW PEG, 25% w/v 1000 MW PEG). Store at 4°C.
- Phenylmethylsulfonyl fluoride (PMSF) solution (0.0174 g PMSF in 100 μl dimethylsulfoxide). Store at -20°C.

Glutathione reductase assay

Equipment

- Spectrophotometer capable of measuring in the UV and visible spectra

Other supplies

- 1 cm spectrophotometer cuvettes, 1.2 ml capacity
- Caps for spectrophotometer cuvettes, or parafilm

Chemicals

All chemicals should be at least reagent grade.

- Glutathione reductase (GR) (e.g. Sigma, from yeast)
- Distilled water

Buffers

- 0.2 M potassium phosphate (K_2HPO_4/KH_2PO_4), pH 7.0, containing 2 mM EDTA disodium salt
- 100 ml of 10 mM Tris solution, titrated to pH 7.0 with HCl

Stock solutions

- 2 mM NADPH solution (0.0167 g NADPH in 10 ml Tris-HCl buffer, see above). Store at -20°C up to 2 weeks.
- 20 mM glutathione disulphide (GSSG) solution (0.1225 g GSSG in 10 ml distilled water). Store at -20°C.

Superoxide dismutase assay

Equipment

- Spectrophotometer capable of measuring in the UV and visible spectra
- Light box apparatus (see Figure 3)

Other supplies

- 1 cm spectrophotometer cuvettes, 3 ml capacity
- Caps for spectrophotometer cuvettes, or parafilm

Chemicals

- Distilled water
- Superoxide dismutase (SOD) (e.g. Sigma, from horseradish)

Buffers

- 10 mM potassium phosphate (K_2HPO_4/KH_2PO_4), pH 6.6
- 50 mM potassium phosphate (K_2HPO_4/KH_2PO_4), pH 7.8

Stock solutions

- Riboflavin solution (10 mg in 100 ml distilled water). Store at 4°C.
- Triton X-100 solution, 0.833% v/v (1 ml Triton X-100 into 119 ml distilled water). Store at 4°C.
- Nitroblue tetrazolium (NBT) solution (0.1472 g in 100 ml distilled water). Store in the dark at 4°C.

  Note: Make fresh every few days.
- EDTA solution (0.75 g EDTA disodium salt in 100 ml distilled water). Store at 4°C.
- Potassium cyanide (KCN) solution (0.1 g KCN in 10 ml distilled water).

  Note: This is a highly toxic compound. Please handle with care.

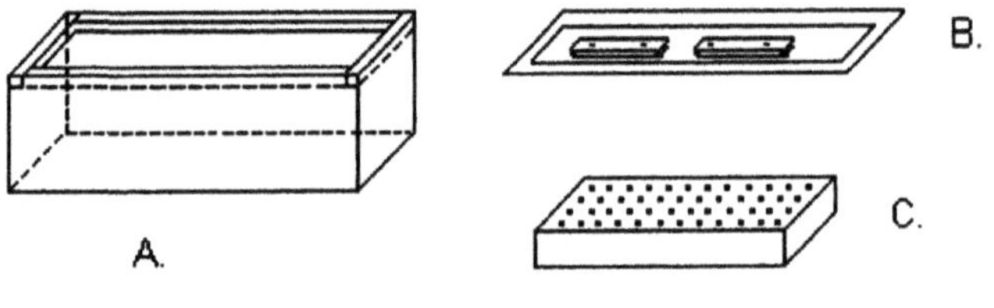

Fig. 3. An example of a light-box apparatus for the photosensitive production of formazin. A = foil-lined light box. The dimensions of the box may vary as needed. An appropriate ratio of length:width:height would be 2:1:1. A typical fluorescent tube light bulb measures approximately 30 cm for every 10 W, with 5 - 10 cm additional for the ballast. Therefore number, intensity and position of the light source(s) will determine the dimensions of the box. B = cover for light box with e.g. 2 15 W full-spectrum fluorescent tube light assemblies mounted on the inside. The light source control (e.g. power cord, power switch) should be externally accessible to enable turning the light on/off while the cover is in place. C = removable incubation tray, which sits under the lights. Seatings on the tray for cuvettes should be no deeper than 3 - 4 mm to enhance light exposure. Actual placement of cuvettes in the tray should be to minimise shadowing. The inner facing of the cover and the exposed areas of the tray should also be foil-lined. The lights should not be turned on until the tray and cover are in position.

Procedure

Sample pre-extraction preparation

All equipment used for extraction and pre-extraction should be cleaned and rinsed well with water of at least de-ionised quality.

1. Lichen thalli collected for assay must be mechanically cleaned as thoroughly as possible of attached materials, such as attached substrate, plant materials, extraneous foreign lichen fragments and debris. Thalli sections which are noticeably parasitised should not be used unless such parasitisation is the focus of the assay. Some lichen species adhere strongly to their substrate. If thallus portions with strongly adherent foreign matter cannot be discarded, then the lichen should be slightly moistened, which will ease removal of extraneous materials with the least degree of tissue damage. The thalli must then be air-dried before proceeding. The time to air dry is dependent on the species of lichen and the lab environmental conditions in addition to the water content, but it may require no longer than 2-3 hours to achieve a stable weight. To ease later preparation, the necessary time frame should be determined beforehand.

   Note: Good experimental design involves controlling the greatest number of extraneous variables. Therefore, pre-treatment procedures should be standardised for all samples. For example, if any samples require moistening for cleaning, then all samples for that experiment should be similarly moistened. The degree of pre-treatment required for a species should be a consideration (when possible) in selecting a species for study. A pendulous or fruticose lichen is much easier to clean than a closely adherent, foliose lichen. The nature of the study is also a consideration in selecting samples. If the goal is to measure the activity of *in situ* material, then it may be more practical to subjectively select thalli from the field that are easier to clean. If the experiment involves laboratory manipulation (e.g. fumigation), then the material should be collected in bulk and cleaned at the same time.

2. Prior to grinding, extracellular phenolics are removed (Fahselt and Jancey 1977).

   Note: In order to reduce the loss of intracellular contents, it is very important to ensure that the lichen material is in an air-dried state and that pure acetone is used. Dried lichens are very hydrophilic, and severe membrane disruption will occur if they are exposed to solvents such as acetone and alcohols while partially hydrated. In a fume

hood, wash approx. 1 g dry weight lichen with 30 - 50 ml cold (4°C) pure acetone for 20 min in a covered container (e.g. 100 ml beaker with watch glass). The amount of acetone used must be sufficient to cover the lichen. At the end of this washing, pour off and discard the acetone and repeat the step. After the second washing, pour off and discard the acetone as before, and allow any remaining acetone on the lichen to evaporate.
An alternative method for measuring GR in lichens that involves a different procedure for lichen pre-treatment (freeze drying of samples and using PVP to bind phenolic compounds) is described in Kranner (1998).

3. The lichen material is ground to a fine powder in liquid nitrogen using a mortar and pestle. The powdered material may be used immediately, or stored (in glass vials) at -20°C for a few weeks (short term) or at -80°C for a longer term. Air-dried material may be stored at -20°C.

Protein extraction

Protease inhibitors are recommended when working with protein extracts to maintain stability of the extract (Deutscher 1990). There are a number of protease inhibitors available (e.g. metallo-, serine, acid and thiol proteases). The selection of protease inhibitors is dependent on what enzyme activity is investigated. Metallo- and thiol protease inhibitors are not recommended for assays of SOD and GR, respectively, as they may interfere with the activities of those enzymes. By default, I recommend PMSF (phenylmethylsulfonyl fluoride), a serine protease inhibitor. It is the most widely used protease inhibitor in studies with yeast (Jazwinski 1990). It can be stored at -20°C for some time. However, once added to the sample buffer, it has an effective half-life of about 100 min at pH 7 and room temperature (Jazwinski 1990), but can be resupplied when the extraction procedure progresses beyond that. It should therefore be added last to the extraction buffer, and temperatures should be maintained at or near 4°C for as long as possible during the extraction.

4. Place 0.3-1.0 g of ground lichen in a 15 ml centrifuge tube. Add to this the 10 mM potassium phosphate stock buffer in the proportion of 1 ml buffer: 0.1 g lichen. Shake the tube to mix contents. **Immediately** add PMSF solution (1 μl PMSF: 1 ml sample/buffer solution; final concentration approx. 1 mM). Remix the contents by shaking the tube. Place the tube in a sonicating water bath for 1 - 2 min to further agitate the mixture. Then incubate the samples at 4°C for one hour. This last step is to ensure hydration of undisrupted cells and organelles, making them more susceptible to rupture by the French press procedure.

5. The sample mixture is placed in a French pressure cell (high pressure bomb) at approximately 75 Mpa for 3 - 4 min, followed by slow release while under pressure. Since much of the lichen material will have settled during incubation, the sample should be shaken immediately prior to transfer. For most species of lichens tested, the total soluble protein content of the extracts was greatly enhanced (2 - 5x) by a second treatment with the French press. Pendulous, fruticose lichens, e.g. *Usnea* or *Alectoria* spp., may require a third treatment and still yield low protein concentrations in the extracts. Collect the material extruded from the French press in a 50 ml centrifuge tube and centrifuge at 21 000 x G at 4°C for 30 min. Conserve the supernatant in a 15 ml centrifuge tube and re-centrifuge at 1000 x G for 3 min to remove any remaining insolubles. Repeat this last step as necessary. The resulting supernatant is the sample protein extract.

   Note: The French press is the recommended procedure. The soluble protein content of the extract is greatly enhanced by this procedure over simple grinding. An alternative technique is to use an ultrasonicating probe.

Protein precipitation

The volume of the protein extract should now be 3 - 7 ml. Although this sample is available for immediate enzyme assay, protein precipitation concentrates the materials, enhancing the assays.

6. Add the PEG solution to the protein extract in a 1:1 (v/v) proportion (final PEG concentration 25% w/v). Thoroughly mix this solution by shaking, and incubate at 4°C overnight. Following incubation, centrifuge the mixture at 3000 x G at 4°C for 15 min to pellet the protein precipitate. Pipette off and discard most of the supernatant, being careful to conserve the pellet. Re-centrifuge the mixture for 5 min, and then discard the remaining supernatant. Resuspend the pellet in the 10 mM potassium phosphate stock buffer (0.5 - 1 ml depending on desired protein concentration) by repeated pipetting. Following resuspension, transfer the solution to a 1.5 ml microeppendorf tube. The sample may then be assayed immediately, or stored short term (2 - 3 weeks) at -20°C or longer term (more than 6 months) at -80°C. The total protein concentration of the sample should be determined spectrophotometrically (recommended: bicinchoninic acid assay). Because of differences in protein extractability of lichens by such factors as species and thallus age, I recommend that the results from enzyme assays be normalised as per protein content rather than per thallus dry weight.

Glutathione reductase assay

The effectiveness of the assay is checked periodically by using appropriate GR standards in place of and added to samples, respectively, preferably GR from yeast.

7. Into a 1.2 ml spectrophotometer cuvette add the following: 50 μl NADPH solution, 50 μl protein extract, 400 μl of 0.2 M potassium phosphate stock buffer, and 500 μl distilled water. Mix by inverting.

8. Place the cuvette in the spectrophotometer, wait until the absorption at 340 nm stabilizes (1- 2 min) and record the background reading. Add 50 μl GSSG solution and invert to mix. Record the initial absorption reading, and continue to record readings every 30 s for 10-15 min. The decrease in absorption can be linear for up to 15 min. Use only the linear portion of the decline to calculate GR (see Troubleshooting).

9. Activity calculations: Using the Lambert-Beer equation, $c=A/(\epsilon d)$, calculate the molar quantity of NADPH oxidised (c), where A is the change in absorption (in nm), ϵ is the extinction coefficient of NADPH at 340 nm (6.2 mM^{-1} cm^{-1}) and d is the spectrophotometer path length (in cm). GR activity is described as a function of the NADPH oxidised to $NADP^+$, and is typically expressed as either units of activity (U), where U=1 μM NADPH oxidized min^{-1}, or as nkats, where 1 katal (kat) = 1M NADPH oxidised sec^{-1}. GR activity is typically normalised for comparison as a function of either dry weight (g) or protein content (mg).

Superoxide dismutase assay

This assay must be done in low-light conditions after the addition of riboflavin. All standards and samples must be exposed to light concurrently. KCN is highly toxic, always wear gloves. The assay proceeds at 25°C.

10. Add the following to a 3 ml spectrophotometer cuvette: 90 μl Triton X-100 solution (0.025% v/v final concentration), 100 μl NBT solution (60 μM final concentration), 150 μl EDTA solution (0.1 mM final concentration), and 50 μl protein extract, SOD standard or 10 mM potassium phosphate stock buffer (blank).

11. For cyanide-inhibition replicates, 20 μl KCN solution (1 mM final concentration) is also added to distinguish cyanide-sensitive SOD activity (Cu,Zn-SOD) from total SOD activity. Troubleshooting for alternative inhibition protocols.

12. Add a sufficient amount of the 50 mM potassium phosphate stock buffer to bring the total volume of the cuvette mixture to 2.99 ml (calculate this beforehand). Cap each cuvette and mix the contents by inversion.

13. Rapidly, under **low-light** conditions, add 10 µl of the riboflavin solution (0.886 µM final concentration) to each cuvette. Mix the cuvettes by inversion (as before).

14. Place the cuvettes in a shallow, foil-lined holding tray, which is then inserted into a light box (Giannopolitis and Ries 1977, Beyer and Fridovich 1987) with (e.g.) 2 - 4 15 W full spectrum light bulbs as a light source (Fig. 3). Turn on the light box light source and illuminate the cuvettes for approximately 8 min. At the end of the incubation period, turn the light source off. Continue to maintain low-light conditions.

Note: Formazin will continue to be produced with additional light exposure until the reactants are used up. The effectiveness of this assay is dependent on removing the light source while differential amounts of product have been produced and the standards are still distinguishable from each other. A general approximation of the necessary timeframe of light exposure is 7 - 9 min. The optimal timeframe is determined by the design of the light box used (shape, reflectance, light source and intensity), and should be pre-determined by trial and error. In general, the reaction should probably not proceed beyond 12 min.

15. The production of formazin in each cuvette is assayed by measuring the absorption at 560 nm with a spectrophotometer. Before measuring the samples, the spectrophotometer should be zeroed with a blank solution containing all the reactants (except inhibitors and sample, but including 50 µl of sample buffer), but not exposed to light. Measurement of the sample and standard mixtures should be made as quickly as possible, and as soon as possible following the light incubation.

16. Activity calculations: SOD inhibition of formazin production is nonlinear. The absorption values are made linear by conversion as: $V/v - 1$, where V is the absorption of the control cuvette (no SOD addition, and so maximum formazin production) and v is the absorption of the sample or standard (Asada et al. 1974, Giannopolitis and Ries 1977). Regression equations are generated using the converted standard values, and the sample SOD concentrations are then calculated. SOD activities are typically reported as units per µg protein. The percentage of total SOD represented by the various forms (i.e. cyanide-sensitive, hydrogen peroxide-sensitive) is also calculated.

Troubleshooting

- Sampling procedure and pre-treatment
  It is very difficult, if not impossible, to grow lichens under greenhouse conditions, and they generally must be collected in the field, often at great distances from laboratory. In order to minimise differences in pre-conditioning, the material for laboratory manipulations (e.g. fumigation studies) is generally collected in sufficient quantities and stored air-dried at -20°C. However, I have found that some lichens, e.g. *Peltigera* spp., respond better to short-term storage (up to 4 weeks) at 4°C than at -20°C. If longer storage is necessary then it should all occur at -20°C. Acetone pre-extraction has been the recommended pre-treatment prior to protein extraction in lichens (e.g. Fahselt and Yancey 1977) to remove extracellular phenolics which might denature proteins. It may not be necessary for use with those lichens with limited secondary compound production, e.g. gelatinous lichens such as *Leptogium* spp. I have found that acetone pre-extraction does interfere with genomic DNA extraction by some techniques.

- Protein extraction (pH of extracting buffer, and disruption technique)
  For the lichens I have tested, within a range of pH 5.6 - 8.8, pH 6.6 comes out as the optimal for the extracting buffer. This is true for most species. However, at least one lichen, *Teloschistes*, extracted better at a slightly basic pH 7.8, with poor extractions at pH 6.6. The French press is the recommended disruption technique. Multiple exposures (2 - 3) increase the amount of protein appreciably. If a French press is not available, then multiple grindings in liquid nitrogen may be of some use, although more material is generally lost during the grindings. Combine this technique with an ultrasonicating probe.

- Protein precipitation (alternative to PEG)
  An alternative to PEG precipitation is the use of dialysis membranes of suitable porosity. Occasionally a sample with visible precipitate will not form a pellet properly during centrifugation. This is apparently species specific and concentration dependent. To recover the precipitate, it may be necessary to vacuum filter the PEG/sample mixture.

- GR assay (non-linear absorption)
  If the absorption becomes non-linear too rapidly (less than 5 min), you may need to increase the concentrations of the reactants, NADPH or GSSG, or reduce the amount of sample used, or dilute the sample mixture. If the reaction takes more than a few min to stabilise before the

addition of GSSG, this is probably due to the utilization of NADPH by other reactions, and additional NADPH should be added.

- SOD assay (standards are too dark)
  Formazin is a purplish colour. The zero standard should be a dark blue-violet. If the standards and samples appear too light, then it may be necessary to increase the light exposure by duration or intensity. If all standards are dark, then the reaction ran too long. Leaving gaps between samples on the tray will reduce shadowing. If standards are not linear following conversion, check the light box to ensure that the light intensity is evenly distributed in the area of the tray.

- SOD assay (lack of variation between total and inhibited SOD)
  I have found that 1 mM KCN is sufficient to inhibit Cu,Zn-SOD activity. Some authors (e.g. Wingsle et al. 1991) have used concentrations of up to 3 mM. The Cu,Zn-SODs should be responsible for a substantial portion of the total SOD activity (30-90%). If little or no variation is found between the total SOD activity and the cyanide-inhibited activity, increase the KCN concentration to 2 - 3 mM.

- SOD assay (quantification of additional alternative forms)
  As noted in the introduction and outline, hydrogen peroxide has been used to inhibit both Cu,Zn-SOD and Fe-SOD activity, while not interfering with Mn-SOD. Various authors have used H_2O_2 additions of 3 mM (e.g. Wingsle et al. 1991) to 10 mM (e.g. Tandy et al. 1990) final concentrations. Effective inhibition by H_2O_2 takes several hours to complete. If this assay is desired, I recommend the addition of 1 µl of 30% H_2O_2 (approximately 6 mM final concentration) to a 50 µl sample aliquot, followed by overnight incubation at 4°C before assaying for SOD activity. Weisiger and Fridovich (1973) and others also describe a method for selectively inactivating Mn-SOD, which involves the addition of a 0.25 volume of ethanol and a 0.215 volume of chloroform to the sample aliquot, stirring and incubating for 15 minutes before assaying.

References

Asada K, Takahashi M, Nagate M (1974) Assay and inhibitors of spinach superoxide dismutase. Agric Biol Chem 38:471-473

Asada K, Kanematsu S, Okaka S, Hayakawa T (1980) Phylogenic distribution of three types of superoxide dismutase in organisms and in cell organelles. In: Bannister JV, Hill HAO (eds) Chemical and Biochemical Aspects of Superoxide Dismutase. Elsevier North Holland, New York, pp 136-153

Beauchamp CO, Fridovich I (1971) Superoxide dismutase: Improved assays and an assay applicable to acrylamide gels. Anal Biochem 44:276-287

Beyer Jr WF, Fridovich I (1987) Assaying for superoxide dismutase activity: Some large consequences of minor changes in conditions. Anal Biochem 161:559-566

Carlberg I, Mannervik B (1985) Glutathione reductase. In: Meister A (ed) Glutamate, Glutamine, Glutathione, and Related Compounds. Methods in Enzymology, vol. 113. Academic Press, London, pp 484-490

Deutscher MP (1990) Maintaining protein stability. In: Deutscher MP (ed) Guide to Protein Purification. Methods in Enzymology, vol. 182. Academic Press, London, pp 83-89

Fahselt D, Jancey RC (1977) Polyacrylamide gel electrophoresis of protein extracts from members of the Parmelia perforata complex. Bryologist 30:429-438

Foyer CH, Hall D (1980) Superoxide dismutase activity in the functioning chloroplast. In: Bannister JV, Hill HAO (eds) Chemical and Biochemical Aspects of Superoxide Dismutase. Elsevier North Holland Inc, New York, pp 380-389

Foyer CH, Halliwell B (1976) The presence of glutathione and glutathione reductase in chloroplasts: A proposed role in ascorbic acid metabolism. Planta 133:21-25

Giannopolitis CN, Ries SK (1977) Superoxide dismutases I: Occurrence in higher plants. Plant Phys 59:309-314

Gries C (1996) Lichens as indicators of air pollution: In Nash III TH (ed) Lichen Biology. Cambridge University Press, Cambridge, 240-254

Hess JL (1994) Free radical scavenging. In: Alscher RG, Wellburn AR (eds) Plant Responses to the Gaseous Environment. Chapman & Hall, London, pp 99-119

Jazwinski SM (1990) Preparation of extracts from yeast. In: Deutscher MP (ed) Guide to Protein Purification. Methods in Enzymology, vol. 182. Academic Press, London, pp 154-174

Kranner I (1998) Determination of glutathione, glutathione disulfide, and two related enzymes, glutathione reductase and glucose-6-phosphate dehydrogenase, in fungal and plant cells. In: Varma A (ed) Mycorrhiza Manual, Springer, Berlin, pp 227-241

Kunert KJ, Foyer C (1993) Thiol/disulfide exchange in plants. In: De Kok LJ, Stulen I, Rennenberg H, Brunold C, Rauser WE (eds) Sulfur Nutrition and Assimilation in Higher Plants. SPB Academic Publishing bv, The Hague, pp 139-151

Nakano Y, Asada K (1981) Hydrogen peroxide is scavenged by ascorbate-specific peroxidase in spinach chloroplasts. Plant Cell Physiol 22:867-880

Nylander W (1866) Les lichens du Jardin du Luxembourg. Bull Soc Bot Fr 13:364-372

Paoletti F, Mocali A (1990) Determination of superoxide dismutase activity by purely chemical system based on NAD(P)H oxidation. In: Packer L, Glazer AN (eds) Oxygen Radicals in Biological Systems. Part B. Oxygen Radicals and Antioxidants. Methods in Enzymology, vol. 186. Academic Press, London, pp 209-220

Pell EJ (1987) Ozone toxicity - Is there more than one mechanism of action? In: Hutchinson TC, Meema KM (eds) Effects of Atmospheric Pollutants on Forests, Wetlands and Agricultural Ecosystems. Springer-Verlag, Berlin, pp 229-240

Scandalios JG (1994) Molecular biology of superoxide dismutases. In: Alscher RG, Wellburn AR (eds) Plant Responses to the Gaseous Environment. Chapman & Hall, London, pp 147-164

Tandy NE, DiGiulio RT, Richardson CJ (1990) Assay and electrophoresis of superoxide dismutase from red spruce (*Picea rubens* Sarg.), loblolly pine (*Pinus taeda* L.), and Scotch pine (*Pinus sylvestris* L.). A method for biomonitoring. Plant Physiol 90: 742-748

Weisiger RA, Fridovich I (1973) Superoxide dismutase: Organelle specificity. J Biol Chem 248:3582-3592

Winglse P, Gardeström P, Hëllgren J-E, Karpinski S (1991) Isolation, purification and subcellular localization of isozymes of superoxide dismutase from Scots pine (*Pinus sylvestris* L.) needles. Plant Physiol 95:21-28

Abbreviations

Cu, Zn-SOD	superoxide dismutase form with copper and zinc in the reaction centre
Fe-SOD	superoxide dismutase form with an iron reaction centre
GR	glutathione reductase
GSH	glutathione (reduced)
GSSG	glutathione disulphide, oxidised form of glutathione
Mn-SOD	superoxide dismutase form with manganese in the reaction centre
NBT	nitroblue tetrazolium
PEG	polyethylene glycol
PMSF	phenylmethylsulfonyl fluoride
SOD	superoxide dismutase

Studying the Effects of Elevated Concentrations of Carbon Dioxide on Lichens Using Open Top Chambers

ZOLTÁN TUBA, EDIT ÖTVÖS, and ATTILA SÓVÁRI

Introduction

The concentration of CO_2 in the atmosphere has increased in the last two centuries from about 280 ppm to the present value of 360 ppm, and is expected to reach more than twice the pre-industrial concentration later this century (Houghton et al. 1990). Besides the well-known implications of global warming, an increase in concentrations might be expected to have direct impacts (through changes caused by photosynthesis and other physiological processes) on plants and vegetation because of its direct use in photosynthesis (Long 1991). Short-term investigations supported the optimistic idea that elevated CO_2 concentrations will be accompanied by increased net primary production. Vegetation may therefore exert a negative feedback on CO_2 levels (Bazzaz and Carlson 1984). However, prolonged exposure to elevated levels of CO_2 can cause down regulation of photosynthesis and thus high CO_2 seems likely to lose its beneficial effect on plant productivity (Gunderson and Wullschleger 1994). However, the response of plants seems to vary greatly between different species, and as a result large changes in community structures and vegetation processes might develop. As lichens are important components of many communities from the tropics to the Polar Regions, and even dominant in some ecosystems, they have to be taken into consideration when making predictions about the effects of global environmental changes on the vegetation (Tuba et al. 1999). Because they are desiccation tolerant (e.g. Tuba et al. 1996) and comprise two or more symbionts, it cannot be assumed that the results obtained from the more widely investigated vascular plants

✉ Zoltán Tuba, Szent István University, Faculty of Agricultural and Environmental Sciences, Department of Botany and Plant Physiology, Gödöllö, 2103, Hungary (*phone* +36-28-522075; *fax* +36-28-410804; *e-mail* tuba@fau.gau.hu)
Edit Ötvös
Attila Sóvári

apply to lichens. Lichens are slow-growing organisms and store significant amounts of sugar-alcohols and lipids as reserve materials and less starch (Fahselt 1994). They contain high concentration of secondary metabolites (Lawrey 1995). These special features of lichens justify research on their responses to elevated CO_2. However, the main difference between lichens and higher plants is that lichens are poikilohydric, and this needs to be taken into consideration when exposing lichens to elevated atmospheric CO_2.

For exposing plants to higher than present-day CO_2 concentration, four different techniques have been developed: Open Top Chamber (OTC) (Last 1986), solardome (Rafarel et al. 1995), wind tunnel (Soussana and Loiseau 1997), and Free Air CO_2 Enrichment (FACE) (Rogers et al. 1992) (for comparative review see e.g. Vourlitis and Oechel 1993). This chapter describes the use of OTCs for exposing lichens to long-term elevated CO_2 concentrations. We developed this lichen fumigation system from the OTCs's system of Ashenden et al. (1992) that has been successfully used in grassland experiments (e.g. Tuba et al. 1998a).

Before describing OTCs in detail, it may be appropriate to review the advantages and disadvantages of FACE systems. Using FACE eliminates the problems arising from the altered microclimate within the chamber (see below), because CO_2 is released by valves or by perforated plastic tubes into the air around the experimental plots. Without walls, however, the vertical profile of the CO_2 concentration is much steeper. Therefore, for epiphytic lichens we recommend the use of OTCs. As wind velocity and CO_2 concentration above the plot can change from second to second, the control system has to react much faster. This partly explains the high costs of the hardware. Running expenses are also much higher due to the increased CO_2 demand.

Outline

The general design of the OTC system is shown in Fig. 1. Fumigation occurs in Plexiglas chambers. The adjusted CO_2 concentration can be maintained within limits of 50 ppm. The CO_2 supplied to the chambers via mass-flow controllers (MFCs) is diluted with air in order to achieve an even distribution. Plastic pipes transfer air sample to an infra-red gas analyser (IRGA) through valves. A computer compares the signal with the set point and regulates the MFCs correspondingly. An irrigation system, not indicated in Fig. 1, must be included in the system.

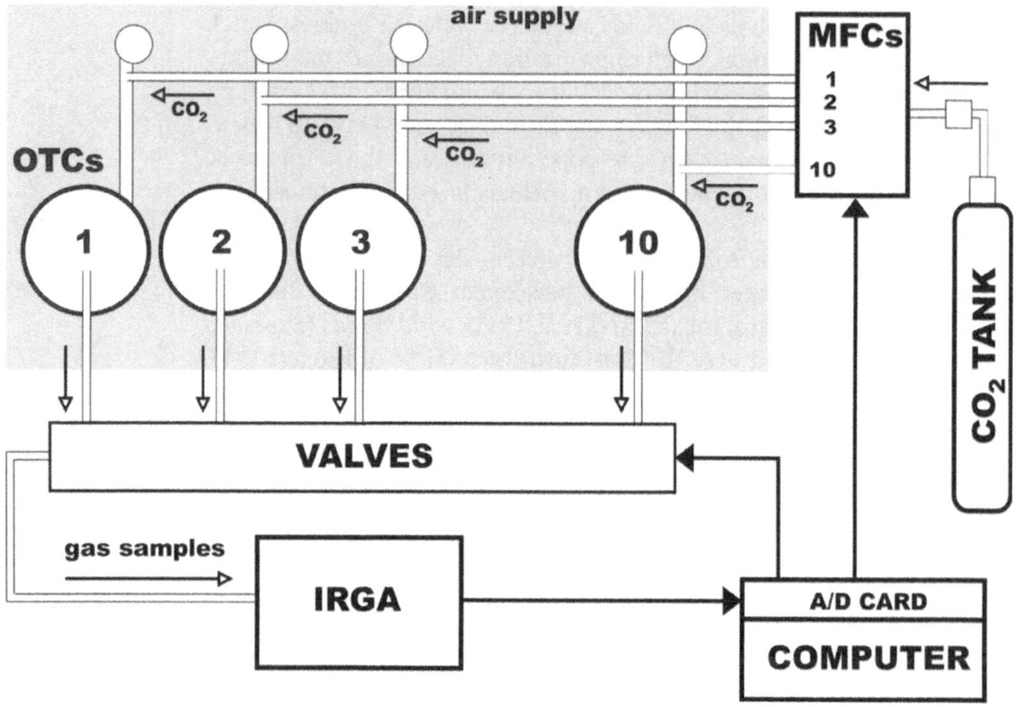

Fig. 1. Layout of the OTC system. Single lines represent data flow, double lines stand for gas pipes.

Materials

Hardware Construction of the hardware of a system with 10 open-top high CO_2 chambers and 10 ambient chambers will require the following equipment:

- **Plexiglas chambers**; these are round, 1 m in height and 1.284 m in diameter, and are braced by stainless steel rings. The system has 20 chambers, half of them without additional CO_2 supply serving as controls with present-day CO_2 concentration. It is also advisable to have unchambered control plots of the same diameter, so researchers will be able to separate experimental results into CO_2 effects and chamber effects.
- **Tank containing liquid CO_2**; while you can store CO_2 in cylinders, the high gas demand makes it highly advisable to obtain a tank. This tank must be equipped with an evaporizer and a pressure-reducing valve.

- **Flexible plastic tubes**; these carry the CO_2 supply and air samples; they should be 4 mm in diameter.
- **Mass-flow controllers** (UFC 1020, Unit Instruments, Ireland); the maximum flow rate must be 2 $dm^3 min^{-1}$ at working gas pressure (0.2 - 0.5 MPa).
- **D/A card**; the card provides the electric input of up to 5 V, and is connected to the personal computer.
- **Personal Computer** (PC); the requirements of the system can be fulfilled even by an AT 286.
- **Axial electric fans**; the fans supply air to the chambers; to achieve complete air exchange every minute within the chambers (1.02 m^3), the capacity of a fan has to be between 25 and 45 $dm^3 sec^{-1}$. This airspeed causes turbulence and satisfactory mixing of CO_2 within the chamber without an internal fan.
- **Aluminium chimney tube**; the air enters the chamber through this tube; the diameter should be 100 mm. Plastic tubing carrying CO_2 enters here in case of the high-CO_2 chambers.
- **Lay-flat tubing**; this tubing distributes air via a perforated encircling the interior of the chamber.
- **Infra-red gas analyser** (IRGA, e.g. WMA 2, Parkinson, UK); the IRGA monitors the CO_2 level in the high-CO_2 chambers one by one.
- **Electric pump**; this carries the air sample to the IRGA, and may be contained within the IRGA. Its capacity ranges from 0.2 to 0.5 $l\ min^{-1}$.
- **Peltier module**; this is in the path of the air sample, and must precede the IRGA. It serves to eliminate water vapour via condensation, which would cause an error in the measurement.
- **Row of electric valves**; these are integrated into the system, and guide the air sample from one chamber to the IRGA. The computer controls these valves.
- **A/D card**; this is used to transform and store the CO_2 concentrations in the PC. The control system has to be installed in a rainproof building or container.
- **Irrigation system**; this comprises flexible plastic tubes, spraying heads, an electric valve and a timer (e.g. DAN Sprinklers, Israel; Hardie, USA; Hunter, USA).

Software The control system is based on a self-developed computer program written in Turbo Pascal. The program is not commercially available, but requests sent to the correspondence address are welcome. The program enables a proportional, integral and differential regulation. It is integral, as an average of the last array of CO_2 concentration data is used in regulation. It is proportional, as MFCs are opened or closed to an extent that is proportional to the deviation of the actual CO_2 concentration from the set point. It is differential, as it considers the error originating from the trend. This feature allows the system not to overshoot the set point. The program also shows the regulative process graphically, and stores CO_2 concentrations in a format easily processed with any common data-handling software. The program reboots and restarts automatically following power cuts with the last parameters being used. The lag-time of the regulation is influenced by the travel time of the CO_2, which ranges from 5 to 20 s, if tubes are 4 mm in diameter and a maximum of 25 m in length. Another factor is the aeration of the chambers. Complete exchange of air with the fans described occurs within 20 - 50 s in the absence of wind. The efficiency of the system at maintaining an elevated CO_2 level is shown in Fig. 2.

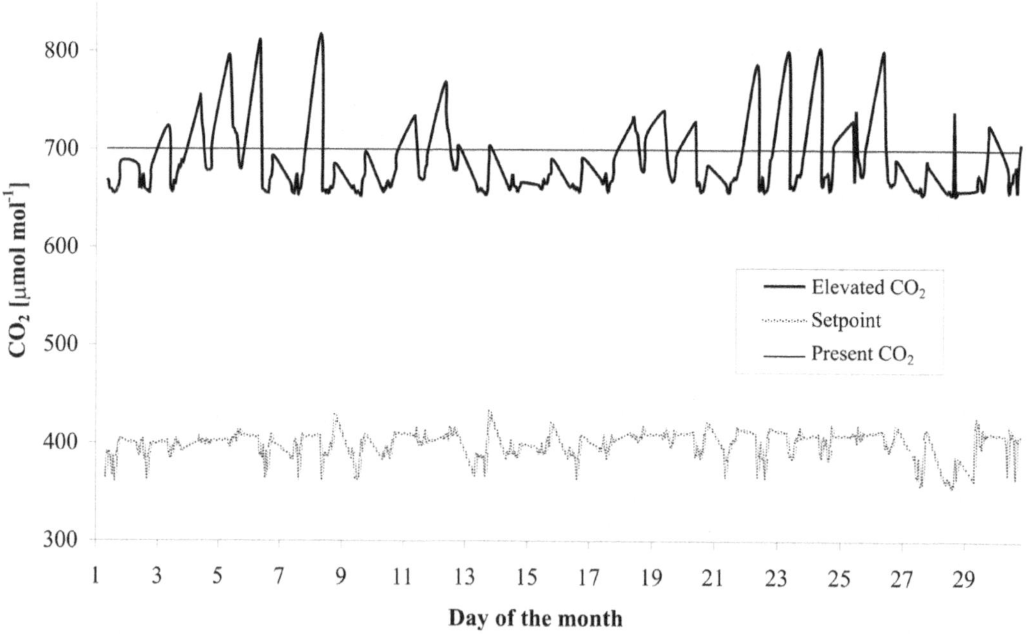

Fig. 2. CO_2 concentrations measured within a high-CO_2 OTC, and an unchambered plot during September 1999.

Procedure

Setting up the open top chambers and the control system

1. Attach chambers to the ground with poles. Do not bury the chamber even centimetres deep into the soil, as this would block the pathway of runoff. For the same reason, do not set the chamber up in a hollow. Long distances between chambers, or between chambers and the control system result in unacceptably long travelling times for air samples. On the other hand, close placement of chambers results in extensive mutual shading. In our experience, the distance between the control system and the farthest chamber should not exceed 25 m; therefore, the distance between chambers can be up to 2 m. **Chambers**

2. Enclose the bundles of plastic tubes in protecting sheaths, and elevate them from the ground using small stands. This will protect the turf between chambers. **Pipelines**

3. Place the end of the air sample tubes in the middle of the chambers (which is the most distant point from the CO_2 source) and bend them so that they point downwards, to avoid sucking up water. Note that liquid water can damage the IRGA!

4. Place all lichens, even terricolous ones, in the chambers with their original substratum. Though one of the purposes of air injection is to achieve turbulence and a uniform CO_2 concentration, CO_2 levels in the upper third of the chamber volume, depending again on wind velocity, can be considerably lower. For this reason, only lichens situated on a similar elevation can be compared. This is a particular problem when exposing epiphytic lichens. **Placement of the test plants**
   Another problem is that the air stream coming from the perforated lay-flat tube dries out adjacent lichens very rapidly. As a consequence, a strip of lichens 10-15 cm wide at the margin of the chamber is lost for experimental purposes. To use expensive experimental area more effectively, shield lichens from the direct air stream with small screens made of Plexiglas reaching not more than 5 cm above thallus surface.

5. Seed grass between chambers, because its albedo is lower than that of bare soil. This will prevent the microclimate of the experimental station from being warmer than the surroundings. **Surroundings of the chambers**

Running the system

6. A computer-controlled row of valves switches to one of the ten air-sample tubes. The air-sample is measured every second by the IRGA 25 times before the valves switch to the following channel. This period is lengthened automatically if the difference between the actual CO_2 concentration and the set point exceeds 10 %. The measurement period is shortened if the valve is at its endpoint and further regulatory action cannot be carried out. Before measuring the channel selected, the tube between valves and IRGA has to be purged. This purge-time is about 6 s for a tube-length of 2 m but purge time should be set after trials and measurements. The software settings have also to be adjusted to different travel times of air-samples from different chambers (with a pump with an air delivery of 30 - 35 $cm^3 s^{-1}$, the travel times range between 15 and 33 s). As a summary, the length of a period used for controlling each channel is 1 - 3 min, and a complete rotation for ten chambers takes 20 min on average.

Results

Once your chambers have been running for the required period, you can start making physiological measurements. Data on acclimation of lichens to elevated CO_2 concentration is sparse (Tuba et al. 1999). However, the following studies indicate the kind of information that can be obtained.

Sonesson et al. (1995) did not find any short-term (or rather rapid) effects of elevated CO_2 on the photosystem II yield of *Cetraria islandica.* On the other hand, Balaguer et al. (1996) report downward acclimation of photosynthesis in *Parmelia sulcata* after one month of exposure. In a long-term experiment *Parmelia caperata* shows practically no acclimation (Balaguer et al. 1999).

However, one has to use the expression "short-term" with care in the case of desiccation tolerant plants. The time spent in dry, "anhydrobiotic" state, must be subtracted from their exposure period to get the time available for acclimation.

In a medium-term experiment (4 months under 700 ppm CO_2) no signs of acclimation of photosynthesis were detected in *Cladonia convoluta* (Csintalan et al. 1997). Total chlorophyll content, soluble sugar and protein-N contents were also unchanged. Though the maximum rate of net CO_2 assimilation rate at optimum water content did not change, elevated CO_2 proved to be beneficial during desiccation and rehydration. On rehydration, the time needed to reach the CO_2 compensation point was halved by high CO_2 concentration and photosynthetic activity was higher during the first 30 minutes. During desiccation, photosynthetic net carbon assimilation was higher and more prolonged under elevated CO_2. As the CO_2 level did not affect the rate of water uptake or loss, this cannot account for the observed beneficial effect. These changes however, were not accompanied by shifts in the chlorophyll fluorescence parameter, RFd690 [variable fluorescence decrease ratio, RFd = (Fm - Ft) / Ft] showing the functionality of the thylakoid membrane.

After fumigating *Cladonia convoluta* with 700 ppm CO_2 for five months Tuba et al. (1998b) measured higher rates of carbon assimilation when CO_2 levels were elevated during desiccation. Desiccation tolerant cryptogamic plants may therefore improve their competitive ability in a future elevated CO_2 environment. Interestingly, elevated levels of CO_2 can alleviate the harmful effects of heavy metals in *Cladonia convulata* (Takács et al. 1999).

In conclusion, it must be emphasized that considerable uncertainties exist concerning the responses of lichens to long-term elevated CO_2 levels, and clearly there is a need for further research in this field.

Troubleshooting

Most frequently problems arise from extreme deviations in CO_2 levels from the set point.

- If the program indicates low CO_2 concentration in all chambers, check
  - the pressure within the CO_2 tank (or cylinder)
  - the state of the pressure reducing valve
  - the piping between CO_2 tank and OTCs
  - the mass flow meter.
- If the program indicates low CO_2 concentration in some chambers, check
  - the CO_2 supply pipe between the certain mass flow controller and the OTC
  - the air sample pipe between the particular OTC and the row of electric valves
  - the axial electric fan and its filter
  - the intactness of the Plexiglas cover of the OTC (whether cracked or leaky).

- If the program indicates high CO_2 concentration, check
  - if there is water in the air sampling circle
  - the CO_2 absorbent within the IRGA is exhausted and needs changing. Soda lime granules turn from blue to yellow after about two months.

Comments

Microclimate within the OTC

Temperature An OTC undoubtedly changes the environment of the test organisms (Lee and Barton 1993). Due to the well-known greenhouse effect, the temperature is higher inside the chamber. The difference rises linearly with radiation, and at 700 W m^{-2} it reaches 3°C. However, the OTC interior is also warmer in the night by about 0.5°C. Temperature regimes are shown in Fig. 3.

Water availability Setting up an OTC will also affect the water relations of a plot. In windy weather, rain will be unevenly distributed within the chamber. At a wind speed of 1.6 m s^{-1} there is practically no rainfall sheltering. At 3.6 m s^{-1} only the edge of the plot is affected. At a wind speed of 4.3 m s^{-1}, half of the plot remains dry (Colls et al. 1988). As soils conduct water at least to some extent, the effect of the chamber is evened out. However, as the terrestrial lichens we have studied do not appear to rehydrate from the substrate, rainfall sheltering will increase variation within the plot.

Relative humidity The OTC also modifies the relative humidity. Adaros et al. (1989) measured an increase of 12 % during daylight and a reduction up to 15 % at night. On the other hand, they were growing vascular plants (broad beans and spring barley), which have high transpiration rates. In OTCs containing lichen thalli the nocturnal reduction in relative humidity can even reach 25 % hindering dew formation. In addition, elevated relative humidity during daylight can be detrimental if the temperature is high. The net result of these effects is that water must be supplied to the OTCs at dawn to give the thalli time to desiccate before temperature rises. As uniform wetting is needed, it is best to apply the water as a spray. On hot days excess water may cause the death of the lichens. If the experimental set-up requires control of the amount and quality of water one can rely entirely on artificially administered water by covering the top of the chamber by a shield of Plexiglas. It is important to mount this shield well above the chamber to allow aeration.

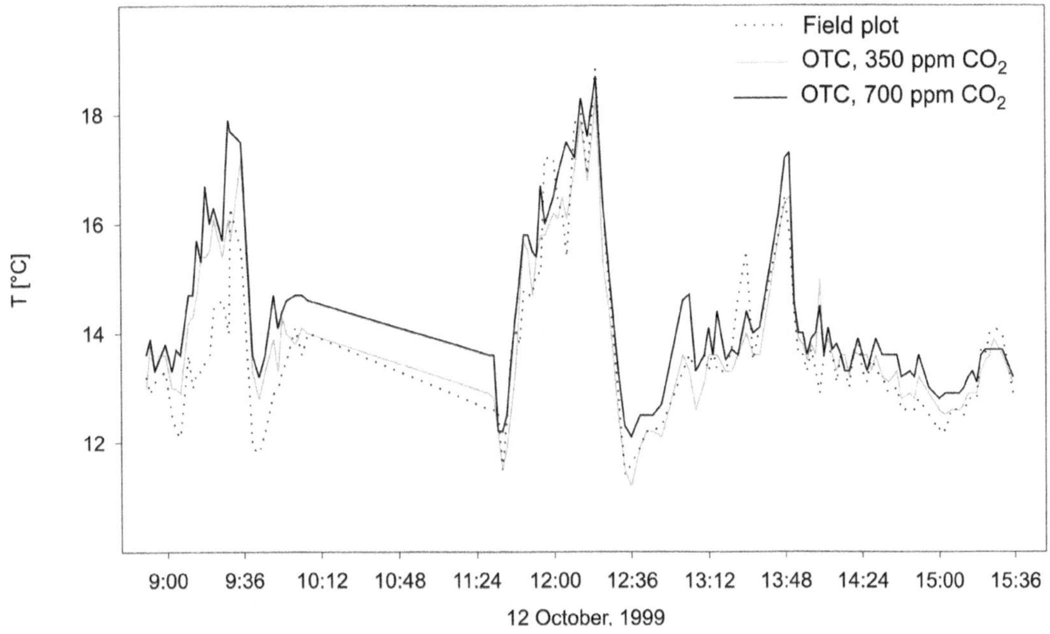

Fig. 3. Temperature values measured within a high-CO_2 OTC, present-day CO_2 OTC, and an unchambered plot.

Energy and gas demand

Electricity consumption is about 1.6 kW for the control system and the ten fans. CO_2 demand depends very much on the current wind velocity. On average, 0.12 kg $CO_2\,h^{-1}$ is released per chamber, implying a total monthly demand of 870 kg CO_2 for the whole system.

Closing the tap and disabling the control system in the night will usually lower CO_2 demand. In experiments with higher plants, the saving reaches 200 kg per month. When working with lichens, this is not advisable. The voluminous biomass of vascular plants growing in the chamber can increase the CO_2 concentration close to the set point (700 ppm in our case) through its respiratory activity. With lichens, this is not the case. As we cannot exclude the possibility that elevated CO_2 increases plant production via decreasing respiration (Bunce, 1994), test organisms should also be fed with CO_2 during the night.

Acknowledgements. We acknowledge financial support by the ECOCRAFT (R&D Programme, EU Brussels), the Hungarian Scientific Research Foundation (OTKA T22723), National Committee for Technical Development (OMFB E0031), Research Development in Higher Education (FKFP 0472/97), ESPACE-Grass Project (R&D Programme, EU Brussels), MEGARICH (R&D Programme, EU-Brussels) and Hung. Acad. Sci. The authors are indebted to Kiskunság National Park (Hungary) and Zoltán Takács and Prof. Richard Beckett for their valuable comments and suggestions.

References

Adaros G, Weigel HJ, Jäger HJ (1989) Environment in open-top chambers and its effects on growth and yield of plants. In: Payer HD, Pfirrmann T, Mathy P (eds.) Environmental Research with Plants in Closed Chambers. E. Guyot SA, Brussels, pp 97

Ashenden TW, Baxter R, Rafarel CR (1992) An inexpensive system for exposing plants in the field to elevated concentrations of CO_2. Technical report. Plant, Cell Env. 15:365-372

Balaguer L, Manrique E, delosRios A, Ascasco C, Palmquist K, Fordham M, Barnes JD (1999) Long-term responses of the green-algal lichen *Parmelia caperata* to natural CO_2 enrichment. Oecologia 119:166-174

Balaguer L, Valladares F, Ascasco C, Barnes JD, de los Rios A, Manrique E, Smith EC (1996) Potential effects of rising tropospheric concentrations of CO_2 and O_3 on green-algal lichens. New Phytol. 132:641-652

Bazzaz FA, Carlson RW (1984) The response of plants to- elevated CO_2. I. Competition among an asseblage at two levels of soil moisture. Oecologia 62:196-198

Bunce JA (1994) Responses of respiration to increasing atmospheric carbon dioxide concentrations. Physiol. Plant. 90:427-430

Colls JJ, Sanders GE, Clark AG (1988) Open-top chamber experiment on field-grown *Vicia faba* L. at Sutton Bonington (1987-1988). In: Bonte J, Mathy, P (eds) The European Community Research Project on Open-Top Chambers. Results on Agricultural Crops. E. Guyot SA, Brussels, pp 43-72

Csintalan Zs, Takács Z, Tuba Z, Proctor MCF, Smirnoff N, Grace J (1997) Desiccation tolerant grassland cryptogams under elevated CO_2: Preliminary findings. Abst Bot 21:309-315

Fahselt D. (1994) Carbon metabolism in lichens. Symbiosis 17:127-182

Gunderson CA, Wullschleger SD (1994) Photosynthesis acclimation in trees to rising atmospheric CO_2: A broader perspective. Photos Res 39:369-388

Houghton JT, Jenkins GJ, Ephramus JJ (1990) Climate Change: The IPCC Scientific Assessment. Cambridge University Press, Cambridge

Last FT (ed) (1986) Microclimate and plant growth in open top chambers. CEC, Air Pollution Research Report 5. EUR 11257

Lawrey JD (1995) Lichen allelopathy, A review. In: Inderjit D, Dakshini KMM, Einhellig FA (eds) Allelopathy. Organisms, processes and applications. American Chemical Society, Symposium Series 582. Washington DC, pp 26-38

Lee HSJ, Barton CVM (1993) Comparative studies on elevated CO_2 using open-top chambers, tree chambers and branch bags. In. Schulze ED, Mooney HA (eds.) Design and Execution of Experiments on CO_2 Enrichment. Commission of the European Communities. E. Guyot SA, Brussels, pp 239-260

Long SP (1991) Modification of the response of photosynthetic productivity to rising temperature by atmospheric CO_2 concentrations - has its importance been underestimated. Plant Cell Environ 14:729-739

Rafarel CR, Ashenden TW, Roberts, TM (1995) An improved solardome system for exposing plants to elevated CO_2 and temperature. New Phyt 131:481-490.

Rogers HH, Prior SA, O'Neill EG (1992) Cotton root and rhizosphere responses to free-air CO_2 enrichment. Crit Rev Plant Sci 11:251-263

Sonesson M, Callaghan TV, Björns LO (1995) Short-term effects of enhanced UV-B and CO_2 on lichens at different latitudes. Lichenologist 27:547-557

Soussana JF, Loiseau P (1997) Temperate grass swards and climatic changes. The role of plant-soil interactions in elevated CO_2. Abst Bot 21:223-234

Takács Z, Csintalan Zs, Tuba Z (1999) Responses of the Lichen *Cladonia convoluta* to High CO_2 Level and Heavy Metal Treatment Z Naturforsch 54c: 797-801

Tuba Z, Csintalan Zs, Proctor MCF (1996) Photosynthetic response of a moss, *Tortula ruralis* (Hedw.) Gaertn. et al. ssp. *ruralis*, and the lichens *Cladonia convoluta* (Lam.) P. Cout. and *C. furcata* (Huds.) Schrad. to water deficit and short periods of desiccation, and their ecophysiological significance: a baseline study at present CO_2 concentration. New Phyt 133:353-361

Tuba Z, Jones MB, Szente K, Nagy Z, Garvey L, Baxter R (1998a) Some Ecophysiological and Production Responses of Grasslands to Long-Term Elevated CO_2 under Continental and Atlantic Climates. Ann NY Acad Sci 851: 241-250

Tuba Z, Csintalan Z, Szente K, Nagy Z, Grace J (1998b) Carbon gains by desiccation tolerant plants at elevated CO_2. Func Ecol 12:39-44

Tuba Z, Proctor MCF, Takács Z (1999) Desiccation-Tolerant Plants under Elevated Air CO_2: A Review. Z. Naturforsch. 54c: 788-796.

Vourlitis G, Oechel WC (1993) Microcosms in natural experiments. In. Schulze ED, Mooney HA (eds.) Design and Execution of Experiments on CO_2 Enrichment. Commission of the European Communities. E. Guyot SA, Brussels, pp 192-210

In Situ Measurement of the Water Content of Lichens

GÜNTER SCHUSTER, SIEGLINDE OTT, ANDRÉ GASSMANN, and JUDITH ROMEIKE

Introduction

The distribution of plants and the development of individuals at a certain location depend to a considerable degree upon microclimate. This is true for phanerogams as well as for lichens. For lichens, however, crucial environmental influences occur on a much smaller scale (Schöller 1991, Canters et al. 1991). The microclimate of a higher plant is defined by the environmental conditions that occur within an area of about 1 m^2 around the plant. However, for lichens important changes occur within a few cm^2. Lichenologists frequently refer to the microclimate of phanerogams as mesoclimate. This difference in scale of the microclimate is the only difference, as the contributing features are identical for both groups of organisms. Air and substrate temperature, solar irradiance, air humidity, water supply and wind speed are the most important parameters. Because the microclimatic observation areas are small for lichens, the sensors also have to be small. Measurement of water availability is a special problem, as it is subject to a very quick change in poikilohydrous thalli (Larson 1979). This chapter describes a new method for measuring water content using thallus impedance. This allows microhabitats to be compared with each other, especially if these measurements are combined with those taken from sensors for solar irradiance, thallus temperature and air temperature. In the field, the method is superior to gravimetric methods that have inaccuracies caused by repeatedly removing the thalli. Thallus water content is determined by wind, irradiation, and temperature, and is

✉ Günter Schuster, Universität Düsseldorf, Botanisches Institut, Universitätsstr. 1, Düsseldorf, 40225, Germany (*phone* +49-211-81-13479; *fax* +49-211-81-12881; *e-mail* schusteg@uni-duesseldorf.de)
Sieglinde Ott
André Gassmann
Judith Romeike

also influenced by substrate type (Ott et al. 1997). For this reason, all these parameters should be measured at the same time.

Water uptake and loss from the lichen thallus

When determining the water content of lichens, always note that water loss from poikilohydric thalli occurs very fast, even if a compact cortex or gelatinous water storage tissues retard the process. Water absorption is a slower process, and depends on the water source, the anatomy of the thallus and its impregnation with more or less hydrophobic substances. Finely distributed droplets, in the form of fog and mist, are optimal for absorption while large raindrops run off (Jahns and Ott 1983). Most lichens can take up water from the air. This process is asymptotic, and saturation is reached much later than expected, often only after days (Jahns 1984). In the thallus, water is mainly stored in the hyphal walls and in gelatinous substances. The capillary regions between the hyphae, necessary for gaseous exchange, are kept free from liquid water by impregnation with hydrophobic substances. Therefore uptake of water through the cortex occurs by diffusion through the walls of the hyphae. Larger quantities of water are stored as surface films on the thallus and between the rhizinae on the lower side (Jahns 1984). As it is impossible to determine how much water is bound to the surface one cannot measure the maximum water content reliably. It is not surprising that the maximum values given in literature obtained by spraying or immersing the thalli vary considerably. However, they are of little ecological and physiological interest as this degree of saturation is hardly ever reached in the natural habitat.

Methods to study water content

After briefly describing the measurement of thallus water status using the classical "gravimetric" method, the main technique described here measures thallus water content using the conduction of electrical currents through the lichen thallus. The currents are the result of a flow of ions in the thallus, and are accompanied by a number of reactions that are incompletely understood. The currents can be measured and are directly proportional to the water content of the thallus. Conductivities in the tissues are usually measured with high frequency alternating currents (Coxson 1991). The practical application of this method, however, is limited, particularly when one measures a large number of lichen thalli simultaneously. Alternating currents are influenced by the dielectric properties of insulating materials and the lichen thallus. This hampers the measurement and makes circuit construction and evaluation difficult. Preventative shields or adjusted records are possible, but it is very time consuming for a routine application. When a large number of measuring points exist in the field it is quite impractical. Every clamp and its wire has to be calibrated individually, and cannot be replaced immediately in the field. After intensive series of tests, we decided to opt for direct current measurements (Schuster et al. 1982). In this set up, the sensors do not need to be calibrated individually and can be replaced at any time.

Subprotocol 1
Gravimetric Determination of Water Content

Procedure

1. Weigh wet thalli.
2. Dry at 110°C for 24 h. It is convenient to use a scale that incorporates specimen drying by infrared (e.g. Mettler PM 100 with Mettler LP 16), as measurements take less time.
3. Determine the dry mass, and express water content as % of dry mass.

Comments

This method is very accurate, but remember that wet thalli can lose a considerable amount of water in seconds as you transfer them in dry air to another container or to the weighing scales. Physiological experiments on photosynthesis overcame this problem by integrating the scales into the experimental set-up (Schroeter et al. 1991). In the field, one or more thalli can be placed into a wire basket and may be repeatedly removed from the substrate for weighing (Hahn et al. 1989). This provides exact results but interrupts the continuity between the lichen and its environment and permanently prevents normal water exchange with the substrate. The method cannot be applied to lichens that grow between dwarf shrubs or that are closely attached to the substrate.

Subprotocol 2
Impedance Measurements

Materials

For the construction of sensors

- 9V alkaline or lithium battery
- Fixed resistors (5 M Ohm) to limit the current (R2 in Fig. 1)
- Variable resistors (value set depends upon the impedance of the volt meter, R1 in Fig. 1)

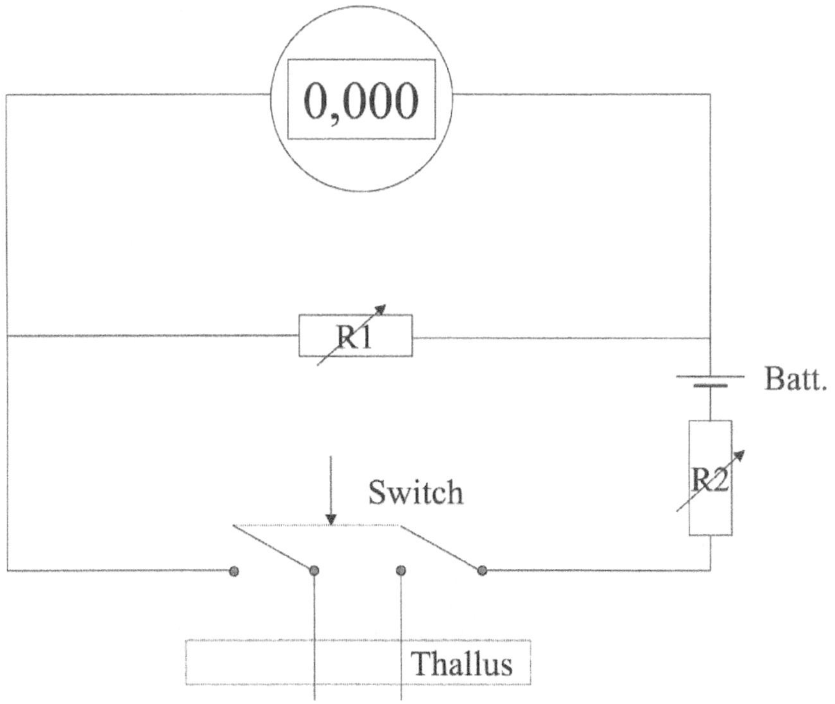

Fig. 1. Circuit diagram for impedance measurements of water content.

- Needle-shaped electrodes
- V2A spring steel wires (0.3 mm in diameter)
- Polyurethane or silicone varnish
- Miniature clamp for electrodes (Fig. 2)
- Thin glass plate to fit the lower part of the clamp to keep the surface smooth and not adsorb water; a cover slip can be used
- Flex (stranded, maximum length 2 m x 0.14 mm^2), must be light and elastic to reduce the tensile stress on the lichen during measurement.
- Digital voltmeter
- Termistor (Siemens B57085 M 473K, 47K)
- Electrolysis bath containing saturated NaCl, capable of delivering 12 V and 1.5 A

For calibration

- Saturated salt solutions to give a range of relative humidities e.g. 12%, LiCl; 33%, $MgCl_2$; 55%, Mg $(NO_3)_2$, 76%, NaCl; 93%, KNO_3; 100%, distilled water (these humidities will be attained at room temperature, approx. 20°C).

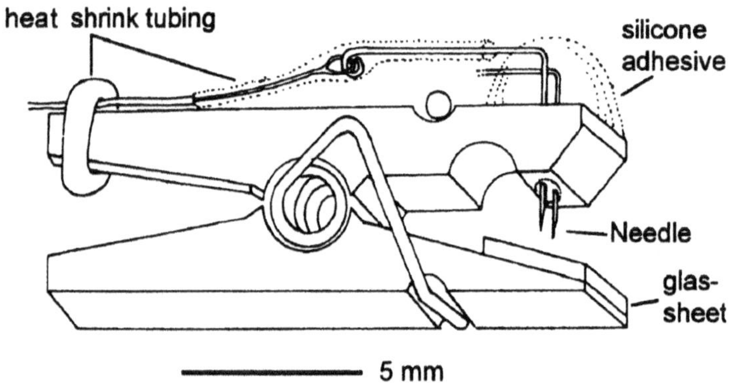

Fig. 2. Miniature clip with electrodes for measuring water content of the thallus. The distance between needles is 1 mm.

Procedure

Construction of sensors

1. Drill a 1 mm hole into the upper part of a miniature clamp (Figs. 2 and 3).
2. Take two spring steel wires and sharpen the points of the needles using an electrolysis bath. Use an electrolysis bath of saturated NaCl solution with a 12 V voltage and a current of approx. 1.5 A. Vary the time of immersion to control the diameter and point of the electrode.
3. Insert the wires into the hole with exactly 1 mm between them and glue them into place with silicone adhesive.
4. Fix the upper part of the wires to the clamp with heat shrink tubing and solder them to elastic flex as shown in Fig. 2.
5. Glue the small glass plate to the lower part of the clamp.
6. Re-sharpen the needles if necessary.
7. Coat the needles and clamps with water repellent varnish (polyurethane or silicone) to stop the formation of moisture bridges between electrodes. Exclude the tips of the needles.
8. Construct temperature sensors in a similar way using the termistor (Fig. 3).

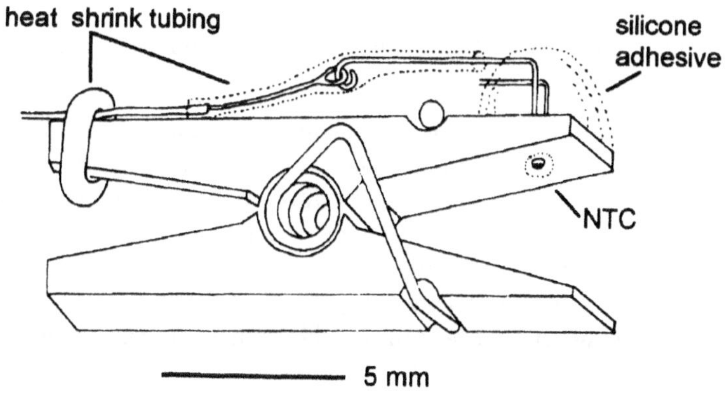

Fig. 3. Miniature clip with thermistor for measuring thallus temperature.

Calibration

Remember that the method only senses the water between the electrodes and assumes that the water content between the needles is representative for the complete thallus. Comparable measurements from different thalli can only be achieved when the electrodes are attached at similar locations. It is important to keep the electrical current loading of the thallus extremely low. This is in the range of a few microamperes in a wet thallus and even lower in the dry thallus. The current is only applied for the duration of the measurement (approx. 3 s). The low currents and the brief exposure, and the inherent electro-chemical environment in the thallus, virtually exclude polarisation effects on the electrodes if intervals of at least 15 min are maintained between measurements.

1. Each lichen species must be calibrated separately using at least ten replicates at the same temperature, e.g. room temperature (20°C). Attach a saturated thallus to the electrode then place it in a closed container with known relative humidity above defined saturated salt solutions as outlined in Materials (also see Jahns 1984). Complete equilibrium with the air humidity is not essential, as the thalli are removed at arbitrary intervals to get a complete set of values for impedance measurements.
2. Adjust the value of the variable resistor - a thallus with 20% water content should produce a reading of approx. 10 mV while a wet thallus should give readings of 1 to 2 volts.
3. Take an impedance measurement for each thallus. The displayed voltage drops quickly in the first 1 - 2 s after the switch is activated. This process depends on the water content. The value after 2 s is taken as the basis for calculating the water content of the thallus.
4. Immediately remove the thallus and take a gravimetric measurement (see Subprotocol 1).
5. Plot a calibration graph of conductance as a function of thallus water content. No values are obtained below 15-20% water content, and higher values from 100% upwards produce very steep, rather inaccurate curves. For ecological measurements these limitations are of little importance as a very high water content, which is species-specific, is rare in nature and most species are physiologically inactive below 20%. The measurements should be reproducible with a standard error of about 10%.

Troubleshooting

Calibration

- Values are obviously inaccurate.
  Discard measurements. A drop of water may have shorted the connection between the needles or a needle may have had insufficient contact with the tissue.

Comments

Although exact research demands a calibration curve for every species, qualitative comparisons in the field can be made with general curves for lichens with similar growth form (e.g. species of *Cladonia* or *Parmelia* respectively) (Jahns 1984). The conductivity of the moist thallus is temperature dependent. Conductivity increases with increasing thallus temperature. Calibration tests in chambers with different temperatures showed that an increase in temperature by 1°C raises the moisture values by 1% on the calibration curve. As a rule of thumb, reduce or increase moisture values by 1% for each 1°C variation in temperature. We have verified this relationship in the laboratory for temperatures from 3°C to 40°C, but our field observations have suggested that the relationship also holds at lower and higher temperatures. It follows that when taking measurements in the field you must make temperature measurements adjacent to lichens. A computer programme that adjusts these values is helpful.

Subprotocol 3
Taking Readings in the Field

The described device is suitable for recording thallus moisture for a period of several days, but the clamps may be used for weeks if conditions include both dry and wet periods and the clamps have a chance to dry out periodically.

Use of data loggers

The sensors should ideally be used with data loggers. We have built one that can record solar irradiance (PAR), air and thallus temperature and the thallus moisture in up to 16 locations simultaneously. For solar irradiance the photosynthetic photon flux density (PPFD) can be measured with a quantum sensor (e.g. LI-Cor 190 SB) or with GaAsP-photodiodes (Hamamatsu) that have been calibrated against a quantum sensor (Schroeter et al. 1995, Pontailler 1990). The parameters are recorded by the relevant sensors, converted to digital values and stored. They can be processed later with a PC. The system is divided into one main appliance and a maximum of 16 boxes with analogue to digital converters (Fig. 4). These are connected to the main appliance with cables of up to 60 m long. Sensors for solar irradiance, air temperature, thallus temperature and water content of the thallus are connected to each sensor box. Three sensors for air humidity can also be connected directly to the main appliance. Power is supplied by solar energy buffered by a 12 V battery. This appliance is particularly suitable for analysing microclimatic gradients in the field.

For construction of the logger and for computer programs for calibration no detailed instructions can be given. If necessary, contact the first author of this paper. In any case, the construction and use of these sensors and this method require some practice that cannot be acquired by the step-by-step advice given in this manual alone.

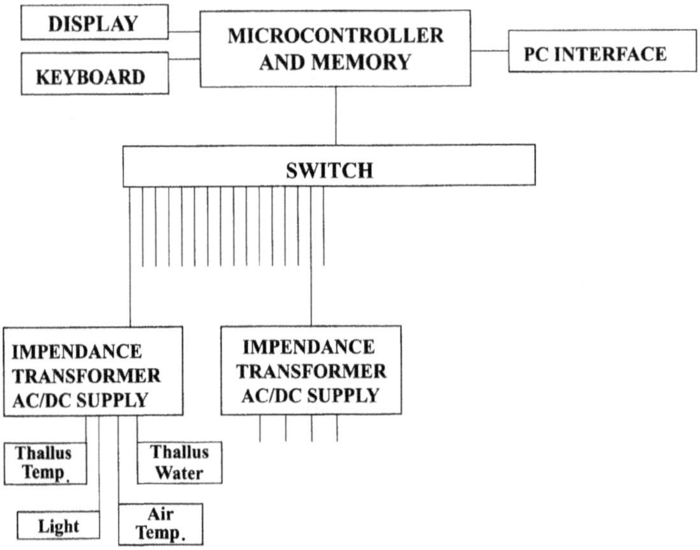

Fig. 4. Diagram of logger for measuring microclimate at 16 locations.

Materials

- Sensors (impedance and thermistor) as constructed in Subprotocol 2
- Data loggers

Procedure

1. In the field, attach both clamps to the lichen thallus. For comparative measurements, thalli of similar diameter should be used, but small variations have no influence on accuracy. Be careful not to tear the tissue of the lichen. If necessary, fix the flex to the substrate close to the lichen. For crustose lichens, use only the upper part of the clamps.
2. Check the contact of the needles by taking one impedance measurement at least once a day.
3. Fix the clamps for temperature measurement with the sensors below the thallus (if possible) or at least not exposed to direct irradiance. We recommend attaching a thermistor to the head of a miniature clamp and pressed to the lower surface of the lichen (Fig. 3).
4. Always record thallus temperature and water content simultaneously, as temperature is needed for calibration.
5. Measure the voltage after 2 s, as voltage falls with time probably as a result of several interacting factors. One possible explanation is the variable rate of ion movement; incipient polarisation of the electrodes may be another. In our experience, however, this is reversible and can be ignored by allowing appropriate intervals between measurements (approx. 15 min). With our self-constructed recorder, the measurement is taken automatically 2 s after the device is switched on.

Troubleshooting

Readings in the field

- Under wet conditions the material of the clamp may absorb water. You will notice that after rain the measurements do not decrease following sunshine.
  Replace the clamps by spares stored over silica gel. Wet clamps can be dried, and if necessary coated again with water repellent varnish.

- The micro-clamps and needles usually do not damage the thallus as they penetrate between the hyphae, and with careful handling do not rupture the tissue. However, some lichens are more sensitive and holes may become larger after one or several days. Check the contact of the needles every day.
- Several days of continuous rain results in polarization and incorrect measurements.

Comments

Both methods have advantages and disadvantages; gravimetric methods give particularly accurate results but do not permit measuring in situ and no long-term and continuous observations. Impedance measurements are effective over a limited range and have a higher standard error. But they are taken in the physiologically relevant range and in the field the discrepancies are negligible when compared with inaccuracies caused by repeatedly removing the thalli for gravimetric measurements. Impedance measurements, used in conjunction with appropriate sensors, are best suited for the comparative characterization of microhabitats, but for physiological research in the laboratory, we recommend gravimetric methods.

References

Canters K J, Schöller H, Ott S, Jahns H M (1991) Microclimatic influences on lichen distribution and development. Lichenologist 23:237-252

Coxson D S (1991) Impedance measurement of the thallus moisture content in lichens. Lichenologist 23:77-84

Hahn S, Speer D, Meyer A, Lange O L (1989) Photosynthetische Primärproduktion von epigäischen Flechten im Mainfränkischen Trockenrasen. I. Tagesläufe von Mikroklima, Wassergehalt und CO2 Gaswechsel zu verschiedenen Jahreszeiten. Flora 173:183-222

Jahns H M (1984) Morphology, reproduction and water relations - a system of morphogenetic interactions in Parmelia saxatilis. Nova Hedwigia, Suppl 79:715-738

Jahns H M, Ott S (1983) Das Mikroklima dicht benachbarter Flechtenstandorte. Flora 173:183-222

Larson D W (1979) Lichen water relations under drying conditions. New Phytol 82:713-731

Ott S, Elders U, Jahns H M (1997) Vegetation of the rock-alvar of Gotland II. Microclimate of lichen-rich habitats. Nova Hedwigia 64:87-101

Pontailler J-Y (1990) A cheap quantum sensor using a gallium photodiode. Funct Ecol 4:591-596

Schöller H (1991) Flechtenverbreitung und Klima. Vegetationsökologische Untersuchungen zur Rolle der Flechten in naturnahen Traubeneichenwäldern des Taunus. Bibl Lichenol 42:1-250

Schroeter B, Kappen L, Moldaenke C (1991) Continuous in situ recording of the photosynthetic activity of Antarctic lichens - established methods and a new approach. Lichenologist 23: 253-266

Schroeter B, Olech M, Kappen L, Heitland W (1995) Ecophysiological investigations of Usnea antarctica in the maritime Antarctic. I. Annual microclimatic conditions and potential primary production. Antarc Sci 7:251-260

Schuster G, Herold K, Jahns H M (1982) Mikroklimatische Messungen an Flechtenstandorten - Neue Meßapparaturen. Herzogia 6:183-200

Suppliers

Parts for the construction of sensors and logger can be bought from local electronic distributors.

Sensors of PPFD (photosynthetic photon flux density) from: LICOR inc, 4421 Superior street, P.O. Box 4425, Lincoln, Nebrasca, 68504 USA; Hamamatsu photonics, 325-6 Sunayama-cho, Hamamatsu City, Shizuoka, Japan (with offices in many countries).

Scales with infrared drying: Mettler PM 100 and Mettler LP16 (www.mt.com)

Determination of the Parameters of Lichen Water Relations

RICHARD P. BECKETT

Introduction

Lichens are poikilohydric, i.e. can lose most of the water from their thalli, and then recover metabolic activity when re-wetted. They share this feature with most bryophytes and many algae, but poikilohydry is absent from the vast majority of higher plants that are termed homoiohydric. Surprisingly little information is available on the basic water relations of lichens; only a few measurements of water potential and its components exist in the literature. Water potential (ψ) is an expression of the chemical potential or free energy status of water (Kramer and Boyer 1995). It is a measure of the driving force that causes water to move into a system, or from one part of a system to another. It is only possible to measure the chemical potential of water in a given system relative to the chemical potential of pure, free water, i.e., of water containing no solutes and bound by no forces. We take the ψ of pure water to be zero under standard conditions of temperature and pressure. The water potential of a plant cell under isothermal conditions has three components:

$$\psi = \psi_\pi + \psi_P + \psi_M \qquad \text{Equation 1}$$

where ψ_π is osmotic potential, ψ_P is turgor potential, and ψ_M is matric potential. Osmotic potential is the contribution made by dissolved solids and is always negative in sign. Turgor potential is the contribution made by pressure exerted by the cell wall and is positive, or zero in plasmolysed cells. Matric potential is the contribution made by water-binding colloids and surfaces in the cell. It is negative in sign, and in living homoiohydric plant cells is usually sufficiently small to be ignored, although in lichen tissues at low water contents it is likely to be important.

Richard P. Beckett, University of Natal, School of Botany and Zoology, Scottsville, Pietermaritzburg, 3209, Republic of South Africa (*phone* +27-33-260-5141; *fax* +27-33-260-5105; *e-mail* beckett@nu.ac.za)

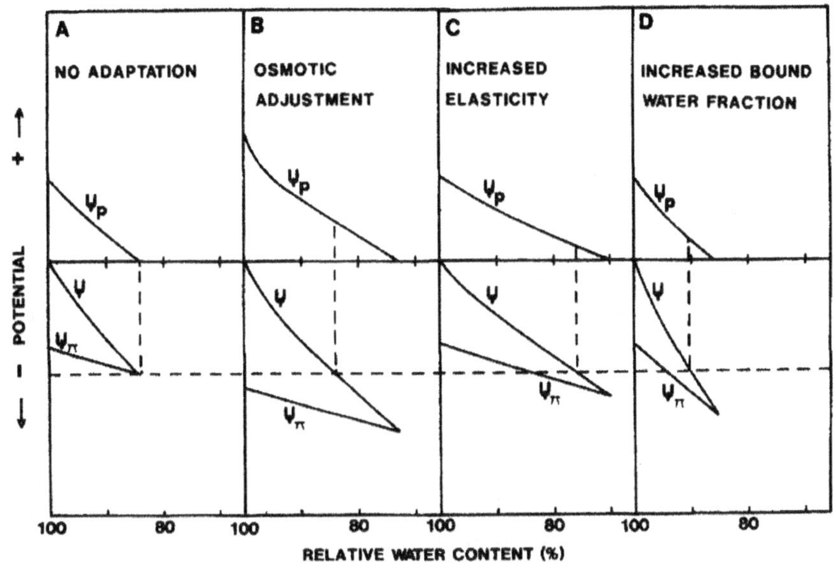

Fig. 1. Representations of the relationships among relative water content and the components of water potential in a tissue. A: no adaptation to drought. B, C, and D: turgor maintenance by osmotic adjustment, increase in elasticity, and increase in bound-water fraction, respectively. The horizontal dashed line indicates the ψ that in A is associated with zero turgor. In B, C, and D the ψ for zero turgor is considerably lower. The vertical dashed lines within each diagram indicate the relative water content and turgor associated with the constant ψ. Note that for constant ψ the relative water content is unchanged in B, decreased in C, and increased in D. Taken from Radin (1983).

Measurement of water potential enables the metabolic activities of lichens to be related to ψ rather than some biologically less meaningful parameter like water content. However, knowledge of ψ and its components is also useful in studies on desiccation tolerance in lichens. Assuming lichens need positive turgor for growth, lichens from xeric environments may benefit from a higher turgor at given water contents (and as a result a lower water content at turgor loss) than more mesic species. In other words, lichens possessing turgor maintenance can lose more water before losing turgor than those that do not. Plants can maintain turgor in several ways (Radin 1983, Fig. 1). First, reducing ψ_π at full turgor will increase ψ_P (Fig. 1). Second, high cell wall stretchiness (and as a result low tissue bulk modulus of elasticity) will also result in higher turgor at a given RWC, and a lower RWC at turgor loss. Third, a low apoplastic water content will also reduce the RWC at turgor loss. It is clearly of interest to compare turgor maintenance and other aspects of the water relations of desiccation sensitive and resistant lichens, or of the same species sampled at different times of the year or from different habitats. In many habitats for much of the time lichens are dry, and as a result metabolically inactive. Although turgor maintenance mechanisms may only be important for a small proportion of the life of a lichen, they may operate for a significant proportion of the time a lichen is metabolically active [for example, look carefully at the data presented in Lange et al. (1990)].

While the measurement of ψ can be made relatively easily using the thermocouple psychrometer, determination of other water relations parameters requires construction of a pressure-volume (PV) isotherm. This involves measuring the relationship between ψ and RWC, then plotting $(-1 / \psi)$ as a function of $(1 - \text{RWC})$. The resulting curve is initially concave, but beyond the region where turgor is lost (i.e. where turgor no longer contributes to ψ) the curve becomes linear. Examples of the parameters that can be obtained from a PV curve include the osmotic potential at full turgor ($\psi_{\pi s}$), the apoplastic water fraction (R_a), a graph of ψ_P as a function of RWC, and the tissue bulk modulus of elasticity (ε_v) (see Fig. 3A and Wenkert et al. [1978], Koide et al. [1989] and Beadle et al. [1993] for more details).

Principles of thermocouple psychrometry

Workers usually determine the water potential of higher plants using the pressure chamber or 'pressure bomb'. This involves severing a leaf or twig, placing it in a sealed chamber with the cut end exposed, then applying pressure until the xylem sap is forced back to the protruding cut surface. The pressure required to force the xylem sap back to the cut surface is equal in magnitude but opposite in sign to the tension (negative pressure) the xylem sap was under originally. However, this technique is clearly unsuitable for lichens, and instead ψ can be determined using the thermocouple psychrometer.

Boyer (1995) provides a detailed description of the theory behind thermocouple psychrometry. Various types of psychrometer exist, but all involve enclosing a small sample in a vapour chamber and using thermocouples to determine the humidity of the air above the sample. At 25°C the relationship between the vapour pressure of a solution and its water potential approximates to:

$$\psi_W = 137.2 \ln (e_W/e_O) \qquad \text{Equation 2}$$

where ψ_w is the water potential of the solution in a thermocouple cup, e_w the vapour pressure of the solution and e_o the vapour pressure of distilled water. Note that if the water potential of the solution in the vapour chamber is -1 MPa then the relative humidity above the solution will be 99.3%. Clearly, humidities above most biological samples will be high, even when their water potential is quite negative. The vapour pressure develops at the surface of the solution in the sample chamber. For plant tissue, this surface is the cell wall.

The principle behind the thermocouple is as follows. If two dissimilar metal conductors are joined to make a circuit, a voltage is generated at each junction. The voltage varies with temperature, but if the temperature is uniform then no current is produced because the voltages in the circuit are the same and oppose each other. However, when the junctions differ in temperature a voltmeter inserted into the circuit will register a current. In thermocouple psychrometry, one junction is held at the temperature of surroundings, i.e. the temperature of the sample chamber in which the sample is located, and is termed the reference junction. The other junction is termed the measurement junction and is exposed to the chamber atmosphere.

Of the various kinds of psychrometer available, perhaps the simplest for measuring the water potential of lichens is the Peltier type (Fig. 2) [see Boyer (1995) for a discussion of other types]. Passing a current through the thermocouple cools the wire and thus condenses water from the atmosphere of the vapour chamber, forming tiny droplets on the thermocouple junction. The current is then switched off, and the droplets allowed to evaporate. As the droplets evaporate, they cool the thermocouple junction, producing a voltage. The drier the air in the chamber, the faster the droplets evaporate, and the greater the voltage produced. The voltage gener-

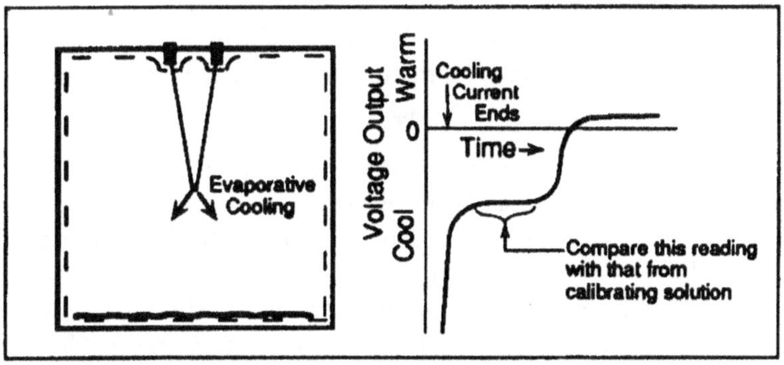

Fig. 2. Peltier psychrometer using current imposed by a voltage source to cool the junction below the dew point (Peltier effect). Turning off the cooling current allows condensate to evaporate from the junction. The output decreases to a semi-stable value that is compared to outputs previously measured with calibrating solutions. Taken from Boyer (1995).

ated is proportional to the vapour pressure of the chamber. Using standard solutions of known water potential will allow the construction of a calibration curve of ψ as a function of voltage. You can then estimate the water potential of any sample by measuring the voltage that the sample produces, then reading off the water potential from the calibration curve. A variant of this, the "Richards" method (Boyer 1995), involves physically placing a droplet of water on the thermocouple. A porous ceramic bead or ring holds the water in place. In both cases, the thermocouple initially produces a high voltage, but this rapidly declines to a semi-stable reading, and this is the value you record (Fig. 2). I recommend using the Richards method for lichens, as it allows measurements of ψ down to at least -10 MPa, but with practice down to -30 MPa, compared with -5 MPa for the method using Peltier cooling. The systematic error in the technique is typically 10% of the reading.

Thallus water content

Different ways of expressing thallus water content exist in the literature (Kershaw 1985). One method is to express the amount of water in lichen material as a percentage of the thallus dry weight, typically derived by drying lichen material at 80°C for 48 hours. However, it can sometimes be helpful, for example when constructing PV isotherms, to present thallus water content as a percentage of the saturating water content, i.e. the "relative water content" (RWC):

$$\text{RWC} = (\text{Fresh weight} - \text{Dry weight})/(\text{Turgid weight} - \text{Dry weight})$$

Equation 3

As may be expected, plots of photosynthesis as a function of water content can show much less variation between different replicates if water content is expressed as RWC (Kershaw 1985). The main problem calculating the RWC of lichens is the estimation of the turgid weight. One possible method is outlined under Materials.

Problems with constructing lichen PV curves

Although lichen PV curves may appear normal, plotting ψ_P as a function of RWC often yields anomalous results (e.g. Fig. 3C). Apparently, initially ψ_P does not change as the RWC decreases. A possible explanation for this is that as lichens dry, they first lose intercellular water. Turgor will only decrease when lichens start losing symplastic water. Although some workers question the significance of intercellular water (see Frey and Scheidegger, this volume chapter 8) its existence is supported by evidence from studies on photosynthesis in lichens (Cowan et al. 1992). Valladares et al. (1993) estimated that in the lichen genus *Umbilicaria* pore space occupies approx. 25% of the thallus volume. It seems reasonable to assume

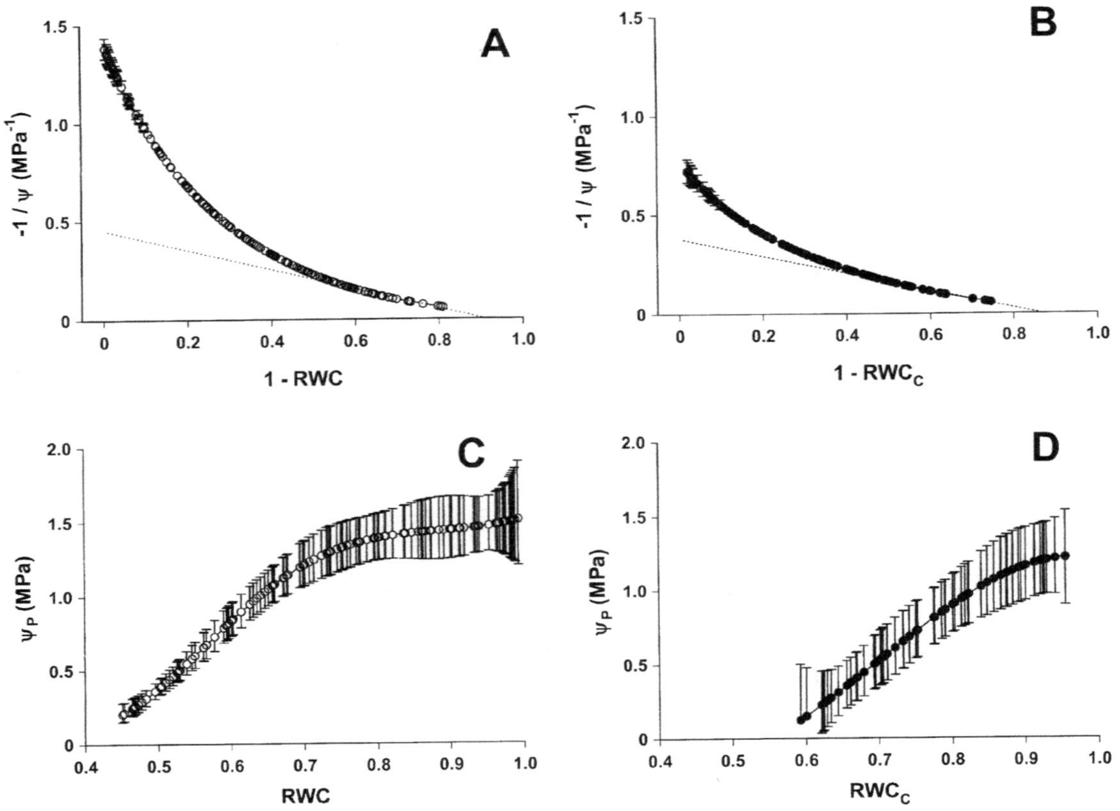

Fig. 3. Pressure volume curve of the lichen *Xanthoria parietina* before (A) and after (B) correction for intercellular water. Turgor potential as a function of relative water content (RWC) plotted before (C) and after (D) correction for intercellular water. Points represent fitted values with 95% confidence limits using the "Spline" program of Hunt and Parsons (1974).

that at least some of this pore space could become filled with water. If uncorrected, this intercellular water will cause inaccuracies in the estimates of the components of ψ.

Unfortunately, no simple method exists to correct for the existence of intercellular water and thus determine the true lichen water content at full turgor. Beckett (1996a) suggested the following approach. Recalculate the values of RWC to exclude intercellular water using this equation:

$$RWC_C = (\text{Fresh weight} - \text{Dry weight}) / (\text{Turgid weight} - \text{Dry weight} - \text{Weight of intercellular water})$$

Equation 4

where RWC_c is the relative water content corrected to exclude intercellular water. The weight of intercellular water is the weight of water that the lichen loses before ψ_P starts to fall. The PV isotherm can be replotted to exclude the points where the values of RWC_c exceed 1 (Fig. 3B), and a new graph of turgor as a function of RWC_c derived showing turgor falling as RWC_c declines (Fig. 3D). It should be noted that the data of Valladares et al. (1993) suggest that a gradation of intercellular pore sizes exists. Water in the smallest pores will only be lost after the lichen has lost some symplastic water, obscuring the point of full turgor. Thus even the corrected graph of turgor as a function of RWC_c (Fig. 3D) is probably influenced by intercellular water. Clearly, no perfect method exists to determine the water content of lichens at full turgor that excludes intercellular water, and therefore errors will occur when calculating the parameters of lichen water relations. However, the approach outlined above and in detail below will probably give sufficiently accurate estimates of these parameters to allow comparisons to be made, for example between different species or single populations sampled at different times of the year.

Proctor et al. (1998), working with bryophytes, suggest an alternative way to estimate intercellular water. As for lichens, in PV analysis of bryophytes these workers found that initially ψ falls more slowly than expected with the decline in RWC. Bryophytes probably contain "external water"; in mosses this may take the form of water trapped between the leaves and the stem. It seems likely that plants will lose this water before losing symplastic water (and hence turgor). Proctor et al. (1998) suggest plotting water content as a function of ψ (Fig. 4A). Any easily lost water will cause a steep upward inflexion of this graph at low values of ψ and data point falling within this region of the graph can be excluded from subsequent analysis. However, a graph of the data obtained from *Xanthoria* plotted in this way

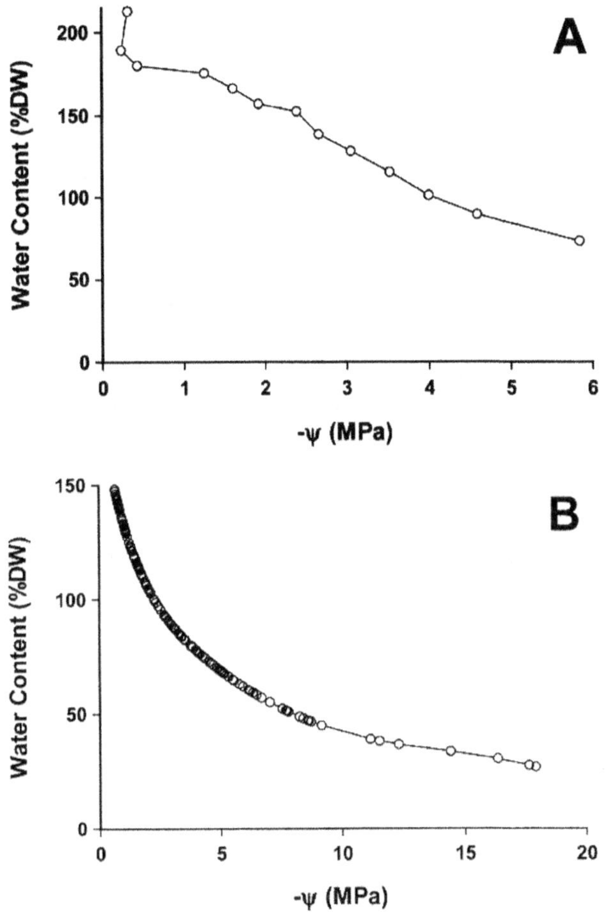

Fig. 4. Water content as a function of ψ in the moss *Polytrichum commune* [**A**, modified from Proctor et al. (1998)] and in the lichen *Xanthoria parietina* (**B**).

gives no indication of any upward inflexion (Fig. 4B), and yet graphs of turgor as a function of RWC derived from this data are clearly anomalous (Fig. 3C). The most likely explanation for this result is that in mosses the water responsible for the sharp upward inflexion is very loosely held and readily lost. While firm blotting will remove most superficial water in lichens, the absence of sharp inflections in graphs of water content as a function of ψ may be a result of water loss from intercellular pores of progressively smaller diameter as a thallus desiccates. Possibly therefore, when working with lichens, the method for correcting for intercellular or external water outlined above may provide better estimates of the components of ψ than that proposed by Proctor et al. (1998).

In future, it seems likely that pressure probes that can be inserted into individual cells will be developed, and these will allow direct determination of ψ_P as a function of RWC in both photo- and mycobiont cells. Although in theory appropriate probes already exist (Boyer 1995), obtaining reliable, reproducible measurements from them is currently extremely difficult.

Materials

Lichen Material

1. If lichens are collected dry, they should be slowly hydrated by storage for 48 h at a relative humidity of 100% (in a desiccator over distilled water) at 20°C and a low light intensity (e.g. approx. 100 µmol photons m^{-2} s^{-1}).
2. Completely hydrate the material by weighing small samples of lichen thallus (e.g. 50 - 100 mg) and gently shake them in approx. 10 ml of deionised distilled water.
3. Remove, carefully blot dry and weigh the samples every 10 - 15 minutes until they reach constant mass (usually approx. 1 - 2 hours).

Equipment

- Thermocouple psychrometer (e.g. Decagon "Tru-Psi")
- Microvoltmeter (e.g. Wescor HR-33T)
- Chart recorder
- Balance weighing to 0.1 mg

Reagents

- Contrad (Merck) or teepol detergents
- NaCl solutions

Water Potentials of NaCl solutions at 20°C (See Lang 1967 for more details)

Concentration (M)	-mPa
0.05	0.232
0.10	0.454
0.20	0.901
0.30	1.349
0.40	1.793
0.50	2.242
0.60	2.699
0.70	3.159
0.80	3.618
0.90	4.087
1.00	4.558
1.50	6.984
2.00	9.591

Procedure

Cleaning the thermocouple

Never directly touch the thermocouple at any stage of the procedure, as it is extremely delicate. However, with routine use the thermocouple and the thermocouple plate will become dirty, and this will affect the accuracy of readings. Clean the thermocouple as follows:

1. Spray the thermocouple with fine jet of deionised distilled water.
2. If the thermocouple is mounted in a metal plate, as for example in the "Decagon" unit, clean the plate surrounding the thermocouple using a matchstick, or an orange wood stick, of the kind used for specimen preparation in electron microscopy. Shave the end of the stick to a point, and place a small piece of cotton wool on the end. Dip the cotton wool in a dilute detergent solution, e.g., "teepol" and, viewing the plate under a dissecting microscope, gently rub the plate to remove contamination.
3. Thoroughly rinse the plate and thermocouple with deionised distilled water, and then dry at 60°C for 1 h.

4. If removable (e.g. for the "Decagon" unit) sample chambers can be soaked overnight in detergent (e.g. 2% "Contrad", but teepol is probably adequate). Next day, thoroughly wash the chambers in tap then deionised distilled water, and dry in an oven.

Carefully read the manufacturer's instructions on these points; "Wescor" in particular offer good advice on thermocouple and chamber cleaning.

Calibration curve Before determining the water potential of biological samples, construct a calibration curve of water potential as a function of microvolts generated by the thermocouple. If this graph is approximately linear, the thermocouple is probably working properly.

5. Prepare a set of standard solutions using sucrose, NaCl or KCl (see Materials) from, e.g., 0 to -20 MPa. NaCl and KCl have the advantage of being less susceptible to microbial attack, but can cause corrosion if left in the sample chamber for too long.
6. Place a disk of filter paper in the base of the chamber and line the walls of the chamber with a rectangular strip of filter paper.
7. Add sufficient solution to moisten the paper thoroughly, leave for 1 h to equilibrate, and then measure.

Equilibration times Even though lichens lack a cuticle, the atmosphere of the sample chamber can take surprisingly long to equilibrate with the water potential of lichen tissue.

8. Prepare samples of lichens at a range of RWCs, and place them in the thermocouple psychrometer.
9. Take readings every 0.5 h for 3 h, and then every 1 h for 9 h.
10. For each sample, plot a graph of ψ as a function of time, and determine the time needed for equilibration at each RWC. For tissue at low RWCs, this can be up to 12 h.

It is important to understand that sample chambers are never completely airtight, and are likely to leak water vapour very slowly. As a result, measurements of ψ as a function of time will typically show an initial rapid rise to less negative values as the sample equilibrates with the chamber atmosphere, followed by a very slow decline as the cups lose water vapour. Incidentally, leakage of water vapour does not affect the accuracy of measurements for the PV curve. Sample RWC is determined immediately after taking a reading, and the equilibration of sample ψ with sample chamber air will be fast compared with leakage rates.

Obtaining raw data needed to construct a PV curve

The Decagon "Tru-Psi" thermocouple psychrometer is well suited to measuring the water potential of lichens as it allows the simultaneous equilibration of ten samples. In addition, the chamber volume is sufficiently large (approx. 4 ml) to take a representative sample of lichen tissue. If you are using the Richards method for wetting the thermocouple, a typical experiment would comprise the following: five sample chambers containing lichen tissue; three standards (distilled water, and one standard higher and one lower than the expected values of tissue ψ); one special chamber with water for wetting the thermocouple, and one empty chamber. The empty chamber is useful for drying the thermocouple while you wait for samples to equilibrate, and you should routinely place this chamber below the thermocouple when you are not taking measurements. To obtain the raw data needed to construct a PV curve proceed as follows.

11. Incubate lichen tissue in deionised distilled water until it is fully hydrated (see above).
12. Place the lichen tissue into a pre-weighed sample chamber. Tissue fresh masses of 50 - 100 mg, weighed to 0.1 mg, are suitable for the Decagon thermocouple psychrometer. Rapidly transfer the chamber to the psychrometer, preferably within 10 s to minimise any loss of water.
13. Allow the sample to equilibrate with the chamber atmosphere for the time estimated above, then read water potential. Make a note of the temperature. It is good practice to read the standards every time you make a set of measurements on lichen tissue.
14. Remove the sample chamber from the thermocouple psychrometer, then rapidly weigh. Allow the sample to lose 2 - 3 mg of water to obtain the next RWC value on the PV curve, then replace in the thermocouple.
15. Repeat stages 13 and 14 until the sample has achieved a RWC of approx. 0.2, increasing the equilibration time as determined above as ψ falls. You may need to take as many as 15 to 20 measurements on each sample.
16. Place the sample cups and lichen samples in an oven at 80°C for 48 h, then allow to cool to room temperature in a desiccator. Determine the dry weight of the lichen and check that the weight of the sample chamber has not changed.

Construction of the PV curve Typing the raw data obtained above into any spreadsheet program will help you calculate the parameters that can be derived from the PV isotherm.

17. For every measurement taken, calculate the ψ and RWC of each sample. Convert the values of ψ measured at temperature T (degrees Kelvin) to their equivalent values at 20°C using the following equation:

    $\psi_{293} = (293/T)\ X\ \psi_T$. Equation 5

18. For each thallus sample, plot the PV isotherm, i.e. $(-1/\psi)$ as a function of (1 - RWC). To carry out statistical analysis of your data it is best to plot PV isotherms for each thallus sample separately. However, to illustrate typical graphs that can be derived from thermocouple psychrometry, Fig. 3A illustrates the combined data from 15 - 20 measurements made on each of 9 replicates of *Xanthoria parietina*.

19. By visual inspection, estimate the linear portion of the PV isotherm. Using simple regression (most spreadsheet programs have a regression function) extrapolate the linear portion, and estimate turgor potential as the difference between this line and raw data points. Plot a graph of turgor as a function of RWC (Fig. 3C).

Deriving parameters from the PV curve

Osmotic potential at full turgor

20. Estimate the weight of intercellular water as the weight of water that the lichen loses before ψ_P starts to fall, estimated from the graph obtained in Step 19.

21. For reasons discussed in the "Introduction", recalculate RWC to exclude intercellular water, using the following equation:

    RWC_C = (Fresh weight – Dryweight)/

    (Turgid weight – Dry weight – Weight of intercellular water)

    Equation 6

22. Replot the PV curve using RWC_c, and excluding any RWC_c values greater than 1.

23. The y-axis intercept of the extrapolated linear portion of the new PV isotherm corresponds to $\psi_{\pi s}$. As indicated above, it is possible that matric potential (ψ_M) becomes important as lichen thalli dry to low RWCs. However, even at low RWCs the PV isotherms of lichens appear

approximately linear. Deviations from linearity would be expected if, for example, ψ_M increased dramatically at low thallus RWCs. Perhaps ψ_M only becomes important at thallus water potentials lower than those that can be measured using thermocouple psychrometry, and will therefore not influence the calculation of parameters of water relations outlined below.

24. Estimate turgor potential as the difference between the extrapolated linear portion of the PV isotherm and raw data points.

RWC at turgor loss

25. Plot turgor as a function of RWC_c and from this graph, estimate the value of RWC_c at which turgor is zero, then correct this back to RWC.

Bulk modulus of elasticity

26. You can calculate the bulk modulus of tissue elasticity from the relationship between turgor and RWC_c. Various methods exist in the literature. Consult Beadle et al. (1993) and Stadelman (1994) for more details.

Cellular location of water

Water in lichens can occur in three cellular locations: inside cells (symplastic water, R_s); in pores within the cell wall (apoplastic water, R_a) and between cells (intercellular water, R_i). R_s, R_a and R_i indicate the proportions of total thallus water in these three locations.

27. As outlined above, R_i can be estimated from the graph of turgor as a function of RWC i.e. the proportion of water that the thallus loses before turgor starts to fall.

28. Apoplastic water occurs in small pores in the cell wall. In higher plants, these are typically from 5 to 10 nm in diameter (Meidner and Sheriff 1976). Plants will only lose this water when the thallus is very dry, and ψ less than -15 MPa. You can estimate R_a by extrapolating the linear portion of the PV isotherm until it intercepts the X-axis (i.e. the RWC at large negative values of thallus ψ).

29. Finally, estimate R_s by subtraction, i.e. $1 - R_a - R_i$.

Proportion of $\psi_{\pi s}$ due to K

30. First, determine the mean thallus intracellular K concentration on a dry weight basis using atomic absorption spectrophotometry. You can then calculate the mean concentration of K as follows:

Mean K concentration (mol l^{-1}) = Mean intracellular thallus K concentration (mol g^{-1} dry mass) $\times$ 1000 / Mean thallus H_2O content at full turgor (g g^{-1} dry mass) $\times$ $(1 - R_a - R_i)$ Equation 7

31. Obtain the osmotic potential of an equivalent concentration of KCl from a list (e.g. thermocouple manual) and express this as a percentage of the osmotic potential derived from the PV isotherm.

A check on the estimate of $\psi_{\pi s}$

32. Campbell (1985) described a method to check the estimate of $\psi_{\pi s}$ derived from the PV curve. Place thallus samples, fully hydrated as above, into the sample chambers, and wrap them in at least three layers of "Parafilm".

33. Immerse the cups in liquid nitrogen for 5 minutes, and then allow them to warm to room temperature (approx. 1 h).

34. Remove the Parafilm, rapidly transfer them to the thermocouple psychrometer, and after an equilibration time of approx. 1 h determine ψ. Assuming freezing ruptures membranes and thus destroys turgor, ψ will equal $\psi_{\pi s}$. This method underestimates $\psi_{\pi s}$ because apoplastic and intercellular water dilute ions and molecules in the symplast. To correct for this, a modification of the equation of Jones and Rawson (1979) can be used:

$$\psi_{\pi s} = \psi_{\pi k}/(1 - R_a - R_i) \qquad \text{Equation 8}$$

where $\psi_{\pi k}$ is the water potential of fully turgid killed lichens.

Troubleshooting

Some lichens e.g. *Ramalina* species display the phenomenon of "over saturation" (Blum 1973) and continue to increase in weight for at least 24 h. This is presumably a result of the hydration of cell wall colloids. It is difficult to define full thallus hydration in these species, and I can only suggest that for consistency you hydrate them for approximately the same time as other, morphologically similar species. One other particular problem arises with maritime lichens. These often contain high intercellular concentrations of NaCl that you need to remove before you can determine ψ and its components (Beckett 1996b,c). However, incubation of these species in deionised distilled water can cause the loss of intracellular ions. I recommend that you hydrate and at the same time wash these species by shaking them in a solution of 0.25 mM $Ca(NO_3)_2$ as recommended by Fletcher (1976).

Occasionally, you may find species in which estimates of $\psi_{\pi s}$ derived from the PV isotherm differ significantly from those estimated by the above method of Campbell (1985) [see for example Beckett (1998)]. Various explanations for this exist, but one possibility is that the cell wall contains pores greater than 10 nm in diameter. Although this would probably have little effect on the estimates of $\psi_{\pi s}$ derived from the PV curve, it would cause underestimation of R_a and thus overestimation of $\psi_{\pi s}$ derived from the freezing method. Unfortunately, it is currently difficult to estimate wall pore size in lichens, although it may be possible to use a recent porosimetric method developed by Valladares et al. (1993).

The three main sources of error associated with thermocouple psychrometers are first vapour pressure disequilibria, second thermal gradients and instability and third changes in ψ due to sample growth and excision [see Koide et al. (1989) for more details]. Vapour pressure disequilibria can be minimised by maximising the ratio of sample surface to chamber volume. In addition, your chamber and thermocouple plate need to be kept scrupulously clean, as salts and other foreign matter on their surface may absorb water vapour. Thermal gradients and instability produce errors in the measurement of ψ by causing the chamber atmosphere to differ in temperature from the sample or by causing electrical potentials to develop within the measuring circuit. Ideally, measurements should be made in a thermostatically controlled water bath, but if this is not possible then carry out measurements in a laboratory not subject to rapid temperature fluctuations. Sample growth would appear to be an unlikely source of error for most lichens, but as a precaution try to always work with mature tissue. Changes in ψ of a tissue will occur if intact cells take up water and solutes released from damaged cells at the cut surface. Minimise excision errors by using samples with a low ratio of cut surface area to sample surface area.

Comments

Shortly after completing the writing of this chapter, Decagon Devices discontinued manufacture of the "Tru-Psi" thermocouple psychrometer, and introduced the "WP4 Water Potential Meter" as a replacement. The WP4 uses the chilled-mirror dewpoint technique to measure the water potential of a sample. The WP4 uses an internal fan that circulates the air within the sample chamber to reduce time to equilibrium. Because both dewpoint and sample surface temperatures are simultaneously measured, the need for complete thermal equilibrium is eliminated, which also reduces measurement times. As discussed above, the "Tru-Psi" Decagon unit allows users to equilibrate multiple samples with the atmosphere of the chambers simultaneously. This is very useful, as long equilibration times are required as lichens reach low water contents. Whether the WP4 is suitable for lichen work will depend on whether it can speed up equilibration times sufficiently to offset the disadvantage that the unit can only equilibrate one sample at a time. Another consideration is that the Tru-Psi's cylindrical sample cups have a diameter of 1 cm, compared with the 4 cm sample cups of the WP4. As plant material should ideally cover the entire base of the cup, in practice this means that the WP4 requires more material. While a larger sample size should provide more accurate values of ψ and RWC for each point on the PV curve, problems could arise if only limited material is available.

The only alternative to the Tru-Psi I am aware of that is currently readily available would be to use the following equipment from Wescor: multiple C-52 sample chambers (or possibly L-51 leaf pschrometers), a PS-10 pschrometer switchbox and a HR-33T dew point microvoltmeter. However, the C-52 sample chamber works on the principle of Peltier cooling. As discussed above, this method can only measure water potentials down to -5 MPa, rather than -30 MPa attainable by the Richard's method. In practice, -5 MPa may be too high for the determination of PV curves for some lichens.

Acknowledgements. I thank the National Research Foundation (South Africa) and the University of Natal Research Fund for financial assistance, Prof M C F Proctor for some extremely useful comments on an early version of this chapter, and Prof M J Savage for help with thermocouple psychrometry.

References

Beadle CL, Ludlow MM, Honeyset JL (1993) Water relations. In: Hall DO, Scurlock JMO, Bolhar-Nordenkampf HR, Leegood RC, Long SP (eds) Photosynthesis and production in a changing environment. A field and laboratory manual. Chapman and Hall, London pp 113-128

Beckett RP (1996a) Some aspects of the water relations of the lichen *Parmotrema tinctorum* measured using thermocouple psychrometry. Lichenologist 28:257-266

Beckett RP (1996b) The water relations of the maritime lichen *Roccella hypomecha* (Ach.) Bory. studied using thermocouple psychrometry. S Afr J Bot 62:122-125

Beckett RP (1996c) Some aspects of the water relations of the coastal lichen *Xanthoria parietina* (L.) Th. Fr. Acta Phys Plant 18:229-234

Beckett RP (1998) Some aspects of the water relations of the desert lichen *Xanthomaculina hottentota* studied with thermocouple psychrometry. S Afr J Bot 64: 346-349

Blum OB (1973) Water relations. In: Ahmadjian V, Hale ME (eds) The lichens. Academic Press, New York pp 381-400

Boyer JS (1995) Measuring the water status of plants and soils. Academic Press, San Diego.

Campbell GS (1985) Instruments for measuring plant water potential and its components. In: Marshall B, Woodward F I (eds) Instruments for environmental physiology: Society for Experimental Biology Seminar Series no. 22. Cambridge, Cambridge University Press pp 193-214

Cowan IR, Lange OL, Green TGA (1992) Carbon-dioxide exchange in lichens: determination of transport and carboxylation characteristics. Planta 187:282-294

Fletcher A (1976) Nutritional aspects of marine and maritime lichen ecology. In: Brown DH, Hawksworth DL, Bailey RH (eds) Lichenology: Progress and Problems. Academic Press, London pp 359-384

Hunt R, Parsons IT (1974) A computer program for deriving growth functions in plant growth analysis. J App Ecol 11: 297-307

Jones MM, Rawson HM (1979) Influence of rate of development of leaf water deficits upon photosynthesis, leaf conductance, water use efficiency, and osmotic potential in sorghum. Physiol Plant 45:103-111

Kershaw KA (1985) Physiological ecology of lichens. Cambridge University Press, Cambridge.

Koide RT, Robichaux RH, Morse SR, Smith CM (1989) Plant water status, hydraulic resistance and capacitance. In: Pearcy RW, Ehleringer J, Mooney HA, Rundel PW (eds) Plant Physiological Ecology: Field Methods and Instrumentation. Chapman and Hall, London, 161-183

Kramer PJ, Boyer JS (1995) Water Relations of Plants and Soils. Academic Press, San Diego

Lange OL, Meyer A, Zellner H, Ullmann I, Wessels DJC (1990) Eight days in the life of a desert lichen: Water relations and photosynthesis of *Teloschistes capensis* in the coastal fog zone of the Namib Desert. Madoqua 17: 17-30

Lang ARG (1967) Osmotic coeficients and water potentials of sodium chloride solutions from 0°C to 40°C. Aust J Chem 20: 2017-23

Meidner H, Sheriff DW (1976) Water and plants. John Wiley, New York

Proctor MCF, Nagy Z, Csintalan Z, Takacs Z (1998) Water-content components in bryophytes: analysis of pressure-volume relationships. J Exp Bot 49: 1845-1854

Radin JW (1983) Physiological consequences of cellular water deficits: osmotic adjustment. In: Taylor HM, Jordan WR, Sinclair TR (eds) Limitations to efficient water use in crop production. American Society of Agronomy, Wisconsin pp 267-276

Stadelmann EJ (1984) The derivation of the cell wall elasticity function from the cell turgor potential. J Exp Bot 35:859-868

Valladares F, Wierzhos J, Ascaso C (1993) Porosimetric study of the lichen family Umbilicariaceae: anatomical interpretation and implications for water storage capacity of the thallus. Am J Bot 80:263-272.

Wenkert W, Lemon ER, Sinclair TR (1978) Water content-potential relationship in soya bean: changes in component potentials for mature and immature leaves under field conditions. Ann Bot 42:295-307

Suppliers

Decagon Devices Inc.
P.O. Box 835, Pullman WA 99163 USA
Sales@decagon.com
http://www.decagon.com
Telephone: 1-800-752-2751 (US and Canada only), 509 332-2756
Fax: 509 332-5158

Wescor, Inc.
Environmental Products Division
459 South Main Street, Logan, Utah 84321 USA
wescor@wescor.com
http://www.wescor.com
Telephone: 435 753-8311
Fax: 435 753-8177

Abbreviations

ψ	water potential
ψ_π	osmotic potential
$\psi_{\pi s}$	osmotic potential at full turgor
$\psi_{\pi K}$	water potential of fully turgid killed lichens
ψ_P	turgor potential
ψ_M	matric potential
PV	pressure-volume
RWC	relative water content
RWC_c	relative water content corrected for intracellular water
ε_V	tissue bulk modulus of elasticity
R_s, R_a *and* R_i	proportions of total thallus water in symplastic, apoplastic and intercellular locations respectively.

Measurement of Lichen Growth

DAVID J. HILL

Introduction

Growth of an organism can be defined as the increase in its living biomass. In lichens, biomass increase is presumed to be manifest in an increase in linear dimension, area or weight, which are convenient to measure (Table 1). However they do have different interpretations. For a lichen, an increase in lobe length, or thallus diameter, may not exactly reflect increase in area or mass; conversely, biomass increase may, for example, occur without an increase in the apparent linear size of a lichen thallus (e.g. a thallus confined in a mosaic may grow only in thickness). The rate of growth, which is expressed as increase per unit time, may give very different rates over different time spans, if it varies with different sized thalli, and, especially, with differing environments. The time span can be predetermined and the size of thallus can be allowed for in the analysis of the data. The growth form of the thallus is also important in deciding what parameter to measure. The type of data analysis planned and the aim of the study may also determine the type of measurement chosen. It should be decided whether the actual direct measurement itself is required or a derived measure, such as a proportional increase or the value of a growth rate coefficient.

Growth and growth rates are influenced by internal and external factors (Table 2) which can be used as a check list for consideration before planning replicated growth measurements in the field.

Thallus growth form

The type of growth measurement chosen will depend partly on the growth form (morphology) of the thallus and the methods most likely to be successful are indicated in Table 3.

David J. Hill, University of Bristol, School of Biological Sciences, Woodland Road, Bristol, BS8 1UG, UK (*phone* +44-117-928-8155; *fax* +44-117-9257374; *e-mail* D.J.Hill@bris.ac.uk)

Table 1. Types of growth measurements and their presentation

Type of growth	Example of measure made	Suggested data presentation	Data presentation of doubtful meaning	Notes
Linear	Linear increase in the direction of growth at thallus margin Increase in radius or diameter	Distance grown and Aplin/Hill coefficients	Increase in radius or diameter as proportion of existing radius or diameter	As only the thallus next to the tip of the lobe or growing point is associated with growth, the proportional increase in radius will decline with increasing radius
Area	Increase in projected area (needs a per unit basis)	Area per unit perimeter	Area × $area^{-1}$ (e.g. new area as proportion of existing area) "Relative Growth Rate (RGR)" (see text) Area × $mass^{-1}$	Since growth is only local to where the growth occurs, normalisation to the whole area (including RGR) will give a value declining with increasing thallus size
Mass	Increase in dry weight (needs a per unit basis)	Mass per unit area	Mass × $mass^{-1}$ RGR = ln(w2/w1)/t	We do not know how much of the mass of a lichen is contributing to the further growth of the thallus, so mass $mass^{-1}$ is difficult to interpret.

Table 2. Examples of factors that may affect growth rate measurements

	External factors	Internal factors
Growth rate may be determined by	Substrate Water availability Nutrient availability Light availability Pollution Biotic factors (e.g. herbivore damage and competition)	Taxon of fungus Taxon of photobiont Development Age Genotype Phenotype
Errors in mass and size can be caused by	Extraneous matter e.g. mineral matter, other organisms Loss of fragments	Loss of dry weight due to decomposition of dead biomass Thallus water content

Growth in **fruticose** lichens is frequently measured as either increase in mass because thalli can be detached relatively easily from the substratum, weighed and replaced for ongoing experiments (Renhorn and Essen 1995; Hyvarinen and Crittenden 1998) or increase in thallus branch length (Vagts and Kinder 1999). With **foliose** lichens, most growth studies have involved measurements of lobe length (Hill 1984, 1992; Armstrong 1995; Honegger et al. 1996), thallus diameter (Hill 1981; Benedict 1990; McCarthy and Smith 1995; Armstrong and Smith 1996), thallus area (Polonyi and Türk 1991) or, with larger detachable thalli, thallus mass (Dennison 1988; Renhorn et al. 1997; Muir et al. 1998). Mass increase can occur all over foliose and crustose thalli and not just at the margin (Hill 1981, 1985). It may be expressed per unit area or per unit protein, chlorophyll, photobiont cell or DNA.

The **umbilicate** lichen *Lasallia pustulata* grows by intercalary growth (Hestmark 1997). The umbilicate form might be expected to be an adaptation to this type of growth, being fixed to the substratum only at the centre. In theory, it would be possible for umbilicate thalli to grow much faster with increasing size, even exponentially but, in practice, radial growth rate in *L. pustulata* appears to be limited and independent of thallus size (Hestmark 1997).

Circular **crustose** lichens grow at the margin and significant subapical and intercalary expansion is not possible because the thallus is fixed over the entirety of its underside to the substratum. Lobed crustose (placodioidal) thalli form circular thalli that are ideally suited to linear and area growth studies (Hill 1981, 1984; Proctor 1983; McCarthy and Smith 1995) owing to their simple geometry. With diffuse crusts, the accurate delimitation of the thallus is a problem but it may be possible to measure thallus area or, perhaps, granule size.

Table 3. Summary of options for measuring growth rates in different types of lichen thalli

Growth form	Linear growth [length × time^{-1}]	Area growth [area × time^{-1}]	Mass growth [mass × time^{-1}]
Fruticose	yes	difficult	yes
Foliose	yes	yes	yes
Umbilicate	yes	yes	yes
Circular crust	yes	yes	difficult
Diffuse crust	difficult	possible	difficult

Time-scale problems

The longer the time-scale the more likely there will be problems of identification of thalli which may also disappear, break up or fuse together. These problems can occur over a period of one or more years in foliose thalli. Follman (1961) was able to use one of the longer time-scales with photographs of crustose species separated by 47 years. Long time-scales have the advantage in that they tend to average out, or obscure, the effects of seasonal or annual variations in environmental conditions. Shorter time-scales, less than a year, are used to investigate the effects of seasonal, monthly, weekly or even possibly daily variations in environmental factors.

When measuring small increments over a short time period, variation of size or mass due to hydration, or presence or absence of debris, can be large compared with the possible growth increment. Growth periods of a month or more do not seem to present difficulties, but variations due to hydration are a major problem for shorter-term measurements down to a week or a few days (Fig. 1). With any time scale, thalli should be measured, photographed or weighed under similar environmental conditions or measurements corrected for effect of variable humidity. A sensor, which measures electrical current through a filter paper (calibrated against

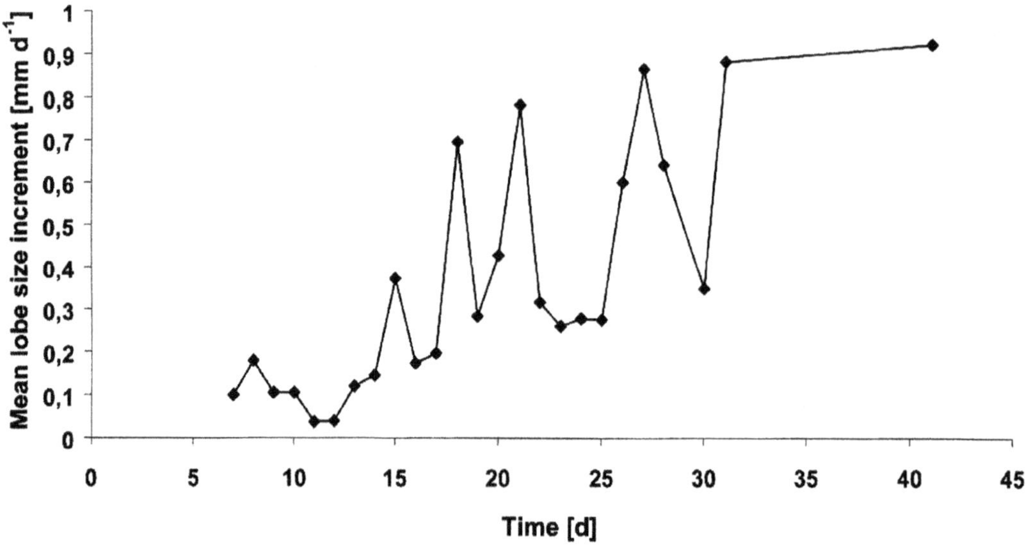

Fig. 1. The daily increment of a lobe of *Xanthoria parietina*. The lower points represent the thallus when dry and the upper points when saturated with water (measured from photographs between 7 October and 10 November 1996 on trunk of *Malus domestica* near Bristol, UK).

lichen thalli), can be used to make a continuous estimate of water content of nearby thalli in the field for growth measurements (Benedict 1990). In general, crusts grow more slowly than foliose and fruticose species but they are more firmly attached to the substratum, so small changes in visible size are more reliably measured. In foliose, and especially fruticose thalli, movement of the lobes with reference to the substratum can frustrate measurement of small changes in lobe length.

Outline

When choosing the most suitable method for measuring growth for a particular project, a number of factors need to be considered. Figure 2 sets out a simplified scheme to assist in this choice.

There are three main approaches for the meassurement of lichen growth: linear measures direct from thalli in the field (Subprotocol 2) or from photographs (Subprotocol 3), area measures from photographs or tracings and mass measures (Subprotocol 4). The photographs and tracings have the advantage that they also provide a valuable permanent record of the lichen thalli and growth, which may be useful for future reference. Techniques for measurements on very small early phases of thallus establishment and measurements for lichenometric studies are also described (Subprotocol 5).

Growth analysis and data presentation is described in Subprotocol 6.

Subprotocol 1
General Protocol for Relocation of Lichen Thalli in the Field

Relocation of lichen thalli, or even a small part of one, can be surprisingly difficult and time consuming in the field. Photographs or drawings (with field notes) and a map of the exact location of the measured lichen thallus can be helpful, or even essential.

Materials

- 35 mm camera and tripod, and for short term measurements either 50 or 90 mm macro lens
- Colour print (or transparency) film

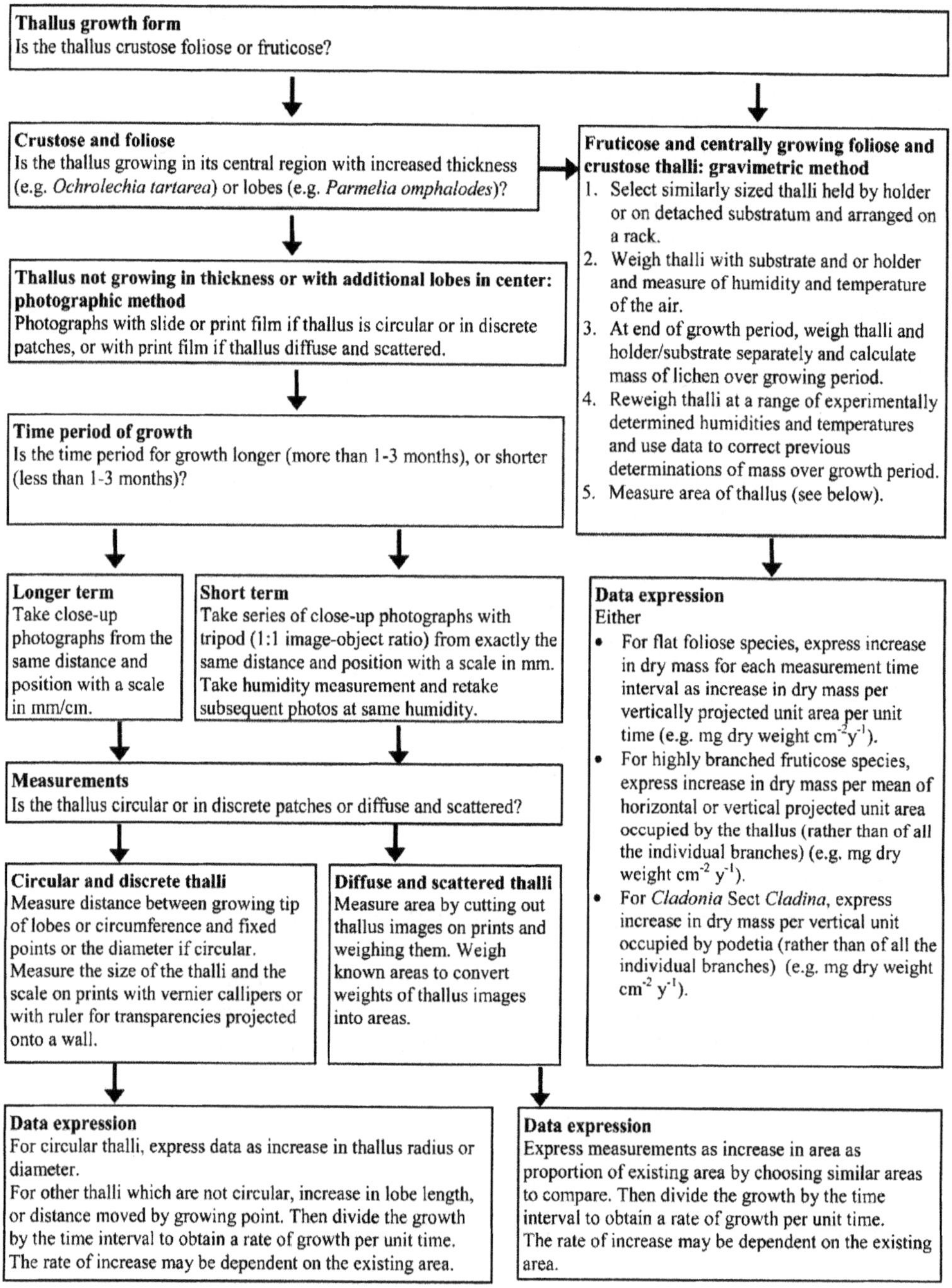

Fig. 2. Flow diagram for choosing suitable method for measuring lichen growth

Procedure

Take the following photographs, marking the position from which the photo is taken, and the key features in the photograph on the map.

1. Whole lichen thallus showing any parts featuring in close-ups.
2. Community of thalli showing the thallus being measured.
3. Tree or rock etc. showing this community.
4. Tree or rock etc. in its location with other land marks.
5. Two or three shots of the locations with the tripod left in situ.

Fewer photos than indicated above may be used if safe to do so, and hand drawn sketches and plans may suffice. Take copies into the field when relocating thalli.

Subprotocol 2
Direct Linear Measurements

Materials

- Clear plastic ruler, vernier callipers or travelling microscope
- Hand lens
- Hardened needle to scratch rock or fine pen with Indian ink (or other water-proof ink) to mark surface
- White gloss paint to paint rock first (optional)

Procedure

1. Select thalli to give required replication (about 30 lobes or thalli are normally sufficient). If growth coefficients are to be calculated, include thalli of representative sizes from the smallest ($>$ 1 mm diameter) to the largest. Always make measurements at similar state of hydration, and with other conditions as similar as possible.

2. Mark rock with reference points scratched into it, e.g. slate (Armstrong 1973), or, paint rock surrounding the thallus margin with white gloss paint, allow to dry then apply marks 1 mm apart moving radially away from the thallus with Indian ink on white gloss paint (Benedict 1990).
3. Measure the diameters of circular lichen thalli in the field. If not perfectly circular take mean of least and greatest diameters (Innes 1986). Calliper measurements are more accurate than those made with a clear plastic ruler (Bull and Brandon 1998).
4. With a hand-lens measure position of tip of lobe (and hence radial increments) to reference marks scratched in a smooth rock surface, or to fine ink marks noting to which mark measurements are made.
5. Repeat measurements at required time intervals.
6. For growth analysis of radial growth of circular thalli, see below.

Troubleshooting

- Results are generally inconsistent.
  Use longer time scale, check that hydration of thalli is similar, check that orientation of the measurement is exactly the same in each sequential measurement.
- Some of the measurements are inconsistent (it may not be possible to correct errors in measurement by checking later).
  If there is clear doubt as to the accuracy of a measurement, omit it in later analyses.

Subprotocol 3
Linear Measurements Using Photographs

There are numerous studies of growth of lichen thalli made by measuring the thalli on photographs taken before and after the growth period. For measurements over short time periods, 35 mm cameras are fitted with macro lens preferably with the use of a tripod (Lawrey and Hale 1977) to detect small changes in size. Special care must be taken to reposition the camera in precisely the same position when retaking the photographs after a growth period. A special tripod can be made that permits very rapid and accurate repositioning of the camera as follows. Construct a cast aluminium box, which is milled to fit round the camera body and into which the camera is screwed precisely, that has three very firmly fixed legs which end in points attached at precise angles, so that the lens points vertically downwards. The points of the three legs can locate 1) into the head of a "posidriv", or Philips screw (fixing the first dimension in space), 2) into the head of a screw whose slot ("bush") is in-line with the first screw (fixing the second dimension) and 3) onto the head of a flat headed nail (fixing the third dimension). The legs are designed to be adjustable to preset lengths which coincide with the plane of focus of the lens(es) when fully extended. The two screws can be adjusted so that the lichen thallus is precisely in plane of optimum focus. The tripod and camera can be rapidly replaced into the screw heads and hence into exactly the same position. Any minor expansion or contraction in the substrate, or tripod, will not prevent the tripod fitting exactly and firmly. The measurements in Figure 1 were obtained using this technique.

For longer time periods (one or more years) a macrolens may not be necessary as the amount of growth would be more substantial except where lichens grow very slowly, for example, in arctic and desert habitats.

Materials

Photographic technique

- 35 mm camera and tripod and for short term measurements either 50 or 90 mm macro lens
- Colour print (or transparency) film
- Hygrometer
- Thermometer
- Strips of graph paper (marked in millimetres)
- "Blu-tack", a detachable mouldable adhesive plastic available from stationers
- Vernier callipers
- Photographic development service

Procedure

1. Select thalli to give required replication (about 30 lobes of thalli are normally sufficient). If growth coefficients are to be calculated, include thalli of representative sizes from the smallest (›1 mm diameter) to the largest.
2. Adjust camera to produce an image of required magnification (x 0.5 or x 1.0 life size for short-term measurements).
3. Place scale in photograph on the same plane as the lichen thallus but without obscuring it.
4. View photograph to be taken to check that enough space is allowed for the lichen to grow, that there are some fixed points to measure to in the prints (see below) and that the scale is in focus with the lichen growing point.
5. Take photographs of the whole thallus or lobe ends.
6. If the whole thallus does not appear on the photograph, measure the diameters in the field. Take care with those that are not circular; I suggest measuring the diameter on same axis as radial growth is being measured or a mean of largest and smallest.
7. Select other thalli to give required replication (about 30 lobes of thalli are normally sufficient). If growth coefficients are to be calculated, select thalli of representative sizes from the smallest to the largest.
8. Record relative humidity and temperature.
9. Select time of repeat visit (with location photographs) to be in similar weather conditions avoiding rapidly changing conditions e.g. recent rain, or sudden increases in atmospheric humidity.
10. Repeat photographs but be sure that the same film is being used and the same focus (distance) settings are used.
11. Record humidity and temperature as before.
12. Develop film and prepare prints from negatives with 5 x - 20 x enlargement. If transparencies have been taken, project them onto a hard white wall.
13. Select a fixed point on the substratum observable in every photograph, e.g. a grain or crack in the rock, or bark, on a radial line through the growing point to be measured.

14. Take measurements using vernier callipers (Hill 1981) or if projected with a ruler from the tip of the lobe or growing point to a fixed point in the substratum.

15. For growth analysis of radial growth of circular thalli, see below.

Troubleshooting

- Photographs are incorrectly aligned due to incorrect repositioning of the camera.
  Parallax errors can be corrected after measurements have been made. See Hooker and Brown (1977) for procedure.
- Environmental conditions were very different causing contrasting thallus hydration resulting in swelling, and/or contraction, to the lobe.
  Some correction can be made by taking measurements of proportional swelling from parts of thallus of size similar to the growing lobe but which are not growing and then using this proportional swelling to correct measurements of the lobe measured.
  These corrections are only approximate however.

Subprotocol 4
Area and Mass Measurements

Areas can be calculated from diameter or radial measurements or measured directly by weighing cut out paper on which tracings of the thalli have been made from photos (Honegger et al. 1996) or by using a Seescan Solitaire image analyser (Smith 1995).

It is possible to trace the outline of the expanding thallus onto clear plastic sheets (Armstrong 1975) but the maximum accuracy achievable is to about 0.5 mm (Benedict 1990), so that periods of growth of less than a year would be difficult. The technique has, however, been done effectively for 3 year intervals of growth in, for example, *Parmelia saxatilis* (Gilbert 1971). As a guide, area measurements should be expressed, where possible, as increase in area per unit length of perimeter.

Mass measurement requires the lichen to be removed from its substratum and this can be difficult except for large foliose or fruticose species. Some authors have made a special frame (Muir et al. 1998) or receptacle (Hyvarinen and Crittenden 1998) for holding the lichen so that it is easier to detach from the substrate without losing or gaining material. Dennison (1988) describes a technique for growing foliose lichens on nylon filaments so that thalli can be removed and weighed. Because the mass of a thallus depends on its water content, the thallus is weighed and the relative humidity and temperature measured at the same time (Renhorn et al. 1997). Known relative humidities can be obtained by using dilutions of sulphuric acid, or saturated solutions of different salts (Lide 1996; Hill and Woolhouse 1969, see also Chapters 14 and 15) exposed in an air-tight chamber. McCune et al. (1996) compared two methods (reference and sacrificial samples) for adjusting for water content.

Materials

Dealing with variation in water content

- Frame, nylon filaments or receptacle for holding lichen (see above)
- Balance
- Enclosed chambers for creating different relative humidities
- Thermometer
- Hygrometer (or wet- and dry-bulb thermometer)
- Saturated solutions of various salts (salts in equilibrium with the saturated solutions) (Table 4)
- Temperature controlled environment (cabinet or room)

Table 4. Relative humidities obtainable with saturated salt solutions in equilibrium with solid salt at 25°C (Lide 1996 p. 15-25 from where details can be obtained for salts giving other humidities and for a formula and coefficients for calculating humidities at other temperatures).

Salt	Relative humidity (%)	Salt	Relative humidity (%)
$LiCl \times H_2O$	11	NH_4NO_3	62
$LiI \times 3\ H_2O$	18	$NaNO_3$	74
$CaCl_2 \times 6\ H_2O$	29	$(NH_4)_2SO_4$	81
$NaI \times 2\ H_2O$	38	KNO_3	92
$Ca(NO_3)_2 \times 4\ H_2O$	51	K_2SO_4	97

Procedure

1. Select thalli that are uniform in size, unless size is a desired factor.
2. Clean off any loose thallus material and debris and substratum.
3. Tie to frame or hook onto nylon threads.
4. Allow samples to equilibrate.
5. Weigh and take measurement of temperature and relative humidity.
6. Take some replicated samples of duplicate material to laboratory.
7. Prepare a series of relative humidity chambers (with different saturated salt solutions (Lide 1996), or dilutions of concentrated sulphuric acid, to cover range of relative humidities in the field) at constant measured temperature. Put in thalli for 12 - 24 h then remove and weigh immediately (within a min). Look up precise relative humidities for each saturated salt solution for the temperature (Lide 1996), or concentration of sulphuric acid. Repeat for each humidity and at different temperatures (e.g. 5°C intervals around those expected in field). When using these relative humidity chambers it is best to minimise the amount of water to be lost or gained as equilibration can be slow if large amounts of water are involved. It is important to check for equilibration by weighing samples at intervals of 12 h until their mass is constant for 3 or more readings to be sure that the process is complete.
8. After equilibration at different relative humidities, place in an oven at 100°C overnight and re-weigh.
9. Calculate weights at different humidities as fractional increase over dry weight.
10. Plot graph of fractional increase in weight versus relative humidity checking effect of temperature.
11. Read off fractional weight increase for field relative humidity and temperature measurement using graphs.
12. Correct weights of thalli in the field to dry weights.
13. Repeat measurement of weights of field samples, measure relative humidity and temperature and look up dry weights from graphs.
14. If possible, express increase in mass as increase in dry weight per unit area.

Troubleshooting

- Thalli weigh less than expected.
  Check that parts of the sample are not becoming detached and lost or being browsed. Photographs of the thalli can help to check the integrity of the samples during the growth measurements.
- Corrected thallus weights appear more variable than expected.
  Check the laboratory determination of the weight-relative humidity curves making sure that the thalli are properly equilibrated over several hours (e.g. several days may be necessary) in the chambers which must be airtight and check that the thalli in the field are properly equilibrated for some hours beforehand with the prevailing relative humidity in the field. Equilibration can be confirmed in the field by repeated accurate weighing every 20-30 min to show that there is no change.

Subprotocol 5
Measurement of Growth in Very Early Stages and for lichenometric studies

When measuring growth of minute thalli, it helps to have the thalli on a detachable substrate. Young thalli can be grown from soredia or thalline fragments on small ceramic discs (Honegger 1996), which are observed microscopically by light microscopy or low-temperature scanning electron microscopy. For corticolous species, the same technique is used but with detachable pieces of bark (Schuster et al. 1985). Soredia or thalline fragments can also be fixed to bark on trees in the field with surgical gauze while observing their growth (Scheidegger et al. 1995).

Lichenometry

Lichenometry is the use of linear growth of a single lichen species to date the length of time a surface has been exposed to the air in a natural setting. A circular thallus, usually of a crustose species such as *Rhizocarpon geographicum*, is presumed to have been growing consistently and the size attained is directly related to the time since it colonised the exposed surface so that the surface exposed the longest has the largest thalli on it. The calibration data published indicate that thalli of *Rhizocarpon geographicum* grow unevenly and large thalli grow more slowly than medium sized ones, although this has not been seen with other lichens such as *Pseudephebe miniscula* in which thalli grow more constantly (Haworth et al. 1986). Brief methodology is given here.

Materials

- Perspex sheet with nest of inscribed circles 5 mm apart
- Clear plastic ruler
- Detailed topographical map to record position of measurement sites
- Computer with curve-fitting statistical software

Procedure

1. Choose species which is abundant in the area of interest and has range of different thallus sizes and grows at the rate commensurate with the study and which also occurs on a series of surfaces of known exposure dates covering the time period of interest.
2. Select an easily defined fixed area for use in the field within which thalli will be located and within each locate Fixed Area Largest Lichen thallus (FALL) (see Bull and Brandon 1998).
3. Measure Largest Inscribed Circle (LIC).
4. Repeat procedure for thalli growing on surfaces of known date and construct a curve of LIC against time using best fitting curve (computer with appropriate statistical software).
5. Read off dates corresponding to each LICs for each FALL in area of interest.

Comments

Because the growth curve on which lichenometry is based is on individual thalli in different locations, it does not usually attempt to describe the growth of any individual over its life span or a part of it. Therefore, lichenometric data should be interpreted carefully if being used for ecological purposes.

Subprotocol 6
Growth Analysis and Data Presentation

The method of data analysis depends on the type of lichen thallus and the purpose of the investigation (Table 1-3). It is important that the investigator is clear as to the reason why he/she applies a particular method of data analysis. Without this clarity, manipulation of the data can obscure rather than clarify the interpretation of the data. It is unfortunate that the term "Relative Growth Rate" (RGR, dimension t^{-1}) has pervaded the literature on lichen growth on the assumption that lichens should grow logarithmically. The RGR concept, originally borrowed from flowering plant growth analysis, has been consequently erroneously applied directly to lichens (Woolhouse 1968, Armstrong 1975) with the result that the calculated value for it declines rapidly as thallus grows larger, without showing any sign of even approximate constancy (until it nears zero), a fact that immediately invalidates the application of the RGR concept. And so it is known that lichens do not, in fact, grow logarithmically. By contrast, the growth of, for example micro-organisms and higher plants, when unrestricted by resources and senescence, has been clearly shown to be approximately logarithmic or to have a logarithmic basis. The method below is for foliose or crustrose thalli.

Growth model for circular thalli

Analysis of the growth of flat circular lichen thalli, based on the functional features of lichens has been made (Aplin and Hill 1979; Childress and Keller 1980) and provides a means for comparing growth in thalli of different sizes:

$$dr/dt = \alpha sr/(r + 2s) \qquad \text{Equation 1}$$

where r is the radius of the circular thallus, t is time, α is the rate constant (dimension t^{-1}) and s is a distance constant (dimension L). For a fuller discussion of the mathematical derivation and its application of this equation see Hill (1981).

It has been tested and found suitable for determining the rate coefficient ($\alpha \times t^{-1}$) for lichen growth from measurements of the radius of circular thalli, separated by a known time period, and applies to any circular thalli such as formed by foliose and crustose species. To describe the observed rate of growth, a distance coefficient ($s \times mm$) in addition to the rate coefficient has to be determined and this relates to the width of a peripheral zone contributing to the measured rate of radial growth (Proctor 1977, Fisher and Proctor 1978, Armstrong 1979, Hill 1981). Innes (1985), after reviewing the then current methods of analysis, concluded that "In the absence of convincing evidence to the contrary, Aplin and Hill/Hill's techniques must be accepted as valid" but since then no other comparable analyses have been proposed. Proctor (1983) was able to apply this technique to *Rhizocarpon geographicum* dating on glacial moraines and, more recently, Armstrong and Smith (1996) demonstrated its applicability to growth in *Parmelia conspersa*. Note that although the Aplin/Hill α for lichens has the same dimension (t^{-1}) as the RGR for micro-organisms and higher plants, values **cannot** be correctly calculated by using the familiar formula:

$$RGR = (\ln x_2 - \ln x_1)/(t_2 - t_1) \quad \text{Equation 2}$$

where x is a measurement of radius, mass or area.

Calculation of growth constants

There are two ways to calculate the values of the Aplin/Hill growth coefficients. The two methods of calculating the values for the growth coefficients, α and s, are summarised in Fig. 3. Refer to this figure when following the methods below.

Procedure

Method a

This is the simplest (though not strictly the most accurate) method.

1. Measure the radius of a population of thalli of different sizes from the very smallest to the largest before and after a growth period.
2. Determine the increase in radius during the growth period.
3. Plot a graph of the increase in radius (Δr) against the mean of the two measurements $(r_1+r_2)/2=r$. - see example in Fig. 4.
4. Obtain the gradient of the straight line which passes through the origin and the smallest thalli ($\Delta r/r$).
5. Divide this value finally by half the time period to give the value for the rate coefficient (α, which has dimension t^{-1}) of lichen growth, see Hill (1981). A calculation is shown in Fig. 4.
6. Take the mean rate of radial growth (mm t^{-1}) of the largest thalli and divide it by the value previously obtained for the rate coefficient, α, and the time period to give the value of the distance coefficient (s which has dimension mm).

Selecting the thalli

- Choose a population of thalli with range in size from the smallest recognisable to the largest. The smallest may not be very circular and largest may not be in complete circles but this does not affect the data.
- Use close-up photography to obtain images of the thalli either as prints or slides.

↓

Taking the measurements

- Measure the size of the thalli and the scale on prints with vernier calipers, or with ruler for transparencies projected onto a wall, before and after the growth period. Measure the largest and smallest diameters of each thallus and mean them.

↓

Plotting graph

- Plot a graph of the increase in radius over a time period of growth (y-axis) against the geometric mean of the radius before and after the growth period (x-axis).

↓

Reading off the graph (or computing the basic mean values)

- Draw a line through the origin and the points representing the smallest thalli and calculate its gradient. This can be done on a computer using regression analysis of the data for the smaller thalli and fixing the line to go through the origin.
- Draw a horizontal line through the points representing the largest thalli and read off the growth increase on the y-axis. This can also be achieved by meaning the growth increase of the larger thalli.

(It is necessary to plot the graph to decide which thalli should be included with the smaller and larger thalli.)

↓

Calculating the growth coefficients

- The rate coefficient (t^{-1}), α, is calculated by multiplying the gradient by 2 and dividing by the time period.
- The distance coefficient (cm or mm), s, is calculated by dividing the mean growth rate of the larger thalli by the rate coefficient obtained from the smaller thalli.
- See Figure 4 for an example.

Alternative more accurate method

- Plot graph of r_2-r_1 (increase in growth over time period) with $\ln(r_2/r_1)$.
- Draw best fitting line through the points furthest away from the origin. This line should be a diagonal to the origin with a negative gradient.
- Find value of intercept on the $\ln(r_2/r_1)$ axis and divide it by the time period and multiply it by 2 to obtain value for α.
- Find value of intercept on other axis to find α s and divide this by obtained value for α to obtain value for s.
- See Figure 5 for an example.

Fig. 3. Flow chart that summarises the method for calculating the rate and distance coefficients for circular thalli.

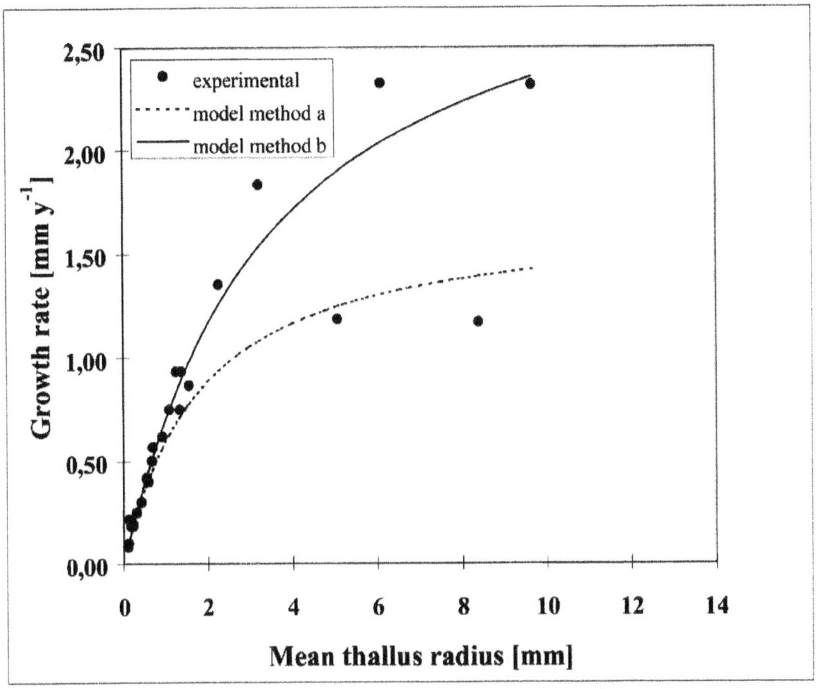

Method a

Statistic	Mean growth rate of larger thalli (over 2 mm radius) [mm y^{-1}]	Gradient between growth rate and size of smaller thalli (under 0.3 mm radius)	Alpha [y^{-1}]	s [mm]
Value	1.69	0.95	1.86	0.909
Standard deviation	0.5408	0.0349	0.065	0.293
Coefficient of variation	0.276	0.044	0.035	0.323

Fig. 4. Growth in *Diploicia canescens* thalli over one year period as a function of thallus size (measured from photographs taken on 8 August 1977 and 3 August 1978 by Dr MCF Proctor, on wall, Prince of Wales Road, Exeter, Devon,UK.) The calculations for growth coefficients (α and s) by the two methods (see text) are superimposed. The lines drawn on the graph assume the values obtained for the coefficients and indicate the differing values obtained for s due to the variation in growth rate of larger thalli.

Method b

This is theoretically a more accurate method for calculating the rate and distance coefficients.

1. Plot the growth observed (r_2-r_1) against ln (r_2/r_1) - for example see Figure 5.
2. Fit the best straight line.

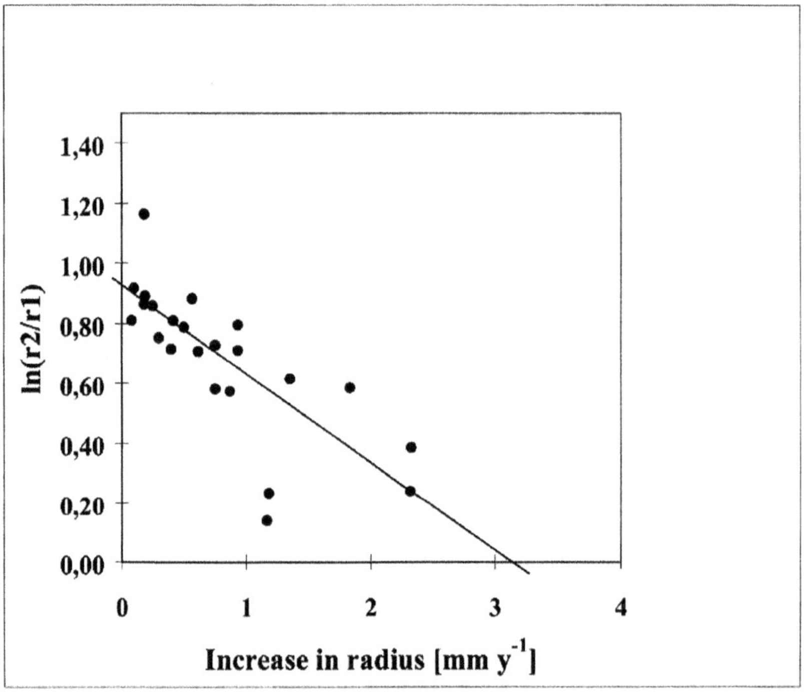

Method b

Statistic	Intercept on $\ln(r_2/r_1)$-axis	Gradient of regression line	Intercept on x-axis [mm]	Alpha [y^{-1}]	s [mm]
Value	0.907	0.281	3.25	1.81	1.79
Standard deviation	0.054	0.053	0.053	0.11	0.153
Coefficient of variation	0.060	0.189	0.016	0.061	0.086

Fig. 5. Example of semi-log plot to calculate growth coefficients by Method b using data in Figure 4.

3. Read off the intercept on the ln (r_2/r_1)-axis and multiply by 2 and divided by the time period to obtain α.
4. Read off the intercept in the (r_2-r_1)-axis and divide by the value for α and the time period to obtain s.

Troubleshooting

Method a

It is important only to use the very smallest thalli (approximately 0.1-1 mm diameter) to obtain the gradient as any larger thalli may cause an underestimate in the rate coefficient. Some of the larger thalli may have been growing much more slowly than others causing a large standard deviation in the mean. Whilst you may wish to include these in the data analysis to represent the whole population, they may be omitted if the values of coefficients from different populations need to be compared. Method b to some extent avoids this problem (see Figure 4).

Method b

The distribution of points on the graph is such that the best fitting line may pass through the points with the highest values, in both the semi-log plot and the graph of growth rate against thallus size, rather than a mid point through all the points. This may be due to the non-normal distribution of "error" on the semi-log plot and/or to some older thalli growing much more slowly than others of the same size due to some inhibition, a well-known feature that needs investigating. For an example see Proctor (1983).

Comments

Interpretation of growth constants

The distance coefficient (s) is related to the width, but is not the actual width, of the peripheral ring, or length of lobe tip, that contributes to the observed growth as mentioned above. Values for the rate coefficient (α) more closely correlate with environmental variables than do measured values for increase in radius as measured as has been shown for *Diploicia canescens* (Hill unpublished). Within a group of similar thalli, values for the rate coefficient are usually much less variable than are the direct measurements of the radial growth of thalli (see example in Figure 4). The growth rate of large thalli may be especially variable possibly because the growth of the larger thalli is affected by more internal factors (represented by both α and s) than contribute to growth in the small slower growing thalli (mainly determined by α).

Formal growth analysis of three-dimensional (e.g. fruticose) thalli has not been successfully attempted yet.

Comments

Even if growth analysis is not attempted, the choice of how to present growth measurements is important but it depends on the nature of the investigation. The idea behind calculating a single figure from many raw measurements is to eliminate the arbitrary factors such as thallus size so that the data is comparable between samples or investigations where the size of thalli is not exactly the same and to reduce the number of data presented. However, it is essential to have a sound proven rationale, or at least a theoretical basis, for the way data are presented. Without this, it is better to leave the measurements in the original form, stating the thallus size (e.g. radius, area or mass).

The linear rate of increase in thallus size (e.g. radius) of a flat circular lichen is absolute and can tell us, for example, the speed at which a lichen can cover a surface. Rate of area increase can be expressed per unit perimeter, even though this would convert the dimensions of the data equivalent to those of linear increase. Mass increase can be expressed per unit area because the derivation of mass is based on light interception on the assumption that lichens are mainly autotrophic organisms.

References

Aplin PS, Hill DJ (1979) Growth analysis of circular lichen thalli. J Theo Biol 78:347-363

Armstrong RA (1973) Seasonal growth and growth rate-colony size relationships in six species of saxicolous lichens. New Phyt 72:1023-1030

Armstrong RA (1975) Studies in the growth rates of lichens. In: Brown DH, Hawksworth DL, Bailey RH (eds.) Lichenology: Progress and Problems. Academic Press, London, pp 309-322

Armstrong RA (1979) Growth and regeneration of lichen thalli with the central portions artificially removed. Env Exp Bot 19:175-178

Armstrong RA (1995) Lobe interactions within the thallus margin and the maintenance of symmetry in the lichen *Parmelia conspersa* (Ehrh. ex Ach.)Ach. Symbiosis 18:127-142

Armstrong RA, Smith SN (1996) Factors determining the growth curve of the foliose lichen *Parmelia conspersa*. New Phyt 134:517-522

Benedict JB (1990) Experiments on lichen growth 1. Seasonal patterns and environmental controls. Arc Alp Res 22:244-454

Bull WB, Brandon MT (1998) Lichen dating of earthquake-generated regional rockfall events, Southern Alps, New Zealand. Geol Soc Am Bull 110:60-84

Childress S, Keller JB (1980) Lichen Growth. J Theo Bio 82:157-165
Dennison WC (1988) Culturing the lichens *Lobaria oregana* and *L. pulmonaria* on a nylon monofilament. Mycologia 80:811-814
Fisher PJ, Proctor MCF (1978) Observations on a season's growth in *Parmelia caperata* and *P. sulcata* in South Devon. Lichenologist 10:81-90
Follman G (1961) Lichenometrische Alterbestimmungen auf Vorchristlichen Steinsetzungen der Polumesischen Osterinsel. Naturwissenschaften 48:627-628
Gilbert OL (1971) Studies along the edge of a lichen desert. Lichenologist 5:11-17
Haworth LA, Calkin PE, Ellis JM (1986) Direct measurement of lichen growth in the Central Brooks Range, Alaska, USA, and its applications to lichenometric dating. Arc Alp Res 18:289-296
Hestmark G (1997) Gap-dynamics recruitment and individual growth in populations of *Lasallia pustulata*. Mycol Res 101:1273-1280
Hill DJ (1981) The growth of lichens with special reference to the modelling of circular thalli. Lichenologist 13:265-287
Hill DJ (1984) Studies in the growth of lichens. Lobe formation and the maintenance of circularity in crustose lichens. Lichenologist 16:273-278
Hill DJ (1985) Changes in photobiont dimensions and numbers during co-development of lichen symbionts. In: Brown DH (ed) Lichen Physiology and Cell Biology. Plenum, New York, pp 303-317
Hill DJ (1992) Lobe growth in lichen thalli. Symbiosis 12:43-55
Hill DJ, Woolhouse HW (1969) Aspects of the autecology and *Xanthoria parietina* agg. Lichenologist 3:207-214
Honegger R (1996) Experimental studies of the growth and regenerative capacity in the foliose lichen *Xanthoria parietina*. New Phyt 133:573-581
Honegger R, Conconi S, Verena K (1996) Field studies on growth and regenerative capacity in the foliose macrolichen *Xanthoria parietina* (Teloschistales, Ascomycotina). Bot Acta 109:187-193
Hooker TN, Brown DH (1977) A photographic method for accurately measuring growth of crustose and foliose saxicolous lichens. Lichenologist 9:65-75
Hyvarinen M, Crittenden PD (1998) Growth of the cushion-forming lichen, *Cladonia portentosa*, at nitrogen-polluted and unpolluted heathland sites. Env Exp Bot 40: 67-76
Innes J (1985) Lichenometry. Prog Phys Geog 9:187-254
Innes J (1986) Dating exposed rock surfaces in the arctic by lichenometry and the problem of the thallus cicularity and its effect on measurement errors. Arctic 19:253-259
Lawrey JD, Hale ME (1977) Natural history of Plummers Island, Maryland XXII. Studies on lichen growth rate at Plummers Island Maryland. Proc Biol Soc Wash 90: 698-725
Lide D R (1996) Handbook of Chemistry and Physics. CRC Press, Boca Raton
McCarthy DP, Smith DJ (1995) Growth curves for calcium-tolerant lichens in the Canadian Rocky Mountains. Arc Alp Res 27:290-297
McCune B, Derr CC, Muir PS, Shirazi A, Sillett SC, Daly WJ (1996) Lichen pendants for transplant and growth experiments. Lichenologist 28:161-169
Muir PS, Shirazi AM, Patrie J (1998) Seasonal growth dynamics in the lichen *Lobaria pulmonaria*. Bryologist 100:458-464
Polonyi S, Türk R (1991) Growth and lobe differentiation of the foliar lichen *Parmelia sulcata*. Phyton 31:81-95

Proctor MCF (1977) The growth curve of the crustose lichen *Buellia canescens* (Dicks.) de Not. New Phyt 79:659-663

Proctor MCF (1983) Sizes and growth rates of thalli of the lichen *Rhizocarpon geographicum* on the moraines on the Glacier de Galorey, Valais, Switzerland. Lichenologist 15:259-261

Renhorn K-E, Essen P-A (1995) Biomass growth in five alectoroid lichen epiphytes. Mitteilungen der Eidgenössischen Forschungsanstalt für Wald Schnee und Landschaft 70:133-140

Renhorn K-E, Essen P-A, Palmqvist K, Sunberg B (1997) Growth and vitality of epiphytic lichens. Response to microclimate along a forest edge-interior gradient. Oecologia 109:1-9

Scheidegger C, Frey B, Zoller B (1995) Transplantation of symbiotic propagules and thallus fragments: methods for the conservation of threatened epiphytic lichen populations. Mitteilungen der Eidgenössischen Forschungsansalt für Wald, Schnee und Landschaft 70:41-62

Schuster G, Ott S, Jahns HM (1985) Artificial culture of lichens in the natural environment. Lichenologist 17:247-253

Smith RIL (1995) Colonization by lichens and the development of lichen-dominated communities in the maritime Antarctic. Lichenologist 27:473-483

Vagts I, Kinder M (1999) The response of different *Cladonia* species after treatment with fertiliser or lime in heathland. Lichenologist 31:75-83

Woolhouse HW (1968) The measurement of growth rates in lichens. Lichenologist 4:32-33

Lichen Compounds

Analysis of Phenolic Products in Lichens for Identification and Taxonomy

H. THORSTEN LUMBSCH

Introduction

The identification of secondary metabolites in lichens is a necessity for the correct determination of lichens in numerous groups, and the presence of substances is often mentioned in taxonomic keys. Numerous review papers have discussed the significance of secondary metabolites in lichen taxonomy (e.g., W Culberson 1969, 1970, 1986; WL Culberson and Culberson 1970; Hawksworth 1976; Brodo 1978, 1986; Leuckert 1985; CF Culberson 1986; Egan 1986; Rogers 1989; Lumbsch 1998a, b) and the chemistry of these substances (e.g., Asahina and Shibata 1954; Shibata 1963; Huneck 1968, 1971, 1973, 1984, 1991; Elix et al. 1984). Our knowledge of the biosynthesis of secondary metabolites in lichenized fungi has been summarised by Mosbach (1969). The biological role of lichen substances was reviewed by Lawrey (1986), and additional information regarding the ability of aromatic lichen substances to protect against irradiation can be found in Rikkinen (1995). Lichens may utilise secondary metabolites as chelating agents (Purvis et al. 1987), as inhibitors of lichenicolous fungi (Lawrey 1995), and to avoid saturation of the medulla by water (Armaleo 1993). Lists recording the distribution of secondary metabolites in different taxa have been compiled by Chicita Culberson and co-workers (CF Culberson 1969, 1970; CF Culberson et al. 1977). Some methods used in the identification of secondary metabolites in lichenized ascomycetes can be found in the literature cited below, but this chapter is restricted to examples of the two most commonly applied techniques, namely thin-layer chromatography (TLC), including high performance thin-layer chromatography (HPTLC) and high performance liquid chromatography (HPLC). Several publications deal with the identification of

H. Thorsten Lumbsch, Universität Essen, Fachbereich 9/Botanik, Essen, 45117, Germany (*phone* +49-201-1832441; *fax* +49-201-1834290; *e-mail* Lumbsch@uni-essen.de)

lichen substances, the most comprehensive being the recent book by Huneck and Yoshimura (1996). The reader is referred to this and the other publications for more detailed information and additional techniques used in lichen chemistry (Santesson 1973; Leuckert 1984; White and James 1985; CF Culberson and Elix 1989). Table 1 presents a list of the classes of secondary metabolites found in lichen-forming fungi.

Table 1. Major classes of lichen substances

1. Acetate-polymalonate pathway		
	I. Fatty acids and related compounds	
	II. Phenolic compounds	
		A. Anthraquinones
		B. Benzyl esters
		C. Chromones
		D. Depsides
		E. Depsidones
		F. Depsones
		G. Dibenzofuranes, usnic acids and related compounds
		H. Diphenyl ethers
		I. Monocyclic compounds
		J. Naphthoquinones
		K. Napthopyran
		L. Tridepsides
		M. Xanthones
2. Mevalonic acid pathway		
	I. Steroids	
	II. Terpenoids	
3. Shikimic acid pathway		
	I. Pulvinic acid derivatives	
	II. Terphenylquinones	

Materials

Equipment

TLC

- Capillary tubes
- Eppendorf tubes
- Merck silica gel 60 F_{254} pre-coated glass-backed TLC plates (layer thickness 0.25 mm) 20 x 20 cm
- Pasteur pipettes
- Pencil
- Safety glasses for UV lamp
- Spray gun
- TLC tanks
- UV lamp

HPTLC

- Identical to equipment described for TLC, with the exception of different plates and tanks
  - Merck silica gel 60 F_{254} pre-coated glass-backed HPTLC plates (layer thickness 0.20 mm, pore size 6 nm) 10 x 10 cm
  - Developing chamber (Camag)
  - Alternatively, a Nanomat (Camag) may be used, but manual application is a more economic alternative.

HPLC

- Cotton swabs or Milipore filter
- Eppendorf tubes
- Pasteur pipettes
- Micropistilles
- HPLC system, preferably connected to a computer for data storage, with a UV detector, coupled with a photometer or photodiode array
- Autosampler and Autosampler vials (optional)

- Reversed phase C 18 column, 5 µm particle size, 250 x 4.6 mm, e.g. Spherisorb 5 ODS 2 column (Kontron)

Solutions

TLC and HPTLC

- Acetone
- Formic acid (96% or greater, 90% or lower will give poor results)
- Glacial acetic acid
- Silicone grease
- Solvent systems:

A	B	B'
180 ml toluene	130 ml hexane	140 ml hexane
45 ml dioxane	80 ml diethyl ether	72 ml methyl tert-butyl ether
5 ml acetic acid	20 ml formic acid	18 ml formic acid
C	E	G
170 ml toluene	75 ml cyclohexane	139 ml toluene
30 ml acetic acid	25 ml ethyl acetate	83 ml ethyl acetate
		8 ml formic acid

Culberson's (CF Culberson and Kristinsson 1970; CF Culberson 1972) standard method employs solvents A, B and C. Solvent C is very stable, reliable and provides the best overall discrimination of lichen substances. Solvent A includes hygroscopic dioxane and thus mixtures that have not been freshly prepared, provide unreliable results. Solvent B and B', which are best for separation of compounds differing in the length of side chains or the number of C-methyl substituents, should be replaced at frequent intervals (every day for solvent B, every week for solvent B'). Non-polar derivatives and compounds with high RF values in Culberson's solvent systems are best discriminated in solvent E (Elix et al. 1988). Compounds with low RF values in Culberson's solvents are more readily separated in solvent G (CF Culberson et al. 1981). This solvent is very stable, in contrast to solvent E, which has to be prepared fresh daily.

HPLC

- Acetone
- Solvent A = Double distilled water containing 1% ortho-phosphoric acid
- Solvent B = Methanol for HPLC
- Benzoic acid standard (10 mg/1000 ml acetone)
- Solorinic acid standard (10 mg/1000 ml acetone). Solorinic acid is not commercially available and has to be isolated from *Solorina crocea* (see e.g. Krog et al. 1980 or Moberg and Holmsen 1982 for illustrations) in which it occurs in large amounts. Solorinic acid is responsible for the strong orange coloration of the lower surface and is easily isolated in a fume-hood:
  - 1. Prepare dry lichen material free of substrate and mortar to powder. Extract with triple amount of acetone.
  - 2. Fill extract into a column and elute column with acetone. The elution of solorinic acid is easily seen by the strong coloration and this fraction should be collected in large Petri dishes.
  - 3. Evaporate the eluate until dry and wash carefully with diethyl ether. For this, the Petri dishes should be inclined and the dry eluate is carefully washed with diethyl ether using a pipette. Be careful not to dissolve the solorinic acid. Discard the liquid.
  - 4. Dissolve the residue with acetone and repeat steps 2-3 twice.
  - 5. Control purity of solorinic acid using HPLC; if several peaks occur, repeat steps 2-3 until the solorinic acid is pure.

Subprotocol 1
TLC

The basic methodology of standardised TLC analysis of secondary metabolites in lichen-forming fungi has been described by CF Culberson and Kristinsson (1970), CF Culberson (1972), and White and James (1985), and the reader is referred to these publications for further information. Lists of chromatographic data for lichen substances are available in book (Huneck and Yoshimura 1996) or computer format (Elix et al. 1987, 1988; Mietzsch et al. 1992, 1993).

Caution: Most solvents used are harmful, and great care should be taken throughout the procedure. Place the solvent tanks in a fume-cupboard, and make all preparations in a fume-hood.

Procedure

1. Place solvents in individual TLC tanks. The lid of the tanks should be sealed with silicone grease in order to minimise evaporation of the solvents. Record name and herbarium number of the lichen material to be examined on data sheets. Carefully separate small fragments (about 5 mg) of the lichens to be examined and place each in a separate Eppendorf tube. Add ca. 0.1 ml acetone or a sufficient amount of acetone to moisten the fragments, and leave them to soak for at least 15 min. For comparison, control substances or lichens known to contain the corresponding metabolite should be run on the same plate to facilitate identification.

2. Number points with a soft pencil 1 cm apart, 2 cm above the base and beginning 2 cm from the edge of the plate to avoid edge effects. Spot extract on TLC plate using a clean capillary tube on the corresponding numbered point. Repeat this process if necessary. To ascertain whether spots are sufficiently intense, it is advisable to check plates under short wavelength UV light.

3. When using solvent A, B or B' a filter paper should be placed at the back of the tank and saturated with the solvent to achieve uniform vapour saturation throughout the tank. Place the TLC plate in the tank with the silica-side facing the filter paper. When using solvent B, B' or C it is necessary to pre-equilibrate the plate for 5 min with 60% formic acid vapour for solvents B and B' or for 10 min with glacial acetic acid vapour for solvent C before elution. Ensure that the plate is not wetted by the liquid acids.

4. Fill tank with appropriate solvent system up to ca. 1 cm height and leave to stabilise for ca. 2 h. Then place the prepared plate into the tank to start elution. The duration of the elution depends on the solvent system and lasts for ca. 30-45 min.

5. After elution to ca. 18 cm height, remove the plate from the tank, mark the front of the solvent with a pencil, and air-dry for ca. 30 min in a fume cupboard.

6. Examine the dried plates initially in daylight for pigments that appear as coloured spots. Record their position and colour. Then examine the plates under short wavelength (λ=254 nm) UV light, where all aromatic substances can be easily observed as dark spots on a fluorescent background. Mark these spots with a pencil to distinguish them from non-aromatic substances.

7. Plates may then be sprayed with water for screening of aliphatic acids. Aliphatic acids produce opaque white spots against a dull background. Subsequently spray the air-dried plates with 10% sulphuric acid. Alternatively, the plates may be directly sprayed with 10% sulphuric acid until wet, left until they just begin to dry and then heated at 110°C in an oven for ca. 5-10 min to develop spots. Overheating should be avoided as this obscures the colours. The RF values and colours of the spots should now be recorded as some colours may alter with time. Extra purple or bluish spots may also appear - not visible before spraying - these are non-aromatic terpenes and steroids. Subsequently examine the spots under long wavelength (λ=350 nm) UV light. Strong spots represent major constituents, while compounds present in minor quantities will produce weak spots.

8. Alternatively, other spraying reagents can be used for special purposes. These include Archer's solution (a 3-methyl-2-benzothiazolone hydrazone hydrochloride solution) (Archer 1978), a stabilised PD reagent (Steiner 1955) or anisaldehyde with sulphuric acid (Leuckert et al. 1979).

9. Calculate the absolute RF value as the ratio of the height of the centre of the corresponding spot and the upper front of the solvent system. Alternatively, the standardised method (CF Culberson 1972; White and James 1985) utilises RF classes determined on each plate by a control mixture of atranorin and norstictic acid so that accurate reproducibility of RF values is not required.

Subprotocol 2
HPTLC

HPTLC is basically the same as standard TLC, but has some advantages, including a shorter elution time (ca. 15-20 min), smaller amounts of lichens and solvents needed, and increased ability to detect compounds present in traces. A detailed description of the application of this method for lichen substances is given by Arup et al. (1993).

Procedure

1. Record name and herbarium number of the lichen material to be examined on data sheets.
2. Carefully prepare small fragments (about 2 mg) of the lichens to be examined each in a separate Eppendorf tube, and add ca. 0.1 ml acetone or a sufficient amount of acetone to moisten the fragments and leave them to soak for at least 15 min. For comparison, control substances or lichens that are known to contain the corresponding metabolite, should be run on the same plate to facilitate identification.
3. Number points with a soft pencil 5 mm apart, 5 mm above the base and beginning 1 cm from the edge of the plate to avoid edge effects. Spot small amounts of the extract (ca. 1 µl) on HPTLC plate using a clean capillary tube on the corresponding numbered point. Repeat this process if necessary. To ascertain whether spots are sufficiently intense, it is advisable to check plates under short wavelength (λ= 254 nm) UV light.
4. When using solvent A, B or B' pour 10 ml of solvent into a conditioning tray of developing chamber to achieve uniform vapour saturation throughout the chamber. Fill the solvent tank with ca. 2-5 ml of an appropriate solvent system, and put the prepared HPTLC plate into the chamber with the silica-side facing downwards. Cover the developing chamber with the glass plate and leave 5 min to precondition.
5. Start the elution by pushing the rod at the side of the chamber that tilts glass strips inwards. The solvent will then rise in the capillary slit and enter the silica gel layer. The duration of the elution is dependent on the solvent system and lasts for ca. 15-20 min.
6. After elution to a height of ca. 8-9 cm, remove the plate from the chamber, indicate the front of the solvent with a pencil, and air-dry the plate for ca. 30 min in a fume cupboard. The examination of plates is identical to that described above in steps 7-8.

Subprotocol 3
HPLC

HPLC is an ideal tool for detecting trace substances, analysing small samples, quantifying phenolic lichen metabolites, and providing structural information from retention characteristics. A standardised method for gradient elution HPLC was developed by Feige et al. (1993) using a UV detector coupled with a photometer. Yoshimura et al. (1994) described the use of a photodiode array detector for HPLC analysis of lichen substances. The latter methodology affords additional capabilities not available with a simple UV detector.

Procedure

1. Carefully clean 10 mg of the lichen sample to be examined. Place it in an Eppendorf tube, crush the material using a micropistille (liquid nitrogen may assist this process), and add 1 ml of acetone if the sample is later injected by an autosampler. If the lichen specimens are too small to provide 10 mg, smaller amounts of lichen material and less acetone can be used to obtain a similar lichen weight/acetone ratio. Leave the extract to soak for 60 min.

2. The extract is then filtered through a Millipore filter into an autosampler vial. As an economic alternative, a pasteur pipette filled with small cotton swabs may be used.

3. Program the run as follows to allow comparability with retention indices available in the literature and computer databases. Analyse samples at 0.7 ml/min. Begin the run with 30% solvent B and continue isocratically for 1 min. Inject 20 µl of sample and increase the concentration of solvent B to 70% over 14 min, then up to 100% B within 30 min, and then keep it at 100% B for a further 18 min. Then decrease solvent system B to 30% and wash the column with 30% B for at least 10 min before starting a new chromatogram. The peaks may be detected at 245 nm and UV spectra (200-400 nm) of each peak should be assessed using a photometer if a UV detector is used. A photodiode array detector allows simultaneous detection of peaks at different wavelengths.

4. The chromatogram itself appears as a series of peaks (as an example see Fig. 1). Instead of using the retention time (RT) it is more advisable

to convert these into retention indices (RI) that remain stable during the life-time of a column. For example, the RI value for the system solorinic acid/benzoic acid is calculated as follows:

$$RI = \frac{R_{t\text{Peak}} - R_{t\text{Benzoic acid}}}{R_{t\text{Solorinic acid}}} x100 \qquad \text{Equation 1}$$

5. The RI values in combination with UV spectral data can be used for the identification of lichen substances using the lists or databases of chromatographic data cited above. However, it is imperative to compare compounds being identified with authentic samples or with extracts from authentic-source lichens.

Troubleshooting

TLC

- Two solvent fronts are visible.
  The solvents are aged and should be replaced by freshly prepared mixtures.
- All substances are present as faint spots.
  The extracts were too dilute, and in a re-run more lichen material or less acetone should be used.
- Substances are present as thick, smearing spots.
  The extracts were too concentrated, and in a re-run less lichen material or more acetone should be used.
- The mixture of lichen substances is very complex and the substances are difficult to separate.
  This problem is quite rare in the analysis of phenolic compounds (exceptions include the stictic acid complex) and is most commonly encountered when studying lichens containing numerous terpenes. In these cases, two-dimensional TLC should be undertaken, as described by Culberson and Johnson (1976).
- The RF values obtained differ from published results.
  The commonest reason is the use of TLC tanks of unusual dimensions. The tank commonly used for the standardised TLC method for lichen products is the Desaga Standard Separating Chamber, which can be obtained from Bodman. Desaga also makes a TLC Separating Chamber that is cheaper but has a very different geometry and should not be mistaken for the Standard Separating Chamber.

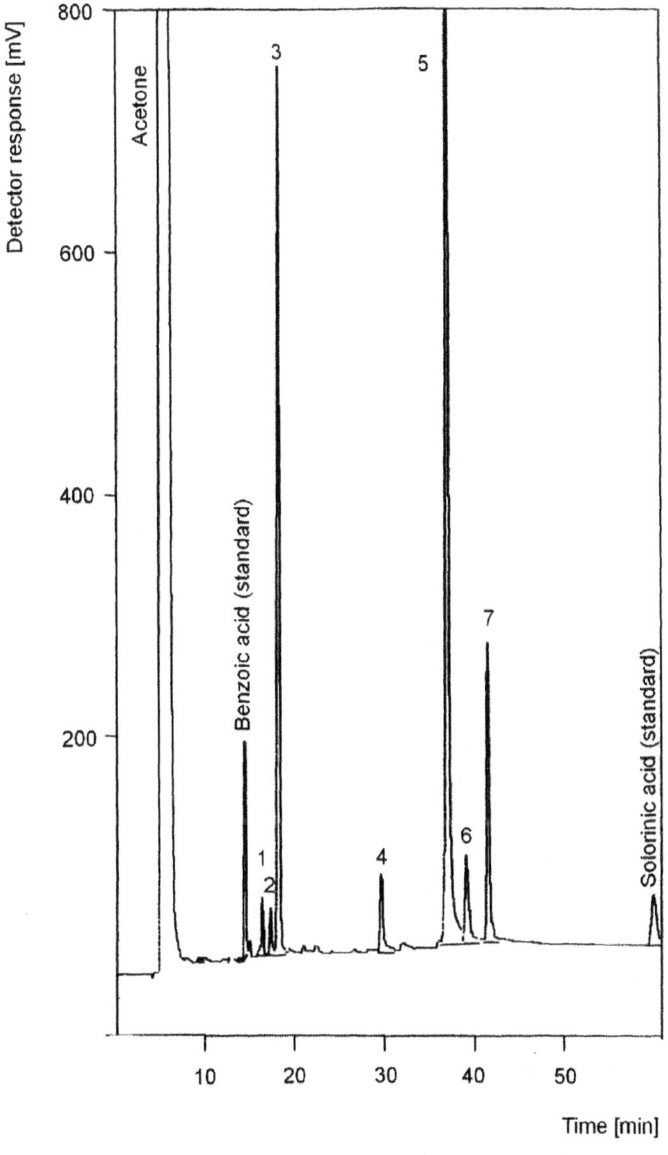

Fig. 1. Example of a HPLC chromatogram of a lichen (*Lecanora epibryon* ssp. *broccha*). The extraction solvent acetone elutes first, the standards benzoic and solorinic acids elute before, respectively after most known lichen substances. *1* = constictic acid, *2* = cryptostictic acid, *3* = stictic acid, *4* = gyrophoric acid, *5* = atranorin, *6* = chloroatranorin, *7* = 2,5,7-trichloro-3-*O*-methylnorlichexanthone.

- It may be important to distinguish atranorin and chloroatranorin using TLC, but this is almost impossible using standard conditions. A special solvent system (30 ml cyclohexane, 15 ml chloroform, 2 ml methyl ethyl ketone) was developed (Ramaut et al. 1978) that separates the two substances on TLC plates. This can also be effected by using solvent E mentioned above.

- Gyrophoric and lecanoric acid may be present, but these are difficult to separate on TLC plate under standard conditions.
  Although easily separated by HPLC, these substances can be reliably distinguished using solvent EA (200 ml diethylether, 2 ml acetic acid) (White and James 1985).

- Depsides with long side-chains are present on the TLC plate and are difficult to separate under standard conditions.
  Most depsides with long side-chains are readily distinguished by the standard HPLC method. However, these may also be distinguished using special TLC solvents, including solvent OH (75 ml ethyl acetate, 20 ml methanol, 5 ml ammonia) as described by Archer (1993). Alternatively, the chromatography of hydrolysis products can prove helpful (Culberson 1972, Esslinger 1994).

- Xanthones are present on TLC plate which are difficult to separate under standard conditions.
  If no HPLC is available, the use of solvent J (Hanko 1983, Leuckert and Knoph 1992) (80 ml dichloromethane, 20 ml acetone) is recommended and is very useful in distinguishing various chlorinated xanthones.

Acknowledgements. Dr. Chicita Culberson (Durham, N.C.) and Professor John A. Elix (Canberra) are thanked for valuable comments on the manuscript.

References

Archer AW (1978) 3-Methyl-2-benzothiazolone hydrazone hydrochloride as a spray reagent for phenolic lichen compounds. J Chromatogr 152:290-292

Archer AW (1993) Identification of orcinol para-depsides in the lichen genus *Pertusaria* by thin layer chromatography. Mycotaxon 46:1-4

Armaleo D (1993) Why do lichens make secondary products? XV Int Bot Congr, Tokyo-Yokohama, Abstr, pp 11

Arup U, Ekman S, Lindblom L, Mattsson J-E (1993) High performance thin layer chromatography (HPTLC), an improved technique for screening lichen substances. Lichenologist 25:61-71

Asahina Y, Shibata S (1954) Chemistry of Lichen Substances. Jap Soc Prom Sci Tokyo
Brodo IM (1978) Changing concepts regarding chemical diversity in lichens. Lichenologist 10:1-11
Brodo IM (1986) Interpreting chemical variation in lichens for systematic purposes. Bryologist 89:132-138
Culberson CF (1969) Chemical and Botanical Guide to Lichen Products. Chapel Hill: Univ N Carolina Press
Culberson CF (1970) Supplement to Chemical and Botanical Guide to Lichen Products. Bryologist 73:177-377
Culberson CF (1972) Improved conditions and new data for the identification of lichen products by a standardized thin-layer chromatographic method. J Chromatogr 72:113-125
Culberson CF (1986) Biogenetic relationships of the lichen substances in the framework of systematics. Bryologist 89:91-98
Culberson CF, Elix JA (1989) Lichen substances. In Harborne JB (ed): Methods in Plant Biochemistry Vol. 1. Plant Phenolics, pp 509-535. London, San Diego: Academic Press
Culberson CF, Johnson A (1976) A standardized two-dimensional thin-layer chromatographic method for lichen products. J Chromatogr 128:253-259
Culberson CF, Kristinsson H (1970) A standardized method for the identification of lichen products. J Chromatogr 46:85-93
Culberson CF, Culberson WL, Johnson A (1977) Second Supplement to Chemical and Botanical Guide to Lichen Products. St. Louis: Am Bryol Lichenol Soc, Miss Bot Gard
Culberson CF, Culberson WL, Johnson A (1981) A standardized TLC analysis of ß-orcinol depsidones. Bryologist 84:16-29
Culberson WL (1969) The use of chemistry in the systematics of the lichens. Taxon 18:152-166
Culberson WL (1970) Chemosystematics and ecology of lichen-forming fungi. Ann Rev Ecol Syst 1:153-170
Culberson WL (1986) Chemistry and sibling speciation in the lichen-forming fungi: ecological and biological considerations. Bryologist 89:123-131
Culberson WL, Culberson CF (1970) A phylogenetic view of chemical evolution in the lichens. Bryologist 73:1-31
Egan RS (1986) Correlations and non-correlations of chemical variation patterns with lichen morphology and geography. Bryologist 89:99-110
Elix JA, Whitton AA, Sargent MV (1984) Recent progress in the chemistry of lichen substances. Progr Chem Organ Nat Prod 45:103-234
Elix JA, Johnston J, Parker JL (1987) Mactabolites. Users Manual. 2nd edn. Canberra
Elix JA, Johnston J, Parker JL (1988) A computer program for the rapid identification of lichen substances. Mycotaxon 31:89-99
Esslinger TL (1994) On the chemistry and distribution of *Cladonia petrophila*. Mycotaxon 51:101-105
Feige GB, Lumbsch HT, Huneck S, Elix JA (1993) Identification of lichen substances by a standardized high-performance liquid chromatographic method. J Chromatogr 646:417-427
Hanko B (1983) Die Chemotypen der Flechtengattung *Pertusaria* in Europa. Bibl Lichenol 19:1-297
Hawksworth DL (1976) Lichen chemotaxonomy. In Brown DH, Hawksworth DL, Bailey RH (eds): Lichenology. Progress and problems, pp 139-184. London

Huneck S (1968) Lichen substances. In: Reinhold L and Liwschitz Y (ed): Progress in Phytochemistry. Vol 1, pp 223-346. Sydney: Intersci Publ, London, New York

Huneck S (1971) Chemie und Biosynthese der Flechtenstoffe. Fortschr Chem Organ Naturstoffe 29:209-306

Huneck S (1973) Nature of lichen substances. In: Ahmadjian V and Hale ME (ed): The Lichens, pp 495-522. New York and London: Academic Press

Huneck S (1984) Fortschritte der Chemie von Flechtenstoffen. Beih Nova Hedwigia 79:793-838

Huneck S (1991) New results in the chemistry of lichens. Symbiosis 11:225-248

Huneck S, Yoshimura I (1996) Identification of Lichen Substances. Berlin, Heidelberg: Springer-Verlag

Krog H, Östhagen H, Tönsberg T (1980) Lavflora. Norske busk- og bladlav. Oslo: Universitetsforlaget

Lawrey JD (1986) Biological role of lichen substances. Bryologist 89:111-122

Lawrey JD (1995) The chemical ecology of lichen mycoparasites. Can J Bot 73 (Suppl 1):S603-S608

Leuckert C (1984) Die Identifizierung von Flechtenstoffen im Rahmen chemotaxonomischer Routineanalysen. Beih Nova Hedwigia 79:839-869

Leuckert C (1985) Probleme der Flechten-Chemotaxonomie - Stoffkombinationen und ihre taxonomische Wertung. Ber Dtsch Bot Ges 98:401-408

Leuckert C, Knoph J-G (1992) European taxa of saxicolous *Lecidella* containing chloroxanthones: identification of patterns using thin layer chromatography. Lichenologist 24:383-397

Leuckert C, Dolling K, Wolters W (1979) Chemische Flechtenanalysen. Herzogia 5:181-185

Lumbsch HT (1998a) Taxonomic use of metabolic data in lichen-forming fungi. In: Frisvad JC, Bridge PD, Arora DK (eds): Chemical Fungal Taxonomy, pp 345-387. New York: Marcel Dekker

Lumbsch HT (1998b) The use of metabolic data in lichenology at the species and subspecific levels. Lichenologist 30:357-367

Mietzsch E, Lumbsch HT, Elix JA (1992) Wintabolites Users Manual. Essen

Mietzsch E, Lumbsch HT, Elix JA (1993) Notice: a new computer program for the identification of lichen substances. Mycotaxon 47:475-479

Moberg R, Holmsen I (1982) Lavar. En fälthandbok. Stockholm: Inetrpublishing

Mosbach K (1969) Zur Biosynthese von Flechtenstoffen, Produkten einer symbiotischen Lebensgemeinschaft. Angew Chem 81:233-244

Purvis OW, Elix JA, Broomhead JA, Jones GC (1987) The occurrence of copper-norstictic acid in lichens from cupriferous substrata. Lichenologist 19:193-203

Ramaut JL, Brouers MB, Serusiaux E, Corvisier M (1978) Separation of mixtures of atranorin and chloratranorin by thin-layer chromatography. J Chromatogr 155:450-453

Rikkinen J (1995) What's behind the pretty colours? A study on the photobiology of lichens. Bryobrothera 4:1-239

Rogers RW (1989) Chemical variation and the species concept in lichenized ascomycetes. Bot J Linn Soc 101:229-239

Santesson J (1973) Identification and isolation of lichen substances. In Ahmadjian V and Hale ME (eds): The Lichens, pp 633-652. New York and London: Academic Press

Shibata S (1963) Lichen substances. In: Modern Methods of Plant Analysis. Vol VI, pp 155-193. Heidelberg: Springer-Verlag, Berlin

Steiner M (1955) Ein stabiles Diaminreagens für lichenologische Zwecke. Ber Dtsch Bot Ges 63:35-40

White FJ, James PW (1985) A new guide to microchemical techniques for the identification of lichen substances. Brit Lich Soc Bull 57:1-41

Yoshimura I. Kinoshita Y, Yamamoto Y, Huneck S, Yamada Y (1994) Analysis of secondary metabolites from lichen by high performance liquid chromatography with a photodiode array detector. Phytochem Anal 5:195-205

Suppliers

Desaga Standard Separating Chamber: Bodman (Cat. No.: 120160; Fax.: 610-459-8036)

Abbreviations

HPLC	high performance liquid chromatography
HPTLC	high performance thin-layer chromatography
TLC	thin-layer chromatography

Investigating the Production of Secondary Compounds in Cultured Lichen Mycobionts

ELFIE STOCKER-WÖRGÖTTER

Introduction

Lichens produce an amazing diversity of compounds, which can be divided into two groups: intracellularly produced primary metabolites, and secondary metabolites that are often deposited extracellularly. Most of the primary products are known from plants. However, a significant percentage of lichen secondary substances (about 630) have been found to be restricted to lichens; 50 - 60 of these secondary products occur in free-living fungi and higher plants (Elix 1996). Culberson and Elix (1989) give an overview of the biosynthetic pathways that lead to the production of typical lichen substances, such as secondary aliphatic acids, polyketide derived aromatic compounds (e.g. depsides, depsidones, dibenzofurans, usnic acids), and derivatives of the mevalonic and shikimic acids.

Lichen secondary products play an important role in routine identification by determining the results of thallus colour tests and standardized thin layer chromatography, TLC (White and James 1985). Information derived from these tests can also be very useful in lichen systematics. Knowledge about their chemical structures and occurrence in various lichen species is mainly derived from extensive studies of thalli collected from the field (e.g. Culberson et al. 1977; Culberson and Elix 1989). TLC and high performance liquid chromatography (HPLC) analyses of lichen secondary products have become an integral tool in modern lichen taxonomy. Many lichen substances have been examined for their biological activity in medicine and agriculture (e.g. Elix 1996, Lawrey 1984, Hawksworth and Hill 1984, Neamati et al. 1997, Sankawa et al. 1982; Umezawa et al. 1983, Yamamoto 1998). Lichens have been used industrially, e.g. for the produc-

Elfie Stocker-Wörgötter, University of Salzburg, Institute of Plant Physiology, Hellbrunnerstr. 34, Salzburg, 5020, Austria (*phone* +43-662-8044-5528 or 5594; *fax* +43-662-645450; *e-mail* elfriede.stocker@sbg.ac.at)

tion of perfumes, litmus and dyeing of wool. Efforts to get satisfactorily high amounts required for applications in medicine, e.g. to grow lichen fungi or lichens in fermenters, green houses and outdoor plantations, with a few exceptions have not been successful. Controlled growth of isolated mycobionts and recombined thallus structures that provide higher amounts of fungal materials open new perspectives of research, e.g. for pharmaceutical use of lichen products and molecular biological studies using axenic materials. Recently, many mycobionts have been isolated and cultured by various investigators (e.g. Crittenden and Porter 1991; Crittenden et al. 1995; Culberson and Armaleo 1992; Culberson et al. 1992; Hamada 1983, 1988, 1989; Hamada et al. 1996; Kinoshita 1993; Kon et al. 1997; Stocker-Wörgötter 1998; Yamamoto et al. 1987; Yoshimura et al. 1994). Most of these studies show that aposymbiontically-grown mycobionts produce secondary products, but often not the same spectrum of lichen substances known from the voucher specimens. Recent results have suggested that the production of lichen compounds depends on the composition of the media, lichenization and differentiation, and various ecological factors, e.g. drought, temperature, and light intensities. However, it seems that no set of conditions exists that will trigger the production of the complete range of secondary compounds that occur in specimens collected in nature. The aim of this chapter is to outline suitable isolation procedures for studying the production of secondary compounds, and how to extract and analyse the compounds. Finally, the manipulation of growth conditions to induce the formation of secondary compounds is demonstrated using examples from the genera *Cladonia*, *Thamnolia* and *Umbilicaria*.

Materials

Lichen Material

As indicated in the introduction, case histories of how to manipulate growth conditions to induce the formation of secondary metabolites will be illustrated using examples from the genera *Cladonia*, *Thamnolia* and *Umbilicaria*. *C. imperialis* Ahti and Marcelli is a tropical species from Brazil, while *C. arbuscula* (Wallr.) Flotow is alpine. *T. vermicularis* (Swartz) Schaerer var. *subuliformis* (Ehrh.) Schaerer is a common alpine European species, while *T. vermicularis* var. *vermicularis* (Sw.) Ach. ex Schaerer was collected in Chile, southern Latin America. *U. virginis* Schaerer is an alpine species collected in Austria, while *U. mammulata* (Ach.) Tuck. is from the Appalachians, eastern North America.

Equipment

- Vacuum desiccator
- Mini-vaporator (Supelco/Sigma)
- Laminar flow bench
- HLPC system including two pumps e.g. Hitachi/Merck
- UV detector or photodiode array detector (190-800 nm)
- Reversed phase C 8 column (4.6 x 250 mm, 5 μm particle size), e.g. Beckman

Reagents

- Cylinder of compressed nitrogen
- HPLC solvent A: 70% H_2O / 30% methanol /1% phosphoric acid
- HPLC solvent B: methanol
- P_2O_5

Culture media

Sabouraud 4% glucose agar

- 10 g polypeptone (casamino acids)
- 40 g glucose
- 18 g agar
- Make up to 1000 ml with double distilled water. Good for Cladoniaceae.

Malt Yeast extract medium

- 20 g malt extract
- 2 g yeast extract
- 20 g agar (Merck)
- Make up to 1000 ml with double distilled water. Good for Cladoniaceae.

Potato Dextrose Agar (PDA)

- Dissolve 39 g PDA in double distilled water, and make up to 1000 ml. Use for isolations in tubes. I recommend PDA from Sigma (P2182). Good for *Umbilicaria*.

Potato Dextrose Agar (modified)

- 39 g PDA
- 20 g mannitol
- 20 g sorbitol
- Dissolve the above in double distilled water, and make up to 1000 ml. Use for cultures in petridishes. Good for *Umbilicaria*.

Sabouraud 2% Sucrose Agar

- 10 g peptone (from meat)
- 10 g peptone (from caseine)
- 20 g sucrose
- 18 g agar (Merck)
- Dissolve the above in double distilled water, and make up to 1000 ml. Use for cultures in petridishes. Good for *Thamnolia*.

Murashige Skoog (MS) Medium (modified)

- 2 g malt extract
- 2 g caseine hydrolysate
- 20 g mannitol
- 40 g sucrose
- 18 g agar
- 1 mg Murashige mineral salts
- Dissolve the above in double distilled water, and make up to 1000 ml. Good for *Thamnolia*.

Procedure

Lichen material

For full details of how to isolate lichen mycobionts, see Chapter 1. For lichen species that are known to produce fruiting bodies only very rarely, I recommend the following modified Yamamoto method (Yamamoto et al. 1998, see also Chapter 2).

1. Excise pieces of thallus approx. 1 cm in diameter.
2. Wash in tap water and then in sterile double distilled water containing a drop of the detergent Tween 80.

3. Carefully clean segments under a dissecting microscope and gently fragment them using a knife homogenizer with 3 ml of sterile water at low speed (15000 rpm for 20 s).
4. Filter the resulting suspension through sieves of different mesh (500 and 150 µm).
5. Pick up small fragments with an average size of 150 µm using inoculation needles or bamboo sticks (Yamamoto 1990).
6. Inoculate agar slants in test tubes with a nutrient medium appropriate for your particular lichen.
7. Grow in the dark to prevent algal growth for 4 to 6 months.
8. Transfer hyphal colonies that grow free of contamination to fresh nutrient media in petri dishes (110 x 15 mm).

Promoting formation of secondary compounds

9. Fragment hyphal colonies in sterile water by using a homogenizer at low speed.
10. Spread aliquots of approx. 5 ml over the solid media in petridishes (110 x 15 mm). This method favours the development of aerial hyphae. Note that although in the intact lichen thallus most of the lichen substances are produced by medullary hyphae, the formation of aerial hyphae appears to be a precondition for the production of secondary compounds (Culberson and Armaleo 1992).
11. Growth conditions that can be varied to promote the formation of secondary metabolites include the following:
    - growth medium
    - light / dark regimes, light intensities (including an occasional burst of high light)
    - application of a cold shock
    - partially desiccating the cultures

    See Results for more details. The best medium for growth can be selected after testing a large variety of recipes. Light-dark regimes can be changed by using different growth chambers and adjusting them partially to conditions that exist in the natural environments of the investigated lichens. Cold shocks can be applied by using refrigerators at different settings of temperature. Desiccation cycles can be simulated by drying cultures in a clean workbench (sterile conditions) over a determined period of time, by using desiccators or long-term cultivation (natural loss of water by evaporation).

12. Cut circular plugs (approx. 1.5 cm in diameter) out from each of the agar plates overgrown by the mycobionts.

13. Dry the samples overnight in a sterile hood (12-15 h), then store them in a vacuum desiccator over P_2O_5.

14. Extract the dried agar discs with attached mycelia 5 times with approx. 3-5 ml acetone (40°C), then evaporate using a mini-vaporator (Supelco/Sigma) under a stream of nitrogen at reduced pressure (modified after Culberson et al. 1992).

15. Just before analysis re-dissolve the extracts in acetone and centrifuge briefly at approx. 7700 g.

HPLC analysis

16. Run a 40 min gradient from 80% to 15% (solvent A) followed by 20 min at 15% solvent A. Detect peaks at 270 nm.

17. Wash the column for 5 min with 5 % solvent A and re-equilibrate for 10 min at 80% solvent A (Narui et al. 1996).

18. Identify peaks using comparisons of voucher specimens and extracts from samples serving as sources of specific compounds. See also Chapter 17.

TLC confirmation

19. Peak identity can also be confirmed by TLC using the three solvent systems described by Culberson and Ammann (1979). See also Chapter 17.

Results

Very little information exists on the production of secondary metabolites in cultured mycobionts. Originally, pure cultures of mycobionts were not believed to form these substances. However, recent research has suggested that it is likely that the production of the whole chemical spectrum of compounds present in intact thalli can be induced in cultured mycobionts. Zorer et al. (1997) reported production of secondary metabolites in natural diaspores (soredia) and soredia-like developmental stages in resynthesis cultures. It seems that culture conditions that promote some form of differentiation, e.g. the formation of resynthesis stages (hyphal network to house the algal partner) or podetia (highly differentiated fungal structures for reproduction) will stimulate secondary metabolite production. The following results are based upon my own observations of *Cladonia*, *Thamnolia* and *Umbilicaria*. These notes may help to decide which environmen-

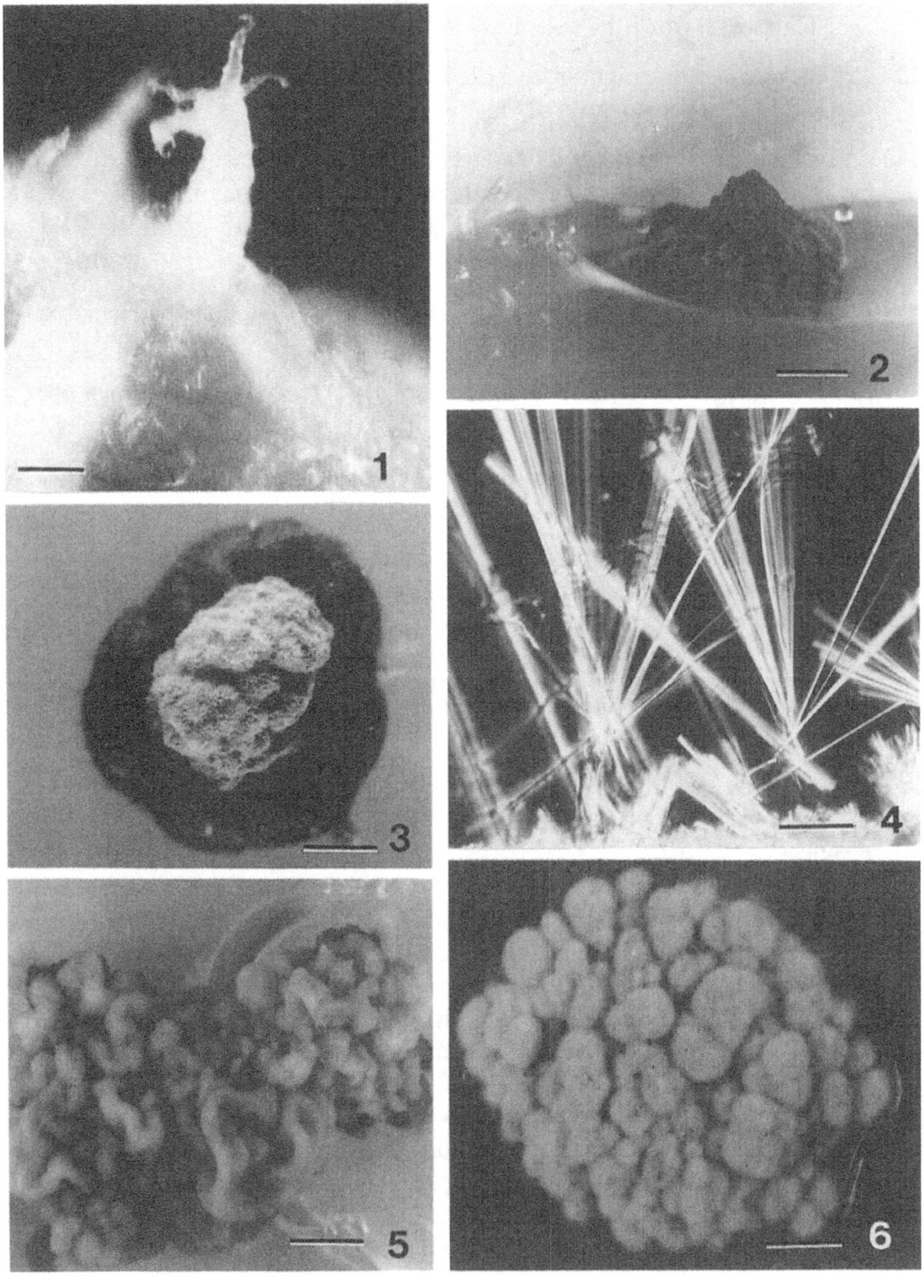
1
2
3
4
5
6

tal conditions should be varied to induce secondary metabolite production. Figure 1 illustrates cultured mycobionts or their products, while Figure 2 shows HPLC chromatograms of secondary metabolites.

Cladonia

On Saboraud 4% glucose agar, cultures of *C. imperialis* formed podetia (Fig. 1.1), while on liquid Lilly and Barnett or MY agar no differentiation occurred (Fig. 1.2). Podetia formation was correlated with the production of fumarprotocetraric, confumarprotocetraric and protocetraric acids, the secondary metabolites present in intact thalli (Fig. 2.A). *C. arbuscula* normally contains fumarprotocetraric, confumar, protocentraric, virensic and usnic acids. Fig. 1.3 illustrates the appearance of the isolated mycobionts grown for 6 months under a "natural" light-dark regime (14:10 hours) and at the same light intensities (60-100 μmol photons $m^{-2}s^{-1}$) that I usually use for culturing isolated photobionts. These conditions induced usnic acid synthesis, but synthesis, and in particular extracellular crystal deposition, was further stimulated by allowing the cultures to partially desiccate. Fig. 1.4 illustrates usnic acid crystals extracted and recrystallised from the cultured mycobiont of *C. arbuscula*.

Thamnolia

Field grown *T. vermicularis* var. *subuliformis* contains baeomycesic and squamatic acids, while var. *vermicularis* contains thamnolic acids. The formation of the appropriate secondary metabolites for each variety was induced in the cultured mycobionts by allowing them to partially desiccate, and growing them at high light intensities (10000-15000 lx) for 24 hours. Exposing the cultures to high illumination was repeated for 3-5 days every 4 weeks. Fig. 1.5 illustrates the appearance of 7-month-old cultures. A typical spectrum of lichen products was formed in desiccated cultures after more than one year (Fig. 2C). In this case the cultures had dried out by slow evaporation.

Umbilicaria

Field grown *Umbilicaria* species are known to contain gyrophoric, lecanoric and umbilicaric acids (Narui et. al. 1996), a result confirmed by ana-

◄

Fig. 1. **1** Cultured mycobiont of *Cladonia imperialis* on Sabouraud 4% glucose agar (3 months in culture). Bar = 1 mm; **2** Cultured mycobiont of *Cladonia imperialis* on MY medium. Bar = 1 mm; **3** Cultured mycobiont of *Cladonia arbuscula* (6 months in culture). Bar = 1 mm; **4** Usnic acid crystals extracted and re-crystallized from *Cladonia arbuscula* mycobiont. Bar = 250 μm; **5** Cultured mycobiont from *Thamnolia vermicularis v. subuliformis* (7 months old). Bar = 1 mm; **6** Cultured mycobiont from *Umbilicaria virginis* (1 year in culture). Bar = 1.3 mm

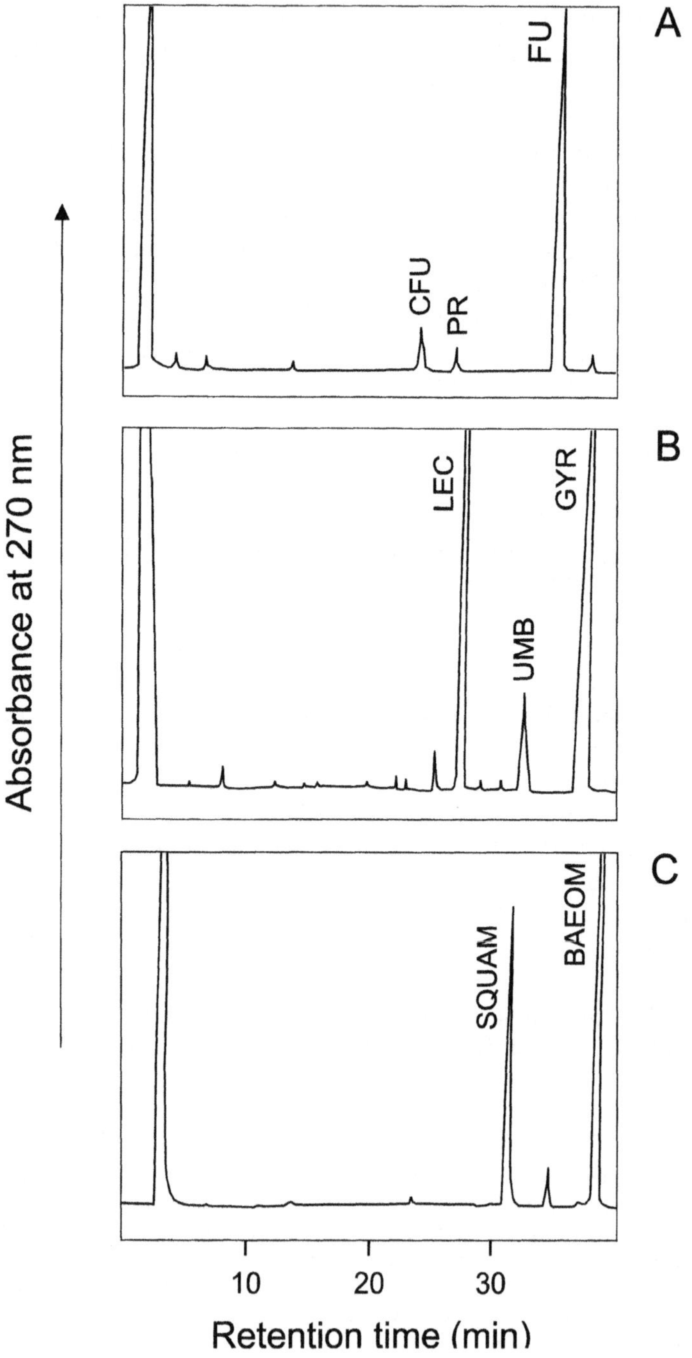
A
FU
CFU
PR
B
LEC
GYR
UMB
C
SQUAM
BAEOM
Absorbance at 270 nm
10
20
30
Retention time (min)

lysis of the intact thallus in the present study. Isolated *Umbilicaria* mycobionts showed the best growth rates on PDA. The production of secondary metabolites only occurred in very dry cultures, when the cultures had stopped growing, after an incubation period of 6 months. Desiccation was achieved by slow evaporation of the medium. Under these conditions, *U. mammulata* produced gyrophoric and umbilicaric acids. Surprisingly, in *U. virginis* these acids were only formed in cultures that, in addition to being desiccated, were regularly refrigerated for 1-2 weeks at -23°C after determined periods, e.g. every 6 weeks. Fig. 1.5 illustrates the appearance of these cultures, while Fig. 2B illustrates the compounds present.

Acknowledgements. I thank the "Fonds zur Förderung der wissenschaftlichen Forschung", Vienna (Austria) for supporting this investigation by the grant P 12789-BIO. I greatly appreciate the help of Chicita Culberson and Anita M. Johnson (Duke University, USA) in identifying lichen products during several stays at the Department of Botany.

References

Crittenden PD, Porter N (1991) Lichen forming fungi: potential sources of novel metabolites. Tr Biotech 9: 409-414

Crittenden PD, David JC, Hawksworth DL, Campbell FS (1995) Attempted isolation and success in the culturing of a broad spectrum of lichen-forming and lichenicolous fungi. New Phyt 130: 267-297

Culberson CF, Armaleo D (1992) Induction of a complete secondary-product pathway in a cultured lichen fungus. Exp Mycol 16: 52-63

Culberson CF, Ammann K (1979) Standartmethode zur Dünnschichtchromatographie von Flechtensubstanzen. Herzogia 5: 1-24

Culberson CF, Elix JA (1989) Lichen Substances. In: Harborne JB (ed) Methods in Plant Biochemistry. Vol. I. Plant Phenolics, Academic Press, London, San Diego, pp 509 - 535

Culberson CF, Culberson WL, Johnson A (1977) Second Supplement to Chemical and Botanical Guide to Lichen Products: The American Bryological and Lichenological Society, Missouri Botanical Garden, St. Louis

Culberson CF, Culberson WL, Johnson A (1992) Characteristic lichen products in cultures of chemotypes of the *Ramalina siliquosa* complex. Mycologia 84: 705-714

Elix JA (1996) Biochemistry and secondary metabolites. In: Nash Th (ed) Lichen Biology, Cambridge University Press, Cambridge, pp 154-180

Fig. 2. **A** HPLC chromatogram of the *Cladonia imperialis* mycobiont grown on Sabouraud 4% glucose agar. FU, fumarprotocetraric acid; PR, protocetraic acid; CFU, confumarprotocetraric acid; **B** HPLC chromatogram of the *Umbilicaria-virginis*-mycobiont grown on PDA. LEC, lecanoric acid; UMB, umbilicaric acid; GYR, gyrophoric acid; **C** HPLC chromatogram of *Thamnolia-vermicularis-v.-subuliformis*-mycobiont grown on MS medium. SQUAM, squamatic acid; BAEOM, baeomycesic acid.

Hamada N (1983) The effect of temperature on lichen substances in *Ramalina subbreviscula.* Bot Mag (Tokyo) 96: 121-126

Hamada N (1988) Depside from isolated lichen mycobiont. II. Lichenologist 20: 294-295

Hamada N (1989) The effect of various culture conditions on depside production by an isolated mycobiont. Bryologist 92: 310-313

Hamada N, Miyagawa H, Miyawaki H, Inoue M (1996) Lichen substances in mycobionts of crustose lichens cultured on media with extra sucrose. Bryologist 99: 71-74

Hawksworth DH, Hill DJ (1984) The Lichen-forming Fungi, Blackie, Glasgow

Kinoshita Y (1993) The production of lichen substances for pharmaceutical use by lichen tissue culture. Nippon Paint Public, Osaka

Kon Y, Iwashina T, Kashiwadani H, Wardlaw JD, Elix JA (1997) A new dibenzofuran, isostrepsilic acid, produced by cultured mycobiont of the lichenized ascomycete *Usnea orientalis.* J Jap Botany 72: 67-71

Lawrey JD (1984) Biology of lichenized fungi. Praeger, New York

Neamati N, Hong HX, Mazunder A, Wang SM, Sunder S, Nicklaus MC, Milne GWA, Proksa B, Pommier Y (1997) Depsides and depsidones as inhibitors of HIV-integrase. Discovery of novel inhibitors through 3D-database searching. J Med Chem 40: 942-951

Narui T, Culberson CF, Culberson WL, Johnson A, Shibata, S (1996) A contribution to the chemistry of the lichen family Umbilicaraceae (Ascomycotina). Bryologist 99: 199-211

Sankawa U, Shibuya M, Ebizuka Y, Noguchi H, Kinoshita T, Iitaka Y (1982) Depside as potent inhibitor of prostglandin biosynthesis: A new active site model for fatty acid cyclooxidase. Prostglandins 24: 21-34

Stocker-Wörgötter E (1998) Culture methods and culture of selected tropical mycobionts and photobionts as exemplified by South American lichens. In: Marcelli MP, Seaward MRD (eds) Lichenology in South America: history, current knowledge and applications CETESB, Sao Paulo, pp 145-154

Umezawa K, Muramatsu S, Ishizuka M, Sawa T, Takeuchi T, Matsushima T (1983) Inhibition of histidine decarboxylase and tumour promotor-induced arachidonic acid release by lecanoric acid analogues. Biochem Biophys Res Comm 110: 739-739

White FJ, James P (1985) A new guide to microchemical techniques for the identification of lichen substances. British Lich Soc Bulletin 57 (suppl): 1-42

Yamamoto Y (1990) Studies of cell aggregates and the production of natural pigments in plant cell culture. Nippon Paint Public, Osaka

Yamamoto Y (1998) Screening of biological activities and isolation of biological active compounds from lichens. Rec Res Devel Phytochem 2: 23-34

Yamamoto Y, Mizuguchi R, Takayama S, Yamada Y (1987) Effects of culture conditions on the growth of Usneaceae lichen tissue cultures. Plant Cell Physiol 28: 1421-1426

Yoshimura I, Kinoshita Y, Yamamoto Y, Huneck S, Yamada Y (1994) Analysis of secondary metabolites from lichen by high performance liquid chromatography. Phytochemical Analysis 5: 197-205

Zorer R, Türk R, Stocker-Wörgötter E (1997) Resynthesis of the lichen *Cladonia fimbriata* from the axenic cultures of the isolated symbionts. In: Kappen L (ed) New Species and Novel Aspects in Ecology and Physiology of Lichens. In honour of O.L. Lange. Bibl Lichenol 67: 85-89

Analysing Lichen Enzymes by Isoelectricfocussing

DIANNE FAHSELT

Introduction

Lichen proteins that most readily lend themselves to analysis by electrophoretic separation are enzymes, particularly those that produce substances capable of reacting with specific staining reagents to form visible products. Presumably due to the nature of cell walls and intercellular matrices, extraction of enzymes from lichens is more difficult than from many other organisms and, thus, some rigorous means of disrupting cells is necessary. Another problem, peculiar to lichens, is the extracellular deposits of phenolics that can denature enzymes unless removed prior to extraction. Of the enzymes and different molecular forms of each enzyme that can be observed, most appear to be those of the mycobiont (Fahselt 1995).

Isoelectricfocussing (IEF) on ultrathin-layer gels is a variant of electrophoresis that seems to provide the best resolution of lichen enzyme bands. Separation can be accomplished using starch gels, but acrylamide-based gels generally produce superior results. Excellent customised gels can easily be prepared at relatively little cost (Winter et al. 1977). Enzymes extracted using the method described here are usually best separated on a pH gradient of approximately 4 - 6, but in some situations another pH range might be preferable. To determine the most appropriate pH range for the enzymes of particular lichens, trial separations may first be made using wide-range pH gels, e.g., 3 - 9.5. For some lichen species, it might also be useful to alter the pH of the extraction buffer or the protein precipitation procedure suggested below.

As a precaution against contamination, lichen material should be cleaned under a microscope to ensure that mosses, mites and other lichens have been removed before extraction of proteins. Thalli will be less prone

Dianne Fahselt, University of Western Ontario, Department of Plant Sciences, London, Ontario, N6A 5B7, Canada (*phone* +01-519-679-2111 ext 86480; *fax* +01-519-661-3935; *e-mail* dfahselt@julian.uwo.ca)

to fragment during the cleaning process if first moistened by spraying lightly with distilled water. As lichenicolous fungi may affect some aspects of lichen chemistry, specimens should also be checked to determine that none are present. Because extractable enzymes tend to be localised in younger parts of the thallus, defined thallus regions should be analysed, and the season of collection should also be standardised for samples being compared. Due to the amount of material required for each sample, the protocol presented here is best suited for macrolichens. Enzyme banding patterns produced using these methods provide a means for characterisation of lichens at the species level or below. They also provide a convenient basis for assessing variability or evolutionary potential in populations. Furthermore, they may provide insight into aspects of lichens biology, such as reproduction, and metabolic processes of the thallus. This chapter describes how to carry out isoelectricfocussing, band visualisation, photography of gels, and band scoring.

Subprotocol 1
Isoelectricfocussing of Proteins

Outline

Figure 1 outlines the protocol.

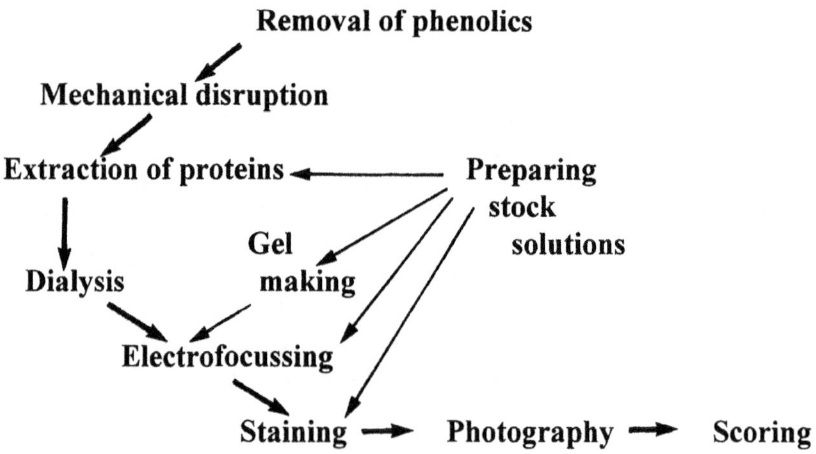

Fig. 1. Summary of operations involved in analysis of lichen proteins using isoelectric-focussing.

Materials

Facilities and equipment

- Fume hood
- Rotary shaker
- Vacuum line or vacuum pump
- Ice machine or other source of crushed ice
- Refrigerated centrifuge, 13 000 g
- pH meter
- Cold room or commercial meat-display case at approximately 4°C
- Magnetic stirrers, 12 mm magnetic stir bars
- Plastic paper clips, various colours
- Gel moulding equipment, glass plates with clamps and neoprene gasket
- Polyacrylamide gel support medium with one side hydrophilic (e.g. Sigma Aldrich)
- Apparatus for horizontal gel electrophoresis with temperature control system (e.g., VWR Scientific)
- High-voltage power supply (e.g. Fisher Scientific), at least 2000 V
- Refrigerating oil bath and circulator (e.g. Fisher Scientific)
- Electrode strips (Amersham Pharmacia)
- Polycarbonate sample application strips
- Electronic balance weighing up to 0.0001 g

Glassware and supplies

- Dewar flask-type vessel approved for liquid nitrogen
- Mortars and pestles
- Ice buckets (2)
- Screw top glass vials
- Wheaton ground glass tissue grinder with teflon pestle (Fisher Scientific)
- Variable speed power drill mounted on stand

- Erlenmeyer flask with side arm
- Dialysis tubing (12 000 - 14 000) MW retention (Fisher Scientific)
- Pasteur pipettes, pro-pipettes
- Centrifuge tubes, 50 ml and 15 ml

Chemicals

All chemicals must be analytical grade.

- Acetone
- Liquid nitrogen
- Ampholites, e.g., Ampholine pH 3.5 - 5, pH 4 - 6, pH 6 - 8 (Sigma Aldrich)
- Protein assay kit (Biorad Laboratories)
- Paraffin oil or immersion oil (VWR Scientific)
- High temperature bath oil (e.g., Fisher Scientific)
- Silicone lubricant (Fisher Scientific)
- Adhesion promoter Silane A174 (Amersham Pharmacia)
- Repel Silane (Amersham Pharmacia)

Stock solutions

Make up these solutions in dark bottles and store them at 4°C.

Acrylamide / BIS stock solution

- 29.1 g acrylamide
- 0.9 g BIS (N, N'-methylene bisacrylamide)
- 75 ml of double distilled (dd) H_2O

Stir the solution until clear, and then make up to 100 ml with dd water and filter. Use within 2 weeks.

Note: Acrylamide and bisacrylamide are dangerous if inhaled or contacted by skin. Wear a dust mask, goggles, lab coat and gloves when handling.

0.05 M Tris buffer, pH 8.5

- 3.1 ml of 0.1 M HCl (8.08 ml concentrated HCl made up to 1000 ml with dd H_2O)
- 5 ml 0.2 M THAM (12.12 g tris hydroxymethyl aminomethane made up to 500 ml with dd H_2O)
- 91.9 ml dd H_2O

Adjust the volume of HCl to give the required pH.

Phosphate extraction buffer

- Stock A (dibasic sodium phosphate)
  - 1.89 g anhydrous disodium phosphate
  - 200 ml dd H_2O
- Stock B (monobasic potassium phosphate)
  - 2.72 g anhydrous monopotassium phosphate
  - 300 ml dd H_2O
- To make an extraction buffer (0.06 M phosphate buffer pH 6.6), mix
  - 149 ml Stock A
  - 251 ml Stock B
  - 0.028 ml mercaptoethanol

  Note: Add in fume hood!

  Check pH; it may be necessary to adjust by adding small amounts of Stock A.

Electrode solutions for pH 4 - 6 gels

- Anode solution (0.1 M glutamic acid in 0.5 M phosphoric acid)
  - 0.44 g glutamic acid
  - 1 ml concentrated phosphoric acid
  - 28.5 ml dd H_2O
- Cathode solution (0.1 M ß-alanine)
  - 0.27 g ß-alanine
  - 30 ml dd H_2O

Ammonium persulphate

Make up a **fresh** 10% solution.

- 0.1 g of ammonium persulphate
- 1.0 ml of dd H_2O

Gels

Gelling solution for one gel, pH 4 - 6

- 10 ml acrylamide / BIS stock solution
- 3.5 ml 87% glycerol (8.7 ml glycerol plus 1.3 ml dd H_2O)
- 0.2 ml Ampholine 3.5 - 5
- 0.9 ml Ampholine 4 - 6
- 0.4 ml Ampholine 6 - 8
- 15.8 ml dd H_2O

Mix the solution well and slowly degas in a 125 ml side-arm Erlenmeyer flask. Degassing the solution for a single gel takes 10 min on a vacuum line, but twice the time is needed for a double batch.

Preparation of gels

1. Place a plastic gel supporting sheet, **hydrophilic** side up, on a glass moulding plate at least 3 mm thick and sized to produce gels that fit the electrophoresis apparatus. Be sure to use the correct side of the gel-support sheet; the hydrophilic side permits a fine spray of water to run off easily as opposed to beading up.
2. Around the edge of a moulding plate place a rubber or neoprene gasket 0.07 mm thick coated with a thin film of silicone lubricant. If gel is being moulded directly on glass without a hydrophilic gel support, the glass must be extremely clean and coated with a product such as Silane A174 to promote adhesion.
3. Place a second 3 mm glass plate, with the inside coated with Repel Silane, on top of the gasket. Sandwich the gasket between plates by placing strong clamps on all sides (Fig. 2), leaving a temporary opening in one corner so the gelling solution can be poured into the mould.
4. **Immediately** pipette 0.15 ml of **fresh** 10% ammonium persulphate into the flask containing gelling solution. Swirl the flask gently to mix and avoid introducing bubbles.
5. **Immediately** pour the gelling solution through a syringe into the gel mould. To prevent bubble formation, hold the gel mould at an angle so that filling proceeds from one corner.
6. Close the gasket, place a clamp over the opening and leave the mould in a vertical position for 1 h to polymerise. The clamps can serve as legs (Fig. 2).

7. Unclamp the mould and lay it horizontally, plastic backing down, for 30 min at 4°C to facilitate gel release.

8. Remove the gel from the completely dismantled mould and rinse with dd H_2O. Use immediately or wrap well in two layers of good quality self-sealing plastic food wrap and store at 4°C. Such gels can be stored for at least two or three months.

Lichen material

Ideally 1 - 2 g of freshly-collected clean, dry lichen is needed for each sample, and a minimum amount is approximately 0.5 g. With no complications, this amount should be sufficient to test for 8 enzymes. Select the terminal 6 - 8 mm of fruticose thalli or peripheral 6 - 8 mm of foliose thalli.

Procedure

Removal of extracellular phenolics

1. Before grinding, place each sample in a separate beaker, cover with acetone, and agitate at low speed on a shaker for 20 min.

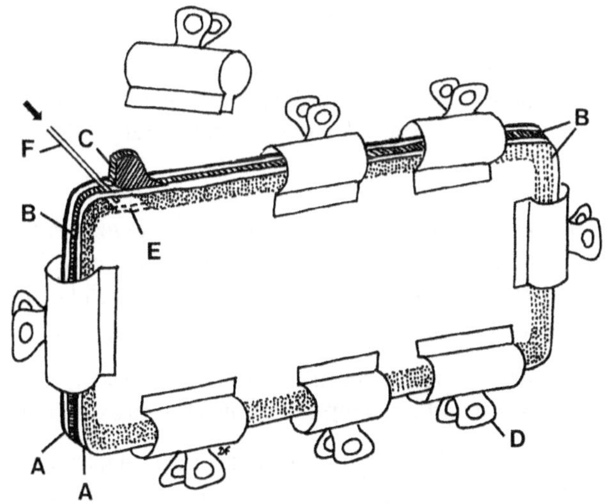

Fig. 2. Gel moulding apparatus. *A* = plate glass; *B* = neoprene gasket; *C* = opening in gasket to allow gelling solution to be inserted into mould; *D* = heavy-duty pressure clips; *E* = upper meniscus of the gelling solution after mould is filled; *F* = syringe; arrow = direction of flow of gelling solution into the apparatus.

2. Pour off acetone as waste and repeat the process 3 more times. Reclaim waste acetone by distillation and use again.

3. Samples are now free of extracellular phenolics and should not be left at room temperature. Freeze immediately and store at -17°C or colder.

4. In addition to material under active investigation, extracts from an already characterised sample should be analysed at the same time, as a standard or reference.

Mechanical disruption of thalli

5. The night before extraction, leave the Dewar bucket as well as mortar and pestles empty in a freezer at approximately -17°C.

6. When starting the extraction procedure, place a small amount of liquid nitrogen in the mortar, with pestle, to cool it further.

7. Place each lichen sample in the mortar and grind **vigorously** in liquid nitrogen. For each sample use two successive applications of liquid nitrogen, both sufficient to cover lichen material.

8. Place the resulting powders in small glass vials with screw top lids, label and store at -20°C. These should probably be analysed within 3 - 4 weeks.

9. Between successive samples, brush the mortar and pestle with a clean brush to remove traces of the last sample.

10. If extraction is not anticipated within a few weeks, freeze acetone-extracted samples directly without liquid nitrogen treatment, then grind in nitrogen either a few days before protein extraction begins or on the same day.

Extraction of proteins from lichens

The extraction of proteins, which is described in the following steps takes approximately 8 h, during which time samples must always be kept at 4°C or less. Begin extraction in the morning, so samples will be ready for dialysing overnight. IEF will occupy all of the following day.

11. Place two mortars and pestles in a freezer, to be used alternately. While one is used for grinding, keep the other at approximately -17°C.

12. Weigh out 0.45 g of lichen powder for each sample; as many as 20 - 24 samples can be run together on the same gel so no more than this should be extracted on the same day.

13. Label a corresponding number of 50 ml centrifuge tubes (e.g. 22) and place in crushed ice. Unbreakable polycarbonate tubes are convenient.

14. First grind the powdered lichen with 1.5 ml of 0.06 M phosphate buffer pH 6.6 in a mortar and pestle for approximately 1 min. Then add an additional 2.5 ml buffer one drop at a time to rinse the mortar walls. Grind further with each addition of buffer.
15. Pour the total of approximately 4 ml of slurry into a motorized 15 ml Wheaton tissue grinder seated in crushed ice, and vigorously grind for 4 min (Fig. 3).
16. Pour the slurry into one of the cold 50 ml centrifuge tubes while the tube remains in ice.
17. Add an additional 5 - 10 ml of buffer drop wise to the sludge on the sides and bottom of the grinder, grind for about 0.5 min with each

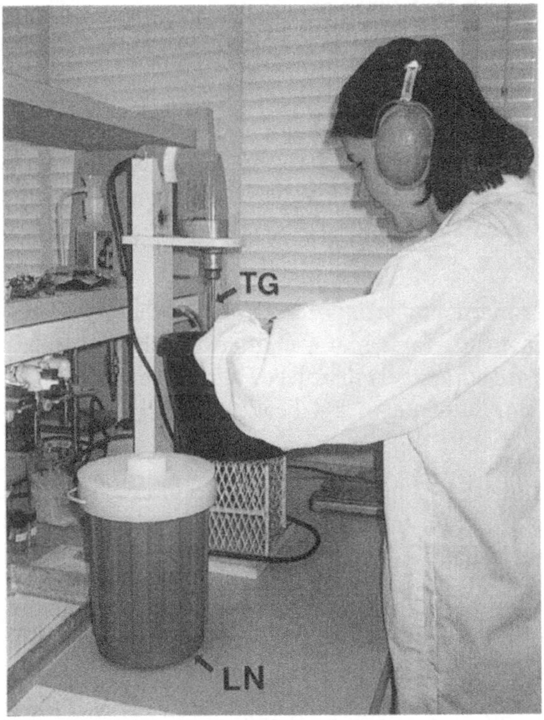

Fig. 3. Standard variable-speed electric drill mounted with a Wheaton tissue grinder for grinding lichen material. Shaft of the grinder (TG) extends into an ice bucket where sample is macerated in a ground glass tube at no more than 4°C. LN = approved vessel for holding liquid nitrogen.

addition, and combine all slurries from each sample in the same polycarbonate tube.

18. Make up two 1000 ml solutions of 1 % glycine in large beakers (10 g glycine in 1000 ml dd H_2O), and cover both. Place on a magnetic stirrer at about 4°C until needed for the next major operation, which is removal of salts through dialysis.

Centrifugation to spin down cell fragments

19. Balance the samples in centrifuge tubes with buffer and spin at 12 400 g for 20 min in the centrifuge at 4°C to remove cell fragments.

20. Place the tubes back in ice and take off the supernatants with pasteur pipettes. If the pellet is loose, a longer period of centrifugation may be used to compact it; otherwise it may be necessary to pass supernatants through coarse (C type) 3 ml sintered glass filter funnels, an operation that must also take place at 4°C.

Precipitation of protein

21. Collect each supernatant in a 25 ml cold graduated cylinder to determine its volume.

22. Use 0.561 g of ammonium sulphate to precipitate protein from each 1 ml of supernatant; calculate the amounts of ammonium sulphate needed for the volumes involved. Weigh ammonium sulphate and add to cooled and numbered 50 ml beakers, along with a small magnetic "flea" or stirring bar in each.

23. Add the supernatant from each sample to the appropriate beaker, and stir magnetically for 2 h at 4°C to ensure that crystals of ammonium sulphate are completely dissolved. With practice it is possible to get the ammonium sulphate in 3-4 beakers mixing simultaneously on one magnetic stirrer.

Protein centrifugation

24. Place numbered 15 ml centrifuge tubes in crushed ice to cool. Larger tubes do not seem to work well at this stage, as the resulting protein pellets are usually not hard enough.

25. Remove stir bars magnetically from beakers containing the ammonium sulphate and pour each extract into a cold tube and spin down at 12 100 g for 40 min at 4°C.

26. Pour off and discard supernatants.

Dialysis

27. While the protein centrifugation is in progress, prepare to dialyse the proteins being precipitated by soaking one 12 cm length of dialyser tubing per sample in hot 10% acetic acid solution for approximately

30 min. Do not overheat, as the tubing will become weak and tend to tear.

28. Rinse the strips of dialyser tubing thoroughly inside and outside with dd H_2O and leave them to soak in 1% glycine.
29. Resuspend each protein pellet in 1 ml cold 0.06 M pH 6.6 phosphate buffer, and remove lumps by repeatedly sucking in and out of a pipette.
30. Tie a knot close to the end of each length of tubing, introduce a protein extract into the tube by pipette and knot the tube at the other end. Position the two knots as close together as possible. This restricts the volume attainable during dialysis and maintains higher protein concentrations.
31. Use coloured plastic paper clips to distinguish samples from one another in tubing, and submerge them in a covered 1 l beaker of 1% glycine. Dialyse on a magnetic stirrer at 4°C for 2 h.

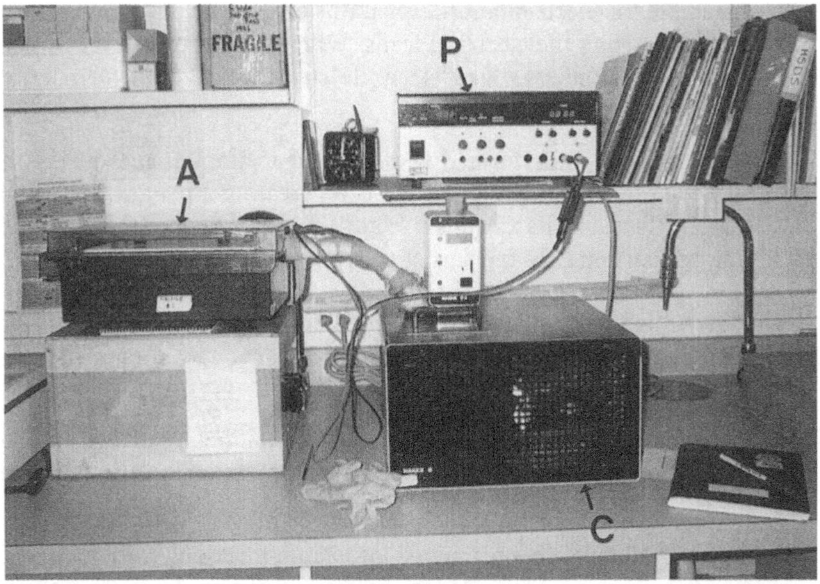

Fig. 4. Electrophoresis apparatus (*A*) connected to power supply (*P*) and cooling bath (*C*). To facilitate cooling, *A* and *C* are placed close together so the length of hose connecting them is minimized, and hose is covered with insulating material.

32. After 2 h, place dialysis tubes in fresh 1% glycine solution. Dialyse overnight at 4°C.

Pre run

33. Turn on the cooling system at least 30 min before electrofocussing, and set to maintain a temperature of 3 - 4°C (Fig. 4).

34. If the electrophoretic apparatus has no grid on which to orient the gel, spread a few drops of paraffin oil into a thin film on the back of a grid or template and apply to the electrophoresis bed. The beds of units marked with a permanent grid require no additional template.

35. Carefully roll a polyacrylamide gel, usually pH 4.0 - 6.5, onto the bed (with or without a template). Start from the middle, so that bubbles beneath the gel support can be more easily eliminated.

36. On two separate pieces of foil (or plastic film) apply anode and cathode solutions to the appropriate electrode strips cut to size. The strips should be evenly wetted by the electrolytic solutions, but apply no more solution than can be absorbed. Use separate scissors to cut strips to the appropriate lengths and clean forceps to apply each of the strips to opposite ends of the gel.

37. Assemble the electrophoretic apparatus (Fig. 4) by positioning the lid and connecting the electrical leads. When the electrodes are in close contact with the saturated electrode strips throughout their length, turn on the power.

38. A 1 h pre-run is performed without lichen samples, and during this step the power is maintained at 1.0 W, creating a pH gradient across the gel [Recall: Volts x Amps = Watts (power)]. Reduce the power setting proportionately for partial gels, i.e., 1/2, 1/3, etc.

Centrifugation to remove debris

39. Pipette dialysed protein extracts into cold numbered 15 ml centrifuge tubes, and spin down at 12 400 g for 40 min at 4°C to remove undissolved debris.

40. Immediately pipette off supernatants containing protein into 5 ml tubes placed in ice and maintain at approximately 4°C until used for IEF.

Loading electrofocussing gel

41. Have 5 x 10 mm holes machined into 2 mm thick polycarbonate strips to serve as re-useable wells for application of samples. Such wells accommodate a larger volume of extract than many other application systems and thus tend to compensate for lower concentrations of protein in sample solutions. If carefully cleaned in detergent and rinsed in

alcohol, such plastic starting gate strips adhere well and do not permit leakage of extract.

42. Use forceps to press sample starting gates firmly into position on the gel, usually near the cathode end (Fig. 5) of the electrical field.
43. In addition to the samples being analysed on each gel, run two applications of a reference extract containing previously characterised enzyme bands.
44. Leave a strip 1.5 cm wide without samples on one edge of the gel to allow an area for pH measurements to be made following the run.
45. If samples on one gel are to be stained for different enzymes, record the positions on the template corresponding to where the gel should be cut.
46. Load 0.25 ml of sample into each sample well.

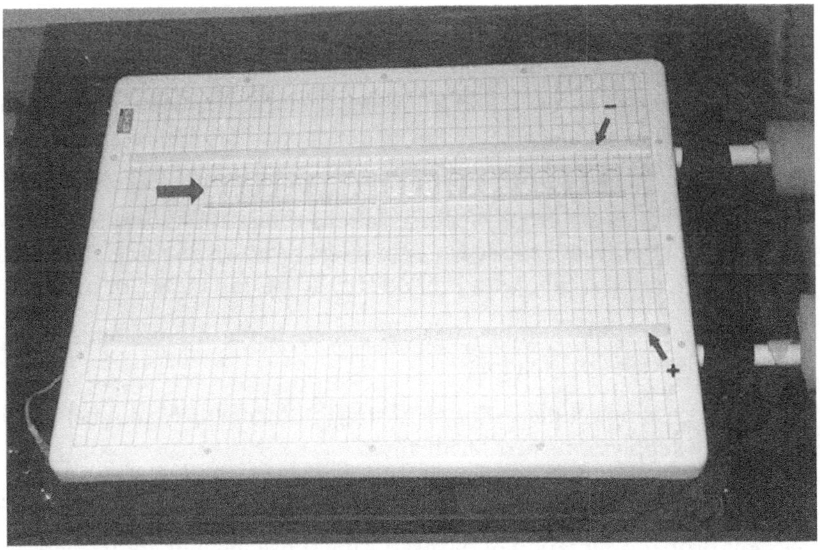

Fig. 5. Polyacrylamide gel set up on bed of electrophoresis apparatus, with electrode strips applied at the cathode (-) and anode (+) ends. The large arrow on the left indicates plastic sample application strips, near the cathode, with reservoirs for up to 23 samples. The shaft of the large arrow extends into space reserved on the left edge of the gel for documenting the pH gradient developed during IEF. Insulated connections to the bath containing coolant appear on the right.

47. Cover unused extract remaining in tubes with "Parafilm" and store in freezer in case further IEF separations are needed for clarification.

48. Close the lid of the electrophoresis apparatus, connect the electrodes and turn on the power.

49. Determine the concentration protein in supernatants following centrifugation using, for example, a Biorad protein assay kit. This is particularly necessary if no stained products are being detected; a protein assay may indicate if the concentration in extracts is too low. Activity of at least some enzymes is usually detectable if the concentration of protein exceeds 1 mg ml^{-1}.

The run

50. For the first 30 min maintain the power close to 1.8 W for a full-width gel, reducing the power proportionately for partial gels.

51. After 30 min increase the power for a full gel to 4 - 6 W. However, the gel should remain at approximately 2 - 3°C, and certainly not more than 4°C. If temperature climbs, as on hot or humid days, lower the voltage, otherwise enzymes may be denatured and fail to stain.

52. One h after the start of a run, turn the power off momentarily and remove the starting gates with clean forceps.

53. Gently absorb excess liquid from the gel surface by wiping toward the cathode with small pieces of thick filter paper. This removes pigmented deposits that could cause streaking on the gel.

54. Protein bands are usually focussed after 3 h, and at this time 5 × 10 mm pieces of the gel are excised at 1 cm intervals down one side. Place each in a covered test tube with 1 ml dd H_2O at 4°C for a few h or overnight.

55. Mix contents of test tubes well, allow them to reach room temperature and then take pH readings.

56. After excising gel pieces for pH testing, re-focus for 10 min before the power supply is turned off finally.

57. Open the apparatus and carefully remove and discard electrode strips.

58. For proper orientation of the gel during subsequent analysis, mark the upper right-hand corner of each gel by cutting off a small piece with a scalpel.

59. Rinse the gel in dd water and transfer to staining container(s).

60. Rinse electrode wires with dd H_2O and dry carefully with lab tissue (e.g. Kimwipes) after every run. Use a separate piece of clean tissue for each electrode.

Subprotocol 2
Isozyme Staining

Materials

Freshly prepared staining solutions should be mixed by the time a run is finished, so that staining can start immediately. Check the pH of all buffers, and adjust if necessary using a pH meter and a magnetic stirrer, before incorporating into staining solutions. Be sure that all glassware and staining containers used have been properly soaked and washed in strong laboratory detergent, as part of the cleaning process.

The following enzymes have been detected electrophoretically in lichens, most in several species. Some stains, such as those for esterase and phosphatase, show activity of many different enzymes, all with an ability to catalyse the same general type of reaction. Other tests, such as those of dehydrogenases are more specific and indicate activity of an enzyme that is only capable of acting on one particular substrate. Substrates and co-factors used in enzyme stains should be stored at -17°C or colder and other reagents can be kept either in a freezer or at 4°C.

for peroxidase (PER) EC 1.11.1.7

- Staining solution: 50 ml ethyl alcohol, 5 ml glacial acetic acid, 12 ml dd H_2O
- 33 ml 3% H_2O_2
- 45 mg 3,3',5,5' tetramethylbenzidine (a substitute for the carcinogen, benzidine)

for catalase (CAT) EC 1.11.1.6

- H_2O_2 solution (5 ml 3% hydrogen peroxide to 95 ml dd H_2O)
- 2% potassium ferricyanide (1 g in 50 ml dd H_2O)
- 2% ferric chloride (1 g in 50 ml dd H_2O)

for alkaline phosphatase (ALP) EC 3.1.3.1

- Staining solution: 1.75 g sodium chloride, 0.02 g magnesium chloride hexahydrate, 100 ml Tris buffer 0.05 M, pH 8.5
- 50 mg sodium napthyl phosphate
- 500 mg polyvinyl pyrolidone (PVP)
- 50 mg Fast Blue RR salt (Sigma Aldrich Chemical Co., F-0500)

for acid phosphatase (ACP) EC 3.1.3.2

- 0.05 M pH 6.0 citric acid buffer (4.7 ml 0.1 M citric acid, 20.7 ml 0.1 M sodium citrate, 74.5 ml dd H_2O)
- 0.359 g phenolphthalein diphosphate
- Ammonium hydroxide, 15 N

for esterase (EST) EC 3.1.1.1

- 0.08 M tris buffer, pH 7.0 (31.2 ml 0.2 M THAM, 56.8 ml 0.1 N hydrochloric acid, 11.9 ml dd H_2O)
- 200 mg fast blue RR salt
- 0.11 g α-napthyl acetate dissolved in approximately 4 ml acetone (rapidly remove any undissolved solute by filtration)

De-staining solution

- 800 ml dd H_2O
- 300 ml ethanol
- 100 ml glacial acetic acid

for carbonic anhydrase (CAN) EC 4.2.1.1

- The materials are the same as for esterase, but Na α-napthyl acetate is substituted for α-napthyl acetate.

for superoxide dismutase (SOD) EC 1.15.1.1

- 0.05 M tris buffer, pH 8.2 (200 ml 0.2 M THAM, 360 ml H_2O, add 0.1 M HCl until pH reaches 8.2)
- 0.013 g 3-(4,5-dimethylthiazol-2-yl)-2,5-diphenyltetrazolium bromide (MPT) (Sigma Aldrich)
- 0.013 g phenazine methosulphate (PMS) (Sigma Aldrich)

for laccase (LAC) EC 1.10.3.2

- 0.06 M sodium phosphate buffer pH 6.6 (the extraction buffer)
- 0.025 g phenylenediamine
- 0.050 g sulphonilic acid

Staining solutions for dehydrogenases

for isocitrate dehydrogenase (IDH) EC 1.1.1.42

- 100 ml 0.05 M Tris pH 8.2 buffer (as for SOD stain)
- 0.0125 g DL-isocitric acid, trisodium salt
- 0.0038 g nicotinamide adenine dinucleotide phosphate (NADP) (Sigma Aldrich)
- 0.0163 g Nitroblue tetrazolium (NBT)
- 0.004 g PMS
- 0.3 g magnesium chloride hexahydrate

for 6-phosphogluconate dehydrogenase (6PG) EC 1.1.1.44

- 100 ml 0.05 M Tris pH 8.2 buffer (as for SOD stain)
- 0.017 g 6-phosphogluconic acid, trisodium salt
- 0.4 g magnesium chloride hexahydrate
- NADP, NBT, PMS as for IDH stain

for glucose-6-phosphate dehydrogenase (G6P) EC 1.1.1.49

- 100 ml 0.05 M TRIS pH 8.2 buffer (as for SOD stain)
- 0.0175 g D-glucose-6-phosphate, monosodium salt
- 0.4 g magnesium chloride hexahydrate
- NADP, NBT, PMS as in stain for IDH

for glutamate dehydrogenase (GDH) EC 1.4.1.4

- 100 ml 0.05 M TRIS pH 8.5 buffer (as for SOD stain, but note the different pH)
- 0.5 g L-glutamic acid
- NADP, NBT, PMS as for IDH stain

for Mannitol dehydrogenase (MAN) EC 1.1.1.38

- 100 ml 0.05 M TRIS pH 8.2 buffer, made up as for SOD stain
- 0.0183 g D-mannitol
- NADP, NBT, PMS as for IDH stain

Procedure

Peroxidase

Method is a modification of Brewer and Sing (1970).

1. Just before staining, add the H_2O_2 and the tetramethylbenzidine to the staining solution and immediately stain.
2. Allow gels to develop at room temperature.
3. Green bands appear instantly but cannot be fixed and preserved. Bands can also overstain, therefore gels must be photographed immediately.

Catalase

The method is a modification of Harris and Hopkinson (1976). Staining according to Brewer and Sing (1970) was less successful.

1. Incubate the gels in the H_2O_2 solution for 15 min.
2. Pour off the H_2O_2 solution.
3. Rinse the gel with water.
4. Mix the ferricyanide solution and the ferric solutions 1:1, and immediately add to the gels.
5. Agitate the container gently for 20 - 30 s.
6. The band should stand out as clear or yellowish zones on a dark blue background.
7. Photograph gels immediately, as bands are very transient.

Alkaline phosphatase

This method is based on Brewer and Sing (1970).

1. Add the sodium napthyl phosphate, PVP and fast blue to the staining solution, and then immediately stain the gel.
2. Reddish brown bands may appear within 1 - 2 h and may intensify with time, although leaving overnight will usually over-stain the gel.
3. Gels with bands should be photographed within a few h, as bands tend to become more diffuse with time.

Acid phosphatase

This method is based on Brewer and Sing (1970).

1. Add the phenolphthalein diphosphate to the citric acid buffer, and immediately stain for 3 h.
2. Decant the staining mixture.
3. Add 2 ml ammonium hydroxide under the fume hood.
4. Immediately cover the staining tray with aluminum foil to create an ammoniacal atmosphere.
5. Red bands usually appear and diffuse quickly, so photograph immediately.
6. If no ammonium hydroxide is added, bands become opaque and white, and they may be photographed against a black background after 3 - 4 h.

Esterase

This method is based on Brewer and Sing (1970).

1. Add the fast blue and the α-napthyl acetate to the buffer and immediately stain.
2. Incubate gel at room temperature; brown bands begin to appear in 15 - 20 min.
3. Remove gel and record bands at this early stage.

4. Continue staining for another 5 min and check the gel for more bands.
5. If bands are faint, a new staining solution may be mixed and the gel incubated for a further 20 - 30 min.
6. Check periodically and do not allow over-staining.
7. When bands begin to diffuse, remove gel from stain, and rinse in dd H_2O. Then place on a shaker covered with dd H_2O and 50 ml of de-staining solution.
8. Make a second drawing after 1 - 2 h. Gel should be left overnight and a third drawing made in the morning. All drawings should be combined into one. Photograph the gel at the time of each drawing.

Carbonic anhydrase

This method is based on Brewer and Sing (1970). The procedure is the same as for esterase, but Na α-napthyl acetate is substituted for α-napthyl acetate.

1. Incubate gel for 2 - 3 h, but watch carefully so as not to over-stain.
2. Reddish-pink band should appear.
3. If bands are poorly developed, try doubling the amount of Na α-napthyl acetate and fast blue RR salt.

Superoxide dismutase

This method is based on Harris and Hopkinson (1976).

1. Add the MPT and PMS to 100 ml of buffer and immediately stain.
2. Clear bands appear on a darker ground.

Laccase

This method is based on Vallejos (1983).

1. Add the phenylenediamine and the sulphonilic acid to 100 ml of buffer and stain immediately.
2. The bands are dark on a lighter background.

Dehydrogenases

Staining for dehydrogenase enzymes involves the use of tetrazolium dyes, and all tetrazolium dye reactions must take place in the dark. Therefore, develop gels in stainless steel staining boxes or containers that are otherwise darkened. All dehydrogenases may not be detectable in a given lichen; sometimes adjusting the pH of assay solutions, e.g., from 8.2 to 8.5, will reveal activity and in some cases, using more concentrated protein solutions is advisable. These staining solutions are based mostly on Brewer and Sing (1970), but Harris and Hopkinson (1976) suggested that substituting MTT (Sigma Aldrich M-2128) for NBT (Sigma Aldrich N-6876) might in some cases be more effective. Gels can be developed overnight in these reaction mixtures, without danger of over-staining. After staining is complete, place gels in de-staining solution (see procedure for Esterase) to reduce background colour.

Comments

Tests may be made for other dehydrogenases by experimenting with different substrates, e.g., shikimic acid, xylose-5-phosphate, along with NADP, NBT and PMS in buffer as shown for IDH.

Subprotocol 3
Photography and Tracing of Gels

Materials

- For black and white prints: Kodak Plus-X film (Fig. 6)
- For color slides: Ektachrome 160 tungsten film

Procedure

1. Set the ASA appropriately, at 120 for Plus-X, and with the shutter speed set at 125, openings of f8, f11 and f16 should be used for each exposure. For best results illuminate gels evenly from below.
2. Clearly label gels to indicate enzyme and run number and display in each photo, along with a metric ruler placed adjacent to one edge of the gel.

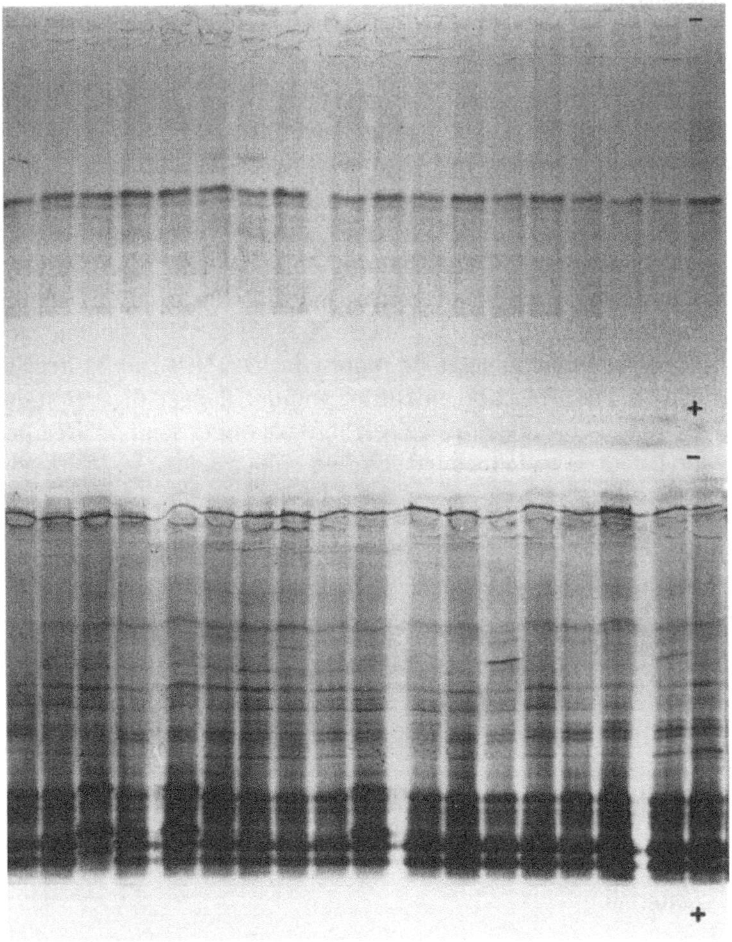

Fig. 6. Gels showing 18 samples stained for a) mannitol dehydrogenase and b) esterase. - = cathode end of gel; + = anode.

3. For enzymes with complex banding patterns, capture images on film at different stages during development.
4. Make accurate tracings on acetate sheets during development of colour. Tracings will sometimes show detail that is not evident on photos, especially for bands that may have been overdeveloped when photos were taken and weak bands that were faint even in the later stages of staining.

5. Wipe any oil off the lower surface of the backing and cover the gel neatly with two independent layers of plastic food wrap film.
6. If carefully sealed and stored in a covered box at 4°C, a gel will be protected from dehydrating appreciably for several weeks.

Subprotocol 4
Scoring of Gels

Procedure

1. The exact position of each band, or electromorph, based on both photographs and tracings, should be carefully indicated on graph paper in relation to the measured pH gradient.
2. Score every band present or absent in each of the samples, and if it is difficult to relate bands in one extract to those in another, re-run problematical samples in adjacent lanes, possibly on a less steep pH gradient than initially.
3. Matrices of presence/absence data can be analysed using similarity coefficients to indicate isozyme diversity within sample sets, such as sets corresponding to populations (Hageman and Fahselt 1990). Enzymes with banding patterns that vary little (e.g., MAN, Fig. 6) may require only one sample per population to reveal all existing variation, while 10 samples are sometimes needed to display isozyme variability in enzymes such as EST (Fig. 6).
4. In addition, gels themselves, or negatives of their photographic images, can be scanned on a densitometer to describe quantitative variation in band intensity within or between samples.

Troubleshooting

- Gel solidifies before it is poured into the mould.
  Check that gelling solution was made up properly; **immediately** after mixing pour it into the gelling apparatus.
- Gel fails to set.
  Check that the ammonium persulphate solution was fresh and that the gelling solution was made up properly.

- No enzyme bands detected in samples.
  - First note whether standards stained successfully. If they did not, the temperature of samples may have reached more than 4°C some time during the procedure. If enzymes were detected in standards and not in experimental samples, other explanations may apply.
  - Changing the pH of extraction buffer or staining solutions might help.
  - If staining for catalase, doubling the amounts of α-naphthyl acetate and Fast Blue RR salt may be effective.
  - When staining for dehydrogenases, substitution of MTT for NBT may produce better results. Also, light may have leaked into the staining container, and if so it must be made light proof.
  - The protein concentration in extracts may not be high enough. If less than 1 mg ml^{-1} it may be necessary to: 1) start out with more lichen material; 2) restrict the capacity of dialysis tubing into which samples are tied; 3) collect samples at a different time of year (Trembley and Fahselt 1999).
- Bands are wavy or "smiling";.
  Very gently clean salts from the platinum electrode wires by scraping with a scalpel to remove residues. Never allow wires to touch anything or to get dusty. Make sure dialysis removes salts from samples prior to electrofocussing.

Acknowledgements. The significant contribution of Dr. C. Hageman to development of these methods is acknowledged, as is the material support of the Natural Science and Engineering Research Council of Canada.

References

Brewer G J, Sing CF (1970) An introduction to isozyme techniques. Academic Press, New York

Fahselt D (1995) Lichen sexuality from the perspective of multiple enzyme forms. Cryptogamic Botany 5: 137-143.

Hageman C, Fahselt D (1990) Enzyme electromorph variation in the lichen family Umbilicariaceae: within-stand polymorphism in umbilicate lichens of eastern Canada. Can J Bot 68: 2636-2643

Harris H, Hopkinson DA (1976) Handbook of enzyme electrophoresis in human genetics. Elsevier Press, Amsterdam New York

Trembley M, Fahselt, D (1999) Effect of sampling time on isozyme banding patterns in *Cladonia rangiferina* and *Umbilicaria mammulata.* Lichenologist 31: 397-402

Vallejos CE (1983) Enzyme activity staining. In: Tanksley SD, Orton TJ (eds)Isozymes in plant genetics and breeding, Part A. Elsevier Press, Amsterdam New York, pp 469-516

Winter A, Ek K, Andersson U-B (1977) Analytical electrofocusing in thin layers of polyacrylamide gels. LKB Application Note 250, LKB, Bromma, Sweden

Suppliers

Amersham Pharmacia Biotec, 1250 University Ave, Montreal, PQ H3B 4H4, Canada, Tel: 1-800-463-5800, Fax: 1-800-567-1008, e-mail: ts-ep@am.apbiotech.com

Biorad Laboratories, 5671 McAdam Rd, Mississauga, ON L4Z 1N9, Tel: 1-800-268-0213, Fax: 905-712-2990

Fisher Scientific Ltd, 112 Colonnade Rd, Nepean, ON K2E 7L6, Tel: 1-800-463-2996, Fax: 1-800-463-2996

Sigma-Aldrich Canada, 2149 Winston Park Dr, Oakville, ON L6H 9Z9, Canada, Tel: 905-829-9500, Fax: 905-829-92929, e-mail: canada@sial.com

VWR Canlab, 2360 Argentia Rd, Mississauga, ON L5N 2Z7, Canada, Tel: 1-800-932-5000, Fax: 1-800-668-6348

Analysis of Lipids in Lichens

IRINA A. BYCHEK-GUSCHINA

Introduction

In all living organisms lipids play several roles, and according to their structures they can be divided into two main groups: the neutral lipids (acylglycerols, sterols, free fatty acids, wax and steryl esters) and polar lipids (phospholipids, glycolipids and betaine lipids). Triacylglycerols act as a compact, easily metabolised and non-hydrated energy store. Waxes are commonly extracellular components such as the surface covering, which function both to reduce water loss and to protect plants from noxious environmental conditions (Harwood 1998). Polar lipids and sterols are important structural components of all cell membranes. Also, there are many examples of what could be termed biologically active lipids (e.g., inositol lipids, sphingolipids, oxidation products). In recent years, scientists have started to realize that lipid metabolism is a key factor in the adaptation mechanisms of many organisms to environmental and anthropogenic stress. In lichens, the importance of their lipids in the response and adaptation to environmental factors such as temperature, elevation, light, high levels of radiation, and sulphur have been studied (Dertien et al. 1977, Piervittori et al. 1995, Bychek and Bychek 1996, Shapiro et al. 1998, Bychek-Guschina et al. 1999). The present chapter will describe methods of lipid analysis that are available for many laboratories and have been used by the author for extraction, separation and quantification of basic lipids.

Irina A. Bychek-Guschina, Institute of Ecology of the Volga River Basin RAS, Togliatti, 445003, Russia (*phone* +7-8482- 489389; *fax* +7-8482-489504;
e-mail bychek@tlt.volga.ru)

Subprotocol 1
Lipid Extraction

The aim of lipid extraction procedures is to extract lipids quantitatively and without contamination by non-lipid components.

Materials

Equipment

- Heating block
- Methylation tubes (16 × 125 mm, with screw caps and Teflon or silicone liners)
- Pestle and mortar
- Centrifuge
- Small volume (e.g. 1 ml) tubes with screw caps and Teflon or silicone liners

Reagents

- Isopropanol (propan-2-ol)
- Methanol
- Chloroform
- Garbus solution (2 M KCl in 0.5 M potassium phosphate buffer, pH 7.4) (Garbus et al. 1963)
- Supply of nitrogen gas

Procedure

1. Cut lichen thalli (about 0.5 g fresh weight) into small pieces.
2. Put the material in a methylation tube, add 3 ml of isopropanol and heat the tissue at 70°C for 30 min. Be sure to seal the tube to prevent evaporation.
3. After cooling, homogenise the tissue using a pestle and mortar in isopropanol and filter the homogenate into a 20 ml centrifuge tube. Perform all lipid extraction procedures in glass using pure solvents, and extract at, or below, room temperature.

4. Re-extract the tissue homogenate in 1.5 ml of methanol and 3 ml of chloroform, and then filter the homogenate to a centrifuge tube, combining the solvents.
5. Rinse the mortar with a further 1.5 ml of methanol and 3 ml of chloroform, filter and combine the solvents.
6. Add 2.5 ml of water and 3 ml of Garbus solution to the extract and mix thoroughly using a vortex mixer. A uniform emulsion should be produced. Allow this to stand, or (better) centrifuge the tubes in a bench centrifuge (at 1000 g for 5 min) to separate the two phases. The green lower phase contains the lipids. Using a Pasteur pipette, carefully remove the upper phase.
7. Wash the lower phase by adding an equal volume of methanol/water (1:1, v/v), then mix and centrifuge to separate the phases.
8. Transfer the lower chloroform phase to an appropriate pre-weighed glass vial with a Teflon-lined screw-cap and evaporate the lipids to dryness under a nitrogen stream at 30°C.

   Note: Rapid evaporation of large amounts of lipid extract can be achieved using a rotary evaporator; however, the samples should not be heated above 40°C.
9. Reweigh the vial and determine the dry weight of total lipids.

   Note: Lipids should not be left for long periods of time in a dry form.
10. Re-dissolve the dry lipid extract in 1 ml of chloroform, flush out the air in the tube with nitrogen, reseal and store in a freezer prior to further analysis.

    Note: If it is necessary to store samples for more than one month, or if unsaturated lipids are present, add 0.05% butylated hydroxytoluene as an antioxidant.

Subprotocol 2
Separation of Total Lipid Extract by Column Chromatography

Usually the lipid extracts from many lichen species contain large amounts of pigments and coloured organic-soluble lichen products. These compounds prevent clear separation of the total lipid extract into individual lipid classes by thin layer chromatography. Because of this, an initial separation of total lipids by column chromatography (Christie 1982) is quite helpful for the subsequent analysis of individual lipids.

Materials

Equipment

- Glass column of 1 cm diameter with glass wool in the lower part to support the adsorbent

Reagents

- Florisil or Silicic acid (60-100 mesh)
- Chloroform
- Acetone
- Methanol
- Supply of nitrogen gas

Procedure

1. Suspend 3.0 g of Florisil or silicic acid in 10 - 15 ml of chloroform and pour into a glass column of 1 cm diameter containing glass wool in the lower part to support the adsorbent. In general, 30 mg of lipids per 1 g of adsorbent is a reasonable load but this can be varied with circumstance.

   Note: Before separation activate the adsorbent by heating overnight at 110 - 120°C.

2. Take an aliquot of total lipids from Subprotocol 1, Step 10, that contains 30 mg of lipids per 1 g of absorbent and dilute this to 2 ml with chloroform, and pipette it carefully onto the column.

3. When no lipids remain above the surface, start the elution procedure as follows: remove the neutral lipids with 8 volumes (24 ml) of chloroform, remove the glycolipids with 25 volumes (75 ml) of acetone and remove the phospho- and betaine lipids with 10 volumes (30 ml) of methanol. The flow rate should be about 1 - 3 ml min^{-1}.

   Note: Never allow the column to run dry, as channels are formed that cause uneven elution of lipids.

4. Evaporate the three fractions obtained under a stream of nitrogen and determine the dry weight of each lipid fraction; re-dissolve them in a known volume (0.3 - 0.5 ml) of chloroform. Store as discussed above (see Subprotocol 1).

Subprotocol 3
HPTLC of Polar Lipids

Although separation of polar lipids may be carried out by various techniques (Cartwright 1993; Henderson and Tocher 1992) two-dimensional high-performance thin layer chromatography (HPTLC) on pre-coated 10 x 10 or 6 x 6 cm plates with silica gel G as adsorbent is one of most useful for the analytical separation of individual lipid classes in lichens.

Materials

Equipment

- TLC tanks
- Hamilton syringe
- Hairdryer
- Spray bottle powered by an aerosol can

Reagents

- Chloroform - methanol - benzene (or toluene) - 28% ammonium hydroxide (65:30:10:6, by volume) **(I)**
- Chloroform - methanol - benzene (or toluene) - acetone - acetic acid - water (70:30:10:5:4:1, by volume) **(II)**
- 0.25% Ninhydrin in acetone
- Solution A for the Dragendorff stain: dissolve basic bismuth nitrate $Bi(NO_3)_3 \cdot 5H_2O$ (1.7 g) in 20% acetic acid (100 ml)
- Solution B: dissolve potassium iodide KI (10 g) in water (25 ml)
- Dragendorff spray reagent: mix 20 ml of solution A with 5 ml of solution B and add 70 ml of water
- TLC plates, pre-coated with silica gel G as adsorbent, 10 x 10 or 6 x 6 cm
- Supply of nitrogen gas

  Note: Only freshly prepared solvent mixtures should be used.
  Caution: Benzene is a highly toxic solvent and can be replaced by toluene.

Procedure

1. Pre-wash TLC plates in chloroform, or chloroform/methanol (1:1, v/v). Dry them in the air and heat at 110°C for 30 - 60 min. Allow to cool.
2. Line two TLC tanks with filter paper and add the solvent mixtures to a depth of 0.5 cm (about 35 - 40 ml of solvents).

   Note: The filter paper should be saturated with solvent and lids should be fitted.
3. Apply approximately 50 μg of lipids (about 25 - 30 μl of separated phospho- and betaine lipid fraction from Subprotocol 2, Step 4) at the lower right-hand corner of the plate, 0.5 - 0.7 cm from each edge for 6 x 6 cm plates, or 1.0 cm for 10 x 10 cm plates. The diameter of the spot should not exceed 0.5 cm.
4. If lipid standards are available, apply the chloroform solution of polar lipids on a separate TLC plate in the same manner.

   Note: If necessary, the lipid solution can be applied repeatedly over the same spot, allowing the solvent to evaporate between applications. Use a Hamilton syringe, a micropipette with a fine tip or a capillary pipette.
5. Evaporate the solvent from the lipid spot and develop the plate in a tank containing solvent mixture (**I**) for the first development.
6. Remove the plate when the solvent front is about 3 mm from top edge.
7. Dry plate under nitrogen for 20 min (or by use of an unheated stream of air from a hair dryer).

   Note: It is important that all traces of the solvent used for development in the first dimension are evaporated from the plate before separation in the second direction.
8. Turn plate 90 degrees clockwise from first development direction and place in tank containing mixture (**II**).
9. Remove plate when solvent front is about 1 - 0.5 cm (for 10 x 10 cm plate) or 0.3 cm (for 6 x 6 cm plate) from top edge.

10. Spray the developed plate with the appropriate lipid detection reagent (a full description of the reagents for identification of lipids on TLC plates is present in Kates 1972 and Christie 1982):
    - For identification of all phospholipids spray the plate with the working reagent for the phosphorus assay (Subprotocol 6): after a few minutes, the phospholipid-containing spots appear as dark blue bands or spots;
    - Lipids with free amino groups (phosphatidylethanolamine, phosphatidylserine and the related lyso compounds) can be detected after spraying the plate with 0.25% ninhydrin in acetone: the amino lipids appear as pink-purple spots within 1-2 h at room temperature or within 15 min if the plates are heated to about 100°C;
    - Phosphatidylcholine, lyso-phosphatidylcholine and diacylglyceryltrimethylhomoserine appear after a few minutes as orange or yellow coloured spots after spraying the plate with the Dragendorff

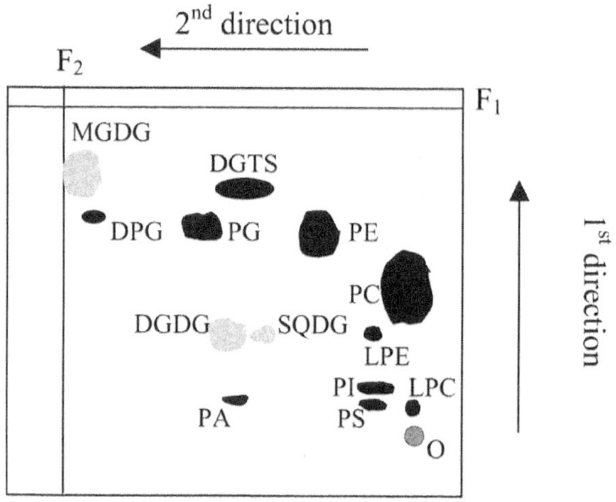

Fig. 1. Schematic two-dimensional HPTLC of polar lipids of lichens. Solvent systems: 1st dimension, development with chloroform / methanol / benzene / 28% ammonium hydroxide (65:30:10:6, by volume) to F_1; 2nd dimension; development with chloroform / methanol / benzene / acetone / acetic acid (70:30:10:5:4:1, by volume) to F_2. Abbreviations: O, origin; PC, phosphatidylcholine; PE, phosphatidylethanolamine; PG, phosphatidylglycerol; DPG, diphosphatidylglycerol; PI, phosphatidylinositol; PS, phosphatidylserine; PA, phosphatidic acid; LPC, lysophosphatidylcholine; LPE, lysophosphatidylethanolamine; DGTS, diacylglyceryltrimethylhomoserine; MGDG, monogalactosyldiacylglycerol; DGDG, digalactosyldiacylglycerol; SQDG, sulphoquinovosyldiacylglycerol. MGDG, DGDG and SQDG appear when total lipid extract is separated. Trace amount of these lipids appear when phospho plus betaine lipid fraction is separated.

stain. If lipid standards are available, the comparison of their retention times and position on the plate with that of the analysed polar lipid mixture may provide additional helpful information for identification of lipids. Figure 1 shows the schematic separation of polar lipids of lichens.

Note: Spray reagents should always be used in fume cupboards. Wear gloves when using the ninhydrin spray.

Subprotocol 4
TLC of Neutral Lipids

The best method for separating and quantifying the neutral lipids of lichens after column fractionation involves using one-dimensional TLC on 10 x 10 or 20 x 20 cm silica gel G plates, and two different sequential solvent mixtures.

Materials

Equipment

- TLC tanks
- Hamilton syringe
- Hairdryer
- Spray bottle powered by an aerosol can

Reagents

- Toluene - hexane - formic acid (140:60:1, by volume) (I)
- Hexane - diethyl ether - formic acid (60:40:1, by volume) (II)
- Sterol stain: dissolve ferric chloride $FeCl_3 \times 6H_2O$ (50 mg) in 90 ml water with 5 ml acetic acid and 5 ml sulphuric acid
- Ester bonds stain, solution A: dissolve 10 g hydroxylamine hydrochloride ($H_3NO \cdot HCl$) in 25 ml water and add 100 ml ethanol, then add 26 ml of a saturated aqueous solution of sodium hydroxide (NaOH) and 200 ml ethanol; filter the mixture
- Ester bond stain, solution B: grind together 10 g ferric chloride ($FeCl_3 \times 6H_2O$) and 20 ml concentrated hydrochloric acid, then shake with 300 ml diethyl ether

- TLC plates, pre-coated with silica gel G as adsorbent, 10 x 10 or 20 x 20 cm
- Supply of nitrogen gas
- Standards of neutral lipids (if available)

Procedure

1. Prepare the plates and a tank as noted in Steps 1 and 2 of Subprotocol 3.
2. For one-dimensional TLC on 10 x 10 or 20 x 20 cm (for large amounts of neutral lipids) plates, use a soft pencil to divide the plate into five columns (i.e., four sample lanes and one standard lane), taking care not to damage the adsorbent layer.
3. Apply 1 - 5 mg of each lipid sample as a streak (0.5 - 1.0 cm wide) at 1.0 or 2.5 cm from the bottom edge of 10 x 10 and 20 x 20 cm plates, respectively. Also apply 10 µg of standard as a single spot.
4. Evaporate the solvents from the lipid streaks and develop the plate in a tank containing solvent mixture (I).
5. Allow the solvents to ascend to 0.5 - 1.0 cm from the top of the plate, remove and dry it in air or under nitrogen.
6. Develop the plate for the second time in the same direction in a tank containing solvent mixture (II) to F_2 (Fig. 2A).
7. Remove the plate, evaporate as in Suprotocol 3a and spray with an appropriate lipid detection reagent:
   - For identification of ester bonds in triacylglycerols, diacylglycerols, monoacylglycerols, steryl and wax esters, spray the plate with solution A, dry, then spray with solution B; lipids appear as purple spots on a yellow background;
   - For identification of free sterols and steryl esters spray the plate with sterol stain solution; the lipids appear as red-violet coloured spots or bands within 2 - 3 min of heating to about 100°C.

     If lipid standards are available, compare their retention times with those of the analysed neutral lipids detected. Figure 2A shows the schematic separation of neutral lipids of lichens.

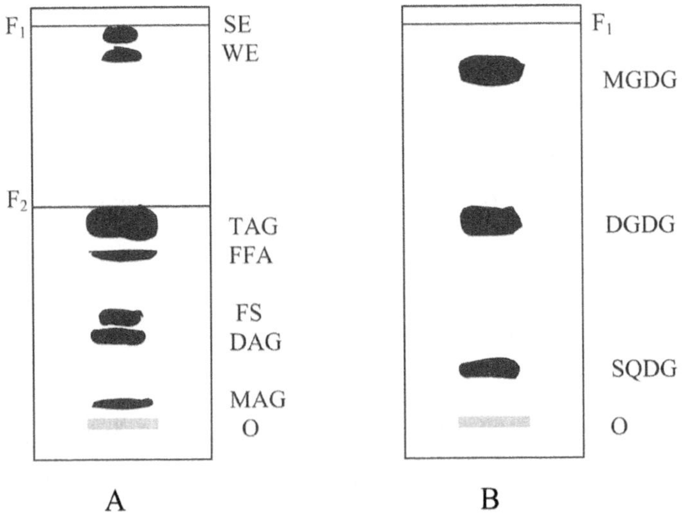

Fig. 2. Schematic one-dimensional separation of neutral lipids (**A**) and glycolipids (**B**). A: Solvent systems: toluene/hexane/formic acid (140:60:1) to F_1 and hexane/diethyl ether/formic acid (60:40:1) to F_2. B: acetone/benzene/water (91:30:8). Abbreviations: O, origin; MAG, monoacylglycerols; DAG, diacylglycerols; FS, free sterols; FFA, free fatty acids; TAG, triacylglycerols; WE, wax esters; SE, steryl esters, MGDG, monogalactosyldiacylglycerol; DGDG, digalactosyldiacylglycerol; SQDG, sulphoquinovosyldiacylglycerol.

Subprotocol 5
TLC of Glycolipids

One-dimensional TLC on 10 x 10 cm silica gel G plates is more useful for separation of individual glycolipids of lichens followed by column chromatography.

Materials

Equipment

- TLC tank
- Hamilton syringe
- Hairdryer
- Spray bottle powered by an aerosol can

Reagents

- Chloroform - methanol - acetic acid - water (170:30:20:7, by volume) **(I)**
- Acetone - benzene - water (91:30:8, by volume) **(II)**
- α-naphthol (0.5 g) in methanol/water (1/1, 100 ml)
- Concentrated sulphuric acid (95 ml) plus distilled water (5 ml)
- TLC plates, pre-coated with silica gel G as adsorbent, 10 x 10 cm
- Supply of nitrogen gas

Procedure

1. Prepare the plates and a tank as in Steps 1 and 2 of Subprotocol 4.
2. Apply 1-3 mg of each lipid sample as a streak (0.5 cm wide) 1.0 cm from the bottom edge of the plate.
3. Develop the plate in a tank containing one from the above mentioned (I or II) solvent mixtures.
4. Remove the plate when the solvent front is about 0.5 - 0.7 cm from top edge, and dry it in air or under nitrogen.
5. For identification of glycolipids spray the plate with α-naphthol solution, air dry and re-spray lightly with 95% sulphuric acid.
6. Heat the plate to 120°C until purple - blue colour develops (Figure 2B).

Subprotocol 6
Determination of Phospholipids After TLC Separation

After separation by HPTLC, phospholipids can be determined by the phosphorus assay that is sensitive to μmolar quantities (Vaskovsky et al. 1975).

Materials

Equipment

- Centrifuge
- Spectrophotometer

- Boiling water bath
- Heating block

Reagents

- Solution A: dissolve 400 mg hydrazinium chloride ($N_2H_4 \cdot HCl$) in 14 ml 4 M HCl
- Solution B: dissolve 10 g sodium molybdate ($Na_2MoO_4 \times 2H_2O$) in 60 ml 4 M HCl
- Stable stock reagent: mix Solution A with Solution B. Heat the mixture in a boiling water-bath for 20 min; after cooling add 14 ml concentrated sulphuric acid, cool and mix. Adjust the volume to 100 ml with water
- 10% H_2SO_4 in methanol
- Nitrogen gas

  Note: If kept in a dark bottle at room temperature, this reagent is stable for 6 months.
- Working reagent: add 26 ml 1 M H_2SO_4 to 5.5 ml stock reagent and adjust the volume to 100 ml with water
- Perchloric acid

Procedure

1. Detect phospholipid spots after development (use plate from Subprotocol 3), by spraying with 10% H_2SO_4 in methanol and subsequent heating at 180°C for 10 - 15 min. Lipids appear as brown spots.
2. Place the plate on a sheet of smooth-surfaced paper or aluminium foil and use a spatula to scrape off each area of silica gel completely.
3. Transfer the scraped spots into Pyrex test tubes.

   Note: As the phosphate assay is extremely sensitive it is essential to use clean test tubes.
4. Pipette 0.05 ml of perchloric acid (72%) into each test tube in a fume cupboard using a glass pipette with an automatic dispenser.
5. Take the blanks from plate areas without lipids.

6. Place marbles of appropriate diameter on top of the tubes and digest the samples by heating in an electrically heated metal block at 180-200°C for 20 min.

   Note: The colour of the solution must change from charred brown to colourless or pale yellow. Take extreme care, as hot perchloric acid may be explosive with organic material.

7. Allow the tubes to cool and add 0.45 ml of working reagent into each tube and mix well.
8. Cap the tubes with marbles and heat them for 15 min in a boiling water-bath; cool the tubes to room temperature.
9. Sediment the silica gel by centrifugation for 10 min at 2000 g and measure the absorbance of the supernatant at 815 - 830 nm against a blank.
10. Prepare a standard curve using 0.1 - 1.1 µg of phosphate. Phospholipid (in µg) in the digested lipid can be determined directly by reading the absorbance from the standard curve. This method is linear to 1.1 µg of phosphate and sensitive to less than 0.1 µg.

    Note: To determine phospholipid phosphorus in total lipid extract or fraction eluted from column or TLC plates, pipette an aliquot of the lipid solution into a test tube. Evaporate the solvent to dryness under a stream of nitrogen. Add perchloric acid, digest and assay the phospholipid as above omitting the centrifugation step.

Subprotocol 7
Determination of Esters After TLC Separation

Ester bonds in triacylglycerols, diacylglycerols, monoacylglycerols, wax and steryl esters, phospholipids, glycolipids and betaine lipids are determined by a modification of the method of Snyder and Stephens 1959 (see also Higgins 1987).

Materials

Equipment

- Desiccator
- Centrifuge

- Spectrophotometer
- Heating block

Reagents

- Stock ferric perchlorate solution: dissolve 5 g of Fe $(ClO_4)_3 \times 6H_2O$ (not yellow) in 10 ml of 70% perchloric acid and 10 ml of distilled water; dilute to 100 ml with cold absolute ethanol; this solution is stable at 4°C for several months
- Working reagent: immediately before use mix 4 ml of stock solution with 3 ml of 70% perchloric acid and dilute to 100 ml with cold absolute ethanol
- Solution A: dissolve 2 g hydroxylamine hydrochloride ($H_3NO \cdot HCl$) in a small volume of water (2.5 ml) and dilute to 50 ml with absolute ethanol
- Solution B: dissolve 4 g NaOH in a small volume of water (2.5 ml) and dilute to 50 ml with absolute ethanol
- Solution C: mix equal volumes of solutions A and B. Filter (or centrifuge) the mixture and use the supernatant as solution C

  Note: Solutions A, B and C should be prepared just before use.
- Solution D: Standard ester solution in chloroform. Triolein 1 μmol ml^{-1} or methylstearate 1 μM ml^{-1} can be used
- Iodine crystals

Procedure

1. After TLC separation (Subprotocol 4), dry the plate and then place it in a sealed desiccator of suitable size containing a few crystals of iodine. The lipids appear as brown - yellow spots or bands. This may take about 5 - 20 minutes depending on the amount of lipids. Outline the lipid spots with a soft pencil and allow the iodine to fade. This may be accelerated by warming the plate. Residual traces of iodine will not interface with the subsequent Snyder and Stephen's method.
2. Scrape the ester bonds containing lipids and a blank into the test tubes.

3. Add 1.0 ml (if large amounts of lipids were separated on 20 x 20 cm TLC plates) or 0.27 ml (if the separation was done on 10 x 10 cm TLC plates) of reagent C, place a marble on top of the tube and heat the mixture to 65°C for 2 min in a heating block or a water-bath.
4. Allow the tubes to cool and add 3.0 ml (for large amounts of lipids) or 0.8 ml (for small amounts of lipids) of working reagent, mix and allow the mauve colour to develop for 30 min.
5. Sediment the silica gel by centrifugation for 5 - 7 min at 4000 g and read the absorbance at 530 nm against a blank of a silica gel spot without lipids.
6. Prepare a standard curve using the ester standard solution D, 0 - 10 μM. The concentration (μM) of tri-, di- and monoacylglycerols and wax and steryl esters can be read directly from the standard curve. Divide the amount of these esters present by 3, 2 or 1, respectively, to determine the amount of each lipid.

Note: The detection limit of this assay is about 0.02 μM of lipid.

References

Bychek IA, Bychek EA (1996) Effect of illumination conditions on lipid and fatty acid composition of the lichen *Parmelia tinctina*. Biochemistry (Moscow) 61: 629-634

Bychek-Guschina IA, Kotlova ER, Heipieper H (1999) Effects of sulphur dioxide on lichen lipids and fatty acids. Biochemistry (Moscow) 64: 61-65

Cartwright I J (1993) Separation and analysis of phospholipids by thin layer chromatography. In: Graham J M, Higgins J A (eds) Biomembrane protocols. I Isolation and analysis. Methods in molecular biology, vol 19. Humana Press, Totowa, New Jersey, pp 153-167.

Christie WW (1982) Lipid analysis 2nd edn. Pergamon Press, Oxford

Dertien BK, De Kok LJ, Kuiper P (1977) Lipid and fatty acid composition of tree-growing and terrestrial lichens. Physiol Plant 40: 175-180

Garbus J, De Luca HF, Loomans ME, Strong FM (1963) The rapid incorporation of phosphate into mitochondrial lipids. J Biol Chem 238:59-63

Harwood JL (1998) What's so special about plant lipids? In: Harwood JL (ed) Plant lipid biosynthesis. Fundamental and agricultural applications. University Press, Cambridge, pp 1-29

Henderson RJ, Tocher DR (1992) Thin-layer chromatography. In: Hamilton RJ, Hamilton S (eds) Lipid analysis. A practical approach. University press, Oxford, pp 65-111

Higgins JA (1987) Separation and analysis of membrane lipid components. In: Findlay JBC, Evans WH (eds) Biological membranes: a practical approach. IRL-Press, Oxford, pp 103-137

Kates M (1972) Techniques in lipidology, 2nd edn. Elsevier, Amsterdam
Piervittory R, Alessio F, Usai L, Maffei M (1995) Seasonal variations in lipids of *Xanthoria parietina* growing at high elevations. Phytochemistry 40: 717-723
Shapiro IA, Ravinskaya AP, Bychek IA (1998) Physiological and lipid changes in lichens due to acute gamma radiation. J Hattori Bot Lab 84: 315-324
Snyder F, Stephens N (1959) A simplified spectrophotometric determination of ester groups in lipids. Biochem Biophys Acta 34: 244-245
Vaskovsky VE, Kostetsky EY, Vasendin IM (1975) A universal reagent for phospholipid analysis. J Chromatography 114: 129-141

Suppliers

TLC pre-coated plates (glass), layer thickness: 0.25 mm (Silica gel: 60 layers), format 20 × 20 or 10 × 20 cm.
BDH (MERCK), Merck Ltd. Hunter Boulevard, Magna Park, Lutterworth, LEICS LE17 4XN, Tel: 0800 22-33-44, Fax: 01455 55-85-86

Methylation tubes (Pyrex tubes with teflon-lined screwcaps): culture tube (16 × 125 mm), Pyrex. BDH (MERCK), Merck Ltd, Hunter Boulevard, Magna Park, Lutterworth, LEICS LE17 4XN, Tel: 0800 22-33-44, Fax: 01455 55-85-86, http://www.merck-ltd.co.uk

Measuring Ergosterol and Chitin in Lichens

LENA DAHLMAN, MARGARETA ZETHERSTRÖM, BODIL SUNDBERG, TORGNY NÄSHOLM, and KRISTIN PALMQVIST

Introduction

Mycobiont biomass dominates in most lichen thalli, and it is generally believed that the fungus acts as a major sink for carbon assimilated by the photobiont, and even nitrogen in cyanobacterial lichens (Rai 1988, Fahselt 1994). However, fungal metabolism within lichens has been poorly studied, and so far there are no studies where fungal respiration, energy or metabolite demand has been separated from photobiont requirements. Even though it has been firmly established that export of assimilates from photo- to mycobiont may be extensive (Richardson and Smith 1966, Tapper 1981, Lines et al. 1989), quantitative measures of carbon and nitrogen distribution patterns between the symbionts of lichens are largely lacking. One important part of such studies would be to find unique cellular components of the myco- and photobiont, that can be related either to their respective metabolism or to their biomass.

Possible fungal markers

There are some prominent components of lichens and fungi that can be used as such markers, namely chlorophyll for the photobiont, and chitin as well as ergosterol for the fungus. Ergosterol is the principal sterol of fungal plasma membranes (cf. Griffin 1994 and references therein) and it has

Lena Dahlman, Umeå University, Department of Ecology and Environmental Science, Umeå, 901 87, Sweden
Margareta Zetherström, Swedish University of Agricultural Sciences, Department of Forest Genetics and Plant Physiology, Umeå, 901 83, Sweden
Bodil Sundberg, Umeå University, Department of Plant Physiology, Umeå, 901 87, Sweden
Torgny Näsholm, Swedish University of Agricultural Sciences, Department of Forest Genetics and Plant Physiology, Umeå, 901 83, Sweden
✉ Kristin Palmqvist, Umeå University, Department of Ecology and Environmental Science, Umeå, 901 87, Sweden (*phone* + 46-90-7869184; *fax* +46-90-7866705; *e-mail* kristin.palmqvist@eg.umu.se)

been demonstrated that this component is also well correlated with basal respiration rates of lichens (Sundberg et al. 1999). However, it is possible that ergosterol may also be a constituent of some lichen photobionts (Goodwin, 1974), so further studies are needed to resolve whether ergosterol is unique for the fungus.

Chitin is the beta (1-4)-linked polymer of N-acetyl-glucosamine, being a cell wall component of most fungi (cf. Griffin 1994). Chitin is synthesised vectorially through the fungal plasma membrane by a transmembrane synthase. This synthase accepts glucosamine precursors from the cytosol, feeding and polymerizing them through the plasma membrane and into the wall (Gooday 1995). Depending on the fungal species, chitin content in relation to other cell wall compounds is highly variable (cf. Griffin 1994), and at least among lichens it appears that chitin content in relation to other cell wall components is higher in species with access to cyanobacterial N_2 fixation (Schlarmann et al. 1990, Palmqvist et al. 1998). Chitin content also varies within species, a trait that can vary with the nitrogen status of the individual thallus (Sundberg et al. 2001).

As for ergosterol, chitin quantification of lichens is still a developing field. Therefore, new applications will become evident as soon as more data is available. For mycorrhizal associations for instance, it has been shown that the ratio of ergosterol to chitin may be used as an indicator of the proportion of metabolically active to inactive fungal biomass in plant roots (Ekblad et al. 1998).

HPLC techniques

This chapter describes how the above two components can be extracted and quantified from intact lichen thalli, starting with material collected from the field. Some simplified guidelines for how these two components can be directly related to lichen biomass or area are also given in Subprotocol 1.

The ergosterol assay (Subprotocol 2) was originally developed for fungi as described by Salmanowicz and Nylund (1988) and results in the extraction of what has been referred to as free ergosterol (Martin et al. 1990), i.e. the ergosterol pool that is most easily extracted from fungal tissue. See also Goad and Akihisa (1997) for a comprehensive survey of sterol analysis methods.

The chitin assay (Subprotocol 3) was developed by Ekblad and Näsholm (1996), and is based on the hydrolysis of chitin and the subsequent production of a fluorescent 9-fluorenylmethylchloroformate (FMOC)-derivative of glucosamine. The method is very sensitive and the resulting derivative is stable for several days. The major problem with this method is working with the FMOC, which is highly toxic, requiring careful handling by an experienced experimentalist. These techniques require HPLC with UV- and fluorescence detection, measuring ergosterol absorption and FMOC-glucosamine fluorescence emission, respectively. Both procedures, as presented here, assume that the investigator is already familiar with HPLC techniques.

We have recently applied the two methods to a range of lichen species (c. 60-70), collected from Antarctica, Northern Canada, Sweden and Tenerife. Both protocols are highly reproducible. However, ergosterol is extremely sensitive to light and some storage conditions. Therefore, ergosterol analysis should be carried out as soon as possible after collection of the lichen, or if this is not possible, the lichen should be stored dry in darkness at, or slightly below, ambient temperature (15-20° C). Drying should be carried out in low light or in darkness, and the thallus must not be sealed in a package until it is completely dry. Freeze-drying and pulverisation should not be carried out until final analysis can be made. Additionally, ensure that freeze-dried samples do not take up humidity from the air after the freeze-drying (store them in a freezer together with silica gel until grinding, and thereafter store them in humidity-proof vials).

Subprotocol 1
Quantification of Thallus Area and Dry Weight and How to Prepare Lichen Tissue for Subsequent Extraction

Outline

Depending on final applications, and the particular species being investigated it may often be convenient to have both a measure of the lichen area and its dry weight, before extracting and quantifying sub-cellular compounds. The area should preferably be measured when the lichen is wet and fully expanded, for instance in conjunction with some physiological characterization. Thereafter, the thallus may be freeze-dried and homogenized to a powder that can be used for sub-sampling and subsequent measures of for instance C:N ratio, concentrations of chlorophyll, Rubisco, chitin and ergosterol (see Chapter 10, for Chlorophyll and Rubisco protocols). If all quantifications are based on dry weight, area or both, the different compounds may subsequently be correlated to each other on an absolute scale. If other measures, such as growth rate, photosynthesis and/or respiration have also been made, these traits may then also be related to the different compounds.

Materials

Equipment

- Leaf area meter (e.g. Li-Cor, Nebraska, USA)
- Balance with precision to the nearest 0.1 or 0.01 mg
- Porcelain mortar, pestle and liquid nitrogen [N_2 (l)] or
- Ball mill with pre-chilled stainless steel cylinders (e.g. Retsch/Brinkmann Mixer mill, Brinkmann Instruments Inc., Canada)
- Freeze dryer with vacuum pump
- Deep freezer (-50 or -80°C) or N_2 (l) storage tank

Procedure

Area determination

1. The area of most foliose and some fruticose lichens can be measured by making a photocopy of the thallus, an alternative, if the lichen is too thick to fit into the leaf-area meter directly. Measure the area when the lichen is hydrated and fully expanded, which will then be a measure of its projected area when being metabolically active. Flatten the lichen gently and place it on the photocopier. Close the lid and make a copy. Check that the copy and the lichen have equal sizes.

2a. Cut the copy from the paper and measure its area in a leaf area meter. Alternatively, weigh the copy and compare with the weight of a known area cut out from the same piece of paper.

2b. Alternatively, lichen area can also be determined with a computer scanner, or any other leaf area device connected to a computer with image analysis software. Many lichens are too thick also for these devices, however. Therefore, making a photocopy may be an alternative also in this case.

Dry weight determination

3. Dry the thallus according to one of the methods described below; the choice of method will depend on the final application and/or future use of the lichen.
   - Air-drying over night is a mild method suitable if the thallus is to be rewetted and used for further metabolic measurements or for field transplantation.
   - Freeze-drying is the best choice when no further physiological measurements will be made, a method that is also better than drying at high temperature in an oven if one wishes to quantify sub-cellular compounds, because oven drying can destroy or modify cell components.

4. Pulverize the dried lichen either in a ball mill or a mortar. In the latter case, a homogenous powder will be obtained if the thallus is placed in a chilled mortar and liquid nitrogen is added. If using a ball mill, it is advisable to chill cylinders and other devices in a freezer or with liquid nitrogen prior to pulverization. This way, excessive breakdown of the different sub cellular compounds (e.g. ergosterol) will, as far as possible, be avoided.
5. Store the dried and pulverized lichen material in sealed vials and in boxes with silica gel in darkness, in a deep freezer, preferably at −80°C or in liquid nitrogen, altough -20°C is also possible for shorter periods.

Subprotocol 2
Ergosterol Extraction and Detection

Outline

Ergosterol extraction is a relatively simple procedure that can be completed in 1-2 hours of laboratory work, starting with pulverised lichen material, involving three major Steps:

- Pulverisation of freeze-dried lichen material and weighing of samples
- Extraction of ergosterol in 99.5% ethanol
- Determination of ergosterol content with HPLC

Materials

Equipment

- HPLC (e.g. Waters, Milford, USA) equipped with
  - PC and/or Integrator
  - One Pump
  - Autosampler (optional)
  - UV-detector (280 nm)
  - ODS ultra sphere column (250 cm x 4.6 mm; particle size 5 μm) (e.g. Beckman, Fullerton, USA)

- Vortex
- Centrifuge for Eppendorf vials
- Balance with precision to nearest 0.01 mg
- Eppendorf tubes (1.5 ml) with screw cap

Chemicals

- Ethanol (99.5%), HPLC grade not necessary
- Methanol, HPLC-grade
- Ergosterol standard, Sigma; E-6510 (Sigma-Aldrich, S. Louis, USA)

Procedure

Note: Ergosterol is sensitive to light and extensive losses may occur both in standards and sample. It is therefore important to avoid light exposure throughout all Steps, and fresh standards should be prepared for each run. If the ergosterol standard turns yellow, re-crystallisation in ethanol is needed. Intact lichens appear to be more resistant than pulverised material to ergosterol losses during storage; therefore freeze-drying and pulverisation should be carried out shortly before analysis.

Extraction

1. Weigh 10-20 mg of freeze-dried and pulverized lichen material into a 1.5 ml screw cap Eppendorf tube. The sample amount is dependent on the lichen species, but 10 mg of lichen powder should be sufficient for a pilot experiment.
2. Add 1 ml of 99.5% ethanol to each lichen sample and vortex vigorously for a few seconds.

Incubation

3. Place the tube on a shaker and incubate the sample for 30 min at room temperature in the dark.
4. Vortex the sample vigorously again, and centrifuge with an Eppendorf centrifuge at maximum speed (c. 20 000 g) for 15 min.

Testing the yield of ergosterol

5. The resulting supernatant contains the free ergosterol. Check the yield of ergosterol from the investigated material by repeated extraction of the pellet with fresh ethanol and measurements of ergosterol content in each extraction. Usually, c. 80% of all free ergosterol is extracted after

carrying out Steps 3 and 4 (cf. Nylund and Wallander 1992). To extract also the bound fraction of ergosterol it is possible to use the procedure of Salmanowicz and Nylund (1988) whereby the sample is heated in KOH. This way, an additional 5% of the total ergosterol pool could be extracted.

Pellet contains chitin

6. Collect all supernatants and transfer to a new Eppendorf tube and save the pellet if chitin analysis will be made. For HPLC analysis at least 50 µl (see next step) is required. The supernatant can be stored in a deep freezer for a maximum of 1-2 days, but it is advisable to perform the analysis immediately after extraction.

Separation and detection

7. Pipette sufficient sample (ca. 100 µl) into a coloured vial, protecting the extracted ergosterol from light. Set the auto injector to 50 µl, and separate the ergosterol on a reversed phase ODS ultra sphere column using methanol as the mobile phase. Set flow rate to 1.5 ml min^{-1} and total analysis time to 12 min. With our system, the retention of ergosterol is 5-10 min. The ergosterol peak is detected with a UV-detector at 280 nm.
8. For quantification of the ergosterol concentration in the lichen samples, produce a standard curve with different amounts of ergosterol. The lichens we have investigated so far contained 0.1-1.8 mg g^{-1} DW ergosterol, so if 10 mg lichen powder is extracted with 1 ml ethanol, the standard curve samples should at least range from 1 to 20 µg ergosterol dissolved in 1ml ethanol. Figure 1 shows chromatograms of an ergosterol standard and a lichen extract.
9. Test the recovery of ergosterol to ensure that it is not destroyed during the extraction. Divide the freeze-dried lichen powder into 10 sub-samples and analyse five of these as described above, and spike five others with known amounts of standard solution. At least 90 % of the added standard should be recovered, otherwise indicate recovery rate in publications. However, losses of ergosterol occurring prior to final extraction will not be detected in this way.

Troubleshooting

Because both ergosterol extraction and its detection are relatively simple procedures there are few sources of errors. However, even if freeze-dried, pulverised lichen material can take up water from surrounding air, parti-

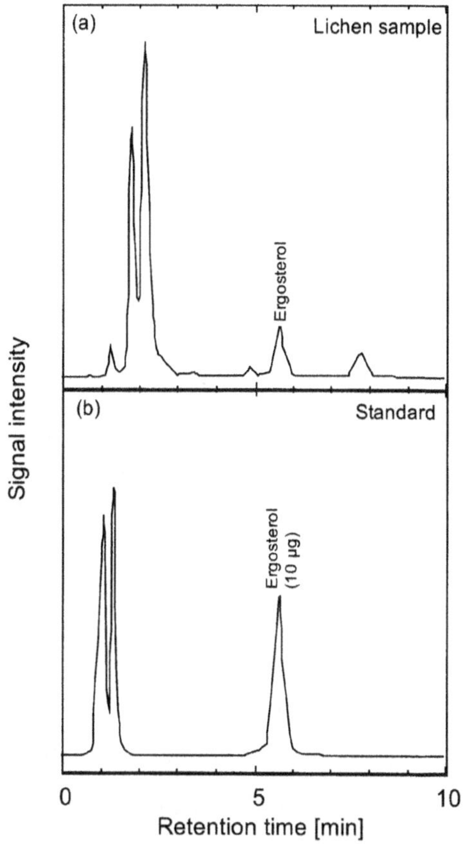

Fig. 1. HPLC chromatogram for an ethanolic extract of pulverised lichen material containing ergosterol (a) and for ergosterol standard dissolved in ethanol (b). Note that the retention time of ergosterol may vary with the particular HPLC system and column. In this case, the retention time of ergosterol was 5-6 min.

cularly if deep-frozen material is exposed to humid and warmer air. This can be avoided by handling the samples over silica gel. Also, clean the lichen thallus thoroughly before grinding, because small amounts of sand, dirt or moss may contribute significantly to the weight of small samples.

Subprotocol 3
Chitin Extraction and Detection

Outline

We have developed a chitin extraction and detection procedure where protein and amino acids are removed with NaOH and repeated washes with water. This is necessary because the FMOC-Cl reacts with all molecules with free amino groups (NH_2) in the sample and because protein and amino acid concentrations are often significantly higher than chitin concentrations. However, some chitin may be lost in this Step (Ekblad and Näsholm, 1996), a loss that can be quantified by spiking a series of known chitin concentrations with a number of otherwise identical samples.

Chitin extraction and detection is a long procedure, starting with pulverised lichen powder or, alternatively, the pellet resulting after extraction of ergosterol (see Subprotocol 2). The procedure comprises four major parts:

- removal of all free amino acids and proteins before chitin hydrolysis (Step 1-8).
- chitin hydrolysis, yielding glucosamine (Step 9-12)
- derivatization of glucosamine with FMOC-Cl (Step 13-15)
- detection and quantification of FMOC-glucosamine with HPLC (Step 16-17).

It is most convenient to carry out the procedure over four days, because several Steps require long incubation. Logical breaks to stop the procedure for the day are included in the protocol.

Materials

Equipment

- HPLC (e.g. Waters, Milford, USA) equipped with
  - Two pumps
  - Gradient controller (Waters 600)
  - PC and/or Integrator
  - Autosampler
  - Fluorescence detector

  - ODS hypersil column (250 cm x 4.6 mm; particle size 5 μm) (e.g. Beckman, Fullerton, USA)
- Balance with precision to 0.1 or 0.01 mg
- Vortex
- Heating block for Eppendorf tubes
- Centrifuge for Eppendorf vials
- Vacuum centrifuge
- Eppendorf tubes (1.5 ml) with screw caps
- Standard Eppendorf tubes (1.5 ml)

Chemicals

- FMOC-Cl (9-fluorenylmethylchloroformate), 15 mM in acetone

  Caution: This chemical is highly toxic and must be handled with care. Note product information given by the supplier.
- n-Heptane
- 0.2 M NaOH
- 6 M HCl
- Borate buffer, 1.0 M, pH 6.3
- Glucosamine standard - G-1514 Sigma-Aldrich
- 5 μM Homocysteic acid, H-2257 Sigma-Aldrich
- Methanol, HPLC-grade (solvent A)
- Acetic acid buffer, 0.8%, pH 4.2 (solvent B)

Triethylamine	1 ml
Acetic acid (glacial)	8 ml
Make up to 1000 ml with ultrapure water (Millipore) and adjust the pH with 10 M NaOH	

Note: Filter solvent B through a 0.45 μm Millipore filter before use.

Procedure

Day 1. Removal of proteins

1. Weigh 10-20 mg of lichen powder into a 1.5 ml screw cap Eppendorf tube. Alternatively, start with the pellet remaining after ergosterol extraction (see Subprotocol 2). According to our observations there is no loss of chitin during the ergosterol extractions but there could be a difference between different species. This can be tested by comparing the chitin content of two samples with or without previous ergosterol extraction.
2. Add 1 ml of 0.2 M NaOH to the pellet, or to the dry material to remove amino acids and proteins that can interfere with the assay. Vortex the tube vigorously to suspend the pellet.
3. Incubate the tube for 6 h at room temperature on the lab bench.
4. Vortex the sample, centrifuge in an Eppendorf centrifuge at maximum speed (c. 20000 g) for 15 min and discard the resulting supernatant.
5. Add another 1 ml of 0.2 M NaOH to the pelleted sample and vortex vigorously to suspend the pellet.
6. Transfer the tube to a heating block and incubate for 17.5 h at 100 °C (do not tighten the cap before the tube has heated somewhat).

Day 2. Washing

7. Vortex the Eppendorf tube vigorously and centrifuge in an Eppendorf centrifuge at maximum speed (c. 20000 g) for 15 min. Discard the resulting supernatant.
8. Add 1 ml ultrapure water (Millipore) to the Eppendorf tube, vortex vigorously to suspend the pellet and centrifuge the tube in an Eppendorf centrifuge at maximum speed (c. 20000 g) for 15 min (at ambient temperature). Discard the supernatant. Repeat this Step three more times. If a large series of samples is being processed simultaneously, these washing Steps will require almost a full day of laboratory work.

Day 3. Chitin hydrolysis

9. To start the hydrolysis, add 1 ml of 6 M HCl to the pellet resulting from the last wash, vortex and incubate at room temperature on the lab bench for 1 h.
10. Vortex the sample, then hydrolyse it by transferring the tube to the heating block and incubate for 7 h at 100 °C (again, do not tighten the cap before the tube has heated somewhat).

11. After hydrolysis, vortex the sample vigorously and centrifuge the tube in an Eppendorf centrifuge at maximum speed (c. 20000 g) for 15 min. The resulting supernatant contains the glucosamine resulting from chitin hydrolysis.

12. Transfer 40 µl of the acid hydrolyse product to a new Eppendorf tube and evaporate this solution in a vacuum centrifuge. This may take between 20-40 min depending on the number of samples and the particular Speed-vac capacity.

    Note: From here it is possible to leave the sample in a freezer over night.

Day 4. Derivatization and analysis

13. Add 200 µl of internal standard (5 µM homocysteic acid) to the tube and incubate on the bench for 1-2 hours to suspend the pellet.

14. In a fume hood, add 250 µl of FMOC-Cl reagent dissolved in acetic acid buffer and 50 µl of borate buffer, and vortex carefully. Incubate for 10 min.

    Note: This reagent is highly toxic. Use protective clothing and gloves and work carefully!

15. Remove excess FMOC-Cl by two-phase partitioning of the sample against 1 ml of n-Heptane. Vortex carefully and discard the upper heptane phase. Repeat this procedure once if hydrolyzed FMOC-OH interferes with FMOC-glucosamine in the HPLC separation. The sample is now ready for injection in the HPLC.

16. Analyse 10-50 µl of the sample by HPLC. The sample volume is dependent on chitin concentration. Separate FMOC-glucosamine on an ODS hypersil column using the following methanol:acetic acid buffer gradient: 0-8 min 40:60; 8-15 min 50:50; 15-20 min 100:0; 20-22 min 100:0; 22-25 min 40:60 (numbers referring to volume proportions of methanol to acetic acid buffer). Set the flow rate to 1 ml min^{-1} and total analysis time per sample to 38 min.

17. The FMOC-glucosamine is detected by fluorescence emission at 330 nm with excitation at 265 nm. The resulting chromatogram (Fig. 2) contains two major glucosamine peaks with approximately 18 and 20 min retention time, respectively. Either of these can be used for the quantification of glucosamine. If an internal standard is used, the response factor of the chosen product (peak one or two) relative to that of the internal standard must be determined. If external standards are preferred, a standard-curve with different concentrations of

glucosamine should be produced instead. Lichens can contain 0.5-100 mg g^{-1} DW chitin. So if 10 mg lichen powder is used, the standard curve should at least range from 5-1000 µg glucosamine.

Testing the yield 18. The above procedure typically extracts 85-90% of total chitin from fungal and lichen material. Yield can be tested both by repeated extraction of the same pellet in fresh HCl (Step 9) and prolonged incubation (Steps 10) until no further chitin can be extracted from the pellet. Too long an incubation may result in losses, so the above procedure appeared to be optimum, but it may be advisable to check the yield for the particular material being investigated.

Troubleshooting

For using HPLC-systems, it is often difficult to give exact proportions of buffers used for gradients and precise retention times. If peaks are not clearly separated, adjustment of buffer proportions and concentrations in the mobile phase might be necessary. As for the ergosterol quantifica-

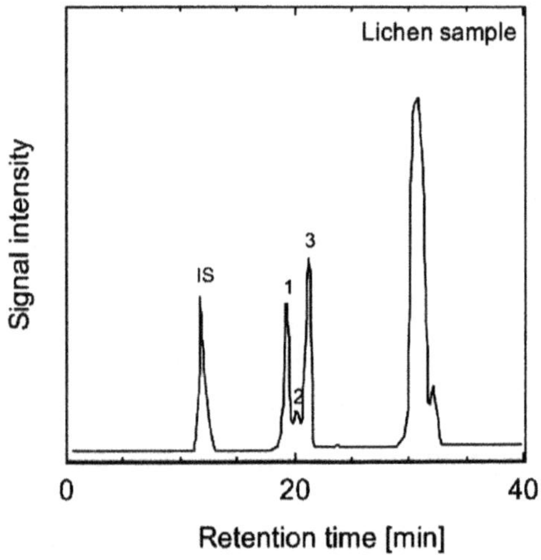

Fig. 2. HPLC chromatogram for FMOC-Cl derivatives of glucosamine extracted from pulverized lichen material. The peaks 1, 2 and 3 are the three peaks formed by glucosamine derivatives; IS = internal standard (homocysteic acid); the small peak to the right of peak 3 comes from impurities in the solution; the big peak to the right is the hydrolysis product of the reagent.

tion, care must be taken to determine sample dry weights as accurately as possible.

Acknowledgements. Alf Ekblad and Jörgen Persson are acknowledged for their pioneer work with lichen ergosterol and chitin analyses. The Swedish Polar Institute and our participating colleagues on the TNW 99 expedition, Jan-Erik Mattson and Martin Westberg, provided material from the Canadian Arctic, Fernando Valladares collected the Antarctic lichens, Karl-Erik Renhorn collected and identified numerous specimens from Sweden and Tenerife. The above protocols have been tested and developed thanks to grants from the Swedish Research Council for Environment, Agricultural Sciences and Spitial Planning to Kristin Palmqvist and Torgny Näsholm, and Center for Environmental Research (CMF, Umeå, Sweden) to Lena Dahlman. We also wish to thank the editors of this book for inviting us to write this chapter and for helping us to improve the protocols.

References

Ekblad A, Näsholm T 1996 Determination of chitin in fungi and mycorrhizal roots by an improved HPLC analysis of glucosamine. Plant Soil 178: 29-35

Ekblad A, Wallander H, Näsholm T 1998 Chitin and ergosterol combined to measure total and living fungal biomass in ectomycorrhizas. New Phyt 138: 143-149

Fahselt D 1994 Carbon metabolism in lichens. Symbiosis 17: 127-182

Goad LJ and Akihisa T 1997 Analysis of sterols. Blackie Academic and Professional, London

Gooday GW 1995 Cell walls. In: Gow NAR, Gadd GM (eds) The Growing Fungus. Chapman and Hall, London, pp 43-62

Goodwin TW 1974 Sterols. In: Stewart WDP (ed) Algal Physiology and Biochemistry. Blackwell, Oxford, pp 266-280

Griffin DH. 1994. Fungal Physiology, Wiley & Sons, New York

Lines CEM, Ratcliffe RG, Rees TAV, Southon TE 1989 A ^{13}C NMR study of photosynthate transport and metabolism in the lichen *Xanthoria calcicola* Oxner. New Phyt 111: 447-456

Martin F, Delaruelle C, Hilbert J-L 1990 An improved ergosterol assay to estimate fungal biomass in ectomycorrhizas. Mycol Res 94: 1059-1064

Nylund J-E, Wallander H. 1992. Ergosterol analysis as a means of quantifying mycorrhizal biomass. In: Norris JR, Read DJ and Varma, A (eds) Methods in microbiology Vol. 24. Academic Press, London, pp 77-88

Palmqvist K, Campbell D, Ekblad A, Johansson H 1998 Photosynthetic capacity in relation to nitrogen content and its partitioning in lichens with different photobionts. Pl Cell Env 21: 361-372

Rai AN. 1988. Nitrogen metabolism. In: Galun M (ed) CRC handbook of lichenology, vol I. CRC Press, Boca Raton, pp 201-237

Richardson DHS, Smith DC 1966. The physiology of the symbiosis in *Xanthoria aureola* (Ach.) Erichs. Lichenologist 3: 202-206

Salmanowicz WE, Nylund J-E 1988 High performance liquid chromatography determination of ergosterol as a measure of ectomycorrhizal infection in Scots pine. Eur J For Path 18: 291-298

Schlarmann G, Peveling E, Tenberge K 1990 The occurrence of chitin in the cell walls of ascomycete mycobionts. Bib Lichen 38: 395-409

Sundberg B, Näsholm, Palmqvist K (2001) The effect of nitrogen on growth and key thallus components in the two tripartite lichens, *Nephroma arcticum* and *Peltigera aphthosa*. Plant Cell Env 24: 517-527

Sundberg B, Ekblad A, Näsholm T, Palmqvist K. 1999 Lichen respiration in relation to active time, temperature nitrogen and ergosterol concentrations. Func Ecol 13: 119-125

Tapper R 1981 Direct measurement of translocation of carbohydrate in the lichen *Cladonia convoluta* by quantitative autoradiography. New Phyt 89: 429-437

Suppliers

Beckman Coulter Ltd., Worldwide Headquarters, 4300 N. Harbor Boulevard PO Box 3100, Fullerton, CA 92834-3100, USA, Internet http://www.beckman.com

Li-Cor Inc., PO Box 4425, Lincoln, NE 68504, USA, Email: envsales@env.licor.com, Internet: http://www.licor.com

Retsch/Brinkmann, Brinkmann Instruments, Inc., Several Worldwide distributors, Internet: http://www.brinkmann.com

Sigma-Aldrich Ltd., P.O Box 14506, St. Louis, MO 63178, USA, Email: sigma@sial.com, Internet http://www.sigma-aldrich.com

Waters Corporate Headquarters, 34 Maple Street, Milford, Massachusetts 01757, USA, Email: info@waters.com, Internet: http://www.waters.com

Analysis of Chlorophylls, Carotenoids, and Tocopherols in Lichens

HARTWIG W. PFEIFHOFER, REGINA WILLFURTH, MARGRET ZORN, and ILSE KRANNER

Introduction

To accurately measure the contents of chlorophylls and carotenoids in lichen tissues, and to determine their metabolic roles, requires careful selection of appropriate techniques and equipment. Despite the progress in the chromatography of pigments it is still frequent practice to measure the amounts of individual chlorophylls and total carotenoids spectrophotometrically in mixed extracts. However, high-performance liquid chromatography (HPLC) techniques have facilitated the separation and quantification of plastid pigments. Fast one-step separations of chlorophylls and carotenoids have been described (e.g. Siefermann-Harms 1988, Pfeifhofer 1989, Thayer and Björkman 1990). Reversed-phase partition chromatography is now most widely used for the routine analysis of pigments in natural extracts (for reviews see Bramley 1992, Young et al. 1997).

In lichens, the presence of acidic secondary metabolites causes major problems, because these compounds can damage pigments during extraction unless suitable precautions have been taken. This chapter describes sample pre-treatment, extraction and the analysis of chlorophylls and carotenoids using HPLC (Subprotocol 1). Additionally, brief instructions for spectrophotometric quantification of chlorophylls and total carotenoids are given (Subprotocol 2). The same extracts prepared for analysis of plastid pigments can also be used for HPLC analysis of tocopherols, which are lipid soluble antioxidants present in membranes of photobionts. A brief description of the tocopherol assay is included in Subprotocol 3. For de-

✉ Hartwig W. Pfeifhofer, Karl-Franzens Universität Graz, Institut für Pflanzenphysiologie, Schubertstraße 51, Graz, 8010, Austria (*phone* +43-316-380-5639; *fax* +43-316-380-9880; *e-mail* hartwig.pfeifhofer@kfunigraz.ac.at)
Regina Willfurth
Margret Zorn
Ilse Kranner

tailed description of a method for determination of the water-soluble antioxidant glutathione in lichens see Kranner (1998). The extract produced in Subprotocol 1 can also be used for determination of ergosterol, which is exclusively present in the mycobiont. For more details about this method see Chapter 21.

Outline

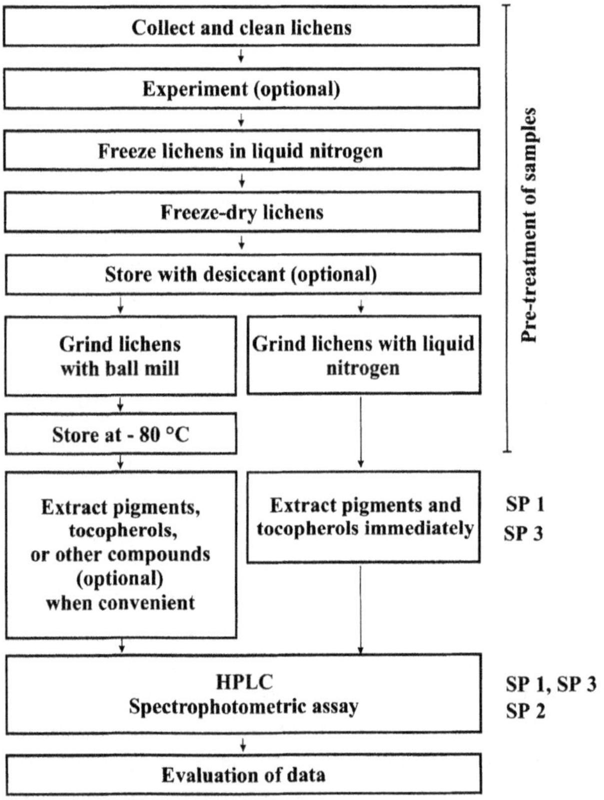

Fig. 1. Outline of procedures described in this chapter. SP, Subprotocol.

Materials

General notes

- All chemicals used should be of analytical grade quality.
- Work in dim light and use dark centrifuge tubes and autosampler vials to avoid pigment destruction by light during extraction and handling. Because of the thermolability of many carotenoids, samples should **not be heated.**

Equipment

- Freeze dryer
- Machine for annealing plastic bags (available in kitchen shops)
- Ball mill for small samples with agate balls (e.g. "dismembrator", Braun)
- Humidity-proof vials with screw caps
- Deep freezer (-25°C or lower, ideally -80°C)
- Refrigerated centrifuge for Eppendorf vials
- Vortex
- HPLC-system consisting of :
  - two high pressure pumps (only one pump needed for Subprotocol 3)
  - an integration and gradient controlling software package
  - a sample injector with a 20 µl dosing loop or autosampler (optional)
  - a RP-18 column (250 x 4.6 mm i.d., Spherisorb S5 ODS-2, 5 µm particle size) and
  - an UV/Vis variable wavelength detector or Diode-array-detector (DAD) (optional).
- Spectrophotometer (required for Subprotocol 2)
- Fluorescence detector (only required for Subprotocol 3)

Reagents

- Liquid nitrogen
- The following extraction media are useful: dimethyl sulfoxide (DMSO), ethanol, acetone, and acetone containing 0.1% N-ethyldiisopropylamine. The extraction efficiency of these media is demonstrated in Figure 2.
- $CaCO_3$ or $MgCO_3$

- Solvent A, acetonitrile : water : methanol = 100 : 10 : 5 (v/v/v)
- Solvent B, acetone : ethyl acetate = 2 : 1 (v/v)

  Note: For preparation of solvents A and B use HPLC-grade reagents, and double distilled or ultra-pure water (Milli Q-Plus, see list of suppliers). Filter both solutions through a 0.2 µm membrane filter before use.

- Methanol, HPLC grade (required for Subprotocol 3)
- Some standards of plastid pigments are commercially available from Sigma and Roth. DHI - Water & Environment provide additional standards of carotenoids (see list of suppliers)

Subprotocol 1
Analysis of Chlorophylls and Carotenoids by High Performance Liquid Chromatography

Procedure

Pre-treatment of the samples

Freeze-drying

We strongly recommend that lichen samples be freeze-dried prior to pigment extraction. Even a minimal water content causes pigment destruction during sample processing due to co-extraction of acidic lichen compounds. Further advantages of the freeze-drying procedure are: First, freeze dried samples can be stored at -25°C for one year at least. Second, lichens are less tough after removing all water and can be ground to a homogeneous powder easily. Thus, problems with disruption of cells are avoided and the yield of compounds (also of enzymes) is usually dramatically enhanced. Third, the same powder can be used for different protocols thus allowing a better statistical correlation of the results.

1. Clean the lichen material, weigh and put it into small paper bags, and close the bags with paper clips.
2. Put paper bags containing the lichens into liquid nitrogen and freeze-dry immediately. Short-term storage for a few days in a deep freezer (preferably -80°C) is feasible.

3. After the freeze-drying is completed, weigh the samples and record their dry weight. **Caution:** Take care not to expose lichens to ambient air, because humidity will cause sample destruction! Transport samples to the balance in a desiccator over silica gel.

4. Put the paper bags containing the lichens into a plastic bag with silica gel, remove the air from the plastic bag by applying a vacuum (if available), and anneal the plastic bag. Repeat the procedure with a further plastic bag to ensure that the samples are not destroyed if one bag leaks. The samples can now be stored in a deep freezer until grinding.

Grinding samples with a ball mill

5. Fill lichen thalli and grinding balls into the sample flask (teflon bomb) of the dismembrator ball mill and cool it with liquid nitrogen before homogenisation to avoid warming of the sample during grinding.

6. Homogenise the sample with the ball mill. The duration of grinding depends on lichen species, sample amount, and size of the sample flask. Usually homogenisation is completed within 7 to 10 minutes.

7. Before opening, let the sample flask adapt to room temperature in a desiccator over silica gel to avoid condensation of water (air humidity) on the lyophilised powder.

8. Transfer the powder to humidity proof vials quickly (less than 2 min). Vials can be stored at -25°C until use. It should be safe to store these vials in a freezer, however, if you want to be on the safe side, put them into plastic bags with silica gel, and anneal the plastic bag as described above. If the samples are treated carefully and never wetted, they can be stored for a long time (probably more than three years), but we recommend immediate analysis.

9. Before analysis, remove the vials from the freezer and let them adapt to ambient temperature to avoid condensation of humidity on the powder. We recommend the vials be placed into a desiccator over silica gel. After withdrawing powder for one sample, close the vial immediately to prevent destruction of the remainder caused by moisture, and put it back into the freezer as soon as possible.

Extraction

Note: Lichens frequently contain high amounts of secondary products, which destroy chlorophylls during extraction. For neutralisation of acids, $MgCO_3$ or $CaCO_3$ is added to the lichen powder. Moreover, the use of organic bases, e.g. N-ethyldiisopropylamine, as solvent additives has proven to be advantageous for pigment extraction from many lichens. Acetone, ethanol (96%), and dimethyl-sulfoxide (DMSO) have been used successfully as solvents. However, it is **not advisable** to extract chlorophylls and carotenoids from lichens with 80 % acetone (see Troubleshooting). Extraction with ethanol can be superior to that with acetone (e.g. Cañas and Pignata 1998). In our laboratory, the following procedure has proven to be most effective for extraction and analysis of chlorophylls and carotenoids from *Pseudevernia furfuracea* (see Figure 2). Nevertheless, we recommend testing which solvent is best for your lichen material.

10a. Put 50 mg lyophilised lichen powder into brown 1.5 ml Eppendorf tubes (micro-centrifuge tubes). Add a spatula tip (approx. 50 mg) of $CaCO_3$.

11a. Extract with 1 ml DMSO.

Note: 100 % DMSO is solid at temperatures below 18°C. Therefore work at temperatures above 20°C. However, when working with powdered lichen material, it is not necessary to treat the sample with temperatures higher than 30°C during extraction. **Caution**: DMSO is an irritant. Work in a fume hood and avoid contact with skin (wear gloves).

12a. Shake samples well for approximately 30 s using a vortex.

Alternative extraction method using mortar and pestle

If no ball mill is available, lichens can alternatively be ground using mortar and pestle, which requires modification (Steps 10b - 12b) of the extraction procedure.

10b. Put 40 - 60 mg lichen material, 50 mg $CaCO_3$, and a small amount of liquid nitrogen into a small mortar and grind with a pestle. Add a spatula tip of quartz sand if lichen thalli are tough and difficult to homogenise.

11b. After evaporation of the liquid nitrogen add 1.0 ml DMSO:ethanol = 2 : 1 (v/v) and grind gently for c. 30 sec.

12b. Transfer the homogenate (using a Pasteur pipette or a pipette with a disposable polypropylene tip) to a dark centrifuge tube, add a further 0.5 ml extraction medium to the mortar to remove residue, and pool with the previous homogenate.

13. Centrifuge for 5 min at 20°C and at least at 20000 g.

14. Decant the supernatant into graduated glass tubes and store them in the dark (cover the glass tubes with aluminium foil).

15. Re-suspend the pellet in 1 ml 96 % ethanol (or 0.75 ml DMSO:ethanol = 2:1, if lichens have been ground using mortar and pestle), vortex vigorously and centrifuge for at least 5 min at 20000 g.

16. Combine the supernatants and repeat Steps 14 and 15 until the pellet is colourless.

    Note: Generally, two to three extraction steps are necessary for quantitative extraction of plastid pigments. Further extraction steps do not enhance the yield of extracted pigments significantly; by contrast, they might be disadvantageous, because extracts become too dilute.

17. Record the volume of the combined supernatants.

HPLC

18. To remove cell debris that would block the analytical column and shorten its lifespan, centrifuge samples (at least at 20000 g for 20 min) prior to injection into the HPLC system, then pipette supernatants into dark autosampler vials. Alternatively, pass extracts through a membrane filter (0.22 µm pore size).

Settings

19. Separate pigments using the following settings:
    - Column: Spherisorb ODS-2 (250 x 4.6 mm i.d.), particle size 5 µm
    - Flow rate: 1 ml/min
    - Gradient: Linear gradient from 10 to 70 % solution B within 18 min, then 4 min at 70 % solution B (until elution of β carotene). Using these settings, adequate resolution, even of the xanthophyll isomers lutein and zeaxanthin, can be achieved (Figure 3)
    - Detection: If plastid pigments in the sample are already identified and a quantitative determination of these is the aim of the study, use a UV/vis variable detector, wavelength set at 440 nm. If the pigment pattern of your lichen species has not been studied so far, or if identification of unknown plastid pigments is desired, use a diode-array detector (DAD). Record the absorbance spectra of all peaks in a range from 250 to 550 nm. UV/Vis detectors are more sensitive than DADs, but the latter allow for recording spectra of compounds and can thus be used in the identification process (see below).

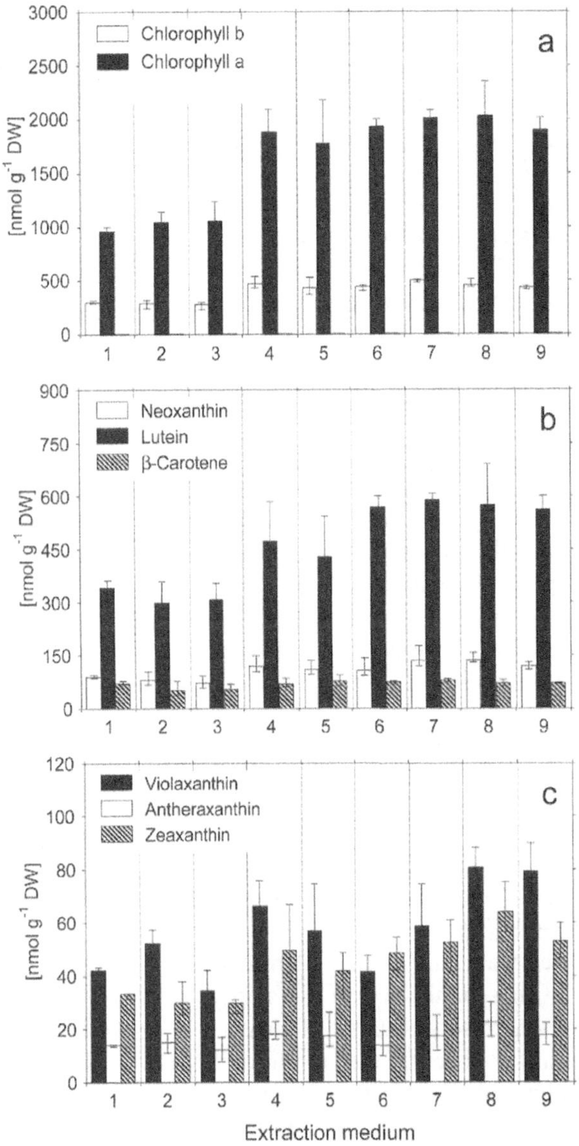

Fig. 2. Extraction efficiency of various solvents used for extraction of the following pigments from the lichen *Pseudevernia furfuracea:* chlorophylls (**a**), the carotenoids neoxanthin, lutein and ß-carotene (**b**), and the xanthophyll cycle pigments violaxanhin, antheraxanthin and zeaxanthin (**c**). Data are means and SD for n = 8-20. Extraction media:

1 pure acetone
2 pure acetone containing 0.1% N-ethyldiisopropylamine
3 96 % ethanol
4 DMSO (room temperature)
5 DMSO (sample heated to 60°C)
6 DMSO followed by washing twice with 96 % ethanol
7 96 % ETOH followed by washing twice with DMSO
8 DMSO followed by washing with DMSO and then pure acetone containing 0.1% N-ethyldiisopropylamine
9 DMSO followed by washing twice with pure acetone containing 0.1% N-ethyldiisopropylamine

20. Equilibrate the column with 10 % B for 10 min before next run.

21. Prepare standards of the following plastid pigments that are usually present in green algae (Goodwin 1980, Goodwin and Britton 1988): chlorophyll a and b, ß-carotene, lutein, violaxanthin, antheraxanthin, zeaxanthin, and neoxanthin. For lichens with cyanobacterial photobionts, standard solutions of chlorophyll a, ß-carotene, zeaxanthin, canthaxanthin and echinenone are required. Dilute a known amount of standard in an appropriate solvent (ethanol, acetone) and analyse as described above.

Note: If the pigment pattern of your lichen species is already known, continue at Step 26.

Identification of unknown compounds

22. Compare retention times in the HPLC trace and spectrum of each compound of the lichen extract with those of authentic standards. The use of a DAD significantly reduces the time for identification, because spectral and chromatographic profiles can be recorded simultaneously. If such a facility is not available, collect fractions containing the separated pigments and check the absorbance properties of the fractions in the range between 250 and 550 nm using a spectrophotometer (use quartz cuvettes).

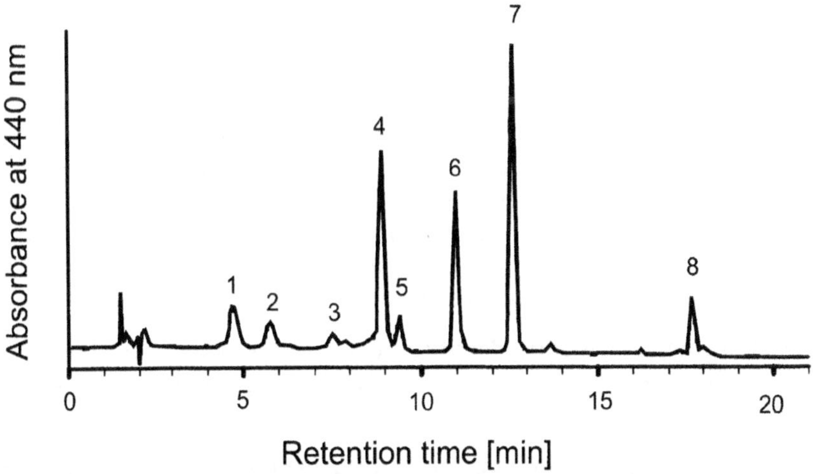

Fig. 3. HPLC separation of chlorophylls and carotenoids of a *Pseudevernia furfuracea* extract. *1*, neoxanthin; *2*, violaxanthin; *3*, antheraxanthin; *4*, lutein; *5*, zeaxanthin; *6*, chlorophyll b; *7*, chlorophyll a; *8*, β-carotene.

23. Compare the position of long-wave absorption bands (usually three bands in the 400 - 500 nm region) of carotenoids with published data (e.g. Davies 1976).

    Note: The absorption spectra of carotenoids are solvent dependent. Davies (1976) did not quote absorbance maxima of carotenoids dissolved in acetonitrile. To our knowledge, however, absorbance properties of carotenoids in the acetonitrile-based HPLC solvents described above are very similar to those observed when using ethanol, methanol or acetone.

24. There are a number of simple chemical tests that are useful for the determination of carotenoid structure; only a few of these are described here.
    - **Epoxides.** The most useful test for identification of 5,6-epoxides is that based on their acid-catalysed isomerisation to the 5,8-form. A typical procedure is to add a drop of concentrated hydrochloric acid to an ethanolic solution of the carotenoid in a spectrophotometer cuvette. 5,6-monoepoxides (e.g. antheraxanthin) are characterised by a hypsochromic shift (i.e. shift of maxima to shorter wavelengths) of 17-22 nm, while 5,6,5',6'-diepoxides show a hypsochromic shift of some 40 nm.
    - **Carbonyl groups.** Add a small portion (20 mg) of solid sodium borohydride to a 95% ethanol solution of the carotenoid in a spectrophotometer cuvette. Monitor spectral changes. If the carotenoid has a carbonyl group that is conjugated with the polyene chain, a hypsochromic shift can be observed.
    - **Hydroxyl groups.** Treat a solution of the carotenoid in chloroform in daylight with a few drops of a saturated solution of HCl in chloroform. Allow the reaction to proceed for about 15 min and monitor spectral changes. Dehydration induced by this treatment introduces a further conjugated double bond into the chromophore. This change results in a spectral shift to higher wavelengths. Additional chemical reactions have been described by Davies (1976).

25. If pigments cannot be identified by the methods described above, you need to apply additional physical and chemical identification methods such as NMR and HPLC-MS.

Quantitative evaluation

26. Calculate the amount of pigment using the results of standards. If the concentration of pigment in the standard solution is unknown (e.g. where collected fractions are to be used either to identify the compounds or for preparing standards), calculate the concentration using

the extinction coefficients for the appropriate wavelength (usually λ_{max}) as shown in Table 1. See Davies (1976) for data of carotenoids not listed in Table 1. When for a given carotenoid no absorbance coefficient is available, a value of $E_{1\,cm}^{1\,\%} = 2500$ in ethanol is useful as an approximation (Davies 1976; Table 1).

27. Test the recovery of plastid pigments to ensure that pigments are not destroyed during extraction. Divide the freeze-dried lichen powder into 10 sub-samples and analyse five of these as described above, and spike five others with known amounts of standard solution. At least 90 % of the added standard should be recovered (otherwise indicate recovery rate in publications).

Table 1. Specific absorbance coefficients of chlorophylls and most common carotenoids according to Hager and Meyer-Bertenrath (1966) and Stransky and Hager (1970).

Compound	Solvent	Specific extinction coefficient at λ max	λ max [nm]
Chlorophyll a	acetone	840	663
Chlorophyll b	acetone	518	645
ß-Carotene	chloroform	2200	464
Violaxanthin	ethanol	2500	441
Lutein	ethanol	2540	446
Antheraxanthin	ethanol	2350	446
Neoxanthin	ethanol	2270	438
Zeaxanthin	ethanol	2480	451
Canthaxanthin	ethanol		474
Echinenone	ethanol		461

Troubleshooting

- Homogenisation of samples: ball mill versus mortar and pestle
  Whenever possible we recommend the use of a ball mill for homogenisation of samples. This has several advantages: there is no loss of material during extraction; samples can be stored for long periods; more accurate data is obtained, particularly where the same powder is used in different assays.

- Extraction medium
  - Frequently, 80 % acetone is used as solvent for spectrophotometric determination of plastid pigments. However, this solvent is unsuitable for extraction of chlorophylls and carotenoids from lichens, because the water that is added to the solvent solubilises lichen acids that destroy the chlorophylls and some carotenoids. We strongly recommend the use of DMSO, pure acetone, ethanol or the other solvents described above.
  - Oxidation during extraction, if this proves to be a problem, may be minimised by addition of anti-oxidants (e.g. butylated hydroxytoluene) to the extraction solvent.

- Large phaeophytin peak appears in chromatogram
  Chlorophyll can be degraded to phaeophytin by co-extracted acidic lichen substances when either organic solvents containing water are used for extraction (see above) or when wet and in most cases also air-dried lichens are analysed. We strongly recommend freeze drying lichens prior to pigment extraction. This is a very simple technique which ensures that you measure the actual chlorophyll content in lichens instead of using protocols which estimate an approximate chlorophyll content by calculating a "phaeophytinization ratio".

- Analytical column
  The performance of octadecylsilyl (ODS) stationary phases used in HPLC is dependent on the manufacturer. Consequently, in certain cases it will be necessary to vary the conditions given above in order to optimise the separation of lutein and zeaxanthin and other carotenoids. Additionally, differences in the selectivity between non-endcapped ODS-1 and endcapped ODS-2 phases must be considered. Thus, some authors (e.g. Thayer and Björkman 1990) prefer ODS-1, if a baseline separation of the xanthophylls lutein and zeaxanthin is desired. In our laboratory we used ODS-2 in combination with an acetonitrile-based mobile phase and achieved an approximate baseline resolution of these two compounds (Figure 3).

Subprotocol 2
Spectrophotometric Determination of Plastid Pigments

If the laboratory is not equipped with an HPLC, chlorophyll a and b and total carotenoids can be assayed using the spectrophotometeric method described by Wellburn (1994).

Procedure

1. Extract the pigments with 100 % DMSO as described in Subprotocol 1 (Steps 10 - 17).
2. Transfer extract to a glass cuvette and measure against a blank (100 % DMSO).
3. Calculate chlorophyll a (C_a), chlorophyll b (C_b) and total carotenoids (C_{x+c}) in $\mu g\ ml^{-1}$ according to the following equations (Wellburn 1994).
   If you use a spectrophotometer with a resolution of 1 - 4 nm:

   $$C_a = 12.19A_{665} - 3.45A_{649} \qquad \text{Equation 1}$$

   $$C_b = 21.99A_{649} - 5.32A_{665} \qquad \text{Equation 2}$$

   $$C_{x+c} = (1000A_{480} - 2.14C_a - 70.16C_b)/220 \qquad \text{Equation 3}$$

   If you use a spectrophotometer with a resolution of 0.1 - 0.5 nm:

   $$C_a = 12.47A_{665.1} - 3.62A_{649.1} \qquad \text{Equation 4}$$

   $$C_b = 25.06A_{649.1} - 6.5A_{665.1} \qquad \text{Equation 5}$$

   $$C_{x+c} = (1000A_{480} - 1.29C_a - 53.78C_b2)/220 \qquad \text{Equation 6}$$

   Note: There are variations in spectrophotometer response as well as a shift of peak maxima of chlorophylls of between 0.5 and 1 nm. By using pure solutions of chlorophyll a and b, the values at 480 nm and at the peak maximum given in the equations above can be re-adjusted, if necessary, for any spectrophotometer.

Subprotocol 3
Analysis of Tocopherols by High Performance Liquid Chromatography

General aspects

Most photosynthetic pigments and tocopherols can simultaneously be determined by a recently developed reversed-phase HPLC method (Garcia-Plazaola and Becerril 1999) which uses a gradient system for separation of compounds, and for detection a UV detector or a DAD, respectively. Here, we describe a method for separation and quantification of tocopherols by isocratic HPLC with fluorescence detection modified after Wildi and Lütz (1996). The major advantage of using a fluorescence detector is, that it is much more sensitive than a UV detector and more so, a DAD, thus allowing for the determination of tocopherol when very little material is available. This is, for instance, the case when analysing lichens grown in axenic cultures. Moreover, tocopherol is only present in the photobiont, and therefore the material available for analyses might be limited.

Sample preparation and the extraction procedure is the same as for chlorophylls and carotenoids. If you use extracts that have been used for pigment analysis, start the procedure at Step 3 of the following protocol.

Procedure

1. Extract tocopherols as described above for pigments (Steps 10 - 17 in Subprotocol 1) or use the same extract (after pigment analyses, the extracts can be stored at -25°C).

   Note: As observed for chlorophylls and carotenoids, a combination of DMSO/ethanol has proven to be significantly more effective for extraction of tocopherols from lichen material than e.g. 100 % acetone (data not shown).

2. To remove cell debris which would block the analytical column and shorten its lifespan, centrifuge samples (at least 20000 g for 20 min) prior to injection into the HPLC system. Alternatively, pass the supernatant through a membrane filter (0.22 µm pore size).

3. Separate tocopherols by isocratic HPLC using the following settings:
   - Column: Spherisorb ODS-2 (250 x 4.6 mm i.d.), particle size 5 µm
   - Eluent: Methanol, flow rate: 1 ml/min
   - Fluorescence detector: Excitation wavelength = 295 nm, emission wavelength = 325 nm.
     The tocopherols elute in the order δ-,γ-,α-tocopherol.

4. Prepare a series of standard solutions with varying concentrations of α-, γ-, and δ-tocopherol in acetone and analyse as described above.
5. Calculate the amounts of tocopherols using the results of the standards.
6. Test the recovery of tocopherols to ensure that they are not destroyed or lost during the extraction. Divide the freeze-dried lichen powder into 10 sub-samples and analyse five of these as described above, and spike five others with known amounts of standard solution. At least 90 % of the added standard should be recovered (otherwise indicate recovery rate in publications!).

Acknowledgements. We thank Prof. A. Keith Cowan (University of Natal, South Africa) for critically reading the manuscript. We acknowledge financial support from the Austrian Academy of Science (APART 428) and the Austrian Science Foundation (P12690-BIO) to Ilse Kranner.

References

Bramley PM (1992) Analysis of carotenoids by high performance liquid chromatography and diode-array detection. Phytochemical Analysis 3: 97-104

Cañas M, Pignata ML (1998) Temporal variation of pigments and peroxidation products in the lichen *Parmotrema uruguense* (Kremplh.) Hale transplanted to urban and non-polluted environments. Symbiosis 24: 147-162

Davies BH (1976) Carotenoids. In: Goodwin TW (ed) Chemistry and Biochemistry of Plant Pigments, vol 2. Academic Press, London New York San Francisco, pp 38 -165

García-Plazaola JI, Becerril JM (1999) A rapid high-performance liquid chromatography method to measure lipophilic antioxidants in stressed plants: Simultaneous determination of carotenoids and tocopherols. Phytochemical Analysis 10: 307-313

Goodwin TW (1980) Algae. In: Goodwin TW (ed) The Biochemistry of the Carotenoids. Vol. 1. Plants. 2nd edition, Chapmann and Hall, London New York, pp 207–256

Goodwin TW, Britton G (1988) Distribution and analysis of carotenoids. In: Goodwin TW (ed) Plant pigments. Academic Press, London San Diego, pp 62–132

Hager A, Meyer-Bertenrath T (1966) Die Isolierung und quantitative Bestimmung der Carotinoide und Chlorophylle von Blättern, Algen und isolierten Chloroplasten mit Hilfe dünnschichtchromatographischer Methoden. Planta 69: 198-217

Kranner I (1998) Determination of glutathione, glutathione disulfide, and two related enzymes, glutathione reductase and glucose-6-phosphate dehydrogenase, in fungal and plant cells. In: Varma A (ed) Mycorrhiza Manual, Springer, Berlin, pp 227-241

Pfeifhofer H (1989) Evidence for chlorophyll b and lack of lutein in *Neottia nidus-avis* plastids. Biochem Physiol Pfl 184: 55-61

Siefermann-Harms D (1988) High-performance liquid chromatography of chloroplast pigments. One-step separation of carotene and xanthophyll isomers, chlorophylls and pheophytins. J Chromatography 448: 411-416

Stransky H, Hager A (1970) Das Carotinoidmuster und die Verbreitung des lichtinduzierten Xanthophyllcyclus in verschiedenen Algenklassen. Arch Mikrobiol 72: 84-96

Thayer SS, Björkman O (1990) Leaf xanthophyll content and composition in sun and shade as determined by HPLC. Photosynth Res 23: 331-343

Wellburn AR (1994) The spectral determination of chlorophylls a and b, as well as total carotenoids, using various solvents with spectrophotometers of different resolution. J Plant Physiol 144: 307-313

Wildi B, Lütz C (1996) Antioxidant composition of selected high alpine plant species from different altitudes. Plant Cell Env 19: 138-146

Young AJ, Phillip D, Savill J (1997) Methods for carotenoid analysis. In: Pessaraki M (ed) Handbook of Photosynthesis, Marcel Dekker, New York, pp. 597-622

Suppliers

Ball Mill

- F. Kurt RETSCH GmbH & Co.KG, Rheinische Str. 36, 42781 Haan, Germany
- B. Braun Biotech International GmbH, Schwarzenberger Weg 73 - 79, P.O.Box 1120, D - 34209 Melsungen; http://www.bbraunbiotech.com/

Ultra pure water: Milli Q-Plus, Millipore Corporation

Filters: Schleicher & Schuell GmbH, Postfach 4, 37582 Dassel, Germany, phone: +49 55561 791-0, fax: +49 5561 79-15-36

Pigment standards can be obtained from

- Sigma, St. Louis 63178, MO, http://www.sigma-aldrich.com (β-carotene, chlorophyll a, chlorophyll b, lutein) and
- Carl Roth GmbH & Co., Schönperlenstraße 1-5, D-76185 Karlsruhe, phone: +49-0721-5606-0, fax: +49-721/5606-149, http://www.carl-roth.de (β-carotene, chlorophyll a, chlorophyll b, lutein, zeaxanthin).
- A good range of plastid pigments (α-carotene, β-carotene, antheraxanthin, chlorophyll a, chlorophyll b, chlorophyll c, lutein, neoxanthin, violaxanthin, zeaxanthin, canthaxanthin, echinenone and others) can be obtained from DHI - Water & Environment, The International Agency for ^{14}C Determination, Agern Allé 11, DK - 2970 Hørsholm, Denmark, phone: +45 4516 9200, fax: +45 4516 9292, E-mail: c14@vki.dk, http://www.c14.vki.dk/

Nucleic acids

Isolation of Nucleic Acids From Lichens

OSCAR F. CUBERO and ANA CRESPO

Introduction

This chapter is a primer for those investigators working with lichens for whom molecular techniques are a relatively new experience, or who are having problems obtaining consistent results due to difficulties in deriving nucleic acids of sufficient purity. We will deal with some of the particular problems of nucleic acid extraction from lichens, focusing on genomic DNA isolation protocols, although we will also make some suggestions about RNA isolation. Some of the published protocols will be considered and a CTAB (Cetyl-trimethyl ammonium bromide) based protocol for lichen DNA isolation will be explained in detail.

Nucleic acid isolation is the first step in any molecular application involving DNA or RNA. It is carried out with minimal laboratory equipment following simple protocols that usually have a similar general layout but differ in key steps affecting the quantity, quality and purity of the resulting product. The most important points to consider when choosing between alternative protocols are those related to the type and quantity of nucleic acid needed for a particular molecular application (RNA, DNA or both), the quantity of sample available, and the type and quantity of cellular components that have to be eliminated during the purification process.

When deriving nucleic acids from lichens we are faced with several major problems. The existence of several genomes in the lichen is usually resolved by the separate culture of the symbionts, using algal-free fungal parts of the thallus or discriminative techniques, such as specific primers in Polymerase Chain Reaction (PCR), or hybridisation with specific

✉ Oscar F. Cubero, Universidad Complutense, Departamento de Biología Vegetal II, Facultad de Farmacia, Ciudad Universitaria, Madrid, 28040, Spain
(*phone* +34-1-3941771; *fax* +34-1-3941774; *e-mail* OSCARHFC1@mi.madritel.es)
Ana Crespo, Universidad Complutense, Departamento de Biología Vegetal II, Facultad de Farmacia, Ciudad Universitaria, Madrid, 28040, Spain

probes. The drawback of having a small quantity of sample available is overcome by using amplification techniques (see Chapters 24 and 25). However, nucleic acid purification presents additional problems due to the persistence of cellular contaminants. Lichen-forming fungi usually produce phenolic compounds that inhibit protein catalysis (thus inhibiting polymerases and restriction enzymes). Additionally, they contain large quantities of polysaccharides, which at high concentration also inhibit enzymatic activity. The major problem with polysaccharide elimination is that they co-precipitate with nucleic acids when using alcohol precipitation (the most common way to perform the final precipitation) and additional purification steps have to be performed.

There are several published protocols for obtaining nucleic acids from lichens or lichen symbionts for molecular applications. Some of them are general protocols designed for fungi (e.g. Lee and Taylor 1990; Bruns et al. 1990), while others are modified protocols that include additional steps to ensure the removal of polysaccharides or phenolic compounds. Some more specific protocols to eliminate polysaccharides include purification using resins or chromatographic columns (as in Armaleo and Clerc 1991), the use of ammonium bromides (Armaleo and Clerc 1995; Crespo et al. 1997) or the use of SiO_2 (Grube et al. 1995).

In general, protocols without non-alcoholic precipitation can yield DNA of sufficient quality for most molecular applications, such as PCR or Restriction Fragment Length Polymorphism (RFLP). However, they eliminate less polysaccharide and in some cases will not be as reproducible and the DNA will not be as durable (sometimes it cannot be used for amplification after as little as several weeks). Protocols including precipitation without alcohol (such as those that use CTAB or SiO_2 precipitations or purification columns) produce nucleic acids with less polysaccharide and give more reproducible results in molecular applications. Commercial protocols that use purification columns are shorter, easier and give excellent results, but they are expensive. Protocols that use CTAB or SiO_2 usually take longer and in some cases have a lower yield. However they are inexpensive and they are preferable if you have to process a high number of samples and a short protocol does not work, or if you need greater reproducibility or durability of your DNA.

CTAB protocol

The protocol we will describe in detail (Cubero et al. 1999) is based on that of Rogers and Bendich (1988) and similar to that described by Armaleo and Clerc (1995). It exploits the ability of cetyl-trimethyl ammonium bromide (CTAB) to prevent the co-precipitation of polysaccharides (Jones 1963) and of polyvinyl polypyrrolidone (PVPP) to eliminate polyphenolic compounds (Pich and Schubert 1993). It has been successfully used to obtain PCR-amplifiable DNA from several groups of lichens and from different types of material, such as lobuli, rhizinae, apothecia, lirella, the central cord of *Usnea*, dissected medulla and lichen cultures. The protocol can be performed in a single day and the DNA obtained is suitable for use for at least three years if stored at -20°C (even if the samples are occasionally unfrozen for use).

Outline

The protocols described here consist of four main parts:

- Tissue and cell disruption. Tissues and cell walls are broken by mechanical disruption and cell membranes are dissolved using detergents under denaturing conditions.
- Chloroform extraction. This is performed twice during the protocol and serves to remove protein and organic compounds.
- CTAB non-alcoholic precipitation. This removes most of the polysaccharides.
- Alcoholic precipitation. This is used to concentrate the nucleic acids before dissolving them in a storage solution.

Materials

Reagents

To prepare the solutions required by the protocol quickly and easily, the following stock solutions should be available in the laboratory:

- 1 M Tris HCl adjusted to pH 8.0 with concentrated HCl (sterilised by autoclaving)
- 0.5 M EDTA (ethylene-diamine-tetra acetate) adjusted to pH 8.0 with NaOH pellets (sterilised by autoclaving)
- 4 M NaCl (sterilised by autoclaving)
- 4% CTAB w/v

To perform the extraction the following solutions are required:

- CTAB extraction buffer (100 mM Tris/HCl pH 8.0, 30 mM EDTA, 1 M NaCl, 1% w/v CTAB)

- CTAB precipitation buffer (40 mM NaCl, 0.5% w/v CTAB)
- Modified TE buffer (1 mM Tris/HCl pH 8.0; 0.1 mM EDTA). Sterilize by autoclaving (TE buffer as described in common protocols is 10 mM Tris/HCl, 1 mM EDTA. As high Tris concentrations can interfere with some molecular applications we use here a dilution of this buffer).
- CI (chloroform: isoamyl alcohol 24:1 v/v)
- Isopropanol
- 70% ethanol

All the solutions have to be prepared with double-distilled water. No special quality is required for the reagents, except for Tris, EDTA and agarose, which have to be suitable for molecular work. Discard Tris solutions with yellowish colour. To visualise the DNA the following reagents and solutions are required:

- Ethidium bromide (this is a hazardous carcinogenic substance and it has to be handled with care)
- Agarose
- TAE buffer. It is usually prepared as a fifty-times (50X) concentrated stock solution (242 Tris base, 57.1 ml glacial acetic acid, 100 ml 0.5 M EDTA, pH 8.0, double-destilled water to 1l), which is autoclaved and diluted before use.
- Loading buffer. 0.25% w/v bromophenol blue, 30% v/v glycerol
- DNA standard. This is a DNA of known concentration. Preparations of lambda phage digested with the enzyme HindIII are the commonest DNA standards as they contain several DNA bands with different concentrations (they can be purchased or prepared from a known concentration of lambda phage). Commercial DNA ladders can also be used.

Equipment and plasticware

- Waterbath
- Micropipettes that cover the range between 5 µl and 1000 µl
- Sterilised plastic tips
- Microcentrifuge and 1.5 ml eppendorf tubes (sterilised)
- Pestles that fit into 1.5 ml eppendorf tubes. Plastic pestles can usually be purchased, but glass pestles can be easily made from glass rods and can be sterilised using a flame.

- Apparatus for gel electrophoresis
- UV transilluminator

Procedure

Sample preparation

1. For mycobiont cultures, place single colonies of about 1 - 2 mm^2 in a 1.5 ml plastic tube and freeze-dry. For field-collected lichens, inspect small dried pieces of thalli (between 3 and 100 mg, although we have successfully obtained DNA from smaller quantities e.g. a couple of *Physconia* rhizinae or small apothecia from *Graphidaceae*) under the dissecting microscope, clear of any contaminating organisms or remaining substrate, and place in 1.5 ml tubes.

Material disruption

2. Several methods can be used to disrupt the thallial structures and cell walls, but best results are obtained from samples frozen in liquid nitrogen. Procedure is as follows. Place dried samples into 1.5 ml tubes and transfer to a container with liquid nitrogen. When the tissue is completely frozen (2 min is enough), remove one tube from the container, place in an insulated rack and approximately half fill the tube with clean liquid nitrogen. The insulated rack keeps the temperature low and avoids spillage of the powdered lichen material when more liquid nitrogen is added. Use a pre-cooled pestle to grind the material until a fine powder is obtained. After grinding, place the tube on ice or in the refrigerator until the process has been performed for all the tubes.

Incubation

3. Add 400 µl of CTAB extraction buffer to each tube and tap the tube with your finger until all the powder has mixed with the buffer.
4. Add PVPP powder directly to each tube to a final concentration of 1% w/v. If you have experience of judging a 4 mg quantity of PVPP, you can estimate the amount to be added.
5. Place all tubes in a water bath at 60 - 70°C for 30 min. Once or twice during this time mix the contents of the tubes by inverting several times.

First chloroform purification

6. After incubation, add 500 µl of CI to each tube and mix by vigorous shaking (do not use a vortex).
7. Centrifuge at 10 000 G for 5 min at room temperature.

8. Measure and transfer the upper aqueous phase to a new tube. During this step the CI will drag the cell debris and most of the proteins to the interface. The PVPP with some polyphenolic compounds attached will also precipitate in this phase. Polyphenolic compounds will also be dissolved in the CI (lower) phase, but the nucleic acids will remain in the upper aqueous phase.

CTAB precipitation

9. Dilute the aqueous phase of each tube with CTAB precipitation buffer three times to lower the NaCl concentration and mix.

10. Centrifuge at a speed of 10 000 G for 5 min at room temperature. The CTAB will precipitate with the DNA attached. After this step you should obtain a small white pellet at the bottom of the tube. If you cannot detect a pellet, even a minute one, centrifuge for longer. If you are working with small quantities of lichen, place the tubes at 4°C for 5 min before centrifugation to help the CTAB-DNA complex to precipitate.

11. Eliminate the aqueous phase and resuspend the pellet in 0.4 ml of prewarmed (at 37°C) 1.2 M NaCl solution. The pellet usually dissolves easily, but if it does not then place the tubes in a water bath at 50°C. If RNA-free DNA is needed for your molecular application (for PCR it is not usually necessary to eliminate the RNA) then proceed in the following alternative manner. Resuspend the CTAB-DNA pellet in 25 µl of prewarmed 1.2 M NaCl solution and add 3 µl of ten-times concentrated (10X) RNAase buffer (if a different concentration is supplied by the manufacturer, then correct the quantity to obtain the final 1X concentration) and 2 µl of RNAase A (10 mg ml^{-1}). Incubate for 30 min at 37°C. Add the remaining 370 µl of prewarmed NaCl.

Second chloroform purification

12. Add 500 µl of CI and proceed as described in Steps 6–8.

Alcohol precipitation

13. Add 0.6 volumes of isopropanol (e.g., 180 µl if you have recovered 300 µl of the aqueous phase) and centrifuge at 13 000 G for 15 min at 4°C. The pellet obtained, if visible, is usually white or translucent.

Final wash

14. Eliminate the aqueous phase and add 500 µl of 70% ethanol to eliminate excess salt. Shake by hand and collect the pellet by centrifugation for 3 min at 13 000 G at 4°C.

15. Drain the pellet (using a micropipette to eliminate the last drops) and dry it at 50°C.

16. Resuspend in 50 μl of prewarmed (at 37°C) modified-TE buffer. The pellet must be completely dry before resuspension.

Quantification

17. DNA quantity is monitored by electrophoresis in 0.6% (w/v) agarose gels stained with ethidium bromide. Cast agarose gels by melting the agarose in TAE buffer until a clear, transparent solution is achieved (the quantity prepared depends on the size of the gel mould).

18. Cool the solution to about 60°C, add ethidium bromide to a final concentration of 0.5 gl ml^{-1} and mix thoroughly.

19. Position the comb of the mould in its place (follow the manufacturer's instructions), pour the agarose into the mould and let the gel solidify.

20. Mount the gel in the electrophoresis tank filled with 1X TAE buffer. Mix 10 μl of the samples of DNA with 3 μl of loading buffer and load the mixture into the slots of the submerged gel. Additionally, load several dilutions of the DNA standard in empty slots. Band fluorescence is directly proportional to its DNA content so that a band with double the fluorescence of a 20 ng DNA standard band contains 40 ng.

The DNA electrophoresis gels also give information about DNA purity. High purity DNA gives well-shaped bands. Protein and polysaccharides

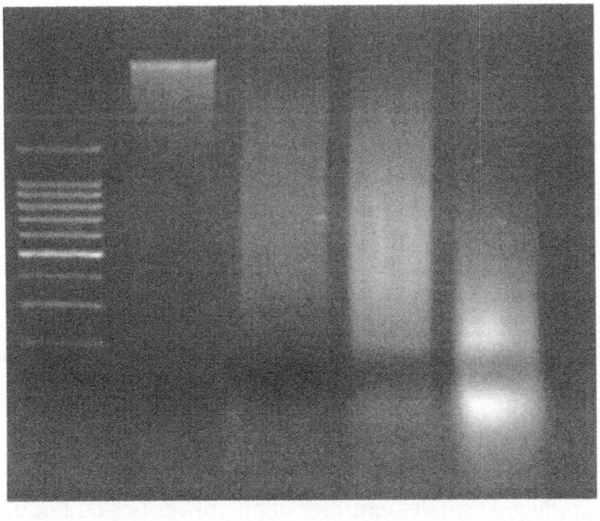

Fig. 1. Agarose gel electrophoresis of genomic DNA obtained as described in the text. Lane 1: Commertial DNA ladder. Lane 2: isolation from a fresh sample. Lanes 3, 4 and 5: isolation from herbarium specimes. DNA from lane 5 was used to amplicy up to 600 bp PCR products.

contaminants produce residual fluorescence in the wells, trails between the wells and the band, and deformation of bands. Excess salts give bands with peaks. Degraded DNA gives smears instead of well-shaped bands. RNA, if present, will appear as a band or smear at the bottom of the gel (see Fig. 1). If different types of RNA need to be separated and identified you must run the RNA in a formaldehyde gel (Sambrook et al. 1989). See Troubleshooting for more details.

Results

This method can give an acceptable yield of undegraded DNA that is free of most polysaccharides and polyphenolics. RNA, mitochondrial DNA and plasmid yield is lower than with other protocols (which do not include CTAB non-alcoholic precipitation) but is enough to perform PCR. A typical DNA isolation gel is shown in Fig. 1.

Troubleshooting

Here we cover some points to check and suggestions that may help if you do not obtain the desired results with this or similar protocols.

- You obtain less DNA than expected.
  Check your tissue disruption. Be sure that you have obtained a fine powder (and not just small lichen fragments) after grinding with liquid nitrogen in Step 2. Try to perform stronger centrifugation steps (increasing the duration or speed of the centrifugation). In Step 13 use cold isopropanol and store the tubes at -20°C for 20 min before centrifugation. Try a shorter protocol without CTAB-precipitation (if your sample does not contain much polysaccharide).
- You obtain no DNA.
  You have probably lost the DNA pellet during Steps 11, 14 or 15. Try to eliminate the aqueous phase carefully with a micropipette. Be sure that you do not suck up the DNA pellet.
- Your final DNA solution is coloured.
  This probably means that some polyphenolics remain. Do not worry if this happens, as small quantities of polyphenolics do not usually interfere with most molecular applications. However, to eliminate them, warm the tubes after adding CI at Steps 6 and 12 for 2 min at 50°C

(first seal the tubes with parafilm and take the necessary precautions to avoid chloroform vapours). Alternatively, the samples can be washed in acetone before the DNA isolation.

- Your sample contains too much polysaccharide.
  In such cases you will obtain a white gelatinous pellet after Step 14 that is difficult to resuspend. To avoid this, centrifuge at a lower speed at Step 10 (however, be aware that this will also reduce the DNA yield). You can also perform a second CTAB precipitation: at Step 11 resuspend in 0.5 ml CTAB extraction buffer (at 60 - 70°C) instead of NaCl and continue from Step 6.
- Your DNA is degraded.
  If you are working with cultures or fresh samples there is no reason why you should obtain degraded DNA. Do not dry your samples with heat. Check all your buffers (mainly their pH).
- You have tried everything and the DNA extraction does not appear to have worked when the extract is used for subsequent applications, e.g. PCR or RFLPs.
  Check for possible mistakes in your molecular application (e.g. change your PCR conditions, check your PCR primers, or be sure that your restriction enzyme is in good condition). Use a different protocol or purify your DNA solution through purification columns.

Comments

RNA isolation

Most protocols designed to obtain DNA also yield reasonable quantities of total RNA. However, RNA tends to be degraded during the isolation process or during storage by RNAases liberated during cell lysis or that are contaminating laboratory equipment. Special care has to be taken with sterilisation of equipment, and RNAase inhibitors should be used during the isolation procedure and storage. RNA can also be specifically isolated (without DNA). There are also protocols to separate or visualise the different types of RNA but they are beyond the objectives of this chapter. For some general protocols and advice, see Sambrook et al. (1989).

Herbarium material

Unfortunately, many problems exist when extracting useful DNA from old, dried, herbarium samples. The problems with this kind of material are related to nucleic acid modification and degradation, and to the accumulation of enzymatic inhibitors during sample storage and death

(Doyle and Dickson 1987; Savolainen et al. 1995; Höss et al. 1996). There is no definite solution for this problem, however some suggestions are available. A protocol that uses PVP (polyvinyl pyrrolidone) or PVPP and several purification steps with phenol or chloroform can be used to eliminate inhibitors. Checking the DNA integrity in an agarose gel will also help to decide which applications are possible. If the DNA is totally degraded it will appear as a big spot at the same position as the RNA in the gel and it will not be of use for molecular applications. If it is partially degraded it will appear in the gel as a smear containing DNA fragments of different sizes. This can be used for PCR, although the smaller the DNA fragments obtained, the lower the possibility of obtaining good results.

References

Armaleo D, Clerc P (1991) Lichen chimeras: DNA analysis suggests that one fungus forms two morphotypes. Exp Mycol 15: 1-10

Armaleo D, Clerc P (1995) A rapid and inexpensive method for the purification of DNA from lichens and their symbionts. Lichenologist 27: 207-213

Bruns TD, Fogel R, Taylor JW (1990) Amplification and sequencing of DNA from fungal herbarium specimens. Mycologia 82: 175-184

Crespo A, Bridge PD, Hawksworth DL (1997) Amplification of fungal rDNA-ITS regions from non-fertile specimens of the lichen-forming genus Parmelia. Lichenologist 29: 275-282

Cubero OF, Crespo A, Fatehi J, Bridge, PF (1999) DNA extraction and PCR amplification method suitable for fresh, herbarium-stored, lichenized and other fungi. Pl Syst Evol 216:243-249

Doyle JJ, Dickson EE (1987) Preservation of plant samples for DNA restriction endonuclease analysis. Taxon 36: 715-722

Grube M, DePriest PT, Gargas A, Hafellner J (1995) DNA isolation from lichen ascomata. Mycol Res 99: 1321-1324

Höss M, Jaruga P, Zastawny TH, Dizdaroglu M, Pääbo S (1996) DNA damage and DNA sequence retrieval from ancient tissues. Nuc Acids Res 24: 1304-1307

Jones AS (1963) Use of alkyltrimethylammonium bromides for the isolation of ribo- and the deoxyribo- nucleic acids. Nature 199: 280-282

Lee SB, Taylor JW (1990) Isolation of DNA from fungal mycelia and single spores. In: Innis MA, Gelfand DH, Sninsky JJ and White TJ (eds) PCR Protocols. Academic Press, San Diego. 282-287

Pich U, Schubert I (1993) Midiprep method for isolation of DNA from plants with a high content of polyphenolics. Nuc Acids Res 21:14

Rogers SO, Bendich AJ (1988) Extraction of DNA from plant tissues. Plant Molecular Biology Manual A6: 1-10

Samsbrook J, Fritsch EF, Maniatis T (1989) Molecular Cloning: A Laboratory Manual. Second Edition. Cold Spring Harbor Laboratory Press. New York

Savolainen V, Cuènoud P, Spichiger R, Martínez MDP, Crèvecoeur M, Manen J (1995) The use of herbarium specimens in DNA phylogenetics; evaluation and improvement. Pl Syst Evol 190: 87-98

Abbreviations

CTAB	Cetyl-trimethyl ammonium bromide
EDTA	Ethylene-diamine tetra acetate
PCR	Polymerase Chain Reaction
PVPP	Polyvinyl Polypyrrolidone
RFLP	Restriction Fragment Length Polymorphism

PCR Techniques and Automated Sequencing in Lichens

ULF ARUP

Introduction

The Polymerase Chain Reaction (PCR) is a fairly simple but powerful technique that allows the amplification of small amounts of genetic material *in vitro* (Saiki et al. 1985). The versatility of this technique has made it a routine method in any molecular work and its application in evolutionary biology has led to important progress in understanding the phylogeny of organisms. Several studies have focused on the phylogeny of lichenized and non-lichenized ascomycetes or bacidiomycetes inferred from data of the nuclear small subunit ribosomal DNA (SSU rDNA; e.g., Gargas and Taylor 1992a, Gargas et al. 1995a, Lutzoni and Vilgalys 1995, Wedin et al. 1998). Studies on the SSU rDNA have also revealed that insertions are apparently more common in lichenized fungi than in many other organisms (e.g., Gargas et al. 1995b). Most studies on lichen-forming fungi were facilitated by the design of specific primers that do not amplify algal DNA (e.g., Gargas and Taylor 1992b). During the last few years, studies on the internal transcribed spacer regions (ITS1 and ITS2) and the large subunit (LSU) rDNA have become more common. The ITS regions and some parts of the LSU are generally more variable than the SSU, and sequences from these regions have been used to study phylogenies within and between genera (e.g., Arup and Grube 1998, Crespo and Cubero 1998, Groner and LaGreca 1997) and at the population level (DePriest 1994).

Ulf Arup, Lund University, Department of Systematic Botany, Östra Vallgatan 14-20, Lund, 223 61, Sweden (*phone* +46-46-222-8970; *fax* +46-46-222-4234; *e-mail* ulf.arup@sysbot.lu.se)

Outline

The principle of PCR is based on the amplification of a template of DNA through thermal cycling involving repeated denaturation of the DNA, annealing of a primer (oligonucleotide) to the single strands of DNA which initiates further addition of nucleotides to form a new double-stranded piece of DNA during the extension step. The amplification is carried out in a thermal cycler. The standard PCR cycling programme may be divided into three major steps which are run for a number of cycles, normally 30 to 35.

1. Denaturation at 94-95°C for 30-45 s
2. Annealing at 50-55°C for 20 s
3. Extension at 72°C for 90-120 s

Optionally, the whole programme can be preceded by a longer denaturation, about five min long. The last cycle can be extended by five to eight minutes, before the temperature is decreased to 4°C. However, not all thermal cyclers can be chilled below room temperatures. An outline of the process, from DNA extraction to sequencing is given in Figure 1.

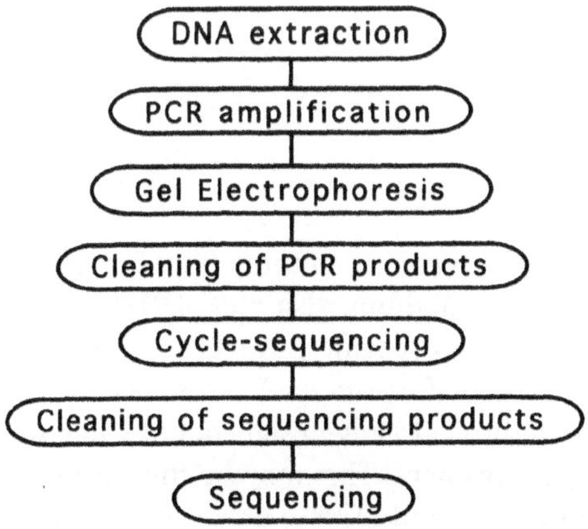

Fig. 1. Outline of procedures described in this chapter.

Subprotocol 1
PCR Techniques

Materials

Equipment

- Thermal cycler
- Bench microcentrifuge
- Plugged tips to prevent contamination

Chemicals

- DNA template (usually 5 % dilution of isolation stock)
- Double distilled sterile water
- 10x Buffer; 10-50 mM Tris-HCl (pH 8,3-8,8). This is usually provided together with the polymerase (and includes other further additives). The amount of buffer and polymerase used in the protocol below works for several brands, but see manufacturer's recommendation for instructions.
- Polymerase; usually between 1-2.5 units per 100 µl reaction volumes
- Desoxynucleosid-triphosphate (dNTP); pH 7.0, 0.8-1 mM
- Primers; 10 µM (see below and Comments)
- $MgCl_2$; 1.5 mM for dNTP concentrations of 0.8 mM (this is often already included in the buffer supplied together with the polymerase. See specific product for details)
- Mineral oil

Primer design

Many of the ribosomal primers are designed to amplify only fungal DNA and can thus be used for isolations containing also algal DNA, which is often the case for lichen extracts. Fungal specific primers can, often successfully, be used together with non-specific primers. In most cases there is no need to use other primers than those already designed, but sometimes it may be necessary to design new primers. This may be the case where already existing primers cannot anneal and initiate extension due to the occurrence of introns. If the primers are supposed to perform well there are a number of useful guidelines to consider when designing them. For design of primers specific to protein-coding genes follow Carbone and Kohn (1999).

The primers should:

- be 22-24 (18-25) nucleotides long;
- have approximately equal proportions of the nucleotides;
- have a C or a G at the 3' end;
- not form hairpin-loops, i.e., complementary folding of the primer itself;
- form no primer dimers, i.e., annealing of the primer pairs to each other;
- include no long series of one nucleotide, e.g., GGGG.

Procedure

All ingredients should be kept on ice during the preparation and gloves should be worn at all times to prevent contamination. The PCR mastermix is prepared in a 0.5 ml Eppendorf tube and here described for a reaction volume of 50 µl and a negative control of 10 µl. In some thermal cyclers 250 µl tubes are used. Use gloves to prevent contamination of the samples. The procedure is described for a standard PCR set-up performed in an ABI Perkin-Elmer thermal cycler and works for some common enzymes. When using other enzymes, the protocol given below must be recalculated according to the manufacturers instructions. Pipetters measuring large volumes with two decimals are used here, but pipetters with only one decimal can also be used.

1. Set up and number an adequate number of tubes.
2. Add two drops of 40-50 µl mineral oil to each of the tubes to prevent evaporation of the sample. This is not needed in thermal cyclers with a heated cover.
3. Add 15.75 µl of double distilled water.
4. Add 20 µl of template DNA.
5. Prepare a master mix with the reagents below and vortex for 2 s. The amount needed is the number of reaction tubes multiplied by 14.25 µl, plus 2.85 µl for control. Allow for some loss during the pipetting.

- 5 µl of 10x Tris-HCl buffer
- 4 µl of dNTPs
- 2,5 µl of each primer
- 0.25 µl of polymerase

Note: When buffer does not contain $MgCl_2$ it must be added to the mix. See product instructions for amount.

6. Add 14.25 µl of the master mix to each tube and centrifuge briefly.
7. Prepare a control tube with 7.15 µl of double distilled water and 2.85 µl of master mix overlaid with one drop, or 20-25 µl, of mineral oil.

   Note: The amount of water and DNA template depends on the concentration of the DNA stock solution. In the protocol above the isolated DNA is dissolved in 25 µl of water and an aliquot of 5 µl diluted in 95 µl of water.
   The optimal amount of polymerase used must be empirically determined. My experience is that the amount of polymerase suggested by the manufacturer is often larger than required.
8. Place the reaction tubes in a thermal cycler and run 30 cycles of a programme with the following parameters: 94°C for 45 s; 52°C for 45 s; 72°C for 90 sec. The programme may include an initial heating phase of 5 min at 94°C and a final phase at 72°C for 5 min. Following the final phase the temperature should be reduced to 4°C.
9. Visualise the PCR product by agarose gel electrophoresis (see Subprotocol 2) or an alternative method.

Clean PCR products

10. 5. Clean the PCR products by using a commercial cleaning kit. Qiagen Spin columns can be recommended for normal cleaning and Qiagen Gel Extraction kit for cleaning of bands cut out from agarose gels. Cleaning can also be carried out following a PEG-based protocol for cleaning (Kusukawa et al. 1990).

Hot-start techniques and touch-down techniques

I recommend using hot-start and/or touch-down techniques to minimise mispriming during the heating phase.

Hot-start techniques

Hot-start techniques prevent the polymerase from taking action before the denaturation temperature is reached. The most simple hot-start technique, which is often sufficient, involves transferring the reaction tubes directly from ice to a preheated block. In some polymerases (e.g., AmpliTaq Gold DNA polymerase from PE Biosystems) an inhibitor has already been added to the polymerase, preventing it from becoming active until after the initial heating phase.

Touch-down techniques

With a touch-down technique the first cycles are carried out at an annealing temperature higher than the calculated optimum and then gradually reduced to annealing temperature or temperatures lower than for optimal annealing.

Annealing temperature

The annealing temperature T_a should be appropriate for the primer pair used. It may be calculated through the melting temperature T_m of the primers: $T_a = T_m - 5°C$. The melting temperature of the primer pair should not differ by too much, preferably by less than 10°C. There are a number of ways to calculate the T_m for primers but one of the most widely used calculation is given in Equation 1:

$$T_m = 2°C \text{ x } (A + T) + 4°C \text{ x } (G + C) \qquad \text{Equation 1}$$

Troubleshooting

The most common problems in amplifying templates of lichen DNA are: There is no or only one weak product formed (no band or a weak band visible on the electrophoresis gel) or formation of more than one product (two or more bands visible on the gel). If optimisation of the PCR does not help, some guidelines are given below. Suggested literature on this subject for lichens includes Ekman (1999), on lichens mainly, and Roux (1995) for PCR in general.

- There is no or only a weak product formed.
  - In herbarium material of lichens the age of the specimen from which the isolation was made is a very important factor. Freshly collected specimens are always best and usually give no problems in the PCR. However freshly collected specimens which are still moist can cause problems in the isolation procedure. My experience with many lichen groups, e.g., *Aspicilia*, *Xanthoria*, *Physcia*, *Buellia*, *Usnea* and some parts of *Lecanora*, is that 1 to 5 years old herbarium material can be used without problems. Material of *Caloplaca* may work well after 15 years in the herbarium, whereas *Multiclavula* amplifies well also after decades of storage.

However, the age of specimens that can be used varies within genera. For example, material of the *Lecanora subfusca* group should not be more than one year old to amplify well. Finally, some groups of Arthoniales are notoriously difficult, and for good results fresh material is usually required. To some degree the result also depends on storage conditions.

- Sometimes even fresh material may be reluctant to amplify. In this case inhibitors may be the cause. Lichens, like higher plants, contain a large number of potential inhibitors that can impair PCR, e.g., polysaccharides, anthraquinones, terpenoid compounds, and carbonised tissues. If you suspect that either of the reasons given above were an inhibitor causing the failure of the PCR, try to isolate from fresher material and/or use another extraction protocol, that completely removes the inhibitors. Ekman (1999) has successfully used the DNA extraction kit DNeasy Plant Mini Kit (Qiagen) that originally was intended for isolation of small amounts of tissue from green plants, but has been shown to work well also for normally problematic material. Try also to use smaller amounts of tissue that do not contain, for example, carbonised parts.
- Other possibilities include reducing the concentration of the template combined with a larger number of cycles. Try also other primers. Some templates are very rich in introns that may block the primer site (see below for primer design). Another powerful method is nested or seminested PCR, which requires very clean working and careful interpretation of the results (see below). Finally, direct PCR may solve the problem (see Subprotocol 3).

- Several PCR products are formed
  - More than one reason exists why PCRs can result in several products. For example, mispriming can occur, and in addition primers can anneal to parts of the genome other than to the target template, or even to the DNA from other organisms, mostly other fungi. If mispriming is the suspected cause, try a higher annealing temperature (increase in steps of 2°C), decrease the magnesium ion concentration, use a more efficient polymerase, use hot-start techniques, touch-down techniques (if possible) or nested PCR. Adding DMSO (dimethylsulphoxide) to the PCR master-mix may also increase specificity.
  - If the primers are not specific for the target sequence more than one product will be formed. Try other primers if possible, or if this does not work, maybe designing new primers may be required.

- One cause of multiple bands formation in PCR on lichens is that the DNA extraction contains contaminations, i.e., DNA other than from the desired species. Almost all extractions from crustose and foliose lichens, and possibly less often from fruticose lichens, are likely to contain DNA from other lichens or non-lichenized fungi. In many cases the contaminant is present in the DNA extract in a concentration much lower than the target DNA. However, even very small amounts of contaminants can be amplified and may sometimes appear on the electrophoresis gel. Isolating from a smaller amount of tissue, e.g., one fruitbody or a section of a fruitbody, may be one solution (Grube et al. 1995, Wolinski et al. 1999). However, when contamination is weak in relation to the desired product, i.e., a weak and a strong band of different size are present on the gel, optimising the PCR should be tried. Very long target products are, for example, often more efficiently amplified when the extension time is prolonged.
- If there are two, or maybe three bands, with more or less equal strength on the gel and the outcome of the PCR does not improve very much from optimising, trying other primers may help. However, the most convenient solution is to extract the bands from the agarose gel and sequence them all. The bands on the gel should in this case be distinct and well separated from each other.

Comments

PCR optimisation

Even though the PCR method is basically rather simple, there are a number of important factors that can be regulated for optimal results. The concentration of some of the reagents in the PCR master-mix as well as the length and temperature of the steps of the PCR programme may be altered to achieve better PCR results. Normally, the protocol described above works well for a variety of DNA templates isolated from crustose, foliose and fruticose lichens.

Several steps may be taken to optimise each PCR. This subject cannot be thoroughly covered in this publication, but for more help see Ekman (1999). Factors worth optimising are: polymerase concentration, $MgCl_2$ concentration, annealing temperature, extension time and number of cycles. Table 1 below indicates how these factors affect PCR.

Table 1. The listed factors are important for PCR and can be regulated in desired direction to give higher specificity, but lower yield, or the reverse.

Higher specificity, lower yield	Lower specificity, higher yield
Increasing annealing temperature	Decreasing annealing temperature
Increasing polymerase concentration	Decreasing polymerase concentration
Decreasing $MgCl_2$ concentration	Increasing $MgCl_2$ concentration
Lower DNA concentration	Higher DNA concentration
Shorter extension time	Longer extension time
Fewer cycles	More cycles (until plateau effect, which normally is c. 40 cycles)

Increasing the DNA concentration in the PCR master-mix to get larger amounts of PCR product is often tempting and can be done successfully, but remember that when increasing the amount of template DNA, all other possible contaminants are also increased in the PCR master-mix. The risk of amplifying undesired DNA is therefore higher. In some rare cases the "normal" amount of DNA template used in the PCR is too large and decreasing the amount of DNA template may give a better yield.

Trying another polymerase sometimes helps because DNA polymerases from different suppliers may behave differently, even if they originated from the same organism.

Extension times between 1 to 2 min should be enough for target molecules up to 2kb long. However, Ekman (1999) states that sometimes even longer extension times may be required for target molecules 2 kb long. However, for relatively short target molecules (<1000 bp) the extension time may often be shortened, in some cases to 30 s. Similar findings have been reported by Mai et al. (1998) who shortened all steps of the PCR cycle to 2 s for denaturation, 5 s for primer annealing, and 8 s for extension and successfully amplified target sequences up to 1.2 kb. One advantage of a shorter annealing time is that mispriming is minimised (Wittwer and Garling 1991, Mai et al. 1998).

Primers and choice of target gene

For phylogenetic studies of the nuclear ribosomal genes many primers are now available. A proposal to standardise the nomenclature and a list of these has been presented by Gargas and DePriest (1996). A large number of primers for amplification of rDNA are also available on the world wide web via http://www.botany.duke.edu/fungi/mycolab/primers.htm. For studies on relationships within genera, families or orders, primers amplifying the SSU and the LSU are recommended because these parts of the ribosomal gene are evolving slowly and variation is rather low. However, a problem with SSU studies is that the variation in this region is sometimes too low to resolve the phylogeny of genera of diverse families (e.g. Mattsson and Wedin 1999, Crespo et al. 1999). Furthermore, the resolution may be good but the statistical support, e.g., bootstrap or jack knife analyses, for a branching pattern can be very low. Therefore, studies of the SSU may have to be extended to include also the LSU, or at least parts of it. The variation of the latter region is larger, especially in the first third of the unit.

For studies on relationships between genera and species I recommend amplifying the ITS regions. The ITS regions can be used also for studies of the variation within species or species groups. However, it has recently been shown that the phylogeny suggested by studies of a single locus may not always reflect that of the species (Arup unpubl., Grube in prep., Kroken and Taylor 1998). There is also growing evidence that more than one allele of nuclear rDNA may be present within a haplotype. This heterogeneity may confound phylogenetic (and population genetic) studies that are based solely on the rDNA locus. Therefore, I strongly recommend using more than one gene, preferably several, for studies at the species level and in studies of genetic variation within species or populations.

In addition to the ITS regions, patterns of intron occurrence or differences in their sequences can be used in population studies (e.g., DePriest 1993). Recently, protein-coding genes (Grube in prep.) and several anonymous loci (Kroken and Taylor 1998) from lichens have been used in studies of species delimitation and species groups. The amount of variation varies between different protein genes and some may not be useful in population studies. Recently, primers amplifying mitocondrial SSU rDNA have been tested on lichenized ascomycetes and the results show that these can be used in phylogenetic studies (Zoller et al. 1999). Another option in population studies is to use randomly amplified polymorphic DNA (RAPD) markers. This technique is widely used in higher plants but has only recently been applied to lichens (Printzen et al. 1999, Murtagh et al. 1999, see also Chapter 25).

Nested and semi-nested PCR

Nested and semi-nested PCR are recommended for amplifying small amounts of template DNA. These techniques have also been considered to be useful for increasing the specificity (Innis et al. 1988), but my experience is sometimes the opposite, and proof-reading enzymes could be important to avoid ambiguous results. Thus, the results of nested and semi-nested PCR should be evaluated with special care. Both, nested and semi-nested PCR, require two consecutive amplifications. The product of the first PCR is used as a template for the second DNA. The first amplification uses one pair of primers and the second either two primers internal to the first ones (**nested**), or only one of the two primers internal to the first pair (**semi-nested**). My personal experience is that nested PCR usually is to be preferred to semi-nested. In the first PCR, priority should be given to efficiency, and in the second, priority should be given to specificity. For the second PCR use a dilution, 0.1-10 %, of the PCR product of the first PCR as template. Nest also the negative control. Because PCR products contain a very high number of replicates, contaminant carry-over to the PCR tubes of the second PCR is a risk. For the cleaning I recommend a gel extraction method which efficiently removes other products possibly present.

Subprotocol 2
Agarose Gel Electrophoresis

The result of the PCR is most conveniently visualised by agarose gel electrophoresis. A small amount of the PCR product is loaded onto a gel, stained with ethidium bromide, and run in an electric field so that negatively charged DNA molecules move toward the anode. The electric mobility of the DNA molecules depends mainly on the size of the molecule, the agarose concentration and the voltage used.

Materials

Equipment

- Electrophoresis tank
- Electric power supply
- UV-light with camera connected to a printer or equipment that saves the information on a floppy disk

Chemicals

- 1x TBE buffer
- Agarose
- Ethidium bromide (10 mg/ml)
- Loading buffer (distilled water:glycerol=1:1, containing 0,2 % bromophenolblue and/or xylenecyanol). The glycerol increases the density of the sample and prevents it from diffusing in the TBE buffer
- DNA size marker

Procedure

1. Prepare a gel tray with a comb adjusted 1-2 mm above the plate.
2. Prepare 1 % agarose gel: weight the agarose in an Erlenmeyer flask and add 1x TBE buffer.
3. Boil in a microwave until the agarose is completely dissolved and cool to about 60°C.
4. Add ethidium bromide, 2-3 µl per 100 ml of agarose, and mix.
5. Pour into the gel tray (avoiding air bubbles) and let it solidify for 40- 60 min.
6. Transfer the tray to the electrophoresis chamber containing 1x TBE buffer enough to cover the gel.
7. Mix 2-10 µl of DNA sample with 2 µl of loading buffer and load the wells, e.g., by placing the loading buffer as small droplets on a piece of parafilm and adding the DNA to this droplet.
8. Run at 80-120 V (10 V per cm of gel) until the indicator dye in the loading buffer has run 2/3 the length of the gel.
9. Place the gel under UV-light to visualise bands and photograph or save on floppy disk.
   Caution:
   - Ethidium bromide is extremely toxic. Always wear gloves when handling this chemical and equipment that has been in contact with it.
   - UV-radiation is dangerous to unprotected eyes and to the skin. Wear gloves and safety mask or goggles.

Modification Staining of the DNA molecules may also be carried out after running the gel by placing it in a solution of ethidium bromide (1 µg/ml) for 10-20 min (depends on thickness of the gel), preferably on a horizontal shaker.

Subprotocol 3
Microslide PCR

Sometimes normal DNA isolation of lichens or lichenicolous fungi is difficult or impossible, especially when the sample is very small. A method offering a way around this problem is direct PCR, where no isolation step is needed but the PCR is carried out directly with a section or small part of the lichen (Wolinski et al. 1997, 1999; Heibel et al. 1999; see also Chapter 25).

Materials

Same as for normal PCR (see Subprotocol 1) plus:

- Microslides (2 x 10 mm large slides cut from a cover slip with a diamond cutter). Microslides should be washed in sterile water, 2 x in 100 % EtOH, air-dried and immersed in a gelatin/chromalum solution (0.25 % gelatin, 0.025 % chromium III potassium sulphate) for 5 s under sterile conditions, or coated with poly-L-lysin, and finally dried on aluminium foil
- Oven

Procedure

Use as far as possible sterilised and appropriately cleaned equipment.

1. Place one to two sections of an apothecium or a thallus, alternatively some soredia, in a small droplet at the end of a microslide. The sections can be made by hand or cut in a cryostat but should be as thin and plane as possible.
2. Dry microslide in a Petri dish for 5 min at 45°C in an oven.
3. Microwave slides for 10 min (at 800 W). This enhances attachment of the specimen to the slides.

4. Remove under stereomicroscope parts of sections likely to inhibit the PCR (e.g., anthraquinone crystals and carbonised tissue) or undesired tissue of algae, lichens or lichenicolous fungi.
5. Place the microslide in a 0.5-ml tube containing 35 µl double distilled water and 14.25 µl of PCR master mix according to the protocol given above for a standard PCR.
6. Overlay with two drops of mineral oil.
7. Place in thermal cycler and run a standard PCR programme.
8. Remove the microslide (these may be stored at 4°C for further PCRs) and continue as described for standard PCR.
   Option: If inhibitors or contaminants are not a problem, or can be easily removed, sections or fragments may be placed directly into the tubes and Steps 1-4 can be left out.

Subprotocol 4
Cycle Sequencing

Direct cycle-sequencing is now the standard method for sequencing of lichen DNA, partly because only nanograms of DNA are needed. The DNA is usually sequenced in both directions, to avoid ambiguities. If the sequences contain a number of ambiguous sites, cloning of the PCR product prior to sequencing is suggested. Thus, the second strand, which has been sequenced in the opposite direction, should verify the first strand. The conservative parts of the nuclear ribosomal DNA, the SSU and the LSU, are often rich in insertions. The expected length of the SSU without insertions is about 1800 bp, but there are examples (*Aspicilia caesiocinerea*) where eight introns were present in the SSU and its total length was 3600 bp (Arup, unpublished). This makes both amplification and sequencing more troublesome because finding suitable amplification primers may be problematic and many more primers for sequencing must be used.

The reaction is carried out in a thermal cycler and is similar to the PCR, but no exponential amplification takes place. It has, like the PCR programme, an initial heating phase (10 s), followed by three main steps that are repeated in 25-30 cycles:

1. Denaturation of the double-stranded DNA at 94-96°C for 10 s;

2. Annealing of primer at 50-55°C for 5 s.
3. Extension of the DNA strand at 60-70°C for 3 to 4 min (Figure 2).

The procedure below is described for the system used by Perkin-Elmer where the probes eventually will be run in a Perkin-Elmer sequencer. Protocols for other brands differ considerably and the procedure must be altered according to the manufacturers instructions. The cycle sequencing preparations are made for each DNA template and primer, which means that if you have six different templates and four different primers for each template, you will prepare 24 different tubes. Each tube contains DNA template, primer, cycle sequencing kit and water according to the manufacturer's recommendations.

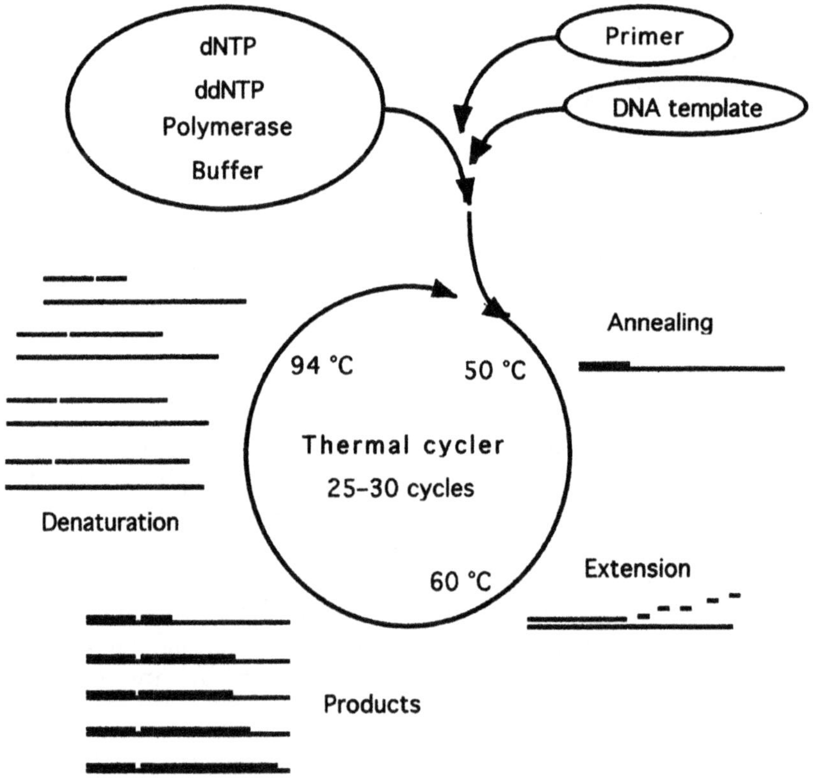

Fig. 2. Overview of the cycle-sequencing process

Materials

Equipment

- Thermal cycler
- Bench microcentrifuge

Chemicals

- DNA template (> 10 ng/μl)
- Double distilled sterile water
- Primers; 1.6 μM
- Cycle sequencing kit (use one that works in the sequencer you will use)

Procedure

The cycle sequencing procedures differ somewhat from each other depending on whether you use labelled primers (e.g., Pharmacia) or labelled ddNTPs (Perkin Elmer). The procedure described below is for a total reaction volume of 10 μl using a cycle sequencing kit where the four different nucleotides are processed in one single tube (Perkin Elmer). To each tube add 4 μl of sequencing kit, 1 μl of primer and enough DNA to give a final concentration of 50-75 ng in 10 μl. Dilute with water to a final volume of 10 μl.

1. Calculate the concentration of cleaned DNA for each template.
2. Calculate for each template how much DNA should be added to give a final DNA concentration of 50-75 ng in 10 μl in each tube. Example: If you want a DNA concentration of 60 ng in10 μl and the concentration of your DNA is 50 ng/μl you need 1.2 μl of DNA. To make up the volume of 10 μl (including 5 μl of sequencing kit and 1 μl of primer) you need to add 3.8 μl of water.
3. Prepare and number tubes for cycle sequencing.
4. Add amount of water as calculated in Step 2 according to the amount of DNA template to all of the tubes (one pipette tip may be used for all tubes).
5. Add 1 μl of primer to the tubes. Use one pipette tip for each primer.
6. Add amount of DNA template as calculated in Step 2 above. Use one pipette tip for each tube.

7. Add 4 µl of cycle sequencing kit to each tube. Use one pipette tip for each tube.
8. Centrifuge the components to the bottom of the tube, place in a thermal cycler and run the programme outlined above. If the thermal cycler has no heated cover, add a droplet of mineral oil to each tube.
9. Store the sequencing product at 4°C.

Note: Sequence reactions with too much DNA generally produce sequences with very strong signal at the beginning, but with very short length of read. Too low concentrations of DNA result in weak signal and short length of read.

Subprotocol 5
Cleaning of Cycle Sequencing Products

Salts, enzymes and unincorporated nucleotides from the sequencing reaction have to be removed by cleaning. A clean sequencing product is very important and, in fact, inadequate template preparation is the most common cause for sequencing problems. There are commercial kits also for this cleaning step, but using the protocol below usually works well. However, it is important that the cleaning reagents are fresh and held at the right concentrations. Wrong concentrations lead to salt remnants or loss of DNA.

Materials

Equipment

- Bench microcentrifuge
- Oven or vacuum centrifuge

Chemicals

- Sodium acetate NaOAc 3M pH 4.4 (store at 4°C)
- Chilled EtOH 95 % (store at -20°C)
- EtOH 80 % (store at 4°C)

Procedure

1. Prepare and number 1.5 ml tubes.
2. Add 1 µl of NaOAc.
3. Add 25 µl of EtOH 95 %.
4. Add sequencing product and vortex 3-5 s.
5. Let the mixture stand at room temperature for 10 min or on ice for 30 min, and centrifuge at 15000 g 25-30 min at room temperature (important!).
6. Wash DNA pellet with 150 µl EtOH 80 % (pipette carefully in and out again).
7. Dry 30-60 min in an oven at 50°C or 10 min in a vacuum centrifuge.
8. Store at 4°C.
9. In the next step the pellets are differently processed depending on the type of Perkin-Elmer sequencer used.

Modifications: Different concentrations (in the range of 70-80 %) and temperatures of the EtOH may be used for the washing in Step 6. Generally, lower concentrations and temperatures remove more of the salt but increase the risk for DNA loss. The best combination must be determined empirically. Washing in Step 6 may also be carried out as follows: Add EtOH to the tube and centrifuge 5 min, and then remove the EtOH. This procedure decreases the risk of losing the DNA.

Acknowledgements. I acknowledge financial support by grant P11806-GEN of the Austrian Science Foundation. I also want to thank M. Grube, B. Spreitzer and anonymous referees for valuable help and suggestions for improvement of previous versions of this chapter.

References

Arup U, Grube M (1998) Molecular phylogeny of *Lecanora* subgenus *Placodium*. Lichenol 30:415-426.

Carbone I, Kohn LM (1999) A method for designing primer sets for speciation studies in filamentous ascomycetes. Mycologia 91:553-556

Crespo A, Cubero O (1998) A molecular approach to the circumscription and evolution of some genera segregated from *Parmelia* s. lat. Lichenol 30:369-380

Crespo A, Gavilan R, Elix, JA, Gutierrez G (1999) A comparison of morphological, chemical and molecular characters in some parmelioid genera. Lichenol 31:451-460.

DePriest PD (1993) Small subunit rDNA variation in a population of lichen fungi due to optional group I introns. Gene 134:67-74

DePriest PT (1994) Variation of the *Cladonia chlorophaea* complex II: Ribosomal DNA variation in a Southern Appalachian population. Bryol 97:117-126

Ekman S (1999) PCR optimization and troubleshooting, with special references to the amplification of ribosomal DNA in lichenized fungi. Lichenol 31:517-532

Gargas A, Taylor J (1992a) Molecular systematics of lichenized and non-lichenized fungi based on rDNA sequences. In: Int Ass Lichens Symp, Abstracts, p 9

Gargas A, Taylor J (1992b) Polymerase Chain Reaction (PCR) primers for amplifying and sequencing nuclear 18S rDNA from lichenized fungi. Mycol. 84:589-592

Gargas A, DePriest PT, Grube M, Theler A (1995a) Multiple origins of lichen symbioses in fungi suggested by SSU rDNA phylogenies. Science 268:1492-1495

Gargas A, DePriest PT, Taylor JW (1995b) Positions of multiple insertions in SSU rDNA of lichen-forming fungi. Mol Biol Evol 12:208-218

Gargas A, DePriest PT (1996) A nomenclature for fungal PCR primers with examples from intron-containing SSU rDNA. Mycol 88:99-124

Grube M Molecular investigation of the *Lecanora rupicola* complex. In prep.

Grube M, DePriest PT, Gargas A, Hafellner J (1995) Isolation of DNA from lichen ascomata. Mycol Res 99:1321-1324.

Groner U, LaGreca S (1997) The 'Mediterranean' *Ramalina panizzei* north of the Alps: morphological, chemical and rDNA sequence. Lichenol 29:441-454

Heibel E, Lumbsch HT, Schmitt I (1999) Genetic variation of *Usnea filipendula* (Parmeliaceae) in Western Germany investigated by RAPDs suggests reinvasion from various sources. Am Journ Bot 86:753-757.

Innis MA, Myambo DH, Gelfand DH, Brow MAD (1988) DNA sequencing with *Thermus aquaticus* DNA polymerase and direct sequencing of polymerase chain reaction. Proc Natl Acad Sci USA 85:9436-9440

Kroken S, Taylor JW (1998) Morphologically cryptic species in the *Letharia* "species pair" and their coevolution with species of *Trebouxia*. In: Grube, M., Wedin, M. (eds) Progress in Molecular Studies of Lichens: Graz, 11-15 August 1998. Programme and Abstracts. 16. Graz

Kusukawa N, Uemori T, Asada K, Kato I (1990) Rapid, reliable protocol for direct sequencing of material amplified by the PCR. BioTechniques 9:66-72

Lutzoni F, Vilgalys, R (1995) *Omphalina* (Basidiomycota, Agaricales) as a model system for the study of coevolution in lichens. Cryptog Bot 5:71-81

Mai M, Grabs R, Barnes RD, Vafiadis P, Polychronakos C (1998) Shortened PCR cycles in a conventional Thermal Cycler. BioTechniques 25:208-210

Mattson J-E, Wedin M (1999) Phylogeny of the Parmeliaceae–DNA data versus morphological data. Lichenol 30:463-472

Murtagh GJ, Dyer PS, McClure PC, Crittenden PD (1999) Use of randomly amplified polymorphic DNA markers as a tool to study variation in lichen-forming fungi. Lichenol 31: 257-268

Printzen C, Lumbsch HT, Schmitt I, Feige GB (1999) A genetic study on the genetic variability of Biatora helvola using RAPD markers. Lichenol 31: 491 – 499

Roux KH (1995) Optimization and troubleshooting in PCR. PCR Methods and Applications 4:185-194

Saiki RK, Scharf S, Faloona F, Mullis KB, Horn GT, Erlich HA, Arnheim N (1985) Enzymatic amplification of ß-globin genomic sequence and restriction site analysis for diagnosts of sickle cell anemia. Science 230:1350-1354

Wedin M, Tehler A, Gargas A (1998) Phylogenetic relationships of Sphaerophoraceae (Ascomycetes) inferred from SSU rDNA sequences. Plant Syst Evol 209:75-83

Wittwer CT, Garling DJ (1991) Rapid cycle DNA amplification: time and temperature optimazation. BioTechniques 10:76-83

Wolinski H, Blanz P, Grube M (1997) Slide PCR als Alternative zu DNA-Isolationstechniken bei Pilzen. Tagung der Gesellschaft für Mykol und Lichenol, Regensburg. Abstracts der Poster, p. 62

Wolinski H, Grube M, Blanz P (1999) Direct PCR of symbiotic fungi using microslides. BioTechniques. 26:454-455

Zoller S, Scheidegger C, Sperisen C (1999) PCR primers for amplification of mitochondrial small subunit ribosomal DNA of lichen-forming ascomycetes. Lichenol 31:511-516

RAPD-PCR of Lichens

H. THORSTEN LUMBSCH and IMKE SCHMITT

Introduction

Randomly amplified polymorphic DNA (RAPD)- polymerase chain reaction (PCR) is a simple and inexpensive tool enabling the study of genetic variation at population level. It can also be applied at the species or subgeneric level to examine phylogenetic questions (e.g., Yoon and Bae 1995, Altomare et al. 1997, Gandeboeuf et al. 1997). However, Rieseberg (1996) showed that a considerable fraction (13%) of RAPD fragments in interspecific comparisons are not homologous. The homology problem will be less important when very closely related organisms are studied, e.g. at an infraspecific level.

While sequences of specific genes provide detailed information about a small portion of the genome, RAPD-PCR provides an estimate of the genetic variation over the whole genome (Hadrys et al. 1992, Hillis et al. 1996). In contrast to PCR with subsequent sequencing, which uses a pair of specific primers, the RAPD method usually employs a single short random primer. With this short primer (usually 10 bp long), anonymous stretches of DNA are randomly amplified. In the subsequent electrophoretic analysis, the presence or absence of bands of randomly amplified polymorphic DNAs is screened. The bands can then be used as markers in studies of genetic diversity. One problem of this method is that the identity of the amplification products is unknown. Because of this, it is not possible to distinguish between target DNA and contaminations, such as lichenicolous fungi or bacteria. Another problem is that, although about 1 to 12 bands are usually amplified, some primers may fail to produce amplification fragments. Still another problem is the reproducibility of the results,

✉ H. Thorsten Lumbsch, Universität Essen, Fachbereich 9/Botanik, Essen, 45117, Germany (*phone* +49-201-1832441; *fax* +49-201-1833118;
e-mail lumbsch@uni-essen.de)
Imke Schmitt

because RAPD suffers from a sensitivity to changes in PCR conditions. Thus, great care should be taken to standardise conditions.

The largest obstacle to using RAPD-PCR with lichenized fungi is that RAPD primers are not fungal-specific. Obviously, an examination of whole lichen thalli or apothecia containing symbiotic algae would lead to uninformative results, because the DNA of both symbionts would be amplified. Recent studies, however, have avoided these complications by using algal-free parts of thalli or apothecia (Heibel et al. 1999, Printzen et al. 1999), or mycobiont DNA extracted from axenic cultures (Murtagh et al. 1998). Amplified fragment length polymorphism (AFLP) has a similar potential to RAPD for population studies of lichens. This method relies on differences at endonuclease restriction sites. This method has the advantage of better reproducibility than RAPD, and a larger number of markers are available, but it has not yet been tested on lichens.

Materials

Equipment

- Computer
- Electrophoresis chamber and voltage source
- Microwave or conventional hotplate
- Polaroid camera or video camera for documentation of gels
- Software for data analysis
- Thermocycler
- UV transilluminator
- Vortex
- Freezing microtome

Chemicals

- Acetone
- Agarose
- Ethidium bromide (10 mg/ml in distilled water)
- Gel load mix (0.25% bromophenol blue, 0.25% xylene cyanole, 30.00% glycerin)
- Mineral oil (some new thermocyclers do not require the use of mineral oil)

- Molecular weight and size marker (e.g., 250 bp ladder)
- 4-15 10mer oligonucleotides as random primers
- Ready-to-go RAPD PCR kit (Pharmacia)/ or PCR core kit / or dNTP mixture (or single dATP, dCTP, dGTP and dTTP), Taq DNA polymerase and PCR buffer (100 mM Tris-HCl, 15 mM $MgCl_2$, 500 mM KCl, pH 8.3)
- Standard DNA (usually *E. coli* DNA)
- Ultra pure water
- TAE buffer, pH 8.0

50 ml	0.5 M Tris solution
14.3 ml	glacial acetic acid
62.5 ml	1M EDTA solution
Make up with water to 250 ml and dilute 1:50 before use.	

Other items

- Disposable gloves
- Eppendorf pipettes
- Parafilm
- PCR tubes
- Tweezers
- Crushed ice
- Polaroid film or printer for video pictures

Procedure

There are two ways of obtaining target DNA from lichen material for RAPD analysis. One possibility is to use only the fungal parts of the lichen material. This may be either by directly using ascospores which is easily done, e.g. in taxa with mazaedia (spore masses on ascomata due to passive spore dispersal). Another possibility is to use sections of algal-free lichen parts, e.g. biatorine apothecia, the central axis of a lichen or the medulla of crustose lichens (Grube, pers. comm.). An alternative to slice PCR is the use of mycobiont DNA extracted from axenic cultures (see Chapters 1-3). The second method is more complex and time-consuming and is not outlined here. A detailed protocol can be found in Murtagh et al. (1998). Because RAPD primers amplify all sorts of DNA, extreme care should be taken to prevent DNA contamination. Use sterile filter pipette tips, autoclaved and acid washed material (including blade) only.

Slice-PCR

1. Cool down freezing microtome and fit blade.
2. Wash lichen material in acetone for ca. 30 min to extract secondary metabolites. Discard acetone and let the material dry for a few minutes. Carefully separate algal-free lichen material, e.g. biatorine apothecia or central axis of a lichen thallus, with sterile tweezers and freeze them in ultra pure water on the freezing microtome.
3. Cut slices of 15-30 μm and transfer 1-3 slices per 0.5 ml Eppendorf tube.

RAPD-reaction

RAPD-PCR can be performed in a total volume of 25 μl.

1. Add water, dNTPs, polymerase, buffer, and primers to PCR tube containing target DNA according to the manufacturer's instructions on your PCR kit. Standard conditions, which may be used, are final concentrations of 0.2-0.4 mM of each dNTP (dATP, dCTP, dGTP, dTTP), 0.2-1 U Taq DNA Polymerase, 2.5 μl PCR buffer, 25 pmol of a single RAPD primer, and a variable amount of sterile water to a total of 25 μl. Ready to go RAPD Analysis Beads (Pharmacia Biotech) may be used as a simplification which require only the addition of sterile water and primers. In this case only 5 μl RAPD primer (diluted to give final concentration of 5 pmol/μl) and 20 μl sterile water need to be added to each RAPD bead. Each specimen should be amplified with about 4-15 RAPD-primers, each in a separate sample. The same DNA will give different banding patterns with different primers. All sorts of random primers can be used and numerous companies offer sets of random primers. No general recommendation can be made regarding the choice of primers. They have to be tested on each species. The following primers have so far been used in studies on lichens: primer 1: 5'-GGTGCGGGAA-3', primer 2: 5'-GTTTCGCTCC-3', primer 3: 5'-GTA-

GACCCGT-3', primer 4: 5'-AAGAGCCCGT-3', primer 5: 5'-AACGCG-CAAC-3', primer 6: 5'-CCCGTCAGCA-3' (Heibel et al. 1999, Printzen et al. 1999) and OPA03: AGTCAGCCAC, OPA11: CAATCGCCGT, OPA18: AGGTGACCGT, and OPAJ04: GAATGCGACC (Murtagh et al. 1998).

2. To control the performance of the PCR reaction, a standard DNA template should be run as well. Various companies offer *E. coli* DNA. To exclude the possibility of an amplification of contaminating-DNA, one sample should be run with no DNA template as a control.
3. Mix contents of the tube by gently vortexing. Centrifuge briefly (few seconds) to collect contents at the bottom of the tube.
4. Add 50 µl of mineral oil.
5. Place samples in thermocycler and start PCR using the following programme. The times in brackets refer to robocyclers (Stratagene; thermocycler with three different heated blocks where the samples are transported to each block). The use of robocyclers is recommended. They will give more reproducible results, because they usually control times and temperatures more precisely than other thermocyclers. All RAPD runs should be done using the same parameters every time to make your results as reproducible as possible.

1 cycle	5 min (6 min)	95°C
45 cycles	1 min (1,3 min)	95°C
	1 min (1,5 min)	36°C
	2 min (2,3 min)	73°C
Soak		4°C

Gel analysis

After amplification, the randomly amplified DNA must be separated and analysed. This analysis can be either performed directly after the PCR reaction or the PCR products can be stored at -20°C until gel electrophoresis. The analysis can be done on agarose gels.

Preparation of agarose gel

1. Boil 1 % agarose gel (e.g. 200 ml TAE buffer, 2 g agarose) in microwave until it is clear.
2. Cool for 2 min at ambient temperature and then add ethidium bromide to a final concentration of 0.1%.

3. Pour solution into electrophoresis form, let solidify for about 30 min and remove comb.

Gel electrophoresis

4. Place gel in electrophoresis chamber and cover with TAE buffer.
5. Prepare 2 µl of gel load mix and 5 µl of each amplification product on parafilm and mix with micropipette.
6. Load gel, placing PCR products with identical primers next to each other.
7. In addition to the PCR products a molecular weight and size marker should be used to allow estimation of the size of the PCR products.
8. Set voltage source to about 80 V and 100 mA and electrophorese for about 1.5 h.

Documentation

To visualize the banding pattern of the randomly amplified DNA place the gel on an UV transilluminator and either photograph the results with a polaroid camera or capture the image with a video camera and print out the picture.

Results

Data analysis

Record presence or absence of bands in a binary (0,1) code in a data matrix. The comparison of bands in different gels is only possible when the electrophoresis conditions are kept to a standard and should be accompanied by the comparison of the molecular weight markers.

Because the presence and absence of bands can hardly be regarded as characters with different character states, it is advisable to use phenetic, i.e. distance matrix methods, for the analysis of the data. Several software packages are available and some are listed below. Mostly the Nei and Li index of genetic similarity (Nei and Li 1979) is applied in the distance calculation. This coefficient scores the number of common bands divided by the total number of bands found in the two samples being compared. The most common phenetic methods for the analysis of the distance values for RAPD data are UPGMA (Michener and Sokal 1957) and neighbor-joining (Saitou and Nei 1987). The branch support can be tested with different methods, bootstrap analysis (Felsenstein 1985) being the most commonly applied one. The phenogram showing the genetic similarity of the samples examined can be drawn with the help of different computer programs, listed below.

Troubleshooting

- No bands, including molecular weight marker, on the gel
  Check whether the voltage source of electrophoresis was correctly set and whether ethidium bromide was added. If this was correct, let the agarose cool for more than 2 min before adding the ethidium bromide. Alternatively, after running, an agarose gel can be soaked in TAE buffer with 1% ethidium bromide stain for 30 min, lightly mixing every few minutes.
- No bands of investigated DNA and standard DNA, but molecular weight marker visible on the gel
  Check whether the right thermocycler program was used and whether the thermocycler is working properly. Alternatively, the wrong reaction volume may have been used. Also check whether water, dNTPs, polymerase, buffer, and primers were correctly added and the dNTPs and buffers were used at the correct concentrations.
- No bands of investigated DNA, but bands of standard DNA and molecular weight marker on the gel
  The quality of the DNA may be poor or the quantity inappropriate; in case of slice PCR thinner sections should be used or less material (in some cases more than 2 slices will decrease the quality of the PCR reaction). Increase of Mg^{2+} and number of cycles may also help (Arup and Grube, pers. comm.).
- Excessive smearing on the gel
  Check whether the primer concentration is correct. Otherwise the DNA quantity may not be appropriate and should be modified as above.
- Only poorly visible bands on the gel
  Check whether the primer concentration is correct. Otherwise the DNA quantity may not be appropriate and should be modified as above.
- Banding pattern is not reproducible using the same template and primer

Check whether the concentration of the DNA used (e.g. number and size of sections) and primers is always the same. Check whether some samples may be degraded or contaminated. Ensure that the same thermocycler is used and the program is always exactly the same.

Comments

RAPD-PCR can be applied to numerous questions regarding populations and species relations. Taxonomic and ecological problems can be adressed by this method. Examples in lichenology include studies on lichen populations reinvading formerly polluted areas (Heibel et al. 1999), correlations of genetic differences of corticolous lichens and their phorophytes (Printzen et al. 1999), and morphological variability within one species (Lohtander et al. 1998). A further important application of RAPD-PCR is DNA sequencing with random primers for multilocus studies (Burt et al. 1994), recently applied in lichenized ascomycetes (Kroken and Taylor 1999).

Acknowledgements. Peter Crittenden (Nottingham) kindly provided us with an at the time unpublished manuscript (Murtagh et al. 1999) for which we are grateful. Dr. Scott LaGreca (Cambridge) and Dr. Martin Grube (Graz) are thanked for valuable comments on the manuscript. HTL wishes to thank Heidi Döring (Bayreuth) for introducing him to molecular techniques.

References

Altomare C, Petrini O, Logrieco A, Bottalico A (1997) Taxonomic relationships among the toxigenic species of *Fusarium acuminatum*, *Fusarium sporotrichoides* and *Fusarium tricinctum* by isozyme analysis and RAPD assay. Can J Bot 75:1674-1684

Burt A, Carter DA, White TJ, Taylor JW (1994) DNA sequencing with arbitrary primer pairs. Mol Ecol 3: 523-525

Felsenstein J (1985) Confidence limits on phylogenies: an approach using the bootstrap. Evolution 39:783-791

Gandeboeuf D, Dupré' C, Roeckel-Drevet P, Nicolas P, Chevalier G (1997) Grouping and identification of *Tuber* species using RAPD markers. Can J Bot 75:36-45

Hadrys H, Balcik M, Schierwater B (1992) Application of random amplified polymorphic DNA (RAPD) in molecular ecology. Mol Ecol 1:55-63

Heibel E, Lumbsch HT, Schmitt I (1999) Genetic variation of *Usnea filipendula* (Parmeliaceae) populations in western Germany investigated by RAPDs suggests reinvasion from various sources. Am J Bot 86: 753-757

Hillis DM, Moritz C, Mable BK (eds) (1996) Molecular Systematics, 2nd edn. Sinauer Associates, Sunderland, Mass.

Kroken S, Taylor JW (1999) A multi-locus approach to reveal speciation, population genetics, and life history changes in the genus *Letharia*. Abstr XVI Int Bot Congr, St Louis/USA: 92

Lohtander K, Källersjö M, Tehler A (1998) Dispersal strategies in *Roccellina capensis* (Arthoniales). Lichenologist 30:341-350

Michener CD, Sokal RR (1957) A quantitative approach to a problem in classification. Evolution 11:130-162

Murtagh GJ, Dyer PS, McClure PC, Crittenden PD (1998) Use of randomly amplified polymorphic DNA markers as a tool to study variation in lichen-forming fungi. Lichenologist 31:257-267

Nei M, Li WH (1979) Mathematical model for studying gene variation in terms of restriction endonucleases. PNAS 76:5269-5273

Printzen C, Lumbsch HT, Schmitt I, Feige GB (1999) A study on the genetic variability of *Biatora helvola* Körb. ex Hellb. using RAPD markers. Lichenologist 31: 491 – 499

Rieseberg, LH (1996) Homology among RAPD fragments in interspecific comparisons. Mol Ecol 5: 99-105

Saitou N. Nei M (1987) The neighbor-joining method: a new method for reconstructiong phylogenetic trees. Mol Biol Evol 4:406-425

Yoon CS, Bae KS (1995) Genetic relationships among *Penicillium* species by characterizing RAPD markers. J Microbiol 33:171-177

Suppliers

There are numerous suppliers for equipment and chemicals necessary for molecular research. Anyone starting with molecular research should contact microbiologists or molecular biologists in his/her department and ask for advice. Usually these departments get discounts from certain companies and it may be advantageous to cooperate with them regarding orders in the beginning.

Some programs and software packages available for data analysis are listed below:

- Paup*
  Software package with numerous applications, including distance methods and bootstraps. Available from Sinauer Associates: orders@sinauer.com

- Phylip
  Software package with numerous applications, including distance methods and bootstraps. Available via Internet: http://evolution.genetics.washington.edu/phylip.html

- RAPDistance
  Program for computing distance matrices in RAPD analyses. Available via Internet: gopher://life.anu.edu.au/molecular/software/rapd.html

- Treecon
  Software package including computation of distance matrices and possibilities to draw dendrograms. Available from the author; information in the Internet: http://bioc-www.uia.ac.be/u/yvdp/treeconw.html

- Treeview
  Program for drawing dendrograms. Available via Internet: http://taxonomy.zoology.gla.ac.uk/rod/treeview.html

Abbreviations

bp	base pair
AFLP	amplified fragment length polymorphism
DNA	deoxyribonucleic acid
EDTA	ethylenediaminetetraacetic acid
dATP	2-deoxyriboadenosintriphosphate
dCTP	2-deoxyribocytidintriphosphate
dGTP	2-deoxyriboguanosintriphosphate
dNTP	2-deoxyribonucleotidtriphosphate
dTTP	2-deoxyribothymidintriphosphate
PCR	polymerase chain reaction
RAPD	randomly amplified polymorphic DNA
TAE buffer	buffer consisting of Tris solution, acetic acid and EDTA
TE buffer	buffer consisting of Tris solution and EDTA
UPGMA	unweighted pair group method with arithmetic mean

Bioindication and Biomonitoring

Biomonitoring Radionuclide Deposition with Lichens

GEORG HEINRICH and KLAUS REMELE

Introduction

Aboveground nuclear tests conducted in the fifties and sixties of the 20th century gave rise to large amounts of ^{137}Cs and ^{90}Sr in the environment. Both radionuclides have physical half-lives of approximately 30 years and are still found in parts of the ecosystem. The long-lived fission products persist, especially in alpine and circumpolar environments characterised by a very slow biological turnover rate (Svoboda and Taylor 1979, Taylor et al. 1985). However, lichens are contaminated by fallout to a higher degree than vascular plants growing in the same habitat (e.g. Gorham 1959). The high surface area to biomass ratio, the slow growth rate and persistence of lichens, and the lack of a well-developed cuticle or wax layers explain their extraordinary capacity for fallout interception and retention. Additionally, lichens are well known as effective accumulators of metals because they take up soluble metal ions very rapidly via an extracellular ion exchange process (Nieboer et al. 1976), and they also trap and intracellularly accumulate metal-rich particles (Richardson 1995).

Measurements of ^{137}Cs in lichens were conducted by Hviden and Lillegraven (1961) in Northern Norway. Svensson and Lidén (1965) noted that 95 % of the total airborne complement of ^{137}Cs deposited by precipitation in Northern Sweden is taken up and retained by lichens. The enhanced accumulation in lichens results in greatly increased body burden of radionuclides in caribou (Canada and Alaska) or reindeer (Northern Scandinavia and Russia) and subsequently in predators such as fox and wolf, and in humans, mainly Eskimos and Lapps. Considerable ana-

✉ Georg Heinrich, Karl-Franzens Universität Graz, Institut für Pflanzenphysiologie, Schubertstraße 51, Graz, 8010, Austria (*phone* +43-316-380-8812; *fax* +43-316-380-9880; *e-mail* georg.heinrich@kfunigraz.ac.at)
Klaus Remele

lytical data are now available on the lichen-caribou or reindeer-man food chain (references cited in Bothmer et al. 1990). Fallout measurements of radionuclides in lichens collected in Europe were conducted shortly before the Chernobyl accident by Eckl et al. (1984, 1986), Kirchmann et al. (1979), Kwapulinski et al. (1985). The reactor accident in Chernobyl in 1986 caused new fallout and a new wave of measurements that also led to several studies on lichens (references cited in Sawidis et al. 1997, Heinrich et al. 1999). After the accident in Chernobyl, the contamination of lichens has remained high in most parts of Europe, and it is difficult to get uncontaminated material.

Sampling plant material for bioindication or biomonitoring has received serious attention, as it was recognised that incorrect procedures may introduce an error exceeding the analytical error by orders of magnitude (Markert 1993, Djingova and Kuleff 1994). A main task of environmental monitoring is to obtain information covering whole regions or countries about the magnitude and the spatial distribution of radioactivity. For this purpose, lichens are used rarely. Information on the contamination of big areas comes more often from measurement of soil, mosses, milk or "grass". A comparison of data recorded by different researchers is often difficult, because in different studies different lichen species are collected at different times and the sample preparation often differs in many ways. Consequently, it is necessary to develop standard methods for sampling lichens in order to produce comparable results.

The techniques commonly used at present for sampling plant material for bioindication or biomonitoring are described in Markert (1993, 1994). The sampling of lichen and moss species for trace-element analysis with special reference to biomonitoring of air pollution is described by Garty (1993, see also Chapter 27), Nimis et al. (1993) and Tuba et al. (1994). Reckel et al. (1999) describe monitoring with lichens using the method of the VDI [VDI- Richtlinie 3799, 1989, 1993)]. The authors give some practical advice, and suggestions on how to work with the data obtained by lichen monitoring. They also refer to Internet resources concerning biomonitoring by lichens.

Two methods can be applied to use cryptogams for bioindication or biomonitoring. "In-situ" or "passive monitoring" uses lichens growing in the survey area. "Active monitoring" or the "transplantation method" uses thalli that are transplanted and exposed at the survey field for a certain period of time (Tuba et al. 1994). Transplantation experiments for measuring radionuclide concentrations with lichens have been carried out only in a few cases however. A useful comparison of instrumental and biological monitoring, dealing with the advantages and disadvantages of the monitoring system and with the demands on an ideal biomonitor can be found in Wittig (1993).

In this chapter we describe methods of how to collect and handle lichen samples and how to measure radionuclides in lichens. Macroautoradiography and the determination of α-, β-, and γ-emitting radionuclides are described in detailed step-by-step protocols.

Subprotocol 1
Collecting and Handling of Lichen Samples

Materials

Lichen species

The choice of which group and which species to use in monitoring will depend on the intention of the project. Lichens can be used for a general screening, monitoring of ecosystem fluxes, with emphasis on a particular emission source, monitoring with particular respect to human health (e.g. analysis of ^{137}Cs in the food chain lichens to humans) and on the ecosystem under investigation.

Not all lichens can be found at each fallout monitoring site in sufficiently large amounts. Most work on contamination of lichens with radionuclides has been carried out using terricolous species, such as *Cladonia* sp. and *Cetraria* sp. The epiphytic lichens *Pseudevernia furfuracea* and *Hypogymnia physodes* occur on many trees in mountainous and subalpine regions, whereas *Cetraria islandica* is a common terricolous lichen.

Often used monitors of radionuclides are *Alectoria ochroleuca*, *Cetraria cucullata*, *C. islandica*, *C. nivalis*, *Cladina arbuscula*, *Cladina stellaris*, *Cladina mitis*, *Cladina rangiferina*, *Cladonia phyllophora*, *Cl. squamosa*, *Cl. furcata*, *Cornicularia divergens*, *Hypogymnia physodes*, *Platismatia glauca*, *Pseudevernia furfuracea*, *Stereocaulon paschale*, *Thamnolia vermicularis*, and *Xanthoria parietina*.

Terricolous lichens are believed to accumulate higher ^{137}Cs concentrations than epiphytic lichens (Eckl et al. 1984). The ^{137}Cs activity of *Cladonia rangiferina*, collected 1993 at the Stubnerkogel, Salzburg, was reported to be 1.6 times higher than that of *Cetraria islandica* and *C. cucullata* and 3.2 times higher than that of *Alectoria ochroleuca* (Hofmann et al. 1995). ^{137}Cs was also monitored using saxicolous lichens of the genus *Umbilicaria*. *U. cylindrica* and *U. deusta* were reported to be good bioindicators of ^{137}Cs, but *U. hirsuta* and *U. murina* were apparently less effective (Kwapulinski et al. 1985).

Procedure

Collecting lichens

A first step in any analytical strategy is the definition of the purpose for which the analysis is performed, and on this basis a sampling procedure has to be developed. In all cases the samples should be taken in a way that permits meaningful comparison of the analytical results according to time and/or the place (Djingova and Kuleff 1994). A checklist of sample collection criteria can be found in Scholz et al. (1994).

The following points have to be considered:

- Can the knowledge desired be obtained by the study planned?
- Does the sampling plan yield a good picture of the real contamination?
- Is the sample big enough to enable sufficiently precise statements to be made about mean or variance of the actual data?
- Is the collection of samples sufficiently recorded?
- Is the sampling error quantified, e.g. by a sufficiently large number of sampling repetitions?
- Is the laboratory/analysis error quantified by repeated analysis of reference samples?
- Is the selection of parameters determined sufficient to obtain the knowledge desired?
- Can the knowledge desired be obtained at smaller cost?
- Are more samples obtained than statistically necessary?

A detailed step-by-step sampling design is difficult to create because the intentions of monitoring can differ. However, always consider the following points.

- To get reliable results, use the same species for monitoring purposes.
- The selected standard species should be ubiquitous and the thalli should be large enough so that you can get enough material.
- Take a reasonably large sample, providing sufficient material is available and it is not a protected species.
- Your "sample" should be comprised of 10-30 sub-samples, each with a minimum mass of 10 g.

- For particular questions, e.g. when studying the biological half-life (T_{bio}) of radionuclides in lichens, always collect the same lichen species from the same place (T_{bio} is usually defined as the time required for a 50% reduction of the environmental radioactivity).
  To avoid any shielding effects of trees, epigeic lichens in the open field should be selected instead of epiphytic lichens (e. g. from a flat area of about 100 m^2) (Hofmann et al. 1993).
- Note that uneven trace element and radioactivity distribution can also be found in terricolous lichen thalli. Collect the whole thallus or parts of the thallus, if desired.
- Note that epiphytic lichens, such as *Pseudevernia furfuracea*, present problems due to stemflow of radioactivity from the tree and uneven distribution of the radionuclides on the tree (Heinrich et al. 1999). Take care when sampling in a forest or underneath trees, because the site of sampling can greatly influence the results of investigations. The distribution of the precipitation between stemflow and throughfall depends mostly on the physical properties of the tree crown and the bark surface and can be influenced by species, individuals, year and season, location, crown area, branch gradient, age of trees, crown overhanging by other trees, event type, climate variation etc. (Turcsányi et al. 1994).

Handling of lichen samples

- Separate each plant carefully from the substrate (tree trunk, stone or soil), and clean off other foreign material (twigs, needles, litter debris) before placing in collection bags to minimise contamination.
- If you use plastic bags for the collection in the field, change to paper bags as soon as possible, because mould fungi can grow on partially hydrated lichens.
- Some authors recommend that lichens should be washed in distilled water (e.g. Kwapulinski et al. 1985). It is accepted that in the case of food chain studies washing of plant material prior to analytical procedures should be avoided. In biomonitoring studies final conclusions about the necessity of preliminary washing of plant material have not been reached yet. We do not recommend this procedure, because of the possible loss of radionuclides. In addition, the fact that various washing procedures are used, makes comparison of results rather difficult.

- Dry lichen material at 100°C for 24 h. Sometimes lichen samples were air-dried at room temperature, ground in liquid nitrogen and then dried again at 105°C for 25 h (Hofmann et al. 1993) or freeze-dried (Smith and Ellis 1990).
- Either measure radioactivity of intact thalli, or pulverise them using a homogeniser. If the load of radionuclides is high, it may be sufficient to fill a Marinelli beaker with the lichen material and press with a plastic ring. Grinding lichens to a fine powder increases the amount of material available for filling beakers, and produces a homogenous powder.
- If you want to carry out transplantation experiments collect lichens in a minimally contaminated area (e.g. Canary Islands) and use whole air-dried lichen thalli to measure the initial activity. Use an aliquot of the lichen material dried at 100°C for correction of activity.
- If ashing is necessary, do not overheat the sample, as volatilisation of elements can occur.

Basis for expressing radioactivity

The radioactivity can be expressed:

- Per unit dry weight: Bq g^{-1} (Bq kg^{-1}), to express results on DW basis, which is in practice the most convenient option.
- Per unit area of ground: Bq m^{-2} (Sloof and Wolterbeek 1992, Heinrich et al. 1999).
- Per unit volume for each sample: Bq cm^{-3}.
- The activities (g^{-1}) are sometimes multiplied by the mean standing biomass of the respective species (g m^{-2}).
- Occasionally, the ^{137}Cs activity is expressed on a surface area basis (Sloof and Wolterbeek 1992). The **leaf area ratio** (LAR= plant surface-to-biomass ratio, $cm^{-3}g^{-1}$) and the stand **leaf area index** (LAI = plant surface-to-ground area ratio) are good measures of the relative quantity of fallout that can be intercepted by the plants. The **adsorption index** (LAR x Exp. time) and the **retention index** (LAI x Exp. time) indicate the potential capacity of a plant species to intercept and retain the fallout. LAR and LAI data of several plant species for comparison of their relative interception potential are given in Svoboda and Taylor (1979).

Subprotocol 2
Detection of Radionuclides by Macroautoradiography

This technique can be used to investigate the distribution of radionuclides within single lichen thalli and to detect "hot spots".

Materials

- X-ray films (Fuji X-Ray film; Kodak Scientific Imaging Film X-OMAT AR; Agfa Structurix D4 FW, Industrial X-Ray film; Hyperfilm MP or Hyperfilm ßmax, Amersham Lifescience).
- Common X-ray film developer (Phenisol, Ilford and D-19, Kodak).
- Stop solution. Ilford and Kodak supply concentrated stop solutions. A general purpose stop solution consists of 0.5 % v/v acetic acid in deionised water.
- Fixing solution (e.g. Hypam rapid fixer, Ilford, Unifix, Kodak). A general purpose fix solution consists of 30% w/v sodium thiosulfate in deionised water.

  Note: All three processing solutions should be allowed to equilibrate to 15-20°C and should not differ in temperature from one another by more than 2°C.

Procedure

1. Glue dried lichen samples on a cardboard.
2. Under safelight illumination, cover the lichens with an auto-radiographic film and a second piece of cardboard, hold firmly together using rubber rings and protect against light with an aluminium foil or a light tight bag. Suitable darkroom light filters include Ilford 902, light brown, and Kodak No 2, deep red. Fit the lamps with 15 W bulbs.
3. Expose for sufficient time (two months) in a refrigerator.
4. Develop the film 5 min in an X-ray film developer, transfer to a stop solution for 30 sec, and fix for 10 min in a fixer for autoradiography.

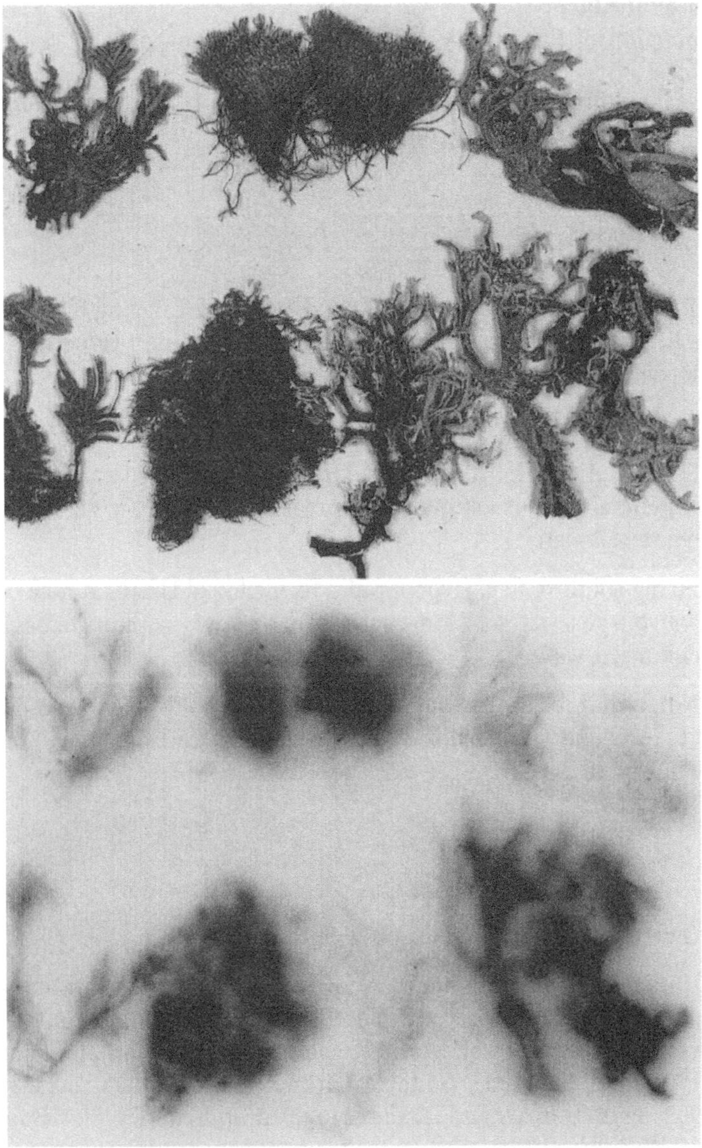

Fig. 1. Autoradiograph of lichens and mosses, collected after the Chernobyl accident and exposed with Agfa Structurix D 7 for two months. Top row, from left to right: *Climacium dendroides* (Graz), *Bryum argenteum* (Graz), *Pseudevernia furfuracea* (Klippitztörl). Bottom row, from left to right: *Climacium dendroides* (Graz), *Amblystegium serpens* (Graz), and two *Pseudevernia furfuracea* samples (Bürgeralpe and Weinebene). The contamination is not uniform, some "hot spots" can be observed.

Figure 1 illustrates the blackening of an X-ray film by moss and lichen samples collected from different sites. The level of blackening and radioactivity [Bq kg^{-1}] measured with a γ-counter are in good agreement.

Subprotocol 3
Determination of γ-Ray Emitting Radionuclides Using NaI-Crystal Detectors

The following **natural radionuclides** can be measured with this procedure:

- ^{7}Be is a short-lived ($t_{1/2}$ = 53d) light radionuclide of cosmogenic origin that has widely been used as a tracer for atmospheric turbulence and to measure the efficiency of uptake into lichens from the atmosphere (Ellis and Smith 1987, Smith and Ellis 1990).
- ^{40}K ($t_{1/2} = 1.3 .10^{9}$ a, 1460 keV) is associated with natural potassium (^{39}K with a relative abundance of 93.1; ^{40}K = 0.0118; ^{41}K = 6.88). The observed ^{40}K concentrations in lichens range from 96 to 392 Bq kg^{-1} (Eckl et al. 1986). In some studies the values of ^{40}K seem to be too high. Therefore we recommend that you should measure potassium by atomic absorption spectrophotometry (AAS) in addition to γ-counting (1 Bq coming from ^{40}K corresponds to 0.0321 g potassium).
- Radionuclides from the decay series (natural radioactive families) can be found not only in soils but also sometimes in lichens:
  - ^{232} Thorium decay series: ^{228}Ac (13 h);
  - ^{224} Ra (3.64 d), ^{212} Pb (10.6 h), ^{208}Tl (3.1 m);
  - ^{238}U-Radium series: ^{226}Ra (1600 a);
  - ^{214}Pb (28.8 m);
  - ^{214}Bi (19.8 m);
  - ^{210}Pb (22 a).

In addition to these natural isotopes there are many **man made radionuclides** (Table 1):

- ^{137}Cs is the most important nuclide and most work has been carried out on this element, because of its long half-life.

Table 1. Physical half-life and spectrum lines of man made fallout radionuclides

Radionuclide	Physical half-life	Spectrum lines (at KeV)
^{137}Cs	30.14 y	661.6
^{134}Cs	2.06 y	569.3; 604.7; 795.8; 1038.6; 1167.9; 1365,1; sum peaks: 1400.5 (604.7+795,8); 1969.8 (604.7+1365.1)
^{103}Ru	39.4 d	497.08
^{106}Ru	371.6 d	511.9; 1050.3
^{110m}Ag	249.9 d	884.7; 937.5; 1505.0; 1562.3
^{125}Sb	2.73 y	427.9
^{144}Ce	248.9 d	133.5
^{95}Nb	30.2 d	765.8
^{95}Zr	64.4 d	724.2; 756.7
^{214}Bi	19.9 m	609.3; 1120.3; 1764.5

Materials

Equipment

- NaI-crystal detector
- Atomic absorption spectrometer (AAS)
- Marinelli beakers of 0.5 l and 1 l
- Plastic tubes of 10 cm length and an outer diameter of 14 mm

Chemicals

- Calibration standard for $^{134,137}Cs$, see Subprotocol 4)
- Calibration standard for ^{40}K (100 g KCl in 1 l distilled H_2O=1637 Bq)

Procedure

Measurement of γ-radioactivity with a LKB-system

Radioactivity is measured using plastic tubes.

1. After determination of the background counts of the instrument (12 h), start with the measurement of the calibration standards for $^{134,137}Cs$ and ^{40}K (30 min) to be sure that the system is giving correct values.

2. Dry the sample at 100°C for 24 h, grind and record the dry weight.
3. Fill the plastic tube to a height of 5 cm. Close the tube with parafilm and analyse each sample for 10 min to 1 h depending on the contamination of the sample. Samples are changed by an autosampler. After 9 samples we recommend that you determine the background counts using an empty tube.
4. Repeat the measurements at least two times to estimate the error of the measurement. One disadvantage is the small volume of the sample in the tubes, especially for slightly contaminated samples. On the other hand, measuring in plastic tubes increases the measuring efficiency up to 400 % compared to a 1 l Marinelli beaker because of the better measuring geometry.

Measurement of γ emitting radionuclides using a Raytest multichannel γ analyser

Radioactivity is measured using plastic tubes and Marinelli beakers.

1. For calibration of the instrument proceed as described above (Step 1).
2. Fill the Marinelli beaker with material (dried at 100°C for 24 h and pulverised as described above) if possible up to the 1 l marker and analyse three times for 0.5 h.
   The Mod MCA-Manuela has manual change of Marinelli-beakers of 0.5 l or 1 l. In addition, it is possible to measure plastic tubes in a well. The apparatus makes a subtraction of the background counts and prints the values of the measured isotopes in Bq kg^{-1} and makes a figure of the spectrum over the 1024 channels (Figure 2).

These systems cost approximately one third of the systems mentioned below. Both are useful for the measurement of the long-lived radionuclides ^{137}Cs and ^{134}Cs. If the radioactivity is low, we recommend systems in which Marinelli beakers can be used.

The ability for the measurement of ^{40} K is limited. The Raytest counter which cannot be cooled, shows a small drift (particularly if room temperature is not stable) and therefore long-time measurements are not recommended. It is better to measure samples for one h, and repeat the measurements twice. In addition to the measurement of ^{40}K with the abovementioned γ-counters, measurement of the ^{39}K content by AAS is recommended.

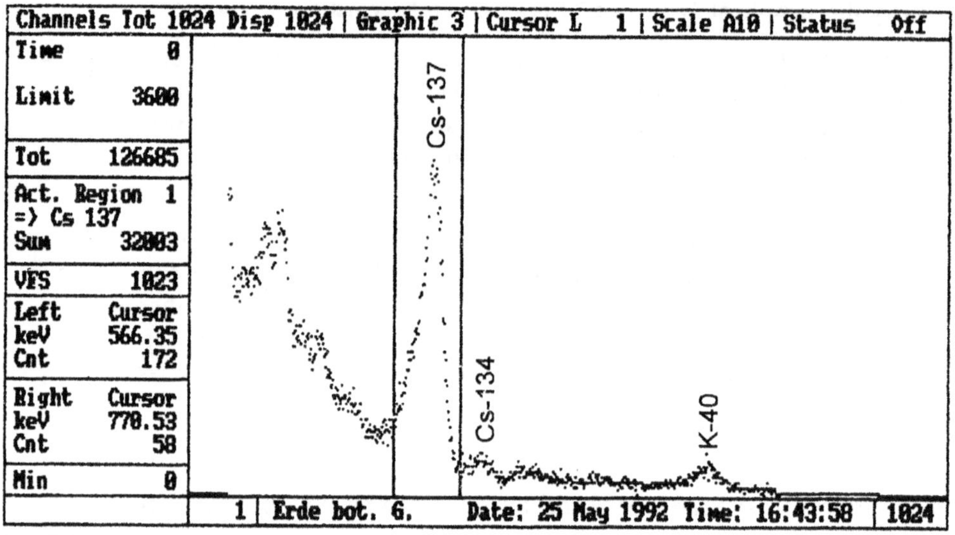

Isotope	Energy range (keV) from	to	Yield	Activity (Bq kg^{-1})
Cs^{134}	777	897	0,028	39,4
Cs^{137}	566	770	0,046	250,7
K^{40}	1340	1585	0,0021	568,1

Fig. 2. A shortened plot of the raytest multichannel analyser of a soil sample from the Botanical Garden in Graz, collected on 25th of May 1992. Note the peaks of ^{137}Cs, ^{134}Cs, and ^{40}K.

Table 2 compares the ^{137}Cs values in *Pseudevernia* and *Hypogymnia* collected in Austria with those from other sites in Europe. The measurements were made shortly after the accident in Chernobyl, showing that the distribution of ^{137}Cs in Europe was not homogeneous.

If the amount of ^{137}Cs exceeds approximately 100 Bq kg^{-1}, the mentioned instruments give correct values. For contamination below 100 Bq kg^{-1} we recommend that you use semiconductor photon detectors. However, these are much more expensive and demand more time and expertise to operate and maintain.

Table 2. ^{137}Cs in kBq kg^{-1} in *Pseudevernia furfuracea* and *Hypogymnia physodes* collected at different sites in Europe.

Pseudevernia furfuracea	^{137}Cs	Hypogymnia physodes	^{137}Cs
Weinebene, Styria	134.6	Weinebene, Styria	79.6
Ischgl, Tyrol	25.1	Oberviechtach, Bavarian Forest	9.5
Choglias Valey, Tyrol	8.1	Bad Kissingen bei Schweinfurt	6.2
Tauplitzalm, Styria	16.2	Isle of Mon, Denmark	3.7
		Tenerife, Canary Island	0.002

Subprotocol 4
Measurements of γ Emitting Radionuclides with Semiconductor Photon Detectors

Materials

Equipment

- semiconductor detectors
  There are some companies that offer a complete line of semiconductor photon detectors for nuclear spectroscopy in the energy range of 1 keV to 10 MeV. We have had good experience with the following detectors:
  - Germanium detector EG & G ORTEC HPGe (High- Purity Germanium) Coaxial Well Photon detector System
  - Germanium well detector, Canberra-Packard, Model 7500SL

Chemicals

- Gamma calibration standards are available from Canberra Packard (Liquid multinuclide solution, 1 l, containing ^{241}Am, ^{57}Co, ^{133}Ba, ^{137}Cs, ^{60}Co, ^{88}Y); and from Czech Metrological Institute, Inspectorate for Ionising Radiation. The calibration standard contains: ^{241}Am, ^{109}Cd, ^{139}Ce, ^{57}Co, ^{137}Cs, ^{203}Hg, ^{113}Sn, ^{85}Sr, ^{88}Y. Certified standard reference material is available from the International Atomic Energy Agency, Vienna, Austria, and from the National Bureau of Standards, Washington. Finally, the Physikalisch-Technische Bundesanstalt, Braunschweig provides standards

Procedure

For sample preparation proceed as described for NaI-crystal detectors. Marinelli-beaker and plastic tubes can be used, the latter only if a well detector system is available. The time of measuring usually is 12 to 24 h. Figure 3 shows a typical γ-spectrum of a lichen sample collected shortly after the Chernobyl accident produced with a semiconductor detector.

Table 3 compares the contents of radionuclides in a lichen sample with that of two higher plants.

Corrections for γ-ray emitting radionuclides

Decay correction The decay correction projects the activity at the time of the count back to the time the sample was collected. This is necessary when there is a long time (relative to the physical half-life) between the sample collection time and the sample count time. A decay correction is useful for the day of collecting the sample and in most cases not for the day of the contamination, because the biological half-life and T_{phys} are different, and T_{bio} is not a constant value.

Table 3. Radionuclides in Bq kg^{-1} in leaves of *Ribes rubrum*, *Acer platanoides* and *Hypogymnia physodes* collected in Graz at 06.11. 1986; measurement uncertainty in %, 1 σ; *man-made radionuclides (not decay-corrected).

Radionuclide	*Ribes rubrum*	*Acer platanoides*	*Hypogymnia physodes*
^{137}Cs	3.452 ± 1	1.665 ± 1	17.427 ± 1
^{134}Cs	1.632 ± 3	648 ± 4	7.677 ± 4
^{106}Ru	2.694 ± 15	1.321 ± 15	4.799 ± 10
^{103}Ru	614 ± 10	329 ± 15	847 ± 15
^{95}Nb	240 ± 20	11 ± 20	203 ± 20
^{144}Ce	555 ± 20	144 ± 20	488 ± 30
^{95}Zr	126 ± 20	96 ± 30	133 ± 30
^{125}Sb	374 ± 20		
^{229m}Te	710 ± 30		977 ± 30
^{110m}Ag		54 ± 20	141 ± 20
sum of activity*	10.379	4.268	32.692
^{40}K	332 ± 10	460 ± 10	124 ± 20

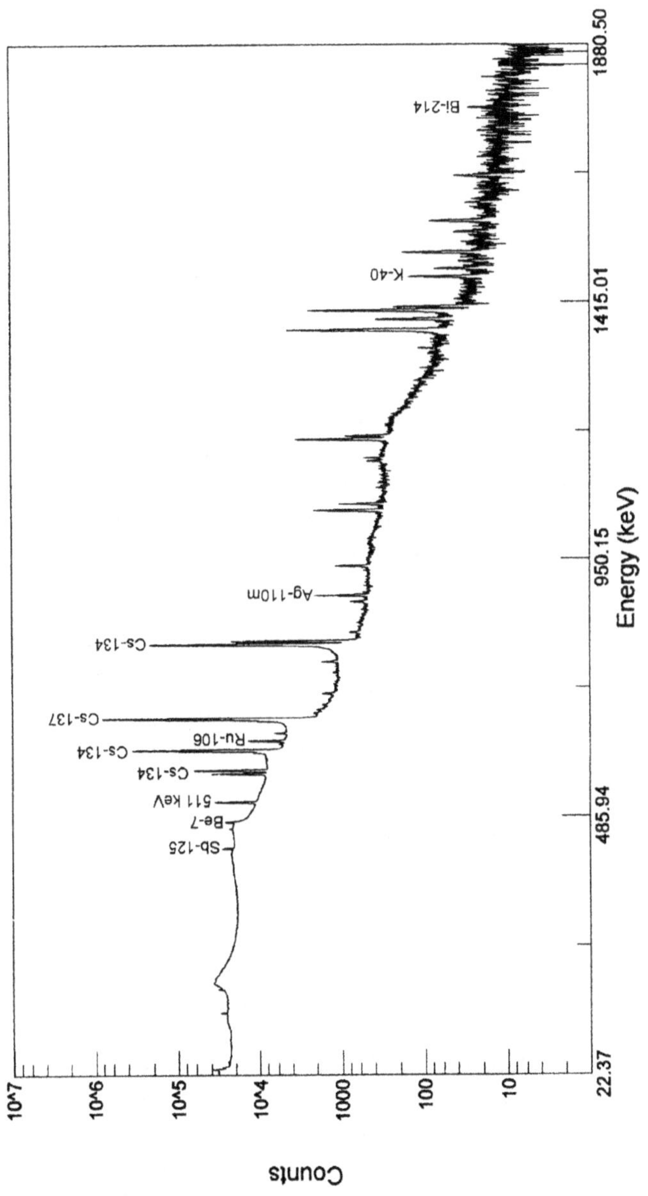

Fig. 3. Gamma spectrum of a 'hot' sample of *Pseudevernia furfuracea* from the Weinebene (1700m) collected on 8th of August 1987, data acquired on 28th of November 1987. Real time=50400 sec. Live time=48892 sec.

Both the time of count and decay-corrected values are presented in the report of some programs (e.g. "Gamma Vision" = Gamma-Ray Spectrum Analysis and MCA Emulation). It is also possible to use the following equation:

$$A = A_0 \times 0.5^{t/\tau} \qquad \text{Equation 1}$$

where A_o is activity at time 0, A is activity at time t, and τ is physical half life.

Note: For t and τ use the same time unit (day or year).

Geometry correction

Problems are caused by the amount of lichens that can be collected. The values of the measurements are only comparable if the Marinelli beakers are always filled to the same height. Note that the filling of the beaker with samples having less volume causes a systematic measuring error because it creates a different measuring geometry. To calculate and correct this possible error, follow this procedure.

1. Measure 100 ml of a standard containing the radionuclides you are interested in.
2. Dilute step by step with 100 ml of distilled water up to a total volume of 1 l. Stir and measure the activity after each dilution step. Always refer to the same weight (1 kg).
3. To obtain a geometrical correction factor, divide the activity per kg measured in 1 l volume by the activity at a certain volume.

Figure 4 demonstrates the influence of the filling height of a Marinelli beaker using a NaI-detector in comparison to a HPGe-Detector. In this experiment the same amount of ^{137}Cs was measured after repeated addition of 100 ml water. With the NaI-detector the values measured are up to 24% higher than the actual activity, depending on the volume, a maximum occurring at 350 ml. Therefore, the activity of a sample of this volume has to be multiplied by the geometrical correction factor 0.806. Using different weights (3, 6, 13, 26, 52, 104 g) of the contaminated lichen *Hypogymnia physodes* (Heinrich et al. 1994), the difference was not as pronounced as recorded for standards (Figure 4). HPGe- detectors can produce more dramatic errors (up to 70 %) than NaI-crystal detectors when measuring volumes smaller than 100 ml.

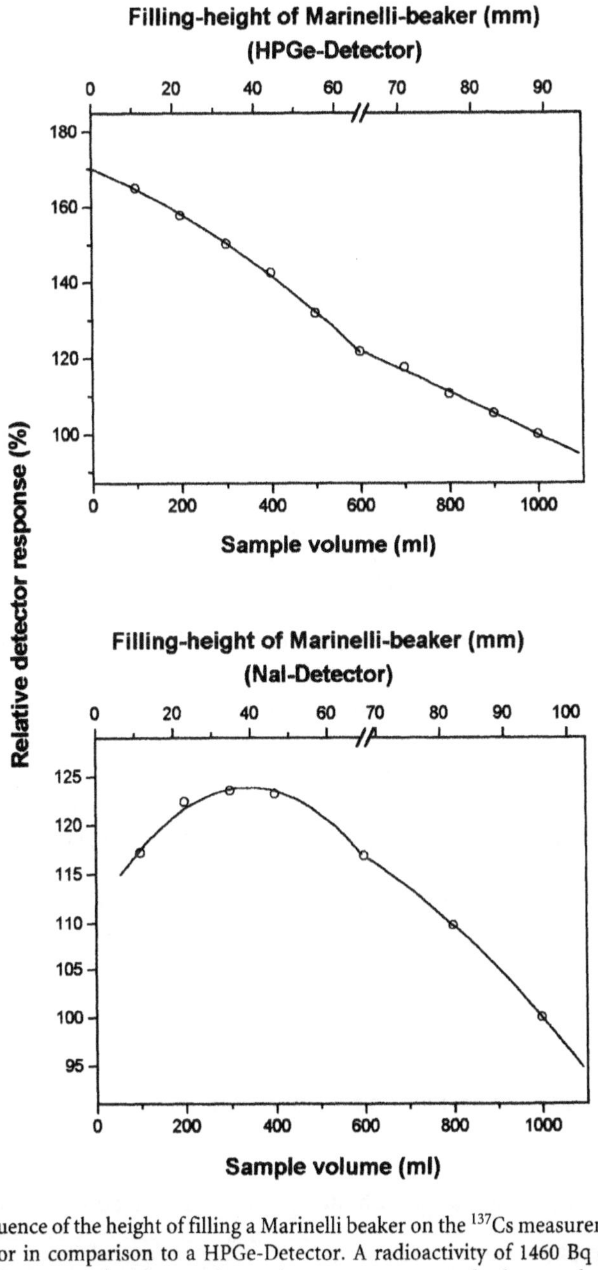

Fig. 4. Influence of the height of filling a Marinelli beaker on the ^{137}Cs measurement using a NaI-Detector in comparison to a HPGe-Detector. A radioactivity of 1460 Bq of ^{137}Cs was recorded after repeated addition of 100 ml water, up to 1000 ml. The sample weight after each dilution was set to 1000 g.

Self absorption

Internal absorption is found when the source (radionuclides present in the thallus) and the absorber (uncontaminated parts of the lichen) are mixed together, so that the γ-rays go through the absorber on the way to the detector. This effect increases with the density and thickness of the sample. The internal absorption of lichens is less than that of soil, and does not vary much between different lichen species compared to different soil samples. Russold (1996) recommends the following corrections for ^{137}Cs: if the plant material shows a density of 0.1 (100g in 1 l Marinelli beaker), divide the activity of the sample by 1.13, if it is 0.5, divide by 1.07. If the soil has a density of 0.8, divide by 1.04, if the density is 1.4, divide by 0.97, and if it is 1.8 divide by 0.93.

Subprotocol 5
Measurement of the β-Ray Emitting Radionuclide ^{90}Sr-Standard Precipitation Method

^{90}Sr in the environment is mainly derived from global fallout following nuclear weapon tests and to a lesser, although increasing extent, from atomic power plants. ^{90}Sr is a pure β emitter with a half-life of 28.5 years. Because β radiators emit radiation in a broad range of energy, their spectra overlap and cannot be separated easily by setting energy windows (as in γ radiation measurement) or by simple calculations. Therefore a chemical procedure is necessary to separate ^{90}Sr from other β radiators. Below, we describe a modified version of the procedure described by Schönhofer et al. (1989). Another technique to separate ^{90}Sr from other radionuclides is given in Wihlidal (1999) and Wihlidal et al. (1998).

Materials

Equipment

- Low-level (low background) scintillation counter
- Muffle furnace
- Centrifuge, preferably with swing out rotor for 50 ml glass-tubes
- Water bath
- AAS

Chemicals

- Nitric acid (65 % HNO_3 : distilled water = 1:1 v/v)
- $SrCl_2$ solution (10 mg Sr ml^{-1})
- $BaCl_2$ solution (20 mg Ba ml^{-1})
- $FeCl_3$ solution (10 mg Fe ml^{-1})
- 25% ammonia solution
- 6 M acetic acid
- 3 M ammonium acetate solution -1.5 M Na_2CrO_4
- $(NH_4)_2CO_3$
- 1 M hydrochloric acid
- Scintillation solution (e.g. Quickszint 400, Zinser)

Procedure

Sample preparation

1. Ash 10-25 g of dried plant material in a muffle furnace at 600°C for 10 h.
2. Weigh 1 g of the ash in a glass centrifuge tube and add 20 ml of nitric acid (HNO_3 65%: distilled water = 1:1 v/v), 5 ml of $SrCl_2$ solution (10 mg Sr ml^{-1}) and 1 ml of $BaCl_2$ solution (20 mg Ba ml^{-1}). Stir for 30 min in a boiling water bath, then centrifuge for 10 min at 5000 g.
3. Add 1 ml $FeCl_3$ solution (10 mg Fe ml^{-1}) to the decanted supernatant, neutralise with 25% ammonia solution and stir for 20 min in a boiling water bath, then centrifuge for 10 min at 5000 g.
4. Decant supernatant and add 6 M acetic acid to the supernatant until the pH is approximately 4, then add 3 M ammonium acetate solution to increase the pH value to 4.75. Add 1 ml 1.5 M Na_2CrO_4 solution to precipitate barium chromate and stir again for 30 min in a boiling water bath. If there is no visible precipitation add 1 ml of the above-mentioned $BaCl_2$ solution. Centrifuge for 10 min at 5000 g.
5. Add 25% ammonia solution to adjust the pH value of the decanted supernatant to about 7 to 8, then add ca 5 g $(NH_4)_2CO_3$ in small portions to prevent the loss of foam from your sample. Stir in a boiling water bath for 20 min and centrifuge for 10 min at 5000 g.

6. After decanting the supernatant, dissolve the precipitated strontium carbonate in 8 ml 1 M hydrochloric acid.
7. If this solution is coloured yellow which causes quenching, another precipitation is necessary. Add 20 ml distilled water, adjust the pH to between 7 to 8 by adding 25% ammonia solution and add again 5 g $(NH_4)_2CO_3$ in small portions. Stir in a boiling water bath for 20 min and centrifuge for 10 min at 5000 g.
8. Dissolve the pellet in 8 ml of 1 M Cl. The solution should now be colourless; otherwise repeat Step 6.
9. To determine the chemical yield, pipette 50 µl of this solution into a 20 ml volumetric flask and fill up with 1 M Cl. Measure the strontium concentration by AAS. A concentration of 15.625 mg Sr ml^{-1} means 100% chemical yield, considering the steps of dilution of the added inactive Sr.
10. Wait for about 4 weeks, so that the ratio ^{90}Sr / ^{90}Y has reached a new equilibrium because ^{90}Y, originating via β-decay from Sr, has been removed through the previous procedures. Then add 12 ml of an appropriate scintillation solution (e.g. Quickszint 400, Zinser; Ready gel, Beckman) to the solution produced at point 7. The result must be a homogeneous solution. Then measure in a scintillation counter.

Measurement of ß-activity

11. Put the samples into an ultra-low-level-scintillation counter e.g. "Quantulus". Although many other scintillation counters exist, the advantage of this apparatus is that it is well shielded by lead; the disadvantage is that it is very heavy. Set the window from channel 100 to channel 900. Measure the blank (8 ml Cl and 12 ml scintillation solution) and each sample in a series of 3 for 4 h. You should find that the efficiency is about 99 %, and that your blanks give about 6.5 counts per minute.

Subprotocol 6
Determination of Radiostrontium Using Crown Ether

Wood et al. (1993) showed that sym-dibenzo-16-crown-5 oxyacetic acid is selective for the extraction of Y^{3+} over Sr^{2+} from aqueous solutions into chloroform.

Other extraction reagents are di-2-ethylhexylphosphoric acid (DEHPA) and tributyl phosphate TBP). Mikulaj and Švec (1993) used TBP to determine ^{90}Sr in milk, plants and soils. Amano and Yanase (1990) presented a method for the measurement of ^{90}Sr and ^{89}Sr in environmental samples by cation exchange using Amberlite CG-120 (100-200 mesh, in a 25 x 2.0 cm column) followed by liquid scintillation counting.

The method described below uses the protocol of Vajda et al. (1992) as slightly modified by Gastberger (1999).

Materials

Equipment

- Low-level (i.e. low background) scintillation counter
- Muffle furnace
- Centrifuge
- Chromatographic glass column (inner diameter 7 mm)

Chemicals

- Crown ether (4,4'-bis-t-butyl-cyclohexano-18 crown-6), Trade name Sr.Spec (100-150 μm)
- 40 % HF
- HNO_3, 65 %, 3 M, 2M, and 0.01 M
- H_3BO_3
- 32 % HCl
- $C_2H_2O_4$
- $Ca(NO_3)_2$
- C_2H_5OH
- 25 % NH_3
- Sr $(NO_3)_2$
- Scintillation cocktail (e.g. Ready gel, Beckman or Quickszint 400, Zinser)

Procedure

1. Add 10 mg of stable strontium as Sr $(NO_3)_2$ in 10 ml 0.01 M HNO_3 solution to each sample.

2. Boil 2 to 5 g of lichen ash in a Teflon beaker to dryness with the following portions of mineral acids: 30 ml 40 % HF, 2 x 30 ml 65 % HNO_3, 30 ml 65 % HNO_3 + 2 g H_3BO_3 and 40 ml 32 % HCl.

3. Add 100 ml HNO_3 to the final residue, warm for at least half an hour to promote dissolution and finally cool and filter through a MN619 1/4 filter paper.

4. Dilute the filtrate with distilled water to 200 ml and add approximately 10 g of $C_2H_2O_4$, and in addition to samples with low calcium content add 1 g of $Ca(NO_3)_2$.

5. Heat the sample to promote dissolution and adjust the pH to 5-6 with 25% NH_3.

6. Keep the solution warm for a few minutes and cool then in a water bath and centrifuge at 5000 g for 10 minutes.

7. Decant the supernatant carefully, and wash the precipitate with c. 100 ml of distilled water and centrifuge again. Repeat the washing procedure.

8. Transfer the precipitate to a glass beaker with a small portion of diluted HNO_3.

9. Destroy oxalates by evaporating to nearly dryness with 2-4 x 20 ml 65 % HNO_3 (the amount of HNO_3 depends on the amount of precipitate).

10. Add 35 ml 3 M HNO_3 to the residue, warm to promote dissolution and cool. The solution is ready for loading onto the chromatographic column.

Chromatographic separation of strontium

11. Fill 2.5 g of the crown ether Sr.Spec into a chromatographic column (inner diameter 7 mm, height 12 cm), wash the column with 100 ml distilled water and 40 ml 3 M HNO_3.

12. Place a filter paper (MN619 1/4) on the top of the column and wait until the 3 M HNO_3 load solution produced at point 10 has passed through the column.

13. Rinse the beaker with 20 ml 3 M HNO_3 and add the solution to the column. Remove the filter and pass another 90 ml of 3 M HNO_3 (in portions of 40, 30, and 20 ml) through the column.

14. Finally, strip the strontium with 15-20 ml distilled water. For cost effectiveness use one column filling for 4-8 measurements. Between single measurements rinse the column with 6 x 100 ml distilled water, 2 x 100 of which are heated up to 40°C before addition onto the column.

Determination of the chemical yield

15. Add 300 mg $C_2H_2O_4$ to the strip solution, heat the solution to promote dissolution and adjust the pH to 9-10 with 25 % NH_3.

16. Filter the solution through a weighed filter paper, wash the beaker with distilled water (add a few drops of 25 % NH_3) and finally with a few ml C_2H_6O.

17. Dry the sample at 50°C for at least one h and then weigh again.

18. Calculate the chemical recovery from the weight of the oxalate precipitation, assuming that the amount of stable strontium originally present in the sample can be neglected.

Preparation of the sample for liquid scintillation counting

19. Place the filter with the strontium oxalate precipitate into a plastic scintillation vial. Add 2 ml of 1 M HNO_3 and dissolve the precipitate.

20. Add 4 ml scintillation cocktail, close the vial, shake thoroughly and centrifuge for 1-2 minutes at low speed to collect the resulting gel at the lower part of the vial.

21. Measure each sample together with a blank. For the blank measurement place an empty filter into a scintillation vial and treat in the same way as the sample. As the gel inside the vial is stable for only 1-2 days, the sample preparation needs to be done immediately before the measurement is started.

Determination of ^{90}Sr activity

22. Measure the samples with a liquid scintillation counter with a special low level mode. Usually, this is done after ^{90}Y has reached equilibrium with ^{90}Sr, i.e. at least 2 weeks after the chemical separation. Alternatively samples can be measured immediately after the chemical separation; in this case the results have to be corrected for ^{90}Y ingrowth originating from ^{90}Sr.

23. Calibrate the counter with a ^{90}Sr solution. Pass about 10 Bq ^{90}Sr and 10 mg stable Sr in a 3 M HNO_3 solution through a chromatographic column filled with 2.5 g supported crown ether. Afterwards, rinse the column with 100 ml 3 M HNO_3 to remove ^{90}Y and finally, strip ^{90}Sr with 20 ml distilled water. The determination of the chemical yield and the preparation for liquid scintillation counting is as described above.

Table 4 shows the ^{90}Sr activities of lichens in Austria. After the Chernobyl accident, the ^{90}Sr level in lichens was high, as expected. Shortly after the accident, the ^{137}Cs-^{90}Sr ratio in *Pseudevernia furfuracea* was 149 and it increased every year, because the loss of ^{90}Sr is faster than that of ^{137}Cs. The T_{bio} for ^{90}Sr was 1.2-1.6 years, whereas a 50% reduction in the ^{137}Cs contamination was reached after approximately 2.5-3 years.

Table 4. Activities of ^{90}Sr, ^{137}Cs in Bq kg^{-1} DW, and the quotient $^{137}Cs/^{90}Sr$ in lichens collected at Weinebene, Styria; measurement uncertainty in %, 1 σ.

Lichens	Date of collection and measurement	^{90}Sr	^{137}Cs	$^{137}Cs/^{90}Sr$
Cetraria islandica	May-91	35 ± 6	9.160 ± 1	233
Cetraria islandica	May-94	6 ± 15	4.540 ± 1	740
Pseudevernia furfuracea	Aug-86	318 ± 3	47.540 ± 1	149
Pseudevernia furfuracea	May-91	44 ± 5	20.910 ± 1	475

Subprotocol 7
Determination of α-Particles Emitting Radionuclides

Because Pu (^{238}Pu and $^{239(240)}Pu$) is usually present only in extremely low concentrations in the biosphere, time consuming radiochemical concentrations and separation procedures are necessary before the various Pu isotopes can finally be determined by α-spectrometry.

Bunzl and Kracke (1994) described an efficient radiochemical separation for the determination of plutonium in environmental samples, using a supported, highly specific extractant.

Materials

Equipment

- α-Spectrometer
- Muffle furnace
- Column

Chemicals

- TEVA Spec™ [methyl - octyl - didecyl - ammonium ion supported on an inert substrate (Amberlite XAD-7 resin)]

- ^{242}Pu tracer
- HNO_3 (conc. and 2 M)
- 0.5 M HNO_3
- $HClO_4$ (conc.)
- 8 M HCl
- 0.5 M HCl
- La $(NO_3)_3 \cdot 6H_2O$
- $NaNO_2$
- Amonia solution (conc.)
- 0.1 M NaOH
- H_2SO_4 conc.
- $(NH_4)_2SO_4$
- KNO_3 (0.1 M)
- Thymol blue (0.04%)

Procedure

Column preparation

1. Before use, swell the resin in deionised water overnight.
2. Prepare the column (18 cm long glass tubes, internal diameter 3.76 mm, bed volume 2 ml with a small glass-wool plug in the bottom) by adding approx. 1 g of the bulk material as a slurry in water. Cover the resin in the column with a piece of filter paper to prevent disruption of the resin bed during sample introduction and connect to the pump by a ground glass joint.

Preparation of the load solution

3. For a sample weight up to 5 g dry weight, add the ^{242}Pu tracer, ash the sample in a muffle furnace at 550°C, dissolve in 2 M nitric acid, and evaporate to dryness in a Teflon-coated dish.
4. Remove traces of organic matter by adding a few ml of concentrated perchloric acid until the colour of the ash is yellowish-white.
5. Take up the residue with 50 ml 2 M nitric acid, and add 10 mg of La (as nitrate) and 50 mg sodium nitrite.

6. After about 15 min add drop-by-drop concentrated ammonia solution while gently heating to co-precipitate Pu with lanthanum hydroxide.

7. Dissolve the precipitate, as obtained after centrifugation (10 min 5000 g), in 20 ml 2 M nitric acid, and adjust the concentration of the nitric acid in the solution again to 2 M (the exact amount of nitric acid required being obtained by titration of an aliquot of the solution with 0.1 M NaOH).

8. Add 50 μl of a solution containing 1 g nitrite in 2 ml water and allow the solution to stand for 15 min.

Column separation of plutonium

9. Feed the load solution to the column via a peristaltic pump at a rate of 1 ml/min. Subsequently, at the same rate, wash the column first with 15 ml 2 M nitric acid and then with 10 ml 8 M HCl to remove traces of iron and thorium. Plutonium is stripped first at a feed rate of 0.5 ml/min with 25 ml 0.5 M hydrochloric acid and then, at the same rate, with 25 ml of a mixture consisting of 0.5 M HCl and 0.1 M HI, containing 0.1 ml of a 35% ammonium sulphite solution.

Preparation of samples for α-spectrometry

10. Collect the eluted solution in a quartz crucible, destroy traces of organic material with 5 ml conc. HNO_3, 3 ml conc. $HClO_4$ and 1 ml 0.1 M KNO_3, and evaporate the mixture to dryness at low heat.

11. Add 0.3 ml conc. H_2SO_4 to the residue and heat the mixture gently until white fume develops. Allow the solution to stand overnight.

12. Add 10 ml distilled water and three drops 0.04% thymol blue.

13. Neutralise the solution with concentrated ammonia until the colour changes to salmon pink (pH 2).

14. Perform the electrodeposition of Pu on a stainless steel disc (2 cm in diameter) at 500 mA for 6 h at pH 2 in ammonium sulphate solution.

15. Add 1 ml conc. NH_4OH before interrupting the current. After the cell is removed, rinse with distilled water and dry.

Modifications

16. If the sample weight exceeds 5 g dry weight, two small modifications are necessary. After taking up the residue from the initial dry ashing process with nitric acid, filter the solution using a membrane filter (Millipore 1.2 μm), treat the black residue separately with concentrated perchloric acid, and combine with the filtrate. Subsequently, add 10 mg lanthanum (as nitrate) and 1 g nitrite.

17. A second modification is required in order to remove ^{210}Po, which, if present in excess, would otherwise interfere with the α-spectrum. Dissolve the $La(OH)_3$ precipitate obtained after centrifugation (see above) in 20 ml 0.5 M nitric acid, and deposit ^{210}Po on an immersed copper disc (c. 2 cm in diameter) while stirring for about 20 h. Adjust the concentration of nitric acid in the solution to 2 M as described above.

Testa et al. (1993) gives alternative methods for separating and determining the α-emitters (^{238}Pu and $^{239(240)}Pu$), and radiochemical techniques for the separation and determination of ß (^{90}Sr, ^{63}Ni and ^{241}Pu) and X emitters (^{55}Fe, ^{59}Ni).

Troubleshooting

The main problem is to collect a representative sample of lichens (see Collecting lichens). There is no point in carrying out a precise analysis if you have used an incorrect sampling technique.

Many potential sources of error exist in the sampling process and the analysis, including the following:

- sampling place is incorrect;
- sampling technique is incorrect, the number of samples and the mass of every sample is too small;
- remember that radionuclides can be very unevenly distributed in lichens; the ^{137}Cs content has been found to be higher in the upper part than in the lower part of certain lichens;
- the ^{137}Cs content varies depending on the position of a lichen on the tree or under a tree;
- radionuclide concentration can increase with altitude; however, after the Chernobyl accident an inversion of this trend was sometimes found, depending on the precipitation of a certain area;
- note that dried lichens can adsorb moisture which results in a change of weight and would produce incorrect results;
- the temperature of the sample changes with time during your analysis (this can cause drift in detectors without cooling system in γ-spectrometry);

- inappropriate or defective instrumentation for the parameters to be determined;
- volatilisation of elements (e.g. too much heat in course of the ashing process).

Comments

The following section lists advantages and disadvantages of lichens as biomonitors.

Advantages of lichens as biomonitors

The main advantages of using lichens as biomonitors are as follows:

- Lichens often have a wide geographical distribution.
- Lichen morphology shows no seasonal variation.
- No seasonal differences in availability and applicability exist.
- Lichens accumulate radionuclides from the air; thus, studying epiphytic lichens allows you to differentiate between airborne and soil borne radioactivity, assuming the latter can be neglected.
- Using lichens, passive and active monitoring is possible.
- Standardised exposure methods for lichens exist.
- A correlation between accumulation and input to the ecosystem exists.
- The vitality of lichens is not affected by radionuclides (at least not in concentrations measured after the Chernobyl accident).
- Some lichen species are easy to sample.
- Lichens grow slowly thus preserving the initial contamination.
- It is possible to measure whole lichen-thalli, and sometimes to separate older and younger parts of the thallus.
- There are no pest and disease-control problems.
- Low costs.
- Lichen samples give good information about recently deposited fall-out.

Disadvantages of lichens as biomonitors

- With very few exceptions, lichens are not eaten by humans, and therefore, unlike crop plants, lichens are not distributed world-wide by man.
- Lichens are not very toxitolerant, particularly not against SO_2, and are therefore not present in areas with high air pollution (centres of towns and industrial areas) and are often not represented in large numbers all over the monitoring area.
- It is often impossible to know the age of parts of the thallus.
- It is very difficult to grow lichens under standardised conditions, and therefore all exposed material has to be obtained from nature.
- In contrast to higher plants (e.g. *Populus nigra*), material with genetic uniformity is not available, i. e. cloning is not possible.
- Lichen identification needs experience.
- Lichens do not reflect the content of radionuclides of the soil (e.g. high concentration of $^{235,\,238}U$ or thorium) as accurately as higher plants (e.g. *Sambucus nigra* or numerous species of Fabaceae).
- Leaking of substances needs to be monitored.
- The radioactivity of lichens measured some years after contamination does not allow the exact calculation of the amount of the initial contamination, because the biological half-life of radionuclides differs in different lichen species as well as in different individuals of the same species.
- It is impossible to determine the exact date of initial contamination, although the presence of only short-lived radionuclides indicates recent contamination.

Comparison of lichen and soil-analysis

The following points can be mentioned (compare Hofmann et al. 1995):

- Lichens are easier to sample and to transport than soil.
- When snow covers the soil, lichens can be picked up from trees or rocks.

- A lichen sample composed of various species can be collected easily, giving a mean value for contamination of an area.
- Information that can be derived on events which happened a long time ago is rather limited using lichens.
- Total radioactivity can be determined only if the whole soil profile is sampled.
- The vertical distribution of the radionuclides in the soil profile gives more information than measuring lichen thalli.

Acknowledgements. We acknowledge financial support by the Österreichische Nationalbank, grant number 5324.

References

Amano H, Yanase N (1990) Measurement of ^{90}Sr in environmental samples by cation-exchange and liquid scintillation counting. Talanta 37: 585-590

Bothmer v St, Johanson K J, Bergström R (1990) Cesium-137 in moose diet; considerations on intake and accumulation. Sci. Total Environ. 91: 87-96

Bunzl K, Kracke W (1994) Efficient radiochemical separation for the determination of plutonium in environmental samples, using a supported, highly specific extractant. J. Radioanal. Nucl., Chem., Letters 186: 401-413

Djingova R, Kuleff I (1994) On the sampling of vascular plants for monitoring of heavy metal pollution. In: Markert B (ed) Environmental sampling for trace analysis. VCH Weinheim, New York, Basel, Cambridge, Tokyo, pp. 395-414

Eckl P, Türk R, Hofmann W (1984) Natural and man-made radionuclide concentrations in lichens at several locations in Austria. Nord. J. Bot. 4: 521-524

Eckl P, Hofmann W, Türk R. (1986) Uptake of natural and man-made radionuclides by lichens and mushrooms. Radiat. Environ. Biophys. 25: 43-54

Ellis KM, Smith J N (1987) Dynamic model for radionuclide uptake in lichen. J. Environ Radioactivity 5: 185-208

Garty J (1993) Lichens as biomonitors for heavy metal pollution. In: Markert B. (ed). Plants as biomonitors. Indicators for heavy metals in the terrestrial environment. VCH Weinheim, New York, Basel, Cambridge, pp 193-263

Gastberger M (1999) Soil-plant-milk transfer of fallout cesium and strontium in Austrian pastures and in an area affected by local fallout from nuclear tests conducted at the Semipalatinsk test site. Thesis Salzburg

Gorham E (1959) A comparison of lower and higher plants as accumulators of radioactive fallout. Can. J. Bot. 37: 327-329

Heinrich G, Oswald K, Müller Hj (1994) Zur Kontamination von Flechten in der Steiermark vor und nach dem Reaktorunglück von Chernobyl. Mitt. Naturwiss. Ver. Steiermark 124: 173-189

Heinrich G, Oswald K, Müller Hj (1999) Lichens as monitors of radiocesium and radiostrontium in Austria. J. Environ. Radioactivity 45: 13-27

Hofmann W, Attarpour N, Lettner H, Türk R (1993) ^{137}Cs concentrations in lichen before and after the chernobyl accident. Health Phys. 64: 70-73

Hofmann W, Gastberger M, Türk R (1995) Bodenflechten der alpinen Höhenstufen als Bioindikatoren für radioaktiven Fallout (Institut für Physik und Biophysik und Institut für Pflanzenphysiologie Univ. Salzburg). Umweltbundesamt Wien, UBA-95-116, 1-24. ISBN 3-85457-258-1

Hviden T, Lillegraven A (1961) ^{137}Cs and ^{90}Sr in precipitation, soil and animals in Norway. Nature (London) 192: 1144-1146

Kirchmann R, Lambinon J, Empain A, Bonnijns-Van Gelder E (1979) Choix et performances de bio-indicateurs pour la surveillance des sites d' implantation d' installations nucléaires. In: 2nd International Symposion of Radiology, Cadarache, 19-22 June 1979, pp 557-589

Kwapulinski J, Seaward M R D, Bylinska, E A (1985) ^{137}Caesium content of *Umbilicaria* species, with particular reference to altitude. Sci. Total Environ. 41: 125-133

Markert B. (1993) Instrumental analysis of plants. In: Markert B. (ed). Plants as biomonitors. Indicators for heavy metals in the terrestrial environment. VCH Weinheim, New York, Basel, Cambridge, pp 65-103

Markert B. (ed) (1994) Environmental sampling for trace analysis. VCH Weinheim, New York, Basel, Cambridge, Tokyo

Mikulaj V, Švec V (1993) Radiochemical analysis of ^{90}Sr in milk, soil and plants by solvent extraction. J. Radioanal. Nucl. Chem., Letters 175 (4): 317-324

Nimis P L, Castello M, Perotti M (1993) Lichens as bioindicators of heavy metal pollution: A case study at La Spezia (N Italy). In: Markert B. (ed). Plants as biomonitors. Indicators for heavy metals in the terrestrial environment. VCH Weinheim, New York, Basel, Cambridge, pp 265-284

Nieboer E, Puckett K J, Grace B (1976) The uptake of nickel by *Umbilicaria muhlenbergii*: a physicochemical process. Canad. J. Bot. 54: 724-733

Reckel S, Löschner M, Stock M (1999) Flechten als Anzeiger der Luftqualität. Biologie in unserer Zeit 29 (6) 364-370

Richardson D H S (1995) Metal uptake in lichens. Symbiosis 18: 119-127

Russold H (1996) Bestimmung der Photonenschwächung mit Monte Carlo Simulation und HPGe Detektoren für Umweltproben in Marinelli-Geometrie. Diploma Thesis Graz 1996

Sawidis T, Heinrich G , Chettri M K (1997) Cesium-137 monitoring using lichens from Macedonia, northern Greece. Can. J. Bot. 75: 2216-2223

Scholz R W, Nothbaum N, May T. W (1994) Fixed and hypothesis-guided soil sampling methods-Principles, strategies, and examples. In: Markert B (ed) Environmental sampling for trace analysis. VCH Weinheim, New York, Basel, Cambridge, Tokyo, pp. 335-345

Schönhofer F, Friedrich M, Buchtela, K. (1989) A new simplified determination method for environmental Sr-90 by ultra low-level liquid scintillation counting. Int. Conf. new trends in liquid scintillation counting and organic scintillators. Gatlinburg, Tennessee, USA, October 2-5, 1989

Sloof J E, Wolterbeek B Th (1992) Lichens as biomonitors for radiocaesium following the Chernobyl accident. J. Environ. Radioactivity 16: 229-242

Smith J N, Ellis K M (1990) Time dependent transport of Chernobyl radioactivity between atmospheric and lichen phases in Eastern Canada. J. Environ. Radioactivity 11: 151-168

Svoboda J, Taylor H W (1979) Persistence of cesium-137 in arctic lichens, *Dryas integrifolia*, and lake sediments. Arctic and Alpine Res. 11: 95-108
Svensson G K, Lidén K (1965) The transport of 137Cs from lichen to animal and man. Health Phys. 11: 1393-1400
Taylor H W, Hutchison-Benson E, Svoboda J (1985) Search for latitudinal trends in the effective half-life of fallout ^{137}Cs in vegetation of the Canadian Arctic. Can. J. Bot. 63: 792-796
Testa C, Desideri D, Meli M A, Roselli C, Queirazza G, Bazzarri S (1993) Radioanalytical procedures for the separation and determination of alpha, beta and X emitters in environmental samples of a nuclear power plant before decommissioning. Sci. Total Environ. 130/131: 403-417
Tuba Z, Csintalan Z, Nagy Z, Szente K, Takács Z (1994) Sampling of terricolous lichen and moss species for trace element analysis, with special reference to bioindication of air pollution In: Markert B (ed) Environmental sampling for trace analysis. VCH Weinheim, New york, Basel, Cambridge, Tokyo, pp 415-434
Turcsányi G, Penksza K, Siller I, Führer E, Tóth S, Kovács M, Büttner S (1994) Sampling in the stemflow and throughfall areas of forests. In: Markert B (ed) Environmental sampling for trace analysis. VCH Weinheim, New york, Basel, Cambridge, Tokyo, pp 449-464
Vajda N, Ghods-Esphahani A, Cooper E, Danesi P R (1992) Determination of radiostrontium in soil samples using a crown ether. J. Radioanal. Nucl. Chemistry 162: 307-323
VDI-Richtlinie 3799 (1993). Messen von Immissionswirkungen. Ermittlung und Beurteilung phytotoxischer Wirkungen von Immissionen mit Flechten: Flechtenkartierung zur Ermittlung des Luftgütewertes
VDI-Richtlinie 3799, Blatt 2 (1989) Messen von Immissionswirkungen. Ermittlung und Beurteilung phytotoxischer Wirkungen von Immissionen mit Flechten: Flechtenexposition Wihlidal H (1999) Analysis of ^{90}Sr in urine and environmental samples. Procedure report -97. Austrian Research Centre, Seibersdorf
Wihlidal H(1999) Analysis of ^{90}Sr in urine and environmental samples. Procedure report 97. Austrian Research Centre, Seibersdorf
Wihlidal H, Sinojmeri M, Lovranich E, Steger F (1998) Sequence analysis of actinides and Sr-90 in urine samples. In: Radioaktivität in Mensch und Umwelt, 30. Jahrestagung des Fachverbandes für Strahlenschutz gemeinsam mit dem Österreichischen Verband für Strahlenschutz, Lindau, 28. Sept.-2. Okt 1998, Bd 1, p 222, ISSN 1013-4506
Wittig R (1993) General aspects of biomonitoring heavy metals by plants. In: Markert B. (ed). Plants as biomonitors. Indicators for heavy metals in the terrestrial environment. VCH Weinheim, New York, Basel, Cambridge, pp 3-27
Wood D J, Elshani S, Du H S, Natale N R, Wai C M (1993) Separation of ^{90}Y from ^{90}Sr by solvent extraction with ionizalble crown ethers. Anal. Chem. 65: 1350-1354

Suppliers

Calibration standards can be obtained from:

- Physikalisch-Technische Bundesanstalt, Bundesallee 100, D 3300 Braunschweig

- Czech Metrological Institute, Inspectorate for Ionizing Radiation. Radiova 1, 102 00 Praha 10, phone: ++4202 67008497, fax: ++4202 67008466
- Intl. Atomic Energy Agency, (IAEA) Post Box 100, Wagramer Str. 5, 1400 Vienna, Austria. Phone: (+43) 1 - 26002 12 88; Fax: (+43) 1 - 26002 96 10; E-Mail: Official.Mail@iaea.org; Internet: www.iaea.or.at

Semiconductor detectors can be obtained from:

- EG & G ORTEC. 100 Midland Road, Oak Ridge, Tennessee, TN 37831-0895 U.S.A. Phone: (423)-482-4411, FAX: (423) 483-0396, E-Mail: INFO_ORTEC@egging.com, Austria: see below, France: (800) 0476907045, Germany (089) 926920, Italy: (02) 27003636, UK (01189) 773003
- Canberra-Packard. Canberra Industries Inc. 800 Research Parkway, Meriden, Connecticut 06450. Tel (203)238-2351. Austria. Canberra Packard Instrument GmbH, Josef Zapf Gasse 2, A 1210 Wien, Austria Tel: (++43-1) 302504, Fax: (++43-1) 305159, Germany. Canberra-Packard GmbH, Hahnstraße 70 , D 6000 Frankfurt 71, Germany. Tel: (++49 69) 663010, Fax: (++4969) 6665921.

Suppliers of scintillation counters are:

- Canberra Packard. See above
- Beckman. Beckman Instruments, Inc. Bioanalytical Systems Group. 2500 Harbor Boulevard, Box 3100, Fullerton California 92634-3100, TWX: 910-592-1260,Telex: 678413. Austria: Beckman Instruments GmbH, Inkustraße 16, A-3400 Klosterneuburg, Tel: (++43 2243) 85656-0
- The low-level-scintillation counter 1220 "Quantulus" can be obtained from EG & Berthold/Ortec/Wallac. Ameisgasse 49-51, A-1140 Wien, Austria, Tel: (++43 1) 914 22 51 9, Fax: (++43 1) 914 22 51 85 or Wallac Oy, P.O. Box 10, 20101 Turku, Finland, Tel: 358-21-678 111, Telex: 623 33

Crown ether can be obtained from:

- EIChroM Industries, Inc. 8205 S. Cass Ave., Suite 107, Darien IL 60559, USA

Biomonitoring Heavy Metal Pollution with Lichens

Jacob Garty

Introduction

Certain characteristics of lichens and bryophytes meet the specifications required for biological monitors. These include large geographical ranges, allowing the comparison of metal content in diverse regions, and a morphology that does not vary with season, thus enabling accumulation to occur throughout the year (Puckett 1988). Lichens integrate long-term deposition patterns and do not reflect necessarily short-term patterns, measured by most atmospheric scientists. The advantages of biomonitoring with lichens over instrumental monitoring are that lichens accumulate most of the elements of the periodic table, are usable at low expense, do not depend on electricity for their operation, do not need treatment and are easy to hide, thus discouraging vandalism. Inexpensive biomonitoring systems also provide much information on the impact of airborne heavy metals on physiological processes. Expensive automatic instrumental air monitoring, on the other hand, provides detailed information on the concentration of gases (e.g SO_2, NO_x, CO, O_3) in the air, whereas airborne heavy metals are often not documented.

Of the large number of lichen species, only a few have been used as biomonitors of trace-element deposition, to assess the degree of atmospheric contamination. However, the study of these few species has yielded hundreds of publications. In this chapter, the term "passive biomonitoring" will be used to denote monitoring that involves analysing lichens in situ. In "active biomonitoring" lichens are collected from an area with clean air, exposed to polluted area and then analysed for metal accumulation.

Jacob Garty, Tel Aviv University, Department of Plant Sciences and Institute for Nature Conservation Research, The George S. Wise Faculty of Life Sciences, Tel Aviv, 69978, Israel (*phone* +972-3-6407468; *fax* +972-3-6409380; *e-mail* garty@post.tau.ac.il)

Lichens accumulate metals both by particulate trapping and via ion exchange. Heavy metal containing particles may be detected in close association with mycobiont hyphae, e.g. in *Caloplaca aurantia* sampled in a contaminated site in Tel Aviv (Garty et al. 1979). The particles appeared in a variety of shapes and sizes and were found to consist of Ti, Cr, Fe, Ni, Mn and Zn. Scanning electron micrographs of *Parmelia chlorochroa* rhizinae produced by Gough and Erdman (1977) show the presence of irregularly shaped foreign matter and fly-ash microspheres.

The assumption that metal-containing particles are enclosed in the thallus was put forward based on the coefficient of variation (CV) obtained by SD/mean values of heavy metal content (Garty et al. 1977). These authors assumed that CVs indicate the mode of aerial dispersion of particles containing heavy metals. Very low CVs were assumed to indicate a low variation due to an even suspension of metal particles; high CVs, on the other hand, indicate a limited, localised deposition of coarse particles. Findings of particulate material in lichens suggest that SEM combined with EDX can be a useful tool for the investigation of different chemical elements incorporated in the thallus. Many lichens display a linear relationship between the Fe/Ti content and trapped particulates (Nieboer et al. 1978). Other lichens exhibited linear relations of Ti and Al (Nash 1989). Bargagli (1989) emphasised the importance of an analysis of Al and/or Ti content and of the normalisation of raw concentrations of metals. Bargagli also recommended a calculation of an enrichment factor (EF) with respect to average concentrations in unpolluted soils or to atmospheric particulates. An estimation of the level of atmospheric pollution should depend on calculated EFs.

The uptake of soluble cations by lichens is initially via an ion exchange process that occurs rapidly and is not inhibited by metabolic poisons (Richardson et al. 1985). The exchange of ions can be shown by displacing the bound cations with hydrogen ions or other metal ions (Nieboer and Richardson 1981; Brown and Beckett 1984).

It is not the intention of this chapter to provide very detailed step-by-step protocols for using lichens to biomonitor heavy metal pollution. Rather, the format of this chapter is more that of a review. In this chapter, the following topics are discussed:

- how to choose the best species to use;
- how to analyse lichens for heavy metals;
- how to use lichens to identify the sources of pollutants by principal component analysis;

- how to assess baseline element levels;
- using transplant experiments to biomonitor pollution;
- which physiological parameters can be used to monitor metal-induced damage.

Procedure

Selection of lichen species applicable as biomonitors

Foliose and fruticose lichens are more suitable for monitoring heavy metal pollution than crustose lichens because they are easier to sample. The foliose lichen *Hypogymnia physodes*, for example, has proved to be very efficient as a monitor of the environmental impact of metal pollution. This lichen was selected to monitor airborne heavy metals in the seventies and the eighties (James 1973; Puckett 1988; Kubin 1990; Richardson 1992; Garty 1992, 1993; Nash 1996; Jacquiot and Daillant 1997). Additional studies using *H. physodes* to biomonitor airborne heavy metals and other chemicals appeared in the nineties, and the species was used to monitor many kinds of industrial activities. Table 1 lists selected studies that have used foliose, fruticose and crustose lichens as biomonitors for heavy metal pollution, with an emphasis on recent publications.

Among corticolous lichens habitats, certain species may exhibit a higher capacity to accumulate metals than others. The epiphytic fruticose lichen *Pseudevernia furfuracea*, for example, collected in Bern, Switzerland, was found to contain 10 504 μg/g Fe whereas the foliose lichen *Parmelia sulcata* contained only 892 $\pm$ 607 μg/g Fe (Garty and Ammann 1987). Morphological and anatomical features probably determine accumulation rates of heavy metals by lichens, by affecting the efficiency of particle entrapment. Nash and Sommerfeld (1981) found that morphological characteristics determine the elemental composition of lichens. Low values for mineral elements in *Rhizoplaca melanophthalma* were related to the fact that this lichen had a relatively thick thallus and consequently a low ratio of surface to volume in comparison with *Acarospora strigata*, *Caloplaca trachyphylla*, *Lecanora frustulosa*, *L. alphoplaca* and *L. novomexicana*. For soluble ions, Puckett and co-workers (1973) showed that different lichen species had varying metal-binding capabilities under laboratory conditions. Individual species of the genera *Umbilicaria* and *Cladonia* differed considerably in their capacity to absorb Cu.

Table 1. Selected studies using foliose, fruticose and crustose lichens as biomonitors for heavy metal pollution, with emphasis on the nineties

Lichens	References
Hypogymnia physodes	Belandria et al. (1991); Kansanen and Venetvaara (1991); Lupsina et al. (1992); Tynnyrinen et al. (1992); Gailey and Lloyd (1993); Takala et al. (1994); Lackovicova et al. (1994); Kytömaa et al. (1995); Jeran et al. (1995); Jovanovic et al. (1995); Makholm and Bennett (1998)
Xanthoria parietina	Bartok and Osvath (1990), Bargagli et al. (1997)
Parmelia sulcata	Sloof and Wolterbeek (1991a,b); Freitas et al. (1997); Reis et al. (1999)
Parmelia caperata	Freitas and Nobre (1997); Rodrigo et al. (1999)
Ramalina lacera	Garty et al. (1997a-d, 1998a,b)
Ramalina menziesii	Boonpragob and Nash (1990)
Ramalina stenospora	Walther et al. (1990)
Ramalina maciformis	Garty (1985); Garty et al. (1995a, 1996b)
Pseudevernia furfuracea	Garty and Ammann (1987); Takala et al. (1994); Quevauviller et al. (1996)
Cladina spp.	Fahselt et al. (1995)
Cladonia spp.	Savidis and Heinrich (1992)
Usnea spp.	Garty and Ammann (1987); Garty et al. (1996a, 1997e)
Lecanora muralis	Seaward (1973, 1974)
Lecanora conizaeoides	Pilegaard et al. (1979); Gailey and Lloyd (1983); Sloof and Wolterbeek (1993)
Caloplaca aurantia	Garty et al. (1977, 1979)
Caloplaca ehrenbergii	Garty (1985)

Preparation of lichen samples for heavy metal determination

The kind of analytical technique you wish to use will determine subsequent specimen preparation. The following instructions apply to the most frequently used methods, e.g. inductively coupled plasma atomic emission spectrometry (ICPAES) or atomic absorption spectroscopy (AAS).

1. Rinse the samples (1 to 2 g) rapidly with double distilled water 2 - 3 times, for 5 s each time. Rapid rinsing will minimize the loss of water-

soluble elements, i.e. K, Mg (Buck and Brown 1979) and Na, known to occur upon the rinsing of desiccated thalli. It is assumed that a rapid, repetitive procedure will not remove particles containing elements like Cu and Pb enclosed by surface hyphae.

2. Air-dry the thalli.
3. Pulverise samples in a mortar with liquid nitrogen. Homogenisation of the lichen material is required to achieve a uniform distribution of metals in the samples.
4. Dry the powder produced in this manner for 24 h at 105°C. Then, digest sub-samples of 250 mg in 10 ml of concentrated analytical HNO_3 in 50 ml test tubes in a heating block for 8 h at a temperature of 120°C.

Markert (1993) describes an additional decomposition system applied recently in trace and ultratrace analysis. High-pressure ashing (HPA) operating at temperatures of around 320°C with a maximum pressure up to 13 MPa was particularly suitable for a complete decomposition of biological samples. In this method, the sample is decomposed in a quartz vessel with HNO_3 within 1 - 2 h.

Instrumental analysis of the elemental content of lichens

In addition to the above-mentioned ICPAES and AAS, the following methods are applicable: instrumental neutron activation analysis (INAA), X-ray fluorescence analysis (XRFA), energy dispersive X-ray fluorescence spectrometry (EDXRF), mass spectrometry with inductively coupled plasma (ICP/MS) and the proton induced X-ray emission technique (PIXE). For some of the methods, e.g. INAA and PIXE, the preparation of pellets is required prior to elemental-content determination. For further details of analytical methods, see the references cited in Table 2. Details about precision, accuracy and concentration-dependence of INAA, XRFA, ICPAES, ICP/MS and AAS are presented in Markert (1993).

Active biomonitoring experiments

Planning To set up a monitoring experiment based on transplanted material, note the following:

- the source of the lichen for transplantation should be in an unpolluted area;

Table 2. References including details on analytical methods for determination of heavy metals in lichens, used in the nineties.

Analytical methods	References
ICPAES	Dillman (1996); Monaci et al. (1997); Bennett and Wetmore (1999)
AAS	Kytömaa et al. (1995); Evans and Hutchinson (1996); Palmieri et al. (1997); Gombert and Asta (1998)
INAA	Sloof and Wolterbeek (1991a), (1993); Markert and Wtorova (1992); Freitas (1995); Sloof (1995); Reis et al. (1996); Freitas and Nobre (1997); Freitas et al. (1997, 1999)
XRFA	Richardson et al. (1995)
EDXRF	Calliari et al. (1995)
ICP/MS	Markert and Wtorova (1992); Vtorova and Markert (1995); Quevauviller et al. (1996); Sarret et al. (1998); Rodrigo et al. (1999); Pablo Valles Mota et al. (1999)
PIXE	Freitas et al. (1999); Reis et al. (1999)

- biomonitoring sites should be in the same climatic region as that of the source (control);
- biomonitoring sites should not differ greatly in their climatic conditions;
- epilithic fruticose lichens transferred together with their substrate (e.g. *Ramalina maciformis* growing on flintstones) should not be located in sites grazed by local animals, e.g. the black goat of the Bedouins in the Negev Desert;
- in all cases, all transplants must be fully exposed to the environment, not covered by roofs and not close to domestic heating stacks, roof-gutters and car parking-lots.

These conditions will assure, to a certain extent, that differences between metal concentrations among samples will indeed reflect deposition differences and not other factors.

As lichen thalli are found in any age and size, the number of samples collected to serve as transplants for massive transplantation should be large enough to expose many thalli in each site. Thalli of all sizes should be used for each site. For small-scale experiments, it is recommended to select healthy thalli of approximately the same size. Analyses of the heavy metal content of transplanted lichens before and after the start-up of industrial plants are mostly conducted with epiphytic lichens.

Biomonitoring procedure

The procedure itself is limited to choosing and pruning the twigs, and then transferring 30 - 40 twigs to each of the biomonitoring sites.

1. For the use of epiphytic lichens growing on twigs, collect terminal lichen-covered twigs with a length of 40 - 60 cm.
2. Secure these 'sticks' with PVC cords and suspend them on peripheral branches of local trees, at 2 to 3 m above the ground. It is important to suspend the twigs on peripheral branches in order to obtain light for the essential physiological processes of the living thalli.
3. The optimal duration of the exposure period is 6 to 9 months. A period of less than 4 months is not enough to accumulate sufficient amounts of airborne heavy metals, whereas an extended period of 12 - 13 months may cause substantial loss of thalli by wind, rain or hail.

Studies of baseline levels

Studies of baseline levels of metal-pollution prior to the start-up of industrial plants, the opening of major motorways or the initiation of other anthropogenic activities are necessary to assess the impact of these activities. Instances of this kind are found in developing countries as well as in old cities and metropolitan areas in Europe and North America. The intention of establishing a new power plant, for example, is announced to the public, that includes the scientific community.

Typical results

As an example, an active biomonitoring study to obtain knowledge on the background level of pollutants was conducted at the construction site of the Maor-David power plant, in the western part of central Israel and in the region encircling this site. For this purpose, we used the epiphytic fruticose lichen *Ramalina lacera*, collected in an unpolluted site in NE Israel with its substrate, carob tree twigs (*Ceratonia siliqua*) and transplanted them to 10 - 30 sites around the power plant (Garty and Fuchs 1982; Garty et al. 1985). The number of sites used for active biomonitoring may vary according to working conditions, budget, etc., as reported in two additional studies using *R. lacera* in the same study area (Garty 1987, 1988). The above-mentioned power plant started to generate electricity in August, 1981, and resulted in an increase of the Cr and Ni content of thalli transplanted in the study area for a one year period: July, 1981 - July, 1982 (Garty 1987) and December, 1981 - December, 1982 (Garty 1988).

Transplant experiments

In the vicinity of both "old" point sources of emission and main motorways, lichens have provided useful knowledge on the source, nature and level of pollutants. Information gathered by analyses of heavy metal content of lichen transplants may be compared with available data on air quality in previous periods with a view to planning urbanisation of rural areas. This information may suggest the need to change the type of fuels being used or to introduce filters in industrial plants etc. Comparative analyses of the elemental content in lichen transplants along roads or along transects from roadsides may provide information on changes in motorcar fuels (e.g. from leaded to unleaded gasoline) and on the increase of traffic.

The duration of the exposure-period of transplants has to be adapted to the intensity of the anthropogenic activity in the relevant study area.

Typical results

In a typical study, Gailey and co-workers (1985) collected small branches covered with *Hypogymnia physodes* and tied them with string onto plastic-covered wire which was then fixed to bamboo poles at a height of 2 m above the ground around a metal foundry in Armadale, Scotland. Batches of the transplants were exposed for eight two-months periods. In the polluted air in the vicinity of the foundry the transplants accumulated greater amounts of Zn than in other parts of the town. The authors found that the Zn content of *H. physodes* transplants ranged between 31 and 202 µg/g, depending on the location of the transplanted material. Gailey and Lloyd (1986) also found that the Fe content of *H. physodes* transplants exposed to an iron-contaminated environment depends on the length of exposure.

In another study, the Pb content of *Ramalina lacera* collected in an unpolluted forest area in Israel and transplanted to a road junction near Tel Aviv was 246 µg/g after 8 months of exposure whereas control samples, left in the forest, contained only 12 µg/g at the end of the experiment. Samples of *R. lacera* transplanted simultaneously to another road junction contained 105 µg/g Pb at the end of the exposure period whereas samples transplanted to a biomonitoring station located 700 m away from this junction contained only 35 µg/g Pb (Garty and Fuchs 1982). These findings are comparable with those of Déruelle and Petit (1983) who found a significant decrease of the Pb content in *H. physodes* at distances of 15 and 600 m from a motorway in France: 488 µg/g and 65 µg/g Pb, respectively.

Differentiation of sources of accumulated metals by principal component analysis (PCA) and related methods

To differentiate between sources of accumulated metals in lichens, Puckett and Finegan (1980) introduced a principal component analysis (PCA) method. This method was carried out for each of the three lichen species *Cetraria cucullata*, *C. nivalis* and *Cladina stellaris*, collected by the authors in the NWT, Canada. PCAs were limited to 17 mineral elements. The first five components in each species explained 88%, 81% and 89% of the variation, respectively, for the above-mentioned lichens. The data of elemental content of *C. cucullata* were standardised to units of variance to eliminate the effects of different scales of measurement and of different concentrations of individual elements. The authors obtained a correlation matrix that referred to elemental correlations in lichens. Two distinct groups of elements were correlated at a 1% level. The first group contained Al, As, Ni, Sc, Ti and V and the second Ca, Fe and Mg. This method enabled the interpretation of 88% of the variation of elemental levels in *C. cucullata* in terms of five variables (components) instead of a large number of variables. Of these five components, the first was found to be highly weighted for Al, As, Ni, Sc, Ti and V, the second for Ca, Fe, Mg and K, the third for S, the fourth for Cu and the fifth for Pb and Mn.

Calvelo et al. in Argentina (1997) compared and standardized elemental concentrations for different lichen species: *Candelariella vitellina*, *Hypotrachyna brevirhiza*, *Parmelia cunninghamii*, *Physcia adscendens*, *Protousnea magellanica* and *Usnea fastigiata*. The information was statistically processed; a cluster analysis was performed using the three principal axes of PCA, which explain 88% of the total variation. The two above-mentioned studies are comparable with that of Bennett and Wetmore (1999) who investigated changes in elemental content of *Cladina rangiferina*, *Evernia mesomorpha*, *Hypogymnia physodes* and *Parmelia sulcata* over 11 years in northern Minnesota, USA. Multivariate analyses of data points revealed two principal components that accounted for 68% of the total variance in the data. These two components, the first highly loaded with Al, B, Cr, Fe, Ni and S and the second loaded with Ca, Cd, Mg and Mn, were inversely related to each other over time and space. The first component was interpreted as comprising an anthropogenic and a dust component, while the second consisted primarily of a nutritional component. Corticolous species accumulated heavy metals to a greater extent than terricolous species.

Physiological parameters as indicators of metal-induced damage

The accumulation of metals in high concentrations often disrupts the physiological processes of lichens. Of the physiological processes that have been investigated in lichens upon metal accumulation, six are probably the easiest to measure and the most sensitive to the interference by metals. It is possible to measure cell membrane permeability and chlorophyll integrity in lichens exposed to field conditions or treated in the laboratory at minimal cost. Analyses of other parameters, e.g. stress-induced ethylene production, spectral reflectance response, photosynthesis and malondialdehyde (MDA) content require special, and rather expensive equipment.

In the following discussion, it is important to realise that lichen injury in polluted environments is not just the result of heavy metals, but also the integrated effect of all pollutants. It is extremely difficult to distinguish the effects of heavy metals from those caused by SO_2, nitrogen compounds, ozone and many other organic and secondary air pollutants. In addition, laboratory studies have suggested that significant synergistic interactions can occur between SO_2 and heavy metal pollutants (e.g. Garty et al. 1992; Kauppi et al. 1998; Punz 1979). Furthermore, it is not always clear which member of the lichen symbiosis is most sensitive to pollutants. In a field experiment, Tarhanen and co-workers (1999) treated thalli of *Bryoria fuscescens* with heavy metal solutions, including a mixture of Cu and Ni or the above in combination with acidity (H_2SO_4, pH 3.0), to study the effects of simulated acid rain and heavy metal deposition on the integrity of membranes. The authors suggested that the different response of membranes is related to the different sensitivity of the algal and fungal partners to the acidity of heavy metal solutions.

Cell membrane permeability and heavy metals

To check the integrity of plasma membranes, a lichen thallus is placed in distilled deionised water for a few minutes (Simon, 1974). In injured thalli electrolyte leakage occurs and this increases the conductivity of the external solution (Puckett, 1976; Alebic-Juretic and Arko-Pijevac, 1989; Garty et al. 1993; Silberstein et al. 1996). According to the laboratory studies of Puckett (1976), exposure of lichens to Ni, Co, Cd and Pb in solution leads to different degrees of damage to lichen cell membranes in accordance with the concentration of the metals. At low concentrations, Puckett observed a loss of K^+ that increased slightly with the increase of concentration until it reached a critical threshold.

1. For an assessment of the integrity of cell membranes prepare samples of 1 g. Quickly rinse the thalli, air dry and keep them overnight in a wet chamber.
2. Immerse the samples in 100 ml of double distilled water for 60 min.
3. Measure the conductivity with a conductivity meter.

Chlorophyll integrity and heavy metals

Laboratory studies have shown that heavy metals can cause chlorophyll degradation. To determine the impact of different metal ions in the presence of certain anions on the integrity of chlorophyll in *Ramalina lacera*, thalli were exposed to different solutions with each of the metal ions Fe^{2+}, Fe^{3+}, Mn^{2+}, Cu^{2+}, Pb^{2+} and Zn^{2+} and the anions SO_4^{2-}, NO_3^- or Cl^- under acidic conditions. Treatments with Cu, Zn, Mn, Pb, Fe(II) and Fe(III) salts under acidic conditions were particularly harmful. The SO_4^{2-} ion combined with K^+ proved to be rather harmful to the chlorophyll in *R. lacera*, over and above Cl^- anions. A $CuCl_2$ treatment led to severe chlorophyll degradation in *R. lacera* (Garty et al. 1992). Chettri et al. (1998) tested the effect of Cu, Zn and Pb on the chlorophyll content of the lichens *Cladonia convoluta* and *C. rangiformis* under controlled conditions. A Cu content up to 1600 $\mu g\ g^{-1}$ dry wt. had no effect on the total chlorophyll content of *C. rangiformis*, whereas Cu contents exceeding 175 $\mu g\ g^{-1}$ in *C. convoluta* led to a decrease of the total content of chlorophyll. Cu effects on chlorophyll were found to diminish in the presence of Pb and Zn in both lichen species.

One way of expressing chlorophyll integrity is to use the "phaeophytization ratio", i.e. the ratio of absorbance at 435 nm to absorbance at 415 nm (Ronen and Galun 1984, see also Chapters 10 and 22). We recommend using DMSO as a solvent for the extraction of photosynthetic lichen pigments, as extraction is simple, rapid and complete (see Barnes et al. 1992 for more details). In addition, the extract is easily stored at low temperatures without degradation. In one study, this parameter was applied under field conditions for lichen-samples transplanted for one year (1980-1981) in biomonitoring stations in northeast and central Israel (Garty et al. 1985). The values obtained for the ratio of absorbance at 435 nm to absorbance at 415 nm decreased in proportion to the amounts of Br, Pb, Fe and Ti. The ratio was inversely correlated with the amounts of Pb, Zn, Cu and S in thalli of *R. lacera* transplanted in the above mentioned study area for another period of one year, 1984-1985 (Garty et al. 1988).

Riga-Karandinos and Karandinos (1998) studied the comparative sensitivity of *Anaptychia ciliaris*, *Lobaria pulmonaria* and *Ramalina farinacea* to chemical pollutants and the impact of pollutants on chlorophyll integrity. Samples of these lichens were collected in the vicinity of a lignite power plant in southern Greece. Multiple linear regressions showed that the amount of S and of the greater part of the metals depended on distance (negative effect) and elevation (positive effect) relative to the location of the power plant. The ratio of chlorophyll *a* / phaeophytin *a* in the three lichens indicated a substantial degradation of chlorophyll. The authors concluded that phaeophytization may be considered a reliable bio-indicator of total environmental stress but not of specific pollutants.

In another study, Bartok et al. (1992) reported on the diminution of assimilation-pigments, dependent on the distance from a polluting chromite dryer. Transplants of *Xanthoria parietina* were placed around a plant of refractory material in the industrial area of Dej, in Cluj county, Romania, which released MgO, Al_2O_3, Fe_2O_3 and CrO_3. The physiological damage observed in the lichen samples coincided with an exaggerated accumulation of Mg, Cr, Fe and Cd.

1. For an analysis of the integrity of the photobiont chlorophyll, use samples of 20 mg of rinsed and air-dried lichens.
2. Extract chlorophyll overnight in the dark in 5 ml of DMSO.
3. Determine the ratio chlorophyll to phaeophytin spectrophotometrically (see Chapters 10 and 22).

The effect of metals on ethylene production in lichens

Ethylene has been termed the "stress hormone", and in higher plants, stress often induces its synthesis. It seems reasonable to assume that heavy metal stress may induce ethylene production in lichens, and several laboratory studies have indeed suggested that this may occur (Lurie and Garty 1991; Garty et al. 1995b). The production of stress-ethylene by thalli of *R. lacera* was measured under controlled conditions upon incubation for 30 min at a low pH; H_2O-treated thalli produced 1.38 nl/g/h ethylene. Additional thalli exposed to 20 mM of either $PbCl_2$, $ZnCl_2$, $MnCl_2$, $CuCl_2$ or $FeCl_2$, produced 3.12, 2.64, 4.12, 7.09 and 17.86 nl/g/h ethylene, respectively (Lurie and Garty, 1991). We obtained a production of 21.8 $\pm$ 6.6 nl/g/h ethylene by *R. lacera* exposed to 20 mM $FeCl_2$ at pH 2.0 for 30 min, whereas a H_2O-treatment yielded only 1.2 $\pm$ 0.4 nl/g/h at pH 7.0 and 4.6 $\pm$ 0.7 nl/g/h at pH 2.0 (Garty et al. 1995b). Large amounts of ethylene were produced by *R. lacera* exposed to $FeCl_3$, $Fe(NO_3)_3$, $FeSO_4$ and $CuSO_4$ whereas treatments with either Pb, Mn or Zn salts yielded lesser amounts of ethylene (Garty et al. 1995b).

Exposure of podetia of *Cladina stellaris* to metal-containing solutions at a low pH yielded large amounts of ethylene in the case of $FeCl_2$ given as 0.5 mM, 5.0 mM and 10 mM solutions at pH 3.5 [6.25, 9.89 and 11.48 nl/g/h, respectively (Kauppi et al. 1998)]. According to the same reference the exposure of *C. stellaris* to $FeSO_4$ under similar concentrations and pH yielded 6.56, 10.16 and 10.48 nl/g/h ethylene, respectively. Lesser amounts of ethylene were produced by *C. stellaris* exposed to either $Fe_2(SO_4)_3$, $FeCl_3$ or $Fe(NO_3)_3$ under the same conditions: 4.75, 4.69 and 4.36 nl/g/h, respectively. According to this study, the exposure of *C. stellaris* to either Cu- or Zn-containing solutions, yielded fair amounts of ethylene depending on the concentration of the metal salts in the solution.

A field study was carried out in Oulu, Finland to test the influence of air-pollution on the amount of heavy metals and other chemical elements and the production of stress-ethylene in the lichens *Usnea hirta* (Garty et al. 1997e) and *Hypogymnia physodes* (Garty et al. 1997f). The accumulation of airborne elements by the two lichens was tested by means of transplantation to several sites in and around Oulu, for a 45 d period. Thalli of *U. hirta* accumulated more K, Ca, Na, Mg, Fe, Zn and Mn in samples exposed in one of the streets with slow traffic in Oulu, than thalli left in a control site out of the city. Lead and Ca were positively correlated with the amount of ethylene produced by the lichen. *Hypogymnia physodes* in comparison, contained more Fe and Mg in streets with slow traffic, than thalli left in the control site. Zinc and Fe exhibited a significant positive correlation with the amount of ethylene produced by the lichen.

1. For measurements of ethylene production, 1 g samples of rinsed and air-dried thalli should be used. Each sample should consist of a few thalli with only one damaged surface each. Avoid using thallus fragments, to minimise the production of ethylene as a result of wounding.
2. Soak the samples in 20 ml of double distilled water for 30 min.
3. Wipe samples gently with filter paper to remove excess moisture and place them in sealed 50 ml Erlenmeyer flasks.
4. After 3 h, withdraw 4 ml of the gas from each flask with a syringe and inject 1 ml into a gas chromatograph equipped with an activated alumina column and a flame ionization detector. The carrier gas is N_2. If a Varian 3350 gas chromatograph is used the N_2 should be injected at a

flow rate of 30 ml/min. The recommended injector temperature for this instrument is 110°C, the column temperature is 110°C and the detector temperature is 150°C.

5. Analyse "controls" i.e. plants from unpolluted habitats that are treated in the same way.

Further details of how to measure ethylene production in lichens are given in Chapter 11.

Heavy-metal content and the spectral reflectance response of lichen thalli

It is possible to measure a "vegetation index" (VI) of the health of plant tissue based on spectral reflectance. Most indices are based on a combination of the ratio of two portions of the electromagnetic spectrum: the red band (R) at 600 - 700 nm, which corresponds to the region of maximum chlorophyll absorption and the near infrared band (NIR) at 700 - 1100 nm which corresponds to the maximum reflectance of incident light in the living vegetation. The basic vegetation index, the "Simple Ratio", is a ratio of the digital value of these two bands. The most widely used VI is known as the normalized difference vegetation index (NDVI; Rouse et al. 1974):

$$NDVI = (NIR - R)/(NIR + R) \qquad \text{Equation 1}$$

The application of the parameter of spectral-reflectance response of lichens exposed to pollutants was reported only in the last decade. Cox et al. and co-workers (1991) found that lichens exposed to Cu concentrations $> 20\ \mu g/g$ exhibited a significant shift of 2 - 3% of the spectral response.

To study the spectral characteristics of lichens exposed to heavy metals either under field conditions or at the laboratory, it is recommended to use a field spectrometer. Measurements of treated thalli may be performed in the laboratory under constant irradiance with light approaching from a fixed angle.

1. Fix the field spectrometer to a 2 nm wavelength of spectral resolution, a scanning width between 400 and 1100 nm, and a field of view (FOV) of 15°.

2. Place the samples in Petri dishes on a black-coated board to minimise the external reflectance or backscatter, and install the spectrometer at 1 m above the sample. A 1000 W quartz sun-simulating halogen lamp should be positioned at a zenith angle of 45°, at a distance of 1 m from the sample.

3. Measure the incident radiation by dividing the spectrum radiance of each sample by the down welling irradiation as measured by a halon reflectance panel.
4. Rotate lichens at 90° between each scan to avoid roughness and shadowing effects. An average spectrum is thus calculated as the mean of four spectra. The reflectance in the red spectral region is determined as the average of reflectance values between 650 and 700 nm. In the same manner, the reflectance in the NIR spectral region is determined as the average of reflectance values between 800 and 1100 nm.
5. Calculate the NDVI values according to the above-mentioned equation. The resulting values range from -1.0 to +1.0.

Typical results

Recently, a series of transplantation studies were carried out to estimate the degree of stress induced by exposure of lichens to pollutants, as expressed by changes in the spectral reflectance response of the thallus (Garty et al. 1997a-d). We compared the alteration of NDVI with the amount of mineral elements, including heavy metals in thalli of *Ramalina lacera* from HaZorea, transplanted in the Ashdod region (SW Israel) for different periods. The exposure of thalli of *R. lacera* to air pollutants in the Ashdod region caused significant shifts in the spectral reflectance of the lichen thallus. Low NDVI values were obtained for lichens exposed for 3–11 months near a sulphur terminal, a chemical plant and oil refineries in Ashdod (Garty et al. 1997a-d). Typical NDVI values of 0.412 ± 0.005 were obtained for *Ramalina lacera* from an unpolluted site and 0.160 $\pm$ 0.008 for samples transplanted to a site contaminated by heavy metals and additional pollutants (Garty et al. 1997d). According to the above-mentioned studies, Pearson correlation coefficients of NDVI and mineral elements accumulated in *R. lacera* transplants indicated an inverse correlation of NDVI and total S, sulphate-S, Ni, V, Pb, Mn, Al, Cr, Fe and Ti and a direct correlation of NDVI, K and P.

Heavy metals and photosynthesis

Photosynthesis in lichens is inhibited by heavy metals. In laboratory studies, Puckett (1976) found that the relative toxic influence of heavy metals on photosynthesis in lichens is Ag, Hg $>$ Co $>$ Cu, Cd $>$ Pb, Ni for short-term exposures and Ag, Hg $>$ Cu $\geq$ Pb, Co $>$ Ni for extended exposures. A photosynthetic decline was observed also in *Cladonia rangiferina* exposed to As (Nieboer et al. 1984) and to anionic forms of U (Boileau et al. 1985).

Using samples of *Hypogymnia physodes*, Punz (1979) investigated the effect of single and combined pollutants such as Pb, NaCl and SO_2 on the gas exchange of the lichen. The author observed synergistic effects in the following combinations: Pb + NaCl, Pb + SO_2 and Pb + NaCl + SO_2.

Nash (1975) demonstrated that the sensitivity of lichen species to metals can vary. Significant depressions in photosynthesis occurred when heavy metal concentrations reached 308 µg/g Zn and 288 µg/g Cd for *Lasallia papulosa* but 178 µg/g Zn and 340 µg/g Cd for *Cladonia uncialis*. Brown and Beckett (1983) found that Zn, Cd and Cu inhibited photosynthesis in lichens containing cyanobacterial photobionts at substantially lower concentrations than lichens containing green algae as photobionts. Nash (1989) concluded that the above mentioned findings of Brown and Beckett may explain the fact that *Peltigera* spp. were found in the control area in the UK but not in an area near a smelter in Pennsylvania, studied earlier by Nash (1972). Beckett and Brown (1983) established a tolerance index based on metal-induced reductions of photosynthesis in members of the lichen genus *Peltigera* from different sites, ranging from heavily metal-contaminated to background levels. The tolerance to Zn was found to be proportional to the Zn content of the thallus. The authors used the following formula to establish a Zn-tolerance index:

$$\frac{\text{Rate of photosynthesis after metal treatment}}{\text{Rate of photosynthesis after water treatment}} x100 \qquad \text{Equation 2}$$

The degree of tolerance did not relate to the presence of other cations or the weight of thallus per unit surface area.

The impact of vehicular pollution on the respiration and net-photosynthesis of different lichen-species was studied by Déruelle and Petit (1983) using infra-red gas analysis (see Chapter 10 for details of this technique). A comparison of lichens from a polluted site, 15 m from a motorway in France, with lichens from an unpolluted site, showed a variable respiratory response. All three lichen species exhibited a decrease of net photosynthesis, ranging from 74% in *Hypogymnia physodes* to 45.5% in *Parmelia caperata* and 42.7% in *Evernia prunastri*. The effect of heavy metals on changes in chlorophyll fluorescence in lichens has been studied only recently (see Chapter 9 for details of this technique). The activity of photosystem II (PSII) in lichens exposed to heavy metals under controlled conditions was monitored by Branquinho et al. (1997a,b) who investigated the impact of heavy metals on chlorophyll fluorescence. The uptake of Pb, particularly in cyanobiont lichens, caused a decrease of PSII photochemical reactions measured as a change in the F_v/F_m ratio (Branquinho et al. 1997b). For the lichens *Lobaria pulmonaria* and

Parmelia caperata, the effect on PSII photochemical reactions could be observed 48 h after Pb uptake, thus confirming that a prolonged exposure to Pb leads to an additional decrease of F_v/F_m. Branquinho et al. (1997a) found the fluorescence parameter F_v/F_m to be applicable to an investigation of the sensitivity of lichens to Cu. *Usnea* spp. were detected as having the greatest sensitivity to Cu uptake as shown by fluorescence studies: a total inhibition of PSII photochemical reactions occurred in *Ramalina fastigiata* when intercellular Cu concentrations exceeded a threshold of ca. 4.0 µmol/g, whereas the concentration of intracellular Cu was already above this apparent threshold in *Usnea* spp. (ca. 4.0 µmol/g). It will be very interesting to see if changes in chlorophyll fluorescence parameters are found in lichens exposed to metals in field situations.

Heavy metals and malondialdehyde (MDA)

Malondialdehyde is a highly reactive aldehyde, generally formed subsequent to the peroxidation of lipids (Turton et al. 1997). Thalli of the lichen *Hypogymnia physodes* transplanted in three sites with different levels of air-pollution in Austria exhibited a positive correlation of their MDA content and the concentration of O_3 in the air (Egger et al. 1994). An additional study using lichen transplants, performed by González and Pignata (1994) in Cordoba, Argentina, revealed a direct relationship between the levels of S, Al, Pb and the MDA content in thalli of the lichen *Punctelia subrudecta*. Of the heavy metals accumulated in *Diploschistes muscorum* in polluted sites in France, Zn was found to be responsible for a significant increase of MDA in the thallus (Cuny 1999).

Comments

Knowledge of the absorption and accumulation of heavy metals in endolithic lichens, particularly in hot-desert ecosystems, is still deficient. The determination of total quantities of heavy metals in endolithic lichens is of interest for those who seek an understanding of the outcome of contaminants in terrestrial food-chains in the desert in which snails, rodents and birds of prey are possibly involved. Knowledge is needed on the capability of endolithic lichens to express stress induced by exposure to heavy metals and other pollutants, by production and release of ethylene. Sufficient data to estimate the quantities of heavy metals leached by forest fires are still unavailable, especially for fires that occur close to industrial cities.

Despite hundreds of articles, reviews and book-chapters dealing with the interaction of lichens and heavy metals, it is quite difficult to establish a biological scale for an assessment of heavy-metal pollution by lichens, as made available for an estimation of SO_2. Moreover, to the best of my knowledge, it has not proved possible to translate elemental levels and physiological alteration, to actual deposition rates.

On the other hand, I have been encouraged by the fact that in the nineties, lichens were applied as air-biomonitors in remote parts of the world: Africa, South America and Antarctica. I am also encouraged by the fact that the vast body of data referring to the capability of lichens to accumulate airborne lead, seems to yield fruitful studies which provide proof to the existence of a definite decrease of Pb in the environment which correlates with the disuse of Pb in gasoline since the 1970's in most western countries.

Acknowledgements. The author thanks Mrs. Rachel Garty-Spitz for her valued contribution to the manuscript.

References

Alebic-Juretic A, Arko-Pijevac M (1989) Air pollution damage to cell membranes in lichens - results of simple biological test applied in Rijeka, Yugoslavia. Water Air Soil Pollut 47:25-33

Bargagli R (1989) Determination of metal deposition patterns by epiphytic lichens. Toxicol Environ Chem 18:249-256

Bargagli R, Nimis PL, Monaci F (1997) Lichen biomonitoring of trace element deposition in urban, industrial and reference areas of Italy. J Trace Elements Med Biol 11:173-175

Barnes JD, Balaguer L, Manrique E, Elvira S, Davison AW (1992) A reappraisal of the use of DMSO for the extraction and determination of chlorophylls *a* and *b* in lichens and higher plants. Environmental and Experimental Botany 32: 85-100

Bartok K, Nicoara A, Victor B, Tibor O (1992) Biological responses in the lichen *Xanthoria parietina* transplanted in biomonitoring stations. Rev Roum Biol - Biol Veget 37:135-142

Bartok K, Osvath T (1990) Influence of pollutants emanated from the Dej cellulose manufactures, over lichens. Contr Bot Cluj-Napoca, 67-72

Beckett RP, Brown DH (1983) Natural and experimentally-induced zinc and copper resistance in the lichen genus *Peltigera*. Ann Bot 52:43-50

Belandria G, Asta J, Garrec JP (1991) Diminutions of fluorine contents in lichens due to a regression of pollution in an alpine valley (Maurienne, Savoi, France) from 1975 to 1985. Rev Ecol Alp Grenoble 1:45-58

Bennett JP, Wetmore CM (1999) Changes in element contents of selected lichens over 11 years in northern Minnesota, USA. Environ Exp Bot 41:75-82

Boileau LJR, Nieboer E, Richardson DHS (1985) Uranium accumulation in the lichen *Cladonia rangiferina*. Part II. Toxic effects of cationic, neutral, and anionic forms of the uranyl ion. Can J Bot 63:390-397

Boonpragob K, Nash III TH (1990) Seasonal variation of elemental status in the lichen *Ramalina menziesii* Tayl. from two sites in southern California: evidence for dry deposition accumulation. Environ Exp Bot 30:415-428

Branquinho C, Brown DH, Catarino F (1997a) The cellular location of Cu in lichens and its effects on membrane integrity and chlorophyll fluorescence. Environ Exp Bot 38:165-179

Branquinho C, Brown DH, Máguas C, Catarino F (1997b) Lead (Pb) uptake and its effects on membrane integrity and chlorophyll fluorescence in different lichen species. Environ Exp Bot 37:95-105

Brown DH, Beckett RP (1983) Differential sensitivity of lichens to heavy metals. Ann Bot 52:51-57

Brown DH, Beckett RP (1984) Uptake and effect of cations on lichen metabolism. Lichenologist 16:173-188

Buck GW, Brown DH (1979) The effect of desiccation on cation location in lichens. Ann Bot 44:265-277

Calliari I, Caniglia G, Nardi S, Tollardo AM, Callegaro R (1995) EDXRF study of lichens as biomonitors and effect of washing procedure on element concentrations. X-ray Spectr 24:143-146

Calvelo S, Baccala N, Arribere MA, Ribeiro Guevara S, Bubach D (1997) Analytical and statistical analysis of elemental composition of lichens. J Rad Nucl Chem 222:99-104

Chettri MK, Cook CM, Vardaka E, Sawidis T, Lanaras T (1998) The effect of Cu, Zn and Pb on the chlorophyll content of the lichens *Cladonia convoluta* and *Cladonia rangiformis*. Environ Exp Bot 39:1-10

Cox JD, Beckett PJ, Courtin GM (1991) Factors affecting spectral responses from lichens. Proceedings of the Eighth Thematic Conference on Geologic Remote Sensing, Ann Arbor, vol 3, Environmental Research Institute of Michigan, MI, USA, pp 1125-1137

Cuny D (1999) Les impacts communautaires, physiologiques et cellulaires des éléments traces métalliques sur la symbiose lichénique, mise en évidence de méchanismes de tolérance chez *Diploschistes muscorum* (Scop.) R. Sant. These pour obtenir le titre de Docteur de l'Université de Lille II. Université de Lille II, Lille, France.

Déruelle S, Petit PJX (1983) Preliminary studies on the net photosynthesis and respiration responses of some lichens to automobile pollution. Cryptogamie Bryol Lichenol 4:269-278

Dillman KL (1996) Use of the lichen *Rhizoplaca melanophthalma* as a biomonitor in relation to phosphate refineries near Pocatello, Idaho. Environ Pollut 92:91-96

Egger R, Schlee D, Türk R (1994) Changes of physiological and biochemical parameters in the lichen *Hypogymnia physodes* (L.) Nyl. due to the action of air pollutants - a field study. Phyton 34:229-242

Evans C, Hutchinson TC (1996) Mercury accumulation in transplanted moss and lichens at high elevation sites in Quebec. Water Air Soil Pollut 90:475-488

Fahselt D, Wu T-W, Mott B (1995) Trace element patterns in lichens following mine closures. Bryologist 98:228-234

Freitas MC (1995) Elemental bioaccumulations in air pollution studies. J Rad Nucl Chem 192:171-181

Freitas MC, Nobre AS (1997) Bioaccumulation of heavy metals using *Parmelia sulcata* and *Parmelia caperata* for air pollution studies. J Radioanal Chem 217:17-20

Freitas MC, Reis MA, Alves LC, Wolterbeek HTh (1999) Distribution in Portugal of some pollutants in the lichen *Parmelia sulcata*. Environ Pollut 106:229-235

Freitas MC, Reis MA, Alves LC, Wolterbeek HTh, Verburg T, Gouveia MA (1997) Biomonitoring of trace-element air pollution in Portugal: Qualitative survey. J Rad Nucl Chem 217:21-30

Gailey FAY, Lloyd O Ll (1983) The use of *Lecanora conizaeoides* as a monitor of the distribution of atmospheric pollution by metals. Ecol Disease 2:215-224

Gailey FAY, Lloyd O Ll (1986) Methodological investigations into low technology monitoring of atmospheric metal pollution: part 2-the effects of length of exposure on metal concentrations. Environ Pollut (Series B) 12:61-74

Gailey FAY, Lloyd OL (1993) Spatial and temporal patterns of airborne metal pollution: the value of low technology sampling to an environmental epidemiology study. Sci Total Environ 133:201-219

Gailey FAY, Smith GH, Rintoul LJ, Lloyd O Ll (1985) Metal deposition patterns in central Scotland, as determined by lichen transplants. Environ Man Assess 5:291-309

Garty J (1985) The amount of heavy metals in some lichens of the Negev Desert. Environ Pollut (Series B) 10:287-300

Garty J (1987) Metal amounts in the lichen *Ramalina duriaei* (De Not.) Bagl. transplanted at biomonitoring sites around a new coal-fired power station after 1 year of operation. Environ Res 43:104-116

Garty J (1988) Comparisons between the metal content of a transplanted lichen before and after the start-up of a coal-fired power station in Israel. Can J Bot 66:668-671

Garty J (1992) Lichens and heavy metals in the environment. In: Vernet J-P (ed) Impact of Heavy Metals on the Environment, vol 2. Elsevier, Amsterdam, pp 55-131

Garty J (1993) Lichens as biomonitors for heavy metal pollution. In: Markert B (ed) Plants as Biomonitors. Indicators for Heavy Metals in the Terrestrial Environment. VCH, Weinheim, pp 193-263

Garty J, Ammann K (1987) The amounts of Ni, Cr, Zn, Pb, Cu, Fe, and Mn in some lichens growing in Switzerland. Environ Exp Bot 27:127-138

Garty J, Fuchs C (1982) Heavy metals in the lichen *Ramalina duriaei* transplanted in biomonitoring stations. Water Air Soil Pollut 17:175-183

Garty J, Fuchs C, Zisapel N, Galun M (1977) Heavy metals in the lichen *Caloplaca aurantia* from urban, suburban and rural regions in Israel (a comparative study). Water Air Soil Pollut 8:171-188

Garty J, Galun M, Kessel M (1979) Localization of heavy metals and other elements accumulated in the lichen thallus. New Phytol 82:159-168

Garty J, Ronen R, Galun M (1985) Correlation between chlorophyll degradation and the amount of some elements in the lichen *Ramalina duriaei* (De Not.) Jatta. Environ Exp Bot 25:67-74

Garty J, Kardish N, Hagemeyer J, Ronen R (1988) Correlations between the concentration of adenosine tri phosphate, chlorophyll degradation and the amounts of airborne heavy metals and sulphur in a transplanted lichen. Arch Environ Contam Toxicol 17:601-611

Garty J, Karary Y, Harel J (1992) Effect of low pH, heavy metals and anions on chlorophyll degradation in the lichen *Ramalina duriaei* (De Not.) Bagl. Environ Exp Bot 32:229-241

Garty J, Karary Y, Harel J (1993) The impact of air pollution on the integrity of cell membranes and chlorophyll in the lichen *Ramalina duriaei* (De Not.) Bagl. transplanted to industrial sites in Israel. Arch Environ Contam Toxicol 24:455-460

Garty J, Harel Y, Steinberger Y (1995a) The role of lichens in the cycling of metals in the Negev Desert. Arch Environ Contam Toxicol 29:247-253

Garty J, Karary Y, Harel J, Lurie S (1995b) The impact of heavy metals on lichens. In: Wilken R-D, Förstner U, Knöchel A (eds) Heavy Metals in the Environment. International Conference, Hamburg, September 1995, vol 1. CEP Consultants, Edinburgh, UK, pp 152-155

Garty J, Kauppi M, Kauppi A (1996a) Accumulation of airborne elements from vehicles in transplanted lichens in urban sites. J Environ Qual 25:265-272

Garty J, Steinberger Y, Harel Y (1996b) Spatial and temporal changes in the concentration of K, Na, Mg and Ca in epilithic and in decomposing detached thalli of the lichen *Ramalina maciformis* and its potential role in the cycling of these elements in the Negev Desert. Environ Exp Bot 36:83-97

Garty J, Cohen Y, Kloog N, Karnieli A (1997a) Effects of air pollution on cell membrane integrity, spectral reflectance and metal and sulfur concentrations in lichens. Environ Toxicol Chem 16:1396-1402

Garty J, Karnieli A, Wolfson R, Kunin P, Garty-Spitz R (1997b) Spectral reflectance and integrity of cell membranes and chlorophyll relative to the concentration of airborne mineral elements in a lichen. Physiol Plantarum, 101:257-267

Garty J, Kloog N, Cohen Y, Wolfson R, Karnieli A (1997c) The effect of air pollution on the integrity of chlorophyll, spectral reflectance response, and on concentrations of nickel, vanadium, and sulfur in the lichen *Ramalina duriaei* (De Not.) Bagl. Environ Res 74:174-187

Garty J, Kloog N, Wolfson R, Cohen, Y, Karnieli A, Avni A (1997d) The influence of air pollution on the concentration of mineral elements, on the spectral reflectance response and on the production of stress-ethylene in the lichen *Ramalina duriaei*. New Phytol 137:587-597

Garty J, Kauppi M, Kauppi A (1997e) The influence of air pollution on the concentration of airborne elements and on the production of stress-ethylene in the lichen *Usnea hirta* (L.) Weber em. Mot. transplanted in urban sites in Oulu, N. Finland. Arch Environ Contam Toxicol 32:285-290

Garty J, Kauppi M, Kauppi A (1997f) The production of stress ethylene relative to the concentration of heavy metals and other elements in the lichen *Hypogymnia physodes*. Environ Toxicol Chem 16:2404-2408

Garty J, Cohen Y, Kloog N (1998a) Airborne elements, cell membranes, and chlorophyll in transplanted lichens. J Environ Qual 27:973-979

Garty J, Kloog N, Cohen Y (1998b) Integrity of lichen cell membranes in relation to concentration of airborne elements. Arch Environ Contam Toxicol 34:136-144

Gombert S, Asta J (1998) The effect of refuse incinerator fumes on the lead and cadmium content of experimentally exposed corticolous lichens. Water Air Soil Pollut 104:29-40

González CM, Pignata ML (1994) The influence of air pollution on soluble proteins, chlorophyll degradation, MDA, sulphur and heavy metals in a transplanted lichen. Chem Ecol 9:105-113

Gough LP, Erdman JA (1977) Influence of a coal-fired power plant on the element content of *Parmelia chlorochroa*. Bryologist 80:492-501

Jacquiot L, Daillant O (1997) Bio-accumulation des métaux lourds et d'autres elément traces par les lichens, revue bibliographique. Bull Obs Myco 12:2-31

James PW (1973) The effect of air pollutants other than hydrogen fluoride and sulphur dioxide on lichens. In: Ferry BW, Baddeley MS, Hawksworth DL (eds) Air Pollution and Lichens. The Athlone Press, London, pp 143-175

Jeran Z, Byrne AR, Batic F (1995) Transplanted epiphytic lichen as biomonitors of air-contamination by natural radionuclides around the Zirovski VRH uranium mine, Slovenia. Lichenologist 27:375-385

Jovanovic S, Carrot F, Deschamps C, Deschamps N, Vukotic P (1995) A study of the air pollution in the surroundings of an aluminium smelter, using epiphytic and lithophytic lichens. J Trace and Microprobe Techniques 13:463-471

Kansanen PH, Venetvaara J (1991) Comparison of biological collectors of airborne heavy metals near ferrochrome and steel works. Water Air Soil Pollut 60:337-359

Kauppi M, Kauppi A, Garty J (1998) Ethylene produced by the lichen *Cladina stellaris* exposed to sulphur and heavy-metal-containing solutions under acidic conditions. New Phytol 139:537-547

Kubin E (1990) A survey of element concentrations in the epiphytic lichen *Hypogymnia physodes* in Finland in 1985-1986. In: Kauppi P, Antilla P, Kenttämies K (eds) Acidification in Finland. Springer-Verlag, Berlin and Heidelberg, pp 421-446

Kytömaa A, Nieminen S, Thuneberg P, Haapala H, Nuorteva P (1995) Accumulation of aluminium in *Hypogymnia physodes* in the surroundings of a Finnish sulphite-cellulose factory. Water Air Soil Pollut 81:401-409

Lackovicova A, Martiny E, Pisut I, Stresko V (1994) Element content of the lichen *Hypogymnia physodes* and spruce needles in the industrial area of Rudnany and Krompachy (NE Slovakia). Ekologia 13:415-423

Lupsina V, Horvat M, Jeran Z, Stegnar P (1992) Investigation of mercury speciation in lichens. Analyst 117:673-675

Lurie S, Garty J (1991) Ethylene production by the lichen *Ramalina duriaei*. Ann Bot 68:317-319

Makholm MM, Bennett JP (1998) Mercury accumulation in transplanted *Hypogymnia physodes* lichens downwind of Wisconsin chlor-alkali plant. Water Air Soil Pollut 102:427-436

Markert B (1993) Instrumental analysis of plants. In: Markert B (ed) Plants as Biomonitors. Indicators for Heavy Metals in the Terrestrial Environment. VCH, Weinheim, pp 65-103

Markert B, Wtorova W (1992) Inorganic chemical investigations in the Forest Biosphere Reserve near Kalinin, USSR. Vegetatio 98:43-58

Monaci F, Bargagli R, Gasparo D (1997) Air pollution monitoring by lichens in a small medieval town of central Italy. Acta Bot Neerl 46:403-412

Nash III TH (1972) Simplification of the Blue Mountain lichen communities near a zinc factory. Bryologist 75:315-324

Nash III TH (1975) Influence of effluents from a zinc factory on lichens. Ecol Monogr 45:183-198

Nash III TH (1989) Metal tolerance in lichens. In: Shaw AJ (ed) Heavy Metal Tolerance in Plants: Evolutionary Aspects. CRC Press, Boca Raton, Florida, USA, pp 119-131

Nash III TH (ed) (1996) Lichen Biology. Cambridge University Press, Cambridge, UK

Nash III TH, Sommerfeld MR (1981) Elemental concentrations in lichens in the area of the Four Corners Power Plant, New Mexico. Environ Exp Bot 21:153-162

Nieboer E, Padovan D, Lavoie P, Richardson DHS (1984) Anion accumulation by lichens. II. Competition and toxicity studies involving arsenate, phosphate, sulphate and sulphite. New Phytol 96:83-93

Nieboer R, Richardson DHS, Tomassini FD (1978) Mineral uptake and release by lichens: an overview. Bryologist 81:226-246

Nieboer E, Richardson DHS (1981) Lichens as monitors of atmospheric deposition. In: Eisenreich SJ (ed) Atmospheric Pollutants in Natural Waters. Ann Arbor Science, Ann Arbor, pp 339-388

Pablo Valles Mota J, Rosario Fernández de la Campa M, Ignacio Garcia Alonso J (1999) Determination of cadmium in biological and environmental materials by isotope dilution inductively coupled plasma mass spectrometry: effect of flow sample introduction methods. J Anal At Spectrom 14:113-120

Palmieri F, Neri R, Benco C, Serracca L (1997) Lichens and moss as bioindicators and bioaccumulators in air pollution monitoring. J Environ Path Toxicol Oncol 16:175-190

Pilegaard K, Rasmussen L, Gydesen H (1979) Atmospheric background deposition of heavy metals in Denmark monitored by epiphytic cryptogams. J Appl Ecol 16:843-853

Puckett KJ (1976) The effect of heavy metals on some aspects of lichen physiology. Can J Bot 54:2695-2703

Puckett KJ (1988) Bryophytes and lichens as monitors of metal deposition. In: Nash III TH, Wirth V (eds) Lichens, Bryophytes and Air Quality. Bibliotheca Lichenologica, Cramer, Berlin and Stuttgart 30:231-267

Puckett KJ, Finegan EJ (1980) An analysis of the element content of lichens from the Northwest Territories, Canada. Can J Bot 58:2073-2089

Puckett KJ, Nieboer E, Gorzynski MJ, Richardson DHS (1973) The uptake of metal ions by lichens: a modified ion-exchange process. New Phytol 72:329-342

Punz W (1979) Der Einfluss isolierter und kombinierter Schadstoffe auf die Flechtenphotosynthese. Photosynthetica 13:428-433

Quevauviller Ph, Herzig R, Muntau H (1996) Certified reference material of lichen (CRM 482) For the quality control of trace element biomonitoring. Sci Total Environ 187:143-152

Reis MA, Alves LC, Freitas M, van Os B, Wolterbeek HTh (1999) Lichens (*Parmelia sulcata*) time response model to environmental elemental availability. Sci Total Environ 232:105-115

Reis MA, Alves LC, Wolterbeek H Th, Verburg T, Freitas MC, Gouveia A (1996) Main atmospheric heavy metal sources in Portugal by biomonitor analysis. Nucl Instr and Meth in Phys Res B109/110:493-497

Richardson DHS (1992) Pollution Monitoring with lichens. Richmond, Slough, England.

Richardson DHS, Kiang S, Ahmadjian V, Nieboer E (1985) Lead and uranium uptake by lichens. In: Brown DH (ed) Lichen Physiology and Cell biology. Plenum Press, New York, pp 227-246

Richardson DHS, Shore, M, Hartree R, Richardson RM (1995) The use of X-ray fluorescence spectrometry for the analysis of plants, especially lichens, employed in biological monitoring. Sci Total Environ 176:97-105

Riga-Karandinos AN, Karandinos MG (1998) Assessment of air pollution from a lignite power plant in the plain of Megalopolis (Greece) using as biomonitors three species of lichens; impacts on some biochemical parameters of lichens. Sci Total Environ 215:167-183

Rodrigo A, Avila A, Gómez-Bolea A (1999) Trace metal contents in *Parmelia caperata* (L.) Ach. compared to bulk deposition, throughfall and leaf-wash fluxes in two holm oak forests in Montseny (NE Spain). Atm Environ 33:359-367

Ronen R, Galun M (1984) Pigment extraction from lichens with dimethyl sulfoxide (DMSO) and estimation of chlorophyll degradation. Environ Exp Bot 24:239-245

Rouse JW, Haas RH, Schell JA, Deering DW, Harlan JC (1974) Monitoring the Vernal Advancements and Retrogradation (Greenwave Effect) of Nature Vegetation. NASA/GSFC final report, NASA, Greenbelt, MD, USA, 371 pp

Sarret G, Manceau A, Cuny D, Van Haluwyn C, Déruelle S, Hazeman J-L, Soldo Y, Eybert-Bérard L, Menthonnex J-J (1998) Mechanisms of lichen resistance to metallic pollution. Environ Sci Technol 32:3325-3330

Sawidis T, Heinrich G (1992) Cesium-137 monitoring using lichens and mosses from northern Greece. Can J Bot 70:140-144

Seaward MRD (1973) Lichen ecology of the Scunthrope Heathlands. I. Mineral accumulation. Lichenologist 5:423-433

Seaward MRD (1974) Some observations on heavy metal toxicity and tolerance in lichens. Lichenologist 6:158-164

Silberstein L, Siegel BZ, Siegel SM, Mukhtar A, Galun M (1996) Comparative studies on *Xanthoria parietina*, a pollution-resistant lichen, and *Ramalina duriaei*, a sensitive species. I. Effects of air pollution on physiological processes. Lichenologist 28:355-365

Simon EW (1974) Phospholipids and plant membrane permeability. New Phytol 73:377-420

Sloof JE (1995) Lichens as quantitative biomonitors for atmospheric trace-element deposition, using transplants. Atm Environ 29:11-20

Sloof JE, Wolterbeek H Th (1991a) National trace-element air pollution monitoring survey using epiphytic lichens. Lichenologist 23:139-165

Sloof JE, Wolterbeek H Th (1991b) Pattern in trace elements in lichens. Water Air Soil Pollut 57-58:785-795

Sloof JE, Wolterbeek H Th (1993) Interspecies comparison of lichens as biomonitors of trace-element air pollution. Environ Mon Assess 25:149-157

Takala K, Olkkonen H, Salminen R (1994) Iron content and its relations to the sulphur and titanium contents of epiphytic and terricolous lichens and pine bark in Finland. Environ Pollut 84:131-138

Tarhanen S, Metsärinne S, Holopainen T, Oksanen J (1999) Membrane permeability response of lichen *Bryoria fuscescens* to wet deposited heavy metals and acid rain. Environ Pollut 104:121-129

Turton HE, Dawes IW, Grant CM (1997) *Saccharomyces cerevisiae* exhibits a yAP-1-mediated adaptive response to malondialdehyde. J Bacteriol 179:1096-1101

Tynnyrinen S, Palomäki V, Holopainen T, Kärenlampi L (1992) Comparison of several bioindicator methods in monitoring the effects on forest of a fertilizer plant and a strip mine. Ann Bot Fennici 29:11-24

Vtorova VN, Markert B (1995) Multi-element analysis of plants of the forest ecosystems in eastern Europe. Biol Bull 4:447-454

Walther DA, Ramelov GJ, Beck JN, Young JC, Callahan JD, Marcon MF (1990) Temporal changes in metal levels of the lichens *Parmotrema praesorediosum* and *Ramalina stenospora*, Southwest Louisiana. Water Air Soil Pollut 53:189-200

Bioindication of Sulphur Dioxide Pollution with Lichens

FRANC BATIČ

Introduction

Bioindicators are organisms that react to environmental pollution with their life functions (Arndt et al. 1987). According to Hawksworth and Rose (1970) and Ferry et al. (1973) lichens can be used as bioindicators in the following methods:

- **Passive bioindication** uses naturally occurring lichens for monitoring purposes.
- **Active bioindication** involves collecting lichens in areas with clean air, moving them to polluted areas then testing for the effects of air pollution.
- **Reactive bioindicators** respond to air pollution and other stress with their life functions (e.g. growth, structure, distribution).
- **Accumulative bioindicators** are organisms that accumulate pollutants without significant damage to their life functions, at least up to a certain dose of pollutant.

The advantage of using bioindicators rather than instrumental monitoring is that the latter can never measure all the abiotic and biotic factors that can affect life. Bioindication can clearly show the effects that particular pollutants and their mixtures are having on biota. Although instrumental monitoring can determine the levels of pollutants in the air, water and soil, the data obtained cannot always indicate when humankind, fauna and flora are at risk. For this, it is necessary to measure the effects on living organisms. The use of plants as bioindicators of environmental conditions

Franc Batič, University of Lubljana, Department of Agronomy, Biotechnical Faculty, Jamnikarjeva 101, Ljubljana, 1000, Slovenia (*phone* +386-1-423-1161; *fax* +386-1-423-1088; *e-mail* franc.batic@bf.uni-lj.si)

is accepted in phytocenology, a study of plant communities (Ellenberg et al. 1992), and also in monitoring water quality. In efforts to monitor air quality, epiphytic lichens were one of the first recognised plant indicators (Ferry et al. 1973), especially in cases where sulphur dioxide was the main air pollutant.

The main advantage of techniques based on bioindication is their relatively low cost, allowing sampling strategies with a relatively high density of sampling points. In remote regions it is often the only technique possible. The main disadvantages in using bioindication are the problems of response standardisation, response reproducibility and quantification, and also determination of their relevance at an ecosystem level.

Toxicity of sulphur compounds

The physiological basis of the great sensitivity of lichens to sulphur compounds lies in the acidic characteristic of sulphur dioxide as the main anthropogenic source of sulphur. It is known that sulphur dioxide, sulphite and sulphate are the most harmful forms for lichens (e.g. Hawksworth and Rose 1970, Nash 1973, Türk and Wirth 1974, Deruelle 1978). Harmful effects of sulphur compounds are connected with proton generation, partly also with oxidation of sulphite to sulphate and free radical formation during the oxidation process. Both partners in lichen symbiosis are affected by air pollution with sulphur compounds. The impact on lichens is much greater when the thallus is hydrated than dry. This is also the reason that mean winter concentrations are more relevant in regions with low temperature in winter when water loss is lower and lichens remain hydrated longer after soaking than in summer. Additionally, the amount of sulphur compounds emitted is much higher in the winter period due to heating and energy production. Sometimes, the transport and dilution of anthropogenic sulphur compounds is smaller in winter than in summer due to the formation of temperature inversions, especially in closed valleys and basins in mountainous regions.

Although emission of sulphur dioxide has been substantially reduced in the more developed countries of the European Community (Posch et al. 1997) and the United States (NAPAP report 1998) there are still large emission sources in Central and Eastern Europe, Russia, China and in many developing countries. Here, the epiphytic lichen flora in particular is still severely endangered, and using lichens as reactive bioindicators may provide much useful information about air pollution.

Lichens as bioindicators

Lichens are more sensitive to sulphur dioxide than higher plants. The main reason for this is probably that the thallus has a very simple structure, lacking a cuticle and an epidermis. In addition, lichens are poikilohydric i.e. they cannot maintain stable water content as the supply of water in the environment changes. That is probably why fructicose and pendulose lichens are more sensitive to pollution than crustose species, as they are more exposed to polluted air, and the substratum on which the lichen is growing (stone, soil, bark, etc.) offers less protection. However, some species probably also show greater inherent tolerance, but the precise mechanisms of this are unknown. Tolerance mechanisms can be of a morphological nature, e.g. the structure of the cortex, production of soredia, hydrophobic lichen substances, all of which are mechanisms of stress avoidance. Some of those species actually appear to have a greater need for sulphur e.g. *Lecanora conizaeoides* Nyl. Ex Crombie and *Scoliciosporum chlorococcum* (Graewe ex Stenh.) Vzda. These species are two of the most resistant lichens to sulphur dioxide in Europe. In general, sulphur dioxide mostly affects lichen species that grow in particular environmental conditions, e.g. those adapted to the physical and chemical bark properties of old trees in old, undisturbed natural forests and are species that occur late in a succession. Lichens of this kind occur in the natural forests of European beech (*Fagus silvatica* L.) and silver fir (*Abies alba* Mill.) (e.g. in Dinaric region of Slovenia) e.g. *Lobaria pulmonaria* (L.) Hoffm., *L. amplissima* (Scop.) Forss., *L. scrobiculata* (Scop.) DC., *Pannaria conoplea* (Ach.)Bory, *Degelia plumbea* (Lightf.) P.M.Jørg. and P.James *Parmeliella triptophylla* (Ach.) Müll.Arg., *Thelotrema lepadinum* (Ach.) Ach., *Menegazzia terebrata* (Hoffm.) Massal., *Gyalecta ulmi* (Sw.) Zahlbr., *Arthonia leucopellaea* (Ach.) Almq. and many other microlichens. Species that are widespread and occur early in the development of lichen communities are less affected. Examples include *Hypogymnia physodes* (L.) Nyl., *Parmelia sulcata* Taylor, *Melanelia glabratula* (Lamy) Essl., *Cladonia coniocraea* auct., *Parmeliopsis ambigua* (Wulfen) Nyl., *Phlyctis argena* (Sprengel) Flotow., *Lepraria incana* (L.) Ach., *Lecanora argentata* (Ach.) Malme, *L. expallens* Ach., *L. carpinea* (L.) Vainio, and *Lecidella elaeochroma*(Ach.) Choisy.

It is important to remember that sulphur dioxide is not the only cause of lichen death in polluted environments. It is difficult to distinguish the effects of sulphur dioxide from those caused by nitrogen compounds or even ozone and other air pollutants (Herzig et al. 1989, Nimis et al. 1991).

The main aims of this chapter are as follows:

- To provide information about the importance of bioindication in air pollution monitoring and to describe the methods for monitoring sulphur dioxide air pollution by lichens;

- To review the advantages and disadvantages of the different methods available for using lichens to monitor air pollution caused by sulphur compounds;
- To provide review guidelines to carry out the most used methods and their interpretation.

Materials

Study area Geographical, meteorological, vegetation and land use history data of the area under study should be acquired. Some of these data can be obtained from national and other libraries; others, especially air pollution and climate data can be provided by natural institutions that measure these parameters within natural and international frameworks. Information about the lichen flora obtained from earlier mapping is particularly useful.

Field work Only well known and easily recognised species can be determined and mapped in the field, while most species must be determined later in the laboratory. For field identification you will need the following:

- Hand lenses (10 x, 15 x, 20 x magnification)
- Camera
- Field bag for carrying samples, materials and tools
- Paper bags of different size for keeping samples
- Soft paper for preserving small and fragile thalli
- Plastic bags for sorting and storing the collected material in paper bags during the sampling, especially necessary in wet conditions
- Pencils and other waterproof writing and labelling materials
- Notebook for recording data
- Field versions of lichen identification keys, especially desirable for beginners
- Map of the investigated area on sufficiently large scale to enable sampling locations to be indicated
- GPS (Global Positioning System), if available
- Suitable knife, hammer and chisel for taking lichens from the substrate

Determination of lichens in the laboratory

- Drying oven (to prevent decay)
- Identification books, chosen according to language, area under study and level of taxonomic knowledge (e.g. Ozenda and Clauzade 1970; Poelt 1969; Poelt and Vezda 1977, 1981; Purvis et al. 1992; Wirth 1995a, 1995b).
- Suitably housed microscope and stereomicroscope
- Reagents for "chemicals tests" necessary for identification (see also Chapter 29)
- Equipment and reagents needed for thin layer chromatography
- Reference herbarium
- Laboratory materials and equipment necessary for cutting and staining

Beginners always need help of better-trained lichenologists. Determination of certain species is possible only by specialists, and such material is best sent for determination to them.

Equipment required for Subprotocol 4

- Liquid nitrogen
- Ball mill, e.g. Fritch vibration micro-puveriser, with zirconia mortar and ZrO_2 balls
- Manual hydraulic press (e.g. SPECAC, art.15011)
- X-ray fluorescence spectrometer (XFR) (e.g. fluorescence spectrometer Philips PW 1410 with rodium tube).

Evaluation of data

After lichen species have been determined, the data field lists and assessments must be produced. Further processing of the data depends on the aim of the study, chosen bioindication method and on the facilities of the researcher. By using computer programs, lists of mapped lichen species and their assessments combined with geographical, climatic and air pollution data can be transformed to indices, zones or otherwise delimited investigated areas. This will require a personal computer with appropriate printers and software.

Subprotocol 1
Mapping Species Distribution

Detailed plotting of the distribution of lichen species on maps is the classical approach to study the effects of air pollutants on lichens. The underlying assumption of this approach is that the number of species increases with distance from the source of sulphur dioxide air pollution. The basic procedures of standardisation when using this method were established by the pioneers of lichen species mapping e.g. Skye (1968), Barkman (1958), Hawkworth & Rose (1970), Gilbert (1970), Deruelle (1978) and many others.

Procedure

1. Make the profile in the area under investigation according to the position of air pollution source, topography of the area and meteorological circumstances.
2. Select the most suitable sites for lichen mapping. Pay special attention to sites/substrata that are very rich in lichens.
3. Decide which lichens you are going to map: epiphytic, epilithic, or terricolous lichens. Epiphytic lichens respond better to air pollution. Therefore you must select proper tree species (phorophytes). Following Barkman (1958), European trees can be grouped according to the bark characteristics (mineral content of bark, pH, wetability, water holding capacity, mechanical bark properties) as follows:
   - bark poor in minerals and acid; these species include the conifers (*Pinus, Picea, Larix, Juniperus, Abies, Cupressus*), birches, (*Betula*), and alders (*Alnus*)
   - bark medium to rich in minerals, slightly acid or neutral pHs; these species include the majority of broadleaves like oaks, sweet chestnut, hop-hornbeam (*Quercus, Castanea, Ostrya*), maples (*Acer*), beech and hornbeam (*Fagus, Carpinus*), ash (*Fraxinus*), and fruit trees and their wild relatives (*Malus, Pyrus, Prunus, Sorbus*);
   - bark rich in minerals, pH neutral or slightly basic; these species include the willows and poplars (*Salix, Populus*), elms, walnut, black locust and horse chestnut (*Ulmus, Juglans, Robinia, Aesculus*).

4. Score all lichens that you are able to determine to the species level in the field and collect the rest of the species for laboratory determination.
5. Collect lichen samples and determine species.
6. Make a list of lichens according to the sites in the investigated profile/area and relate it to the level of measured air pollution, if data are available.
7. Compare list of lichens found in the investigated area with former mapping, where available.

Results

The best results are obtained when epiphytic lichens are mapped on a transect on the same phorophyte. This is rarely possible, therefore usually lichens collected from the same type of bark and tree crown are compared as indicated in the Procedure. According to Hawksworth and Rose (1970) observations on lichens can be grouped into trees with eutrophic and non-eutrophic bark regardless of the above classification. Although both classifications have been used for a long time and used by several workers, no comparative study has been done to evaluate the two groupings. It is clear that a high input of nitrogen and phosphorus compounds masks to a great extent the natural chemical bark characteristics. However, differences among barks of trees due to other chemical and physical bark characteristics persist as does tree crown type (personal observations). Probably, grouping trees into those possessing eutrophic or non-eutrophic barks is too simple. Input of calcareous dust plays another important role, similar to eutrophication, and the successional stages of the managed forests, and the history of the land use at the site are also very important.

Mapping studies carried out in Slovenian forests (Mayrhofer et al. 1996, Grube et al. 1995) have shown that some tree species have very specific epiphytic lichen floras (for example, very species rich, very species poor, specific species always present). This flora does not depend on their age, forest stand structure and history, a fact that must be considered when comparing results. These species are common beech and white hornbeam, sycamore, silver fir, black pine and Scots pine (*Fagus sylvatica, Carpinus betulus, Acer pseudoplatanus, Abies alba, Pinus nigra, Pinus sylvestris*). Studying the distributions of each species separately, and subsequently integrating your data will provide best results. Recent lichen

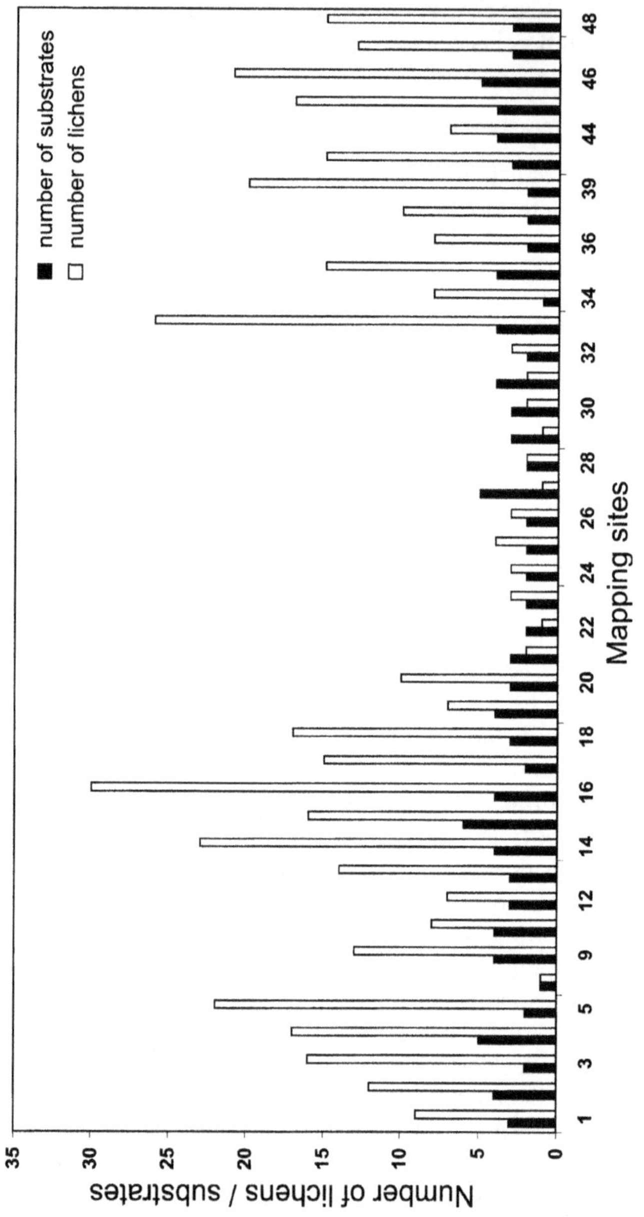

Fig. 1. Relationship between epiphytic lichen species number and number of tree species (y) in a profile across a highly polluted valley (x) in central Slovenia (Mrzlica-Jatna); sites from 22 to 33 are in the vicinity of a thermal power plant (note that very few lichens are present); data from Vidergar-Gorjup 1998.

mapping in a relatively clean air area (Julian Alps) and in highly polluted central parts of Slovenia (Zasavje) confirmed that while the availability of a proper substrate, e.g. phorophyte, affects the number of epiphytic lichens, sulphur dioxide has a larger effect. In Figure 1 the number of epiphytic lichen species is presented according to the number of available substrates at a site in a transect over the Sava valley in the Zasavje district of Slovenia. This district is highly polluted with sulphur dioxide from a thermal power plant situated at the bottom of a deep and steep valley surrounded by hills with very diverse vegetation types. Although the number of phorophytes varies, the decline in epiphytic lichen species number is always the largest where the transect crosses the polluted area (Vidergar-Gorjup 1998).

In Figure 2 epiphytic lichens were mapped along an altitudinal profile in the Julian Alps where data from the lichen mapping was supported by measurements of sulphur dioxide in air using instrumental monitoring (Gomišček 1997). The number of substrates (Norway spruce, European beech, European larch) is more or less the same within the profile, but the number of lichen species present is inversely proportional to the mea-

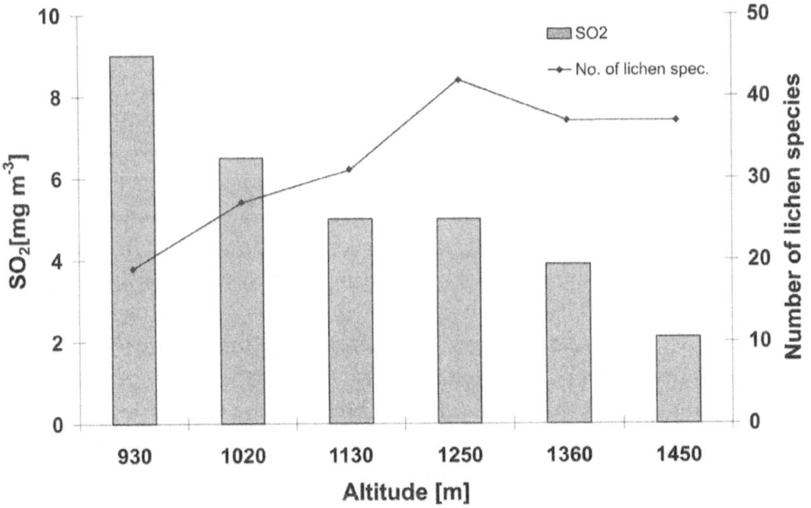

Fig. 2. Relationship between epiphytic lichen species number (line) and sulphur dioxide concentration (bars) in the air along an altitudinal profile at Kurji vrh in the Julian Alps in Slovenia. Site at 980 m is situated at the bottom of a rather polluted valley and has less epiphytic lichens while sites at higher altitudes have more lichens and lower concentrations of sulphur dioxide; data from Surina 1998.

sured sulphur dioxide, as illustrated in Figure 2 (Surina 1998). In all these cases it is clear that sulphur dioxide air pollution is the main cause for the decline of lichen species number and that such an approach can be used to biomonitor sulphur dioxide.

Subprotocol 2 Percentage and Frequency Mapping

Apart from mapping distribution, the occurrence of lichens, the percentage cover and the frequency of single species or lichen thallus types can be used to determine the level of air pollution (Jones, 1952). The approach is particularly useful in highly polluted areas where the majority of lichen species have disappeared. For example Gilbert (1970) mapped the highly tolerant epiphytic lichen species *Lecanora conizaeoides*. He found that in highly polluted areas (80 - 130 µg SO_2 m^{-3}) this species represented 80 - 95% of the epiphytic lichen cover while in less polluted areas (< 60 µg SO_2 m^3) it occupied only 60 - 80% of the total lichen cover.

Another approach in mapping is to assess the cover and frequency of the three major lichen thallus types i.e. crustose, foliose and fructicose lichens. This approach is particularly useful in broad-scale mapping and where non-lichenologists are involved, when species determination is impossible or too time consuming (Fenton 1960, Batič 1991, Batič and Mayrhofer 1996). This method is based on the assumption that crustose lichens are more resistant to air pollution than foliose, and foliose more than fructicose. In general this is true, although some crustose species are much more sensitive to sulphur dioxide and other air pollutants than foliose or even some fructicose species (Barkman 1958; Hawksworth and Rose 1970). Lichen life-form mapping is a very simple method for determining air pollution levels and was often used for teaching purposes with school children, or in broad scale forest decline inventories where a general impression about air quality can be obtained by this method, especially on high density grids and by repetition of mapping at various time intervals (Batič 1991, Batič and Kralj 1995, Batič and Mayrhofer 1996).

From lichen survey studies and "zone mapping", an indicator value for several epiphytic lichen species has been established. The indicator value is most often related to sulphur dioxide concentration in air or just to overall air pollution. In determining the most suitable species to use as indicators, Hawksworth (1973) recommends that the following points should be noted:

- The ecology and distribution of the species must be well known in the area under investigation. The species should have been present in the area before it became affected by air pollution. Therefore old records about the species distribution are very useful, and should be compared with the present distribution if possible. Unfortunately, there are few areas in which the distribution of lichens was well known both before and after becoming polluted. The indicator value of a lichen species changes with climate and some other environmental circumstances, which makes comparison of results difficult.
- The species selected as indicators must still occur on the same substrate after the onset of pollution. This requirement is often difficult to meet because changing land use resulted in a change of available substrates. For example, deciduous forests were often transformed to conifers, drastically changing the epiphytic lichen vegetation.
- The species should be present in the same environmental circumstances in clean nearby areas or on more favourable substrates in the polluted areas. This allows the possibility of the lichen spreading should a reduction of air pollution occur.
- The species must be sensitive to sulphur dioxide, as proved by transplant experiments.
- Skye (1968) recommended avoiding the selection of species that are very tolerant or very sensitive to sulphur dioxide. The best results are obtained by using a group of well-known species, ranging from sensitive to resistant.
- The chosen species must be easily recognised, even by non-specialists.

There are few species that meet all these requirements worldwide, and it is becoming evident that indicator species must be chosen for each area. With certain precautions, the groups of indicator species suggested by Hawksworth and Rose (1970) could be used for screening sulphur dioxide air pollution, but their scaling should be locally tested and certain species should be omitted and additional new ones included.

Procedure

When starting with lichen mapping one must know how to determine lichens; for beginners, complete species mapping is a very hard task. The choice of which method to use (complete flora mapping, indicator species mapping or thallus type mapping) will depend on the abilities of those doing the mapping and the aims of the study.

1. Investigate carefully the area where you are going to monitor air pollution by lichens. Special care should be paid to the position of pollution source, available tree species, topography, and climate parameters (wind direction, occurrence of temperature inversion, etc.).

2. Establish the observation plots with comparable phorophytes and site characteristics and indicate them on the map. They can be set up in profiles or in grids of different density. Note the following:
   - In flat areas consider wind direction (transects taken in the wind direction must be much longer than on the lee side of air pollution source).
   - In hilly or mountainous areas try to have a profile according to the altitude.
   - Grid size (1x 1 km, 2 x 2 km, 4 x 4 km, 16 x 16 km) and the number of plots in a transect depend on the method used and aim of the study. Plots can be established completely randomly or with special care for selected site/trees with expected rich lichen flora. The more mapping plots the better, but a balance must be struck between accuracy on the one hand, and cost and time on the other.
   - Pay particular attention to epiphytic lichens, as they are better indicators of air pollution than epilithic and terricolous species.

3. Prepare a species list for each site, after identification in the laboratory if needed. Note the following:
   - Do not collect very rare species.
   - Preparing a complete lichen inventory is very time consuming and demands well-trained lichenologists. It is better to select very sensitive or relatively resistant epiphytic lichens and map them by using similar procedures to those used for zone mapping.
   - For teaching purposes in work with school children or in large scale inventories map only lichen thalli types (crustose, foliose and fruticose) using methods described in Batič and Mayrhofer (1996).

4. Make distribution maps for all or, if this cannot be done for selected indicator species and compare the results with former investigations and measurements of air pollutants if available.

Subprotocol 3
Zone-Mapping and IAP Maps

From single species mapping and from mapping of all lichen species in chosen areas several investigators have constructed zones according to lichen richness, cover and frequency. Such zone-maps have been constructed for many towns, individual industrial plants, provinces and even whole countries [see Hawksworth (1973) for citations].

Five zone system

The number of zones designated differs among researchers, but most often five zones are recognised on the basis of lichen species occurrence, their cover and frequency. These are as follows:

1. An inner zone, often termed a "lichen desert" where no lichen can be found or only the most tolerant; these often comprise sterile thalli of crustose species;
2. An "inner struggle zone", with a few resistant foliose lichens with low cover and frequency, fructiose species are absent;
3. An "outer struggle zone" in which the most resistant fruticose species appear, especially in sheltered positions;
4. A "transition zone", where lichen species composition is normal according to expectation but signs of air pollution impact are detectable (changed frequency and cover of tolerant and sensitive species, signs of pollution damages);
5. A "normal", unaffected zone.

Later it was realised that phrases such as "lichen desert", "struggle zone", and "normal zone" should be avoided because these terms were used in different circumstances by different authors and caused confusion in interpretation.

Ten zone system

Hawksworth and Rose (1970) devised one of the most elaborate zone mapping systems. They devised a ten zone system for eutrophic and non-eutrophic bark, indicating concentrations of sulphur dioxide from 0 to >170 $\mu g\ SO_2\ m^{-3}$, estimating the mean winter SO_2 air pollution in England and Wales using epiphytic lichens (Hawkworth and Rose 1970, Hawksworth 1973).

Index of Atmospheric Purity

At about the same time as the above zone systems were being proposed, other workers devised IAP (Index of Atmospheric Purity) maps (DeSloover, 1964 LeBlanc and DeSloover 1970; Trass 1971). The IAP is time consuming to estimate, but provides a numerical assessment of the degree of air pollution. It is based on the number of lichen species, their cover and frequency and tolerance to air pollutants at the investigated site. Maps can be drawn by computer, and then compared with data from instrumental monitoring. The first studies of this kind suggested that good correlations between IAP and sulphur dioxide existed. However, later it was realised that other air pollutants are involved. More sophisticated computer techniques were used to correlate lichen data with several measured air pollutants at the observation plot or nearby air quality measuring stations (Herzig et al. 1989, Nimis et al. 1991). Similar IAP maps have also been constructed from the mapping of lichen life-form types where observations of the occurrence, cover and frequency of thalli types were carried out separately for tree base, trunk and crown. Figure 3 gives an ex-

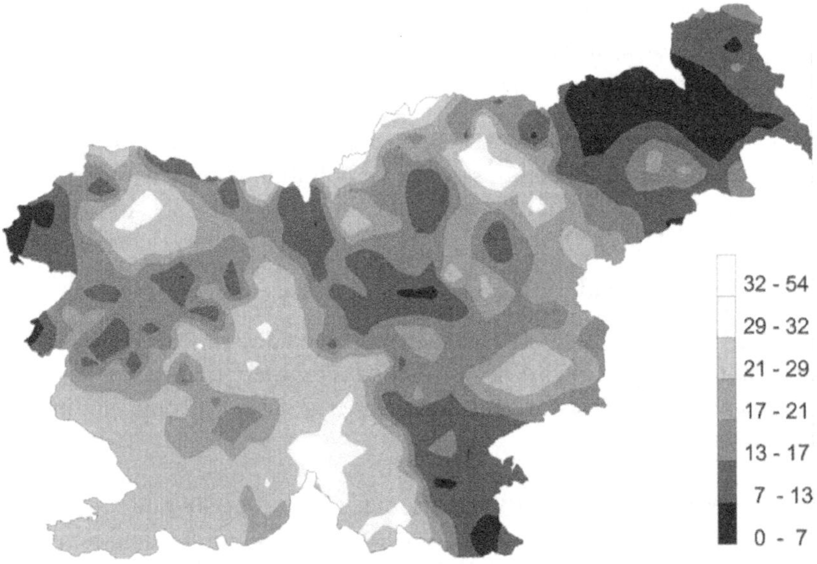

Fig. 3. Lichen map of Slovenia based on the IAP values as obtained from observing lichen thallus types, their frequency and cover within 4x4 km grid in 1991, at regularly forest dieback inventory on different types of trees. The IAP values are between 0 and 54 (where the value 0 means a plot without lichens and very polluted air and the value 54 means very rich lichen vegetation and very clean air), and are divided into 7 classes according to the percentile values (10, 30, 50, 70, 90, 95%); data from Jeran et al. 1996.

ample of such an IAP map for Slovenia from regular forest decline inventories (Batič and Mayrhofer 1996, Jeran et al. 1996).

The most detailed mapping system of this kind was a form of passive biomonitoring with lichens as part of an integrated biological measuring system for monitoring air pollution in Switzerland proposed by Herzig et al. (1989) and Liebendörfer et al. (1988). The procedure of lichen mapping is similar to those described in Subprotocols 1 and 2 for total species mapping and percentage and frequency mapping, but with some modifications.

Procedure

1. Analyse 10 trees supporting epiphytic lichens in a radius of 250 m at each observation plot. Note the following points with respect to tree selection:
   - Only observe autochthonous tree species, typical for the site (e.g. *Tilia platyphyllos*, *T. cordata*, *Acer platanoides*, *Fraxinus excelsior*, *Quercus petraea*, *Q. robur* in most of West, South and Middle European countries)
   - Only observe free-standing trees that are not growing in shade.
   - Only observe vertical trees (with declination less than 10°), and with stem circumferences between 100 and 280 cm.
   - Avoid trees with damaged bark or obviously sprayed with manure or biocide.
2. On each tree assess epiphytic lichens by laying a grid with ten boxes on the trunk between 120 and 170 cm of the stem height (Figure 4). The height of this grid is 50 cm, while the width is adjustable according to the stem size. Place the grid such that the most luxurious lichen cover occurs in the middle of the grid.
3. For each species assess the following parameters:
   - **Frequency** F: assessed at each of the partial fields in the grid; scale: 0-10.
   - **Cover** C: percentage of cover within the grid; scale: 0, 1, 2, 3, 4, 5.
   - **Vitality** V: evaluation of health and growth condition of each species on the base of macroscopic-morphologic appearance of thallus by using 3 scale states: very well, medium and poorly developed.
   - **Injuries** S: descriptions of visible injuries in forms of necrosis/chlorosis in 3 stages: none, slight, very injured.

4. After mapping, calculate the **Q** value, an index of the tolerance of the species to air pollution. It is calculated as average number of lichen species co-occurring with the lichen in the investigated area. High Q values imply that the lichen is sensitive to air pollution, while low values mean that the lichen is tolerant. Q value for each species must be determined for each area separately because there are several factors that contribute to sensitivity or tolerance of the species.

5. On the base of such lichen assessments, calculate the IAP (Index of Atmospheric Purity) by using equation 1.

$$IAP = \Sigma \frac{Q \times C \times F}{V \times S} \qquad \text{Equation 1}$$

6. Calibrate IAP values against measured main air pollutants (SO_2, NO_3^-, Cl^-, dust, Pb, Cu, Zn, Cd) and find the best fitted correlation factors with main pollutants; group the IAP values into five classes delimiting

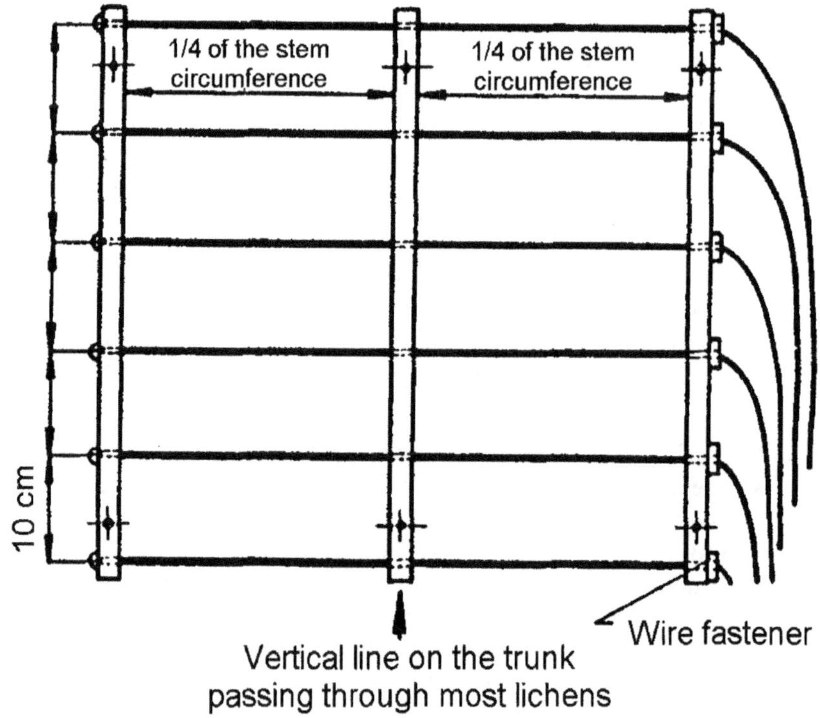

Fig. 4. A 10-field frequency net for lichen assessment after Herzig et al. (1989).

emission and lichen zones: critical pollution - lichen desert, high pollution - inner struggle zone, middle pollution - outer struggle zone, little pollution - transition zone, very little pollution - normal zone (Liebendörfer et al. 1988).

Subprotocol 4
Transplant Studies and Total Sulphur Content Measurements

To prove that sulphur dioxide is having a harmful effect on lichens, they can be transplanted to a polluted environment, for example close to sites where instrumental monitoring is being carried out. Hawksworth (1973) reviews older studies, and these methods are still useful. These approaches can be used in areas without proper natural phorophytes, and also as a substitute for lichen species mapping over large areas to save time and money. Usually more tolerant species are used, e.g. *Hypogymnia physodes*,

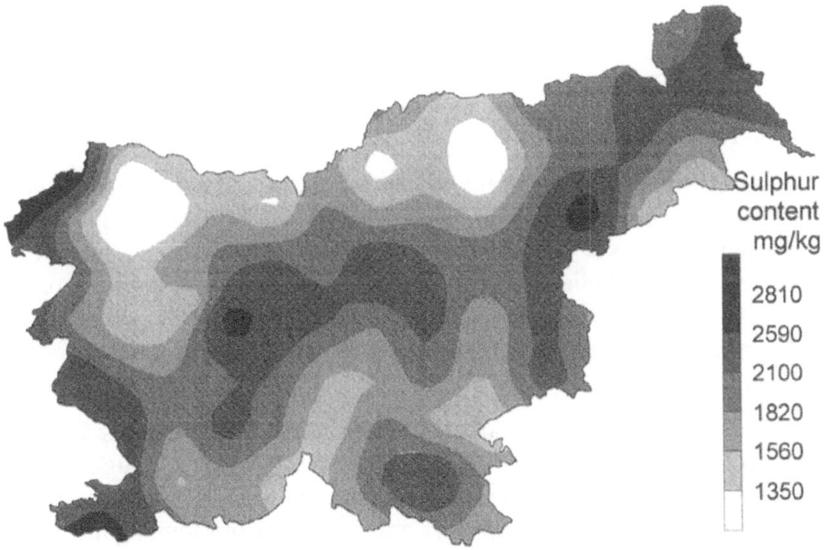

Fig. 5. The pattern of sulphur dioxide concentrations obtained from lichen data on a national scale, divided into 7 classes according to the percentile values (10, 30, 50, 70, 90, 95%). The epiphytic lichen *Hypogymnia physodes* (L.) Nyl. was sampled from, as far as possible, the most similar sites and tree species within the Slovenian 16 x 16 km bioindication grid. This grid was established by the Slovenian Forestry Institute within a research programme dedicated to forest decline studies [data from Jeran et al. (1996)].

Parmelia sulcata and *Cladonia* sp. The same lichen species are also often analysed for total sulphur content, either collected in nature or taken from the transplant experiments. The results are often very difficult to interpret. While alive, lichens can be used as sulphur accumulators. They are very good tools for short-term experiments to monitor sulphur dioxide emissions or abatement of air pollution. When severely damaged or even dead, lichens lose accumulated sulphur quickly and can no longer be used for this purpose (Batič, unpublished results). Figure 5 gives an example of the use of epiphytic lichens as sulphur accumulators (Jeran et al. 1996) in a study carried out in Slovenia for screening heavy metal and radionuclide air pollution. Thalli of the tolerant ubiquitous epiphytic lichen species *Hypogymnia physodes* were collected from nature using a 16 x 16 km sampling grid. Sulphur accumulation in this lichen is in good agreement with expected air quality. In areas with clean air lichen thalli contain less sulphur. The agreement of sulphur content data with IAP values is not perfect, because other pollutants affect lichen growth and contribute to lower IAP values in addition to sulphur dioxide.

Procedure

1. Define precisely the main goal of the study; e.g. screening sulphur and other air pollutants in naturally occurring thalli or in transplants.
2. Select the most convenient species for transplantation. Enough material must be collected from unpolluted areas. The choice of species will depend on local climate, lichen flora and available substrates for exposure. In middle and north Europe *Hypogymnia physodes* is one of the most frequently transplanted species. Schönbeck (1969) recommends transplanting lichens together with their substratum, e.g. a tree branch.
3. Transplant lichen into polluted environment. Select sites for exposure as for mapping studies or as described in Chapter 26. Enough material should be exposed at the site, depending mostly on number and types of analyses, length of exposure and frequency of sampling during the exposure. The longer the exposure period, the more material must be exposed, and accordingly frequency of sampling could be higher.

4. Sample the exposed lichen material and analyse thalli for total sulphur content by using the method described by Jeran et al. (1993) as follows:
   - moisten lichens in the laboratory with distilled water and remove adhering bark particles
   - freeze dry lichen samples in liquid nitrogen
   - grind frozen lichen samples in a zirconia mortar with a ZrO_2 ball
   - take about 200 mg of dry lichen powder to press it into tablets using Manual hydraulic press
   - analyse total sulphur content by using X-ray fluorescence spectrometry

Troubleshooting

- Fieldwork should be done in short time and ensure that the same group of well trained people are involved in the assessments.
- Do not start complete lichen flora mapping, or a large project like the Swiss integrated zone mapping (Herzig et al. 1989) with beginners. These methods are very accurate, but only well trained lichenologists are able to collect and determine all species in the area or assess species properly according to the Swiss method (Subprotocol 3).
- When you have no data about air pollutant measurements be careful when explaining and delimiting zones of air pollution impact based merely on your mapping data, or based only on literature.

Acknowledgements. All participant students, Natalija Vidergar-Gorjup, Tadeja Trošt and Boštjan Surina, Barbara Kruhar and Katja Primožič, are thanked for their contribution to this topic, especially for allowing me to publish some of their results. The same thanks are due also to my friend, Dr. Zvonka Jeran for her long term collaboration in bioindication of air pollution with epiphytic lichens in Slovenia, and for the results she allowed to be included in this article. I am most thankful to Prof. Dr. Helmut Mayrhofer, Karl-Franzens Universität Graz, Austria, his colleague Dr. Martin Grube and students (Marietta Koch, Johannes Prügger and Ursula Zuppan), for their help in improving lichen research in Slovenia as well as for their contribution in lichen mapping in Slovenia and in bioindication methods based on lichens. I would like to thank for technical assistance Gabrijel Leskovec and Boris Turk (Biotechnical Faculty, University of Ljubljana), Dr. A.R. Byrne for correcting my English. I would like to thank the Slovenian Ministry of Science and Technology and Ministry of Agriculture, Food and Forestry for financial support, and last but not least the Slovenian foresters and Slovenian Institute for Forestry who accepted lichen bioindication methods for monitoring air pollution in forests and enabled me to work in this field of research.

References

Arndt U, Nobel W, Schweizer B (1987) Bioindikatoren Möglichkeiten, Grenzen und neue Erkenntnisse. Ulmer, Stuttgart

Barkman JJ (1958) Phytosociology and ecology of cryptogamic epiphytes. Van Gorcum, Assen

Batič F (1991) Bioindikacija onesnaženosti zraka z epifitskimi lišaji. Gozdarski vestnik 49:248-254

Batič F, Kralj A (1995) Bioindication of air pollution in forests by epiphytic lichens. Zbornik gozdarstva in lesarstva 47:5-56

Batič F, Mayrhofer H (1996) Bioindication of air pollution by epiphytic lichens in forest decline studies in Slovenia. Phyton 36:85-90

DeSloover JR (1964) Végétaux épiphytes et pollution del'air. Revue Quest Sci 25:531-61

Deruelle S (1978) Les lichens et la pollution atmosphérique. Bull Ecol 9:87-128

Ellenberg H, Weber HE, Düll R, Wirth V, Werner W, Paulissen D (1992) Zeigerwerte von Pflanzen in Mitteleuropa. Goltze, Göttingen

Fenton AF, (1960) Lichens as indicators of atmospheric pollution. Ir Nat J 13:153-159

Ferry BW, Baddeley MS, Hawksworth DL (1973) Air pollution and lichens. The Athlone Press, University of London

Gilbert OL (1970) Further studies on the effect of sulphur dioxide on lichens and bryophytes. New Phytol. 69:605-629

Gomiščeк S (1997) Zaključno poročilo. Onesnaženost zraka v Triglavskem narodnem parku in njen vpliv na gozd in vode - Atmosfera in narodni park. Final Report, S2-6241. Ministry of Science and Technology, Ljubljana, pp. 133

Grube M, Batič F, Mayrhofer H (1995) Contribution to the lichen flora of Slovenia I. The Snežnik area. Herzogia 11:189-196

Hawksworth DL, Rose F (1970) Qualitative scale for estimating sulphur dioxide air pollution in England and Wales using epiphytic lichens. Nature 227:145-148

Hawksworth DL (1973) Mapping studies. In: Ferry BW, Baddeley MS, Hawksworth DL (eds) Air pollution and lichens. The Athlone Press, University of London, London, pp 38-76

Herzig R, Liebendörfer L, Urech M, Ammann K, Buecheva M, Landolt W (1989) Passive biomonitoring with lichens as a part of an integrated biological measuring system for monitoring air pollution in Switzerland. Int J Env Anal Chem 35: 43-57

Jeran Z, Smodiš B, Jačimovič R (1993) Multielement analysis of transplanted lichens (*Hypogymnia physodes* (L.) Nyl.) by instrumental neutron activation analysis. Acta Chim Slov 40:289-299

Jeran Z, Jačimovič R, Smodiš B, Batič F (1996) The use of lichens in atmospheric trace element deposition studies in Slovenia. Phyton 36: 91-94

Jones EW (1952) Some observations on the lichen flora of the tree boles with special reference to the effect of smoke. Rev Bry Lich 21: 96-115

LeBlanc F, DeSloover J (1970) Relation between industrialization and the distribution and growth of epiphytic lichens and mosses in Montreal. Can J Bot 48:1485-96

Liebendörfer L, Herzig R, Urech M, Ammann K (1988) Evaluation und Kalibrierung der Schweizer Flechten-Indikationsmethode mit wichtigen Luftschadstoffen. Staub 48: 233-238

Mayrhofer H, Koch M, Batič F (1996) Beiträge zur Flechtenflora von Slowenien. II: Die Flechten des Pohorje. Herzogia 12: 111-127

NAPAP (1998) NAPAP bienial report to the congress: An integrated assessment. US national acid precipitation assessment program, Silver Spring, MD
Nash III TH (1973) Sensitivity of lichens to sulphur dioxide. Bryologist 76:333-339
Nieboer E, Tomassini FD, Puckett KJ, Richardson DHS (1977) A model for the relationship between gaseous and aqueous concentrations of sulphur dioxide in lichen exposure studies. New Phytol 79:157-162
Nylander W (1866) Les lichens du Jardin du Luxembourg. Bull Soc Bot Fr 13:364-372
Nimis PL, Lazzarin G, Gasparo D (1991) Lichens as bioindicators of air pollution by SO_2 in the Veneto region (NE Italy). Stud Geobot 11: 3-76
Ozenda P, Clauzade G (1970) Les Lichens. Étude biologique et flore illustrée. Masson et Cie Éditeurs, Paris, pp.801
Poelt J (1969) Bestimmungschlüssel europäischer Flechten. Cramer, Lehre, pp757
Poelt J, Vzda A (1977) Bestimmungschlüssel europäischer Flechten. Ergänzungsheft I. Cramer, Vaduz, pp 258
Poelt J, Vzda A (1981) Bestimmungschlüssel europäischer Flechten. Ergänzungsheft II. Cramer, Vaduz, pp 390
Posch M, Hetteling JP, de Smet PAM, Downing RJ (1997) Calculation and mapping of critical thresholds in Europe. Status report 1997. Coordination centre for effects. RIVM Report No. 90-6960-069-2 , Bilthoven, Netherlands, pp 163
Purvis OW, Coppins BJ, Hawksworth DL, James PW, Moore DM (1992) The lichen flora of Great Britain and Ireland. The British Lichen Society, London, UK, pp 710
Schönbeck H (1969) Eine Methode zur Erfassung der biologische Wirkung von Luftverunreinigungen durch transplantierte Flechten. Staub 29: 14-18
Skye E (1968) Lichens and air pollution. Acta Phytogeogr Suec 52:1-123
Surina B (1998) Lichen mapping in forests from northern part of Triglav national park. Graduation Thesis, University of Ljubljana, Ljubljana
Trass H (1971) Poleotolerantnost lišajnikov. In: Vimba E (ed.) Mater.VI Simpos Mikol Likhenol Pribalt Rep 1, pp 66-70, Riga
Türk R, Wirth V (1974) Der Einfluss des Wasserzustandes und des pH-wetes auf die SO_2 Schädigungvon Flechten. Verhandl.Ges. Ökol., Erlangen
Vidergar-Gorjup N (1998) Bioindication of air pollution in Zasavje. Master of Science Thesis. University of Ljubljana, Ljubljana
Wirth V (1995a) Flechtenflora. 2. Auflage, Ulmer, Stuttgart
Wirth V (1995b) Die Flechten Baden-Württenbergs; Teil 1, Teil 2. Ulmer, Stuttgart

Biodiversity and Information Systems

Management of a Lichen Herbarium

WALTER OBERMAYER

Introduction

Lichen herbaria store preserved specimens of lichenized (and often lichenicolous) fungi. The main tasks of lichen (and all other) herbaria are as follows:

Taxonomic, chemical and molecular studies

Taxonomic studies must be at least in part based upon herbarium specimens, of which the so called type-specimens are the most important. It is on these that descriptions of new taxa are based, and they serve in perpetuity as the reference for these names. For a new species to be validly published, the herbarium in which the type specimens are lodged must be specified. Locations, contents, acronyms etc. of the world's public herbaria can be found in "Index Herbariorum", compiled by Holmgren et al. 1990. A searchable internet-version is available at: http://www.nybg.org/bsci/ih/.

Most secondary lichen substances remain stable in storage, and hence herbarium specimens are reliable subjects for chemical investigations. The ability of lichen thalli to accumulate pollutants means that older specimens may provide evidence of former environmental conditions.

Improved methods in molecular studies have also led to an increasing use of herbarium specimens (preferably not older than 10 years) as important subjects for taxonomic works. When old herbarium collections (e.g. from the last century) are examined, lichen phylogeny and the long term population dynamics of lichens will become promising fields of study.

Walter Obermayer, Karl-Franzens-Universität Graz, Institut für Botanik, Holteigasse 6, Graz, 8010, Austria (*phone* +43-316-380-5658; *fax* +43-316-380-9883; *e-mail* walter.obermayer@kfunigraz.ac.at)

Floristic data Herbarium specimens are an essential basis for the preparation of checklists or floras of particular geographical regions. They can also provide an indication of changes over time in the distributions of species.

Identification Due to the rather complicated morphology of lichens, descriptions (and keys to taxa) are difficult to make or, if done, often are insufficient for a certain recognition of taxa. Illustrations can overcome this problem to some extent, but are unavailable in many cases. Thus, properly identified herbarium specimens are an indispensable tool for a correct identification of unnamed taxa.

Experimental Details

Preparation of lichens

Carefully prepared herbarium specimens are of vital importance for the quality of the whole herbarium. Although the 'Herbarium Handbook' of Forman and Bridson 1998 gives helpful advice on preparing phanerogams, collecting and preserving of lichens is dealt with only cursorily. [Note: While finishing the manuscript a book on 'managing the modern herbarium' (Metsger and Byers 1999) has been published]. General hints regarding the preparation of lichens can be found in, for example, Wirth 1995, Moberg and Holmsen 1992 and Hawksworth 1974. A brief summary of methods for preparing lichens for deposition in a herbarium is given below.

Collecting lichens and preliminary field-preparations Before collecting lichenized fungi (or indeed any biological specimens), the potential rarity of the species should be considered. Many countries require formal collecting permits, and these frequently come with a range of restrictions. In countries where some species are legally protected, collecting of rare, vulnerable or threatened species may be highly restricted or prohibited. Regardless, no collecting should deplete an entire stand.

For epiphytic lichens, the substratum must also be considered, and trees should not be damaged or killed. Depending on the growth habit of lichens and the type of substratum, different collecting-methods are appropriate. Easily removable (large foliose, umbilicate or fruticose) lichens must be collected with their attachment organs (but usually without substratum). Small foliose and all crustose lichens have to be removed with part of the substratum. Stout knifes or caulking irons are used for bark, wood or soil; chisel and hammer help to remove lichens from hard substrate (e.g. rocks). Fragile lichen species (e.g. Caliciales, or many fruticose taxa if totally dried) should be carefully wrapped in soft paper (e.g. uncoloured toilet paper). Rocks with crustose lichens (especially when wet) must be processed in the same way to avoid mutual abrasion.

Preparation in the laboratory

In order to minimise space-requirements, freshly collected (often slightly to totally wet) specimens of foliose or fruticose lichens should be very slightly pressed and dried between uncoloured paper. With great care, rock substrates may have to be made thinner with a chisel or other specialised cutting equipment (see Figure 1), because too thick specimens unnecessarily can cause additional space problems. Specimens on bark, rock and especially soil are usually fixed to stiff card with wood glue. Additional protection against pressure (especially necessary in case of brittle lichens and/or substrate, e.g. delicate Caliciales, unpressed fruticose lichens or specimens on friable or earthy substrate) can be achieved by glueing small wooden rods (or rings of very strong cardboard) around the substrate or by putting it into small shatter-proof boxes. Never close the boxes before the lichen or the substrate or the glue is totally dry; mould fungi may destroy the whole specimen!

Foliose or fruticose lichen thalli and any substratum with lichens on each side must never be stuck directly onto cards. If a sample consists of many small pieces, these can be placed on cardboard between two layers of soft paper, which may prevent displacement. Sometimes it is necessary to group them within a small packet, which can be glued onto the card (see Figure 2a). Arranging packets inside each other should be reduced to a minimum as it slows down the access to the material.

After mounting on card, most lichens are usually placed into folded envelopes (see Figure 2b). Both cards and envelopes need to be of a long-lasting (acid-free) archival quality. It is advisable to print the herbarium name (or acronym) on the outside of every packet and on the cardboard. This prevents confusion when handling specimens from many different herbaria. At least for lichens on rocks and for any fragile material, specimens should be covered with a layer of soft paper (to protect the envelope, the label and the lichen). Duplicates of labels, placed inside the packets, are sometimes used for the same purpose.

For how to handle specimens for determination/examination see "Herbarium Problems".

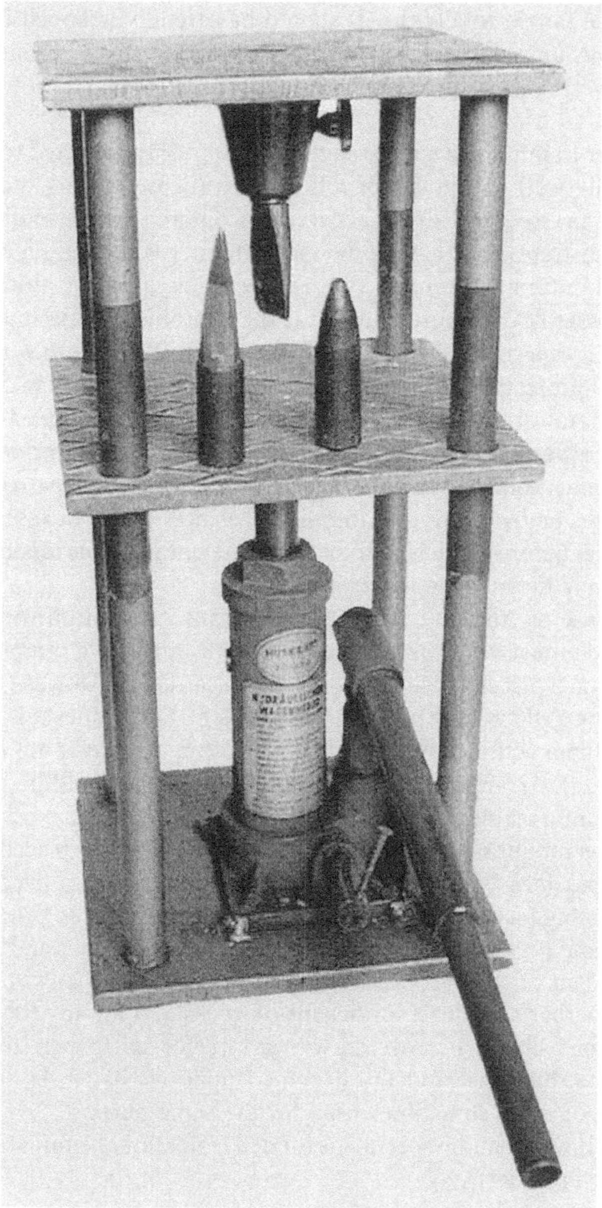

Fig. 1. Home-made "rock-hacker" with interchangeable chisels and using a typical car "jack".

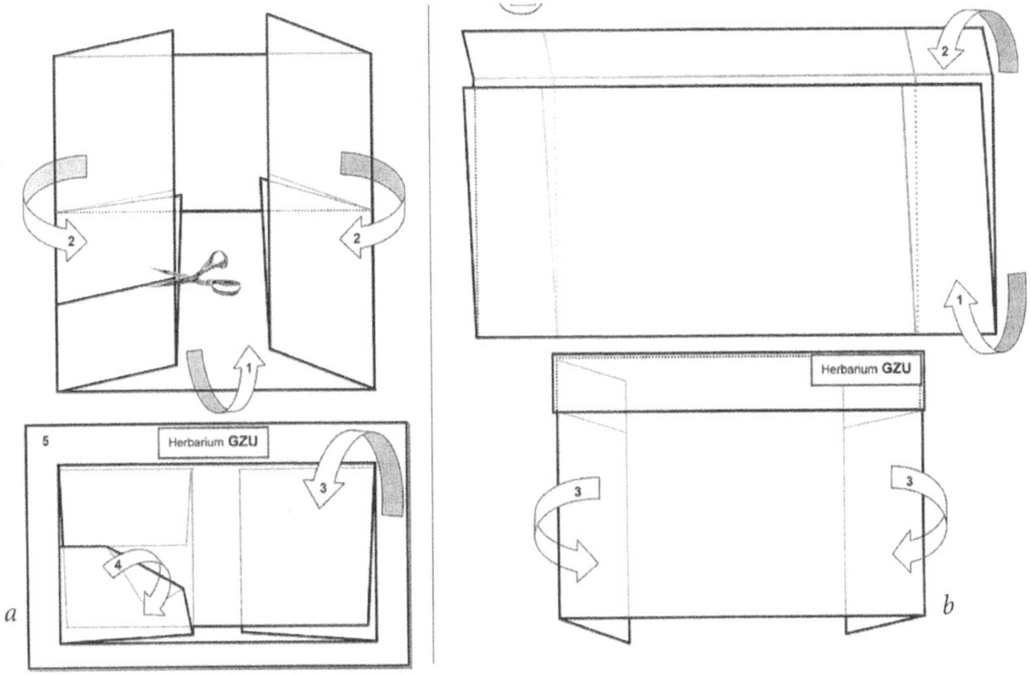

Fig. 2. Suggested design and fold-order for lichen herbarium envelopes. a. Inner envelope, folded (1-4) and glued onto the card (5). This would contain the specimen, or pieces of specimen. b. Typical, folded (1-3) herbarium envelope, within which card and lichen (as per 2a) are placed.

Labelling

Badly labelled or unlabelled herbarium specimens are useless and should be discarded.

The most important data for labels are:

- The name of the taxon
- The location; location details typically include the political region, e.g. country and the precise location, followed by geographical co-ordinates; the actual distance from a large nearby town is often useful, and is preferable to a distance along a road, which may change in the event of road realignment; if the latter is specified, it should be indicated as "road-distance"

- Ecological notes, including substrate
- Date of collection; because different methods of indicating dates exist, it is recommended to give the month in roman numerals (or in letters)
- The collector's name, including collection number
  To make every single specimen unique, the collector should add a successive number (or a non repeating abbreviation of date and field number).
- Reducing subsequent investigation to a minimum, further information on the specimen (e.g. microscopical or chemical features) have to be added, either by writing it directly on the cardboard (this should only be done by the collector him/herself or by preparing annotation slips which are loosely attached to the specimen. Thus, particularly type material can be (and must be!) saved.

Incorporation

After specimens have been decontaminated (see below), they can be incorporated into the herbarium. It is probably best to store lichen-envelopes flat, mounted on a sheet of paper. However, other methods, such as vertical filing of individual packets, are employed in various institutions. The advantage of rapid access to vertically filed specimens is more than offset by the disadvantages. For example, specimens can be easily damaged, and problems arise as a result of different envelope sizes or space may not exist for large envelopes, while small packets tend to slide under larger ones. Horizontally filed specimens can be easily retrieved, and of course any size of envelope can be stuck onto the sheets. However, handling is probably slower and extra costs for folders and cardboard sheets are involved. In case of a flat storage, envelopes or packets may be glued (or even pinned with needles) to sheets or kept loose. As loose packets on sheets tend to fall off, some form of 'fixing-method' is recommended. Depending on the size of the sheets and the envelopes, 2 - 8 packets can be fixed onto a single sheet. Several herbaria prefer to mount only one single envelope on a sheet, which has the advantage that enough space is provided for directly visible annotations, and that envelopes never have to be removed from the sheets. However, specimens stored in this way use much space. Depending on the thickness (and weight) of the specimens, 2 - 5 sheets may be placed inside a stronger and slightly larger folder. Alternatively, a few herbaria use big cardboard-boxes, which may be stacked horizontally. These boxes provide maximum protection for the specimens, but handling is much less efficient. Where envelopes are glued to sheets but need to be removed for some reason, the sheet is torn from the packet but not *vice versa* to avoid damage (e.g. causing a hole at the back side of the envelope).

Arrangement

Because the classification of lichenized and lichenicolous fungi at the family level and above fungi is in a strong state of flux, I recommend arranging lichen taxa alphabetically by genus and species names respectively. Within each species, a geographical classification may be useful. For example you could use a different colour of species cover for different local region/country/continent. Using an additional striking colour for type-covers will allow easier handling of this most valuable component of the herbarium. In some herbaria, types are stored separately as well as important (mostly old) personal collections or exsiccata material. In case of the latter, I recommend avoiding a separate storage, at least of the numerous recent exsiccata, because it dramatically slows down loan management when it is necessary to search the same taxa in many different places.

Fully databased herbarium information also offers the possibility of keeping specimens in the same order that they are accessioned. This method has the advantage, that all specimens keep their original place even if there are taxonomic changes or space extensions. But again, loan management and identification work with comparison-material is made much more difficult.

Herbarium Problems

Decontamination / herbarium conditions

Deep freezing (below -20°C for three days) is the most widely used method for decontaminating incoming herbarium material. In addition, the whole herbarium unit should be fumigated periodically (e.g. every second year) although the need for this varies very much with local conditions such as humidity and the prevalence of certain insect pests. Nowadays very volatile gases (e.g. hydrogen phosphide) are used, which minimise health problems, but require a carefully sealing of every single room. Naturally this process is only undertaken by professionals!

Although lichens are known to be rather resistant to pests, soralia and algal layers of some nitrophilous taxa (e.g. Physciaceae and Teloschistaceae) can be entirely devoured. The main culprits are bark lice [*Liposcelis* spec. div. ("book lice")] and also skin beetles [dermestids, e.g. *Anthrenus museorum* ("museum beetle") and *Anthrenus verbasci*]. These insects mostly live and breed inside a herbarium, whilst others (e.g. mites) are carried in with freshly collected material. If specimens are not decontaminated, the mites can continue feeding for a while (until they and the lichens are dried up). In very damp conditions, the most serious damage can occur through an attack on the specimen label and the glues, rather than on the specimen itself. Among others, silverfish (*Lepisma* spec. div.) and several bark lice (see above) may cause such damage.

Best protection against mould growth (and pest feeding) is provided by storage in constant environment rooms with constant temperature (20–23°C) and low humidity (40-60%).

Storage space

The efficiency of a herbarium is much influenced by the structure of the storage space. Important cost factors are the time needed for inserting and retrieval of specimens and the required building space. Herbarium cases not higher than a person can reach from the floor and nearby working tables (equipped with good microscopes) can help to optimise labour efficiency, both of herbarium staff and scientists (in smaller herbaria this may be only one person). Nowadays building space is often saved by the installation of "compactor systems", using mobile cases running on tracks.

Material acquisitions versus space-problems

Active herbaria not only have the duty to house old material but should also try to acquire freshly collected samples, which represent invaluable genetic resources in the future. Material from the area where the herbarium is situated is often procured by a local staff. Species from distant parts of the world are frequently acquired via exchange with other herbaria, e.g. by means of issuing exsiccata material or duplicate collections.

Although storage of duplicate material is said to involve extra resource consumption for maintaining the same amount of scientific information, the morphological (and chemical) spectrum of species is much better demonstrated with several collections of the same taxon.

However, to expand collections always seem to be in permanent need of additional space. Therefore there are few herbaria in the world that do not suffer from a space shortage. New and improved storage systems (see above) may partly be of help, but in many cases only extensions or even new herbarium buildings can solve the problem. The need to defend the importance of herbarium collections to politicians and senior managers is very much a way of life for most herbarium curators.

Examination and sampling

Caution: Handle specimens with greatest care!

Once dried, a specimen should not be rehydrated with tap water, as chlorides and fluorides can alter its colour or chemical composition (even the purest water may cause a small change in thallus colour). Chemical spot tests with standard reagents such as potassium hydroxide (KOH, "K"), calcium- (or potassium-) hypochlorite ("C") and paraphenylendiamine ("PD"), must never be made directly on the whole specimen. These tests may destroy not only the lichen samples but also envelopes and labels. Many valuable specimens have been damaged by sometimes even famous scientists due to thoughtless handling of these chemicals. Spot tests are easily done on a white background with a tiny fragment of lichen dipped in a small drop of reagent. In case of PD, place a few small crystals into a drop of pure water or alcohol before adding the lichen. Any sampling of thallus-fragments for morphological, chemical or molecular studies has to be undertaken with extraordinary diligence: Especially for TLC and HPLC examination and for genetic analyses, the amount of removed material has to be kept as small as possible and the sampling-position must be indicated. Please note, that herbaria often prohibit this kind of removal from type material. Any new information which has been gathered from the specimen should obviously be added to the label.

Loan Management

All material sent by post, for example on loan or the return of a loan, as a gift or as an exchange, must be carefully packaged. Dried lichen thalli are usually very brittle, and easily damaged or even destroyed, and therefore must be protected from crushing, shaking or mutual abrasion. It is best to keep each envelope under a slight pressure inside the parcel. To avoid damage, never send specimens, particularly type material, in padded bags, but always in stiff cardboard boxes. Loan forms, or similar documents for gifts and exchange, which have to be acknowledged and sent back by the borrower, must be included. The necessary data for such documentation are:

- Address (including herbarium acronyms) of both sender and recipient
- Date of dispatch
- Loan-number (preferably containing the year)
- Detailed list of specimens; the list of specimens should include at least the names and numbers of specimens and a unique detail for every single specimen, e.g. running herbarium number or collection number etc.
- Loan period
- Advice for handling

It may also be useful to make copies of the borrowed specimen labels, or other records and place them in the herbarium collection where specimens have been removed; this alerts other users that some specimens may be absent.

It is strongly recommended that incoming material from other herbaria or returned loans are deep frozen before further handling or reincorporation. Before acknowledging the enclosed form, the material must be checked against the given list. In the case of an incoming return of loan, determination/confirmation slips (hopefully added to the specimen) should be glued to the outside of the envelope to assist the correct filing on reincorporation of the specimen into the herbarium. The determination slip (with the names of taxon, revisor and date of revision) must be as small as possible to leave space for further revisions. Hand-written determination slips should be avoided.

Working steps for loan management

Outgoing loan

1. Search for requested material in the herbarium.
2. Carefully remove specimen from the herbarium sheet.
3. Copy the labels and place the copies at the site of the removed specimens, or, in case of an electronic management system, make the appropriate database entries.
4. Print loan forms and add them to the specimens and files respectively.

5. Pack the specimens.
6. Prepare the shipping papers and mail the parcel.

Outgoing return of loan

1. Check the borrowed specimens against original documentation.
2. Add (small !) determination/confirmation/annotation labels to every specimen (should already have been done by the person requesting the loan).
3. Print return-of-loan forms and add them to the specimens.
4. Pack the specimens.
5. Prepare the shipping papers and mail the parcel.

Incoming loan

1. Check the condition of the package and specimens, and compare the documentation with the actual content.
2. Send acknowledgement-form to the sender, retaining one copy, or acknowledge the receipt by e-mail.
3. Deep-freeze the parcels.
4. Hand over the specimens to the scientists and remind them to handle the specimens with great care and to keep to the terms and conditions of the loan.

Incoming return of loan

1. Check the condition of the package and specimens, and compare the documentation with the actual content.
2. Send acknowledgement-form to the sender, retaining one copy, or acknowledge the receipt by e-mail.
3. Deep-freeze the parcels.
4. Remount any damaged specimens.
5. Glue revision and confirmation labels on the outside of every envelope.
6. Remove the stored copies of labels and/or make the database entries and corrections.
7. Reincorporate the material into the herbarium.

Computer-Assisted Herbarium Management

Table 1. Some database systems for herbarium management and their internet address

FLORIN Information System is designed to deal with a wide range of data about plants: taxonomy and nomenclature, geographic distribution (incl. distribution maps generated automatically), herbarium and living collections, detailed information about plants collected in the wild, bibliographic data, plant images, etc.	http://www.florin.ru/florin/
The PANDORA taxonomic database system is designed for biodiversity research projects, such as floras or monographs, and is the official database used at the Royal Botanic Garden Edinburgh (RBGE) for taxonomic data sets. It can also be used for maintaining catalogues of collections such as herbarium specimens and a herbarium label printing system is included.	http://www.rbge.org.uk/pandora.home
SysTax is a botanical information system based on ORACLE. It supports a wide range of systematical work in botany and can also be used for the administration of botanical gardens, herbaria and other plant collections.	http://www.biologie.uni-ulm.de/systax/index.html
TRACY is a system for the management of herbarium collections. It is not a huge, all inclusive database tool. Rather it was specifically designed to facilitate rapid entry of specimen data by relatively un-skilled operators, and to provide mechanisms that allow complex queries of the data to be carried out with a minimum of training.	http://www.csdl.tamu.edu/FLORA/input/inputsys.html
The University of California Davis Herbarium Management System serves 4 functions: 1. Maintains herbarium incoming and outgoing shipments. 2. Creates herbarium labels and maintains that information. 3. Maintains herbarium library (books and journals) collections. 4. Maintains herbarium support society (e.g. creates mailing labels; membership lists)	http://herbarium.ucdavis.edu/herbaccess/databaseinfo.htm
BIOTA (The Biodiversity Database Manager) helps manage specimen-based biodiversity and collections data by providing an easy-to-use graphical interface to a fully relational database structure. Specimen loan management system, label-printing and label text export facilities are supported.	http://viceroy.eeb.uconn.edu/Biota
KE EMu - Electronic Museum management system for Museums, Art Galleries, Herbaria, and Botanic Gardens	http://www.ke.com.au/emu/index.html

Table 1. Continuous

MUSE - The KUNHM MUSE Project is an effort to provide software for the curation of natural history collections. It is designed explicitly to manage natural history collections and is based upon the experience of curators and collections managers. Built in taxonomic dictionaries are available for several taxonomic disciplines	http://www.biodiversity.uno.edu/muse/
BRAHMS (Botanical Research And Herbarium Management System) has been developed to support three closely related and overlapping activities: 1. the curation and management of botanical collections in herbaria (general accessions management, loans, exchanges, labels, determination slips and lists, visitor services, internet services, etc.). 2. the production of taxonomically oriented outputs (revisions, monographs, taxonomic checklists, taxonomic synopses, etc.). 3. the production of geographically oriented outputs (geographic checklists, floras, biodiversity surveys, etc.).	http://www.brahms.co.uk/

Database-systems for herbarium management

The need to manage extensive amounts of data has forced many herbaria to develop their own database systems. Some of these programs are now freely available via the internet. A collection of software developed for the purpose of databasing biological objects can be found under: http://www.bgbm.fu-berlin.de/TDWG/acc/Software.htm. The catalogue also contains database programs especially written for herbarium management (see Table 1).

Most of the cited programs also create and print out formatted labels from the database, probably their most common use. Some herbaria have developed one-off packages to service their particular needs, and these may be linked to other data-handling or -mapping programs. Databased label information allows retrieval of specimens using more criteria, which is of particular interest for lichen herbaria. For example, specimens often contain several species in addition to the one under whose name they are stored (e.g. lichenicolous fungi and their host, or two sparse thalli of rare, interwoven taxa, which can not be separated).

When starting a project to database all herbarium specimens, one should consider that within a herbarium the percentage of correctly determined species (and subspecies and varieties) varies from 100 % to less than 10 %, for example in some genera of Verrucariales or in leprarioid lichens. As even many genera of macrolichens (e.g. *Usnea*, *Bryoria*, *Cetre-*

Table 2. List of some lichen herbaria with access to their database via the internet (A comprehensive list of botany related URL's is kept under: http://www.botany.net/IDB/botany.html (Brach, A.R. & S. Liu (1996–)).

ASU (Tempe, U.S.A.)	http://mgd.NACSE.ORG/cgi-bin/qml2.0/arizona/arizonaHerb.qml
BG (Bergen, Norway)	see below under "Norwegian lichen database"
ESS (Essen, Germany	http://www.uni-essen.de/botanik/herb-bot.htm
HBG (Hamburg, Germany)	http://www.rrz.uni-hamburg.de/biologie/ialb/herbar/hbg_l2.htm
MIN (Minesota, U.S.A.)	http://www.tc.umn.edu/~wetmore/Herbarium/HERBHOME.htm
Norwegian Lichen Database. The database compiles six independent databases from four herbaria (BG, hb. Holien, O, UPS) and is able to create distribution maps.	http://www.toyen.uio.no/botanisk/bot-mus/lav/soklavhb.htm
O (Oslo, Norway) (lichen types; see also above under "Norwegian lichen database")	http://www.toyen.uio.no/botanisk/bot-mus/lav/sok_ltyp.htm
S (Stockholm, Sweden)	
species list of lichens	http://www.nrm.se/kbo/saml/lichen.html.en
type-database	http://www.nrm.se/kbo/saml/lavtyp.html.en
Olof Swartz' lichen types (scanned images of specimens and labels)	http://linnaeus.nrm.se/botany/kbo/sw/welcome.html.en
Erik Acharius' lichen types (scanned images of specimens and labels)	http://linnaeus.nrm.se/botany/kbo/ach/welcome.html.en
TSB (Trieste, Italy)	http://www.univ.trieste.it/cgi-bin/g/bot/leggi
UPS (Upsala, Sweden) (see also above under "Norwegian lichen database")	http://www.evolmuseum.uu.se/fytotek/
US (Washington, U.S.A.)	http://nmnhgoph.si.edu/gopher-menus/U.S.NationalHerbariumLichenTypeSpecimens.html
private herbarium of Professor M.R.D.Seaward (U.K.)	http://www.brad.ac.uk/acad/envsci/infostore/herbarium/LICHEN.html

Table 3. Additional useful internet links for (lichen-)herbarium-curators

Taxonomy, nomenclature, literature, identification:	
Search Recent Literature on Lichens	http://www.nhm.uia.no/botanisk/lav./RLL/RLL.HTM
DALI: Database of Ascomycete Literature	http://www.mycology.net.dali/search.html
The Authors Database (with their abbreviation)	http://www.herbaria.harvard.edu/Data/Author/author.html
International Code of Botanical Nomenclature	http://www.bgbm.fu-berlin.de/iapt/nomenclature/code/tokyo-e/default.htm
Index to "Index of Fungi"	http://NT.ars-grin.gov/fungaldatabases/databaseframe.cfm
Dictionary of Botanical Epithets	http://www.winternet.com/~chuckg/dictionary.html
Searchable List of Names in the Parmelioid Genera (Lichens)	http://persoon.si.edu/parmeliaceae/
Lichen determination keys available on INTERNET	http://www.bgbm.fu-berlin.de/bgbm/staff/wiss/Sipman+H/keys/default.htm
Online identification of lichens	http://www.mycology.net/lias/index.html
Lichens of the British Isles: A list of published maps (M.R.D.SEAWARD)	http://www.brad.ac.uk/acad/envsci/infostore/herbarium/database.htm
Geography (important tools on creating label texts, on searching data of localities or on drawing distribution maps):	
How far is it? This service uses data from the US Census and a supplementary list of cities around the world to find the latitude and longitude of two places, and then calculates the distance between them (as the crow flies). It also provides a map showing the two places, using the Xerox PARC Map Server (see below)	http://www.indo.com/distance/
The Xerox PARC Map Viewer is a World-Wide Web HTTP server that accepts requests for a World or USA map and returns an HTML document including an image of the requested map. Each map image is created on demand from a geographic database.	http://mapweb.parc.xerox.com/map/color=1/features=alltypes/ht=45.00/iht=450/iwd=800/

Table 3. Continuous

Versamap for Windows is suitable for drawing maps of continents or large countries	http://www.versamap.com/webdoc10.htm
The GEOnet Names Server (GNS) provides access to the National Imagery and Mapping Agency's (NIMA) database of foreign geographic feature names.	http://164.214.2.59/gns/html/index.html
Geographic Names Information System (United States and Territories)	http://mapping.usgs.gov/www/gnis/gnisform.html
Coordinates of worlds bigger cities	http://www.gwdg.de/~unolte/AVG/lexikon/tab13b.html
Miscellaneous useful links and access to e-mail addresses and homepages:	
Index Herbariorum !! (updated every few months)	http://www.nybg.org/bsci/ih/ih.html
Lichenologists' and Mycologists' E-mail Address List	http://www.mycology.net.index.html
IAL (International Association of Lichenologists). New entries, cancellations and changes of address, phone and fax number, e-mail, homepage and scientific interests are personally done by the lichenologist)	http://www.botany.hawaii.edu/lichen/default.htm
Electronic discussion group devoted to lichens.	send mail "subscribe LICHENS-L Your Name" to: listproc@hawaii.edu
English German dictionary	http://dict.leo.org/
EURODICAUTOM (a multilingual terminological database which contains technical terms, abbreviations, acronyms and phraseology)	http://eurodic.ip.lu/cgi-bin/edicbin/EuroDicWWW.pl
Searchable database of Universities around the world	http://geowww.uibk.ac.at/univ/

lia, Lobaria etc.) include chemically defined taxa, many determinations (particularly those without any chemical investigations) should be accepted with some hesitation.

Maintaining the database involves not only adding new data but also the rather time-consuming and expensive task of keeping the whole system up to date. This includes correction of revised taxa, installing new software updates and upgrades, and shifting to more advanced database-systems.

Herbaria on-line

As internet data transfer becomes more advanced, more herbaria are offering online access to their database systems. An incomplete compilation of important lichen herbaria that are online is presented in Table 2. Lists of stored taxa and very detailed data (including label text etc.) can be requested. Type databases are especially useful for taxonomists and strong efforts should be undertaken to connect all such databases to a single network. Some institutions are already striving to connect database systems of different biological collections for large geographical areas, e.g. "NatureWeb" for central Europe (http://www.natureweb.at/), and "BioCISE" for the European community (http://www.bgbm.fu-berlin.de/biocise/).

Other uses of the internet

In addition to the cited internet addresses for database systems (Table 1) and online-herbaria (Table 2), Table 3 presents some further useful links for curators of lichen herbaria.

Acknowledgements. I would like to express sincere thanks to Dr. Gintaras Kantvilas and to anonymous referees for many valuable hints and advice and to the former also for the linguistic improvement of the manuscript.

References

Foreman L, Bridson D (1998) The Herbarium Handbook, 3rd Edition. Kew, RBG.

Hawksworth DL (1974) Mycologist's Handbook, an Introduction to the Principles of Taxonomy and Nomenclature in Fungi and Lichens. Commonwealth Mycological Institute, Kew Surrey England.

Holmgren PK, Holmgren, NH, Barnett LC (1990) Index Herbariorum Part I The herbaria of the world Edition 8. Regnum Vegetabile Vol 120

Metsger DA, Byers SC [eds.] (1999) Managing the modern Herbarium An inter-disciplinary approach. Society for the preservation of natural history collections. Elton-Wolf Publishing, Vancouver, Canada.

Moberg R, Holmsen I (1992) Flechten von Nord- und Mitteleuropa, Ein Bestimmungsbuch. Gustav Fischer Verlag, Stuttgart, Jena, New York

Wirth V (1995) Flechtenflora, Bestimmung und ökologische Kennzeichnung der Flechten Südwestdeutschlands und angrenzender Gebiete. Verlag Eugen Ulmer, Stuttgart

Documentation of Lichens by Macrophotography

VOLKMAR WIRTH

Introduction

While written descriptions of lichens can characterize such features as spore length or width, they are less satisfactory for describing colour or habit. Therefore it is very helpful to append a photograph of the gross morphology of a lichen to the description. The traditional advantage of drawings over photographs is that photographs taken at high magnification have a small depth of field which can be overcome by drawing the object. However, a series of partially focused macrophotos (or microphotos) give all the information required. Modern software can even scan and automatically combine the most in-focus regions from several partially focused source images taken at different focussing points (e.g. "Auto-Montage" by Synoptics Ltd.). The main problems associated with the macrophotography of lichens are their small size and their lack of contrast.

High quality photographic equipment will not guarantee a good picture in a broader sense. Even in scientific photography, selection of subject area, picture composition and aesthetics are important for the impression that a picture produces. It is more difficult to give advice on these aspects than for merely technical points. The photographer may have a natural feeling for what makes a good photograph. More likely, he or she will have to learn by trial and error, by critical judging of the prints or slides, or by comparing his or her work with good pictures taken by other photographers. Practical criticism from an experienced photographer is a sure shortcut to developing a successful technique.

Volkmar Wirth, Staatl. Museum für Naturkunde, Erbprinzenstr. 13, 76133 Karlsruhe, Germany (*phone* +49-711-8936202; *fax* +49-711-8936100; *e-mail* vwirth@gmx.de)

Materials

The basic equipment includes the camera body, macro lenses, tripod, auto extension tubes, auto bellows and electronic flash units.

Camera

The single lens reflex camera is best for the macrophotography of lichens. The subject matter appears through the lens exactly as it will on the film. Usually, the 35-mm camera, delivering slides or negatives with an area of 24 x 36 mm, is sufficient. Good quality films rated at speeds of 100 ASA or less have more than enough resolving power to allow excellent reproduction at the dimensions usually required by scientific publications. Large format cameras, for example those delivering 6 x 6 cm negatives can be useful for overview pictures rich in details, e.g. for fruticose lichen vegetation or lichen mosaics on tree trunks or large rock faces. However, they are often not suited for pictures of small lichen individuals. The film area may exceed the subject area, and the area surrounding the subject often gives no additional information, is out of focus or disturbs the composition of the picture. Choose a camera that will allow you to select the aperture manually and automatically adjusts exposure time. Make sure that you can override the automatic exposure to compensate for expected under- or over-exposure of important subjects parts. This will be required, for example, when you are photographing a black lichen thallus on white limestone substrate or white filaments of a beard lichen on a dark background. Modern single lens reflex cameras allow "through the lens" (TTL) metering of natural or flash lighting, so that changing lenses, bellows or filters creates no problems for estimating exposure times. Most cameras today are equipped with an automatic focusing system, but make sure you can switch this off because it does not work properly at close distances. The viewfinder usually is equipped with a focusing screen with a central microprism spot or a central horizontally oriented split-image spot. In macrophotography the split image spot is inconvenient as it does not work at close distance and appear often as a dark spot. For macrophotography a mat Fresnel field with no spot or special Fresnel fields for high-magnification applications are best. In some cameras the focusing screen is interchangeable.

Lenses

No single macro lens is capable of delivering optimum results for the entire range of magnifications usually needed. For magnifications of up to 0.5 x or (with some firms also) 1 x ("close up" photography) ordinary macro lenses are available which fit the camera body without any additional equipment. They can focus subjects at infinity, but offer highest resolutions in the macro range. If these macro lenses have focal lengths of 50 or 60 mm they can also be used for normal photography, e.g. landscape shots. For close-up photos their often relatively low maximum apertures of f 1:4 or f 1:3.5 are of no consequence. Macro lenses with longer focal length (e.g. 80 to 105 mm) have the advantage that they provide a greater "working" distance, i.e. distance from the end of the lens to the subject. This may make it easier to arrange suitable lighting or position of the tripod. For magnifications higher than 0.5 or 1 x (strictly, the term "macrophotography" should be used only for magnifications greater than 1 x) it is necessary to use extension accessories such as extension tubes or auto bellows. Macro lenses are available that are constructed for use only in conjunction with these accessories. They can extend the magnification up to 20 x and more. The higher the magnifying power of the lens, the shorter the focal length. For example, Olympus provided an 80 mm macro lens for magnifications from 0.5 to 2 x, a 38 mm macro for magnifications from 2 to 6 x, and a 20 mm macro for 6 to 12 x. The latter magnification corresponds to a subject area of 3 x 2 mm. Minolta macro lenses exist with focal distances of 25 and 12 mm which allow magnifications from 3.2 to 20.5 x. Canon offers 35 mm and 20 mm macro lenses with magnification from 2 to 10 x. At magnifications of 15 to 20 x the working distance is extremely short and positioning and illumination of the subjects becomes very difficult.

Extension tubes, auto bellows, and telescopic auto tubes

As discussed above, ordinary macro lenses can only be used at high magnifications when connected to extension tubes, bellows or a telescopic tube. Telescopic tubes and bellows offer continuous extension, i.e. a continuous change of subject distance and thus magnification. Olympus were the first to offer a telescopic auto tube. This can be a useful and lightweight accessory for hand-held shooting. Unfortunately, it does not allow use of all magnifications when used in conjunction with the different macrolenses offered. The maximum extension is too short to reach the lowest magnification of the "next" most powerful macrolens. This problem (which can be reduced by adding further extension tubes) also occurs with relatively short and handy bellows. However, you can avoid this problem by using conventional bellows made by the same manufacturer as your lenses; in this case the focussing rail will be long enough to ensure overlapping of the magnification ranges of the macro lenses. Bellows are usually unsuitable for hand-held shooting because they do not offer automatic diaphragm linkage. However, bellows made by Novoflex and Leica have a double-cable release, allowing automatic stop-down to the preselected lens aperture at the moment of exposure.

Tripod

Macrophotography often involves using high magnifications and small lens apertures, and the resulting exposure times often exceed one second. Using natural lighting at magnifications of greater than 0.5 x requires a tripod to provide a satisfactory depth of field, although you can hand hold the camera if you are using an electronic flash. Even with flash, the chance of moving the camera out of focus increases considerably with increasing magnification. Taking photographs at magnifications of 1 x or more requires the use of a tripod unless you can securely rest your camera or arms. The only exceptions may be subjects that have very large depths. Here, irrespective of the precise focus set, it is highly likely that some important elements can be portrayed sharply, e.g. the podetia of fruticose species like reindeer lichens. The tripod must be strong enough to support a camera that may have a heavy extension system, so cheap and light tripods are inadequate. The tripod should allow the camera to be positioned close to the soil surface by spreading wide the tripod legs, although only a few products allow this. Obtaining sharp photographs at magnifications higher than c. 4 x in the field is often rather time consuming and difficult even with a good tripod. Even a tiny movement will take the subject out of focus. It is advisable in these cases to take the subject into the laboratory, particularly if a copy stand is available.

Electronic flash lighting

There are advantages of using natural lighting for photographing lichens. For example, subjects such as fruticose or beard lichens on twigs look better in front of a naturally illuminated background, whether this be a landscape or the sky. However, natural lighting can create rather sharp contrasts in your subject. Flash lighting makes you independent of ambient conditions, and enables you to control lighting angle and contrast (see below, and Figures 1-6). The effects can be tested easily by shooting series of pictures with different lighting conditions. You will need to use auxiliary lighting if working indoors, or outdoors on a cloudy day or in the shade, otherwise very long exposure times will be needed. Using long exposure times creates several problems. First, it is easy for the camera to move slightly during exposure, particularly if the tripod is resting on soft ground. Secondly, beyond certain exposure times colours recorded by the film shift towards blue or blue-green regions of the spectrum (the Schwarzschild effect). These problems can be solved by using electronic flash. Three main kinds of flash units exist: a single spot flash, twin flashes (two flash units fixed by brackets on both sides of the camera or the lens) and the ring flash. The ring flash fits conveniently around the front of the camera lens and illuminates the subject from all sides; some products allow a differential illumination by cutting off segments of the ring. When using modern cameras, it is as easy to estimate the correct exposure when using flash as for natural light. Pioneered by Olympus, the TTL "off-the-film" centralised control flash system measures correct flash exposure directly at the film plane from inside the camera. This will give the correct flash exposures automatically, regardless of lens apertures. Using electronic flash equipment from the same manufacturer as the camera should guarantee perfect TTL exposure control. Even some units that estimate exposure control using a sensor can offer TTL exposure control with an adapter.

a

b

Fig. 1. Crustose lichen (*Acarospora sinopica*). *a*: with single flash unit from above left, shadows help to pronounce surface unevenness and cracks; *b*: twin flash lighting with left flash closer to the subject diminishes shadow effects but still produces enough relief

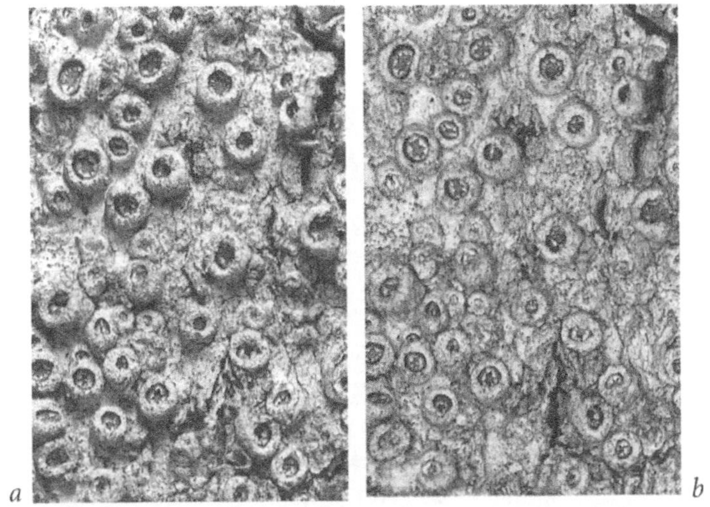

Fig. 2. Crustose lichen (*Thelotrema lepadinum*). *a*: with single flash unit from above; *b*: with ring flash lighting. The difference is striking; the shadowless picture does not give any impression of the plasticity of the fruiting body

Using a stereomicroscope

If you wish to use a reflected light microscope, it is often possible to attach your single lens reflex camera directly to the microscope, and thus continue to enjoy TTL light metering. The stereomicroscope usually does not offer the possibility of choosing different apertures.

Procedure

Subject

Selecting the right subject is very important for the final quality of the picture. Apart from aesthetic considerations, choosing the right subject can help you to minimise problems associated with depth of field or satisfying illumination. A crustose lichen on an even flat surface is better suited to show the characteristics of the lichen than one on a very uneven relief. When photographing podetia of *Cladonia* or the fruiting bodies of *Baeomyces* or *Dibaeis* you should choose specimens that have more than one of these structures in a suitable plane of focus.

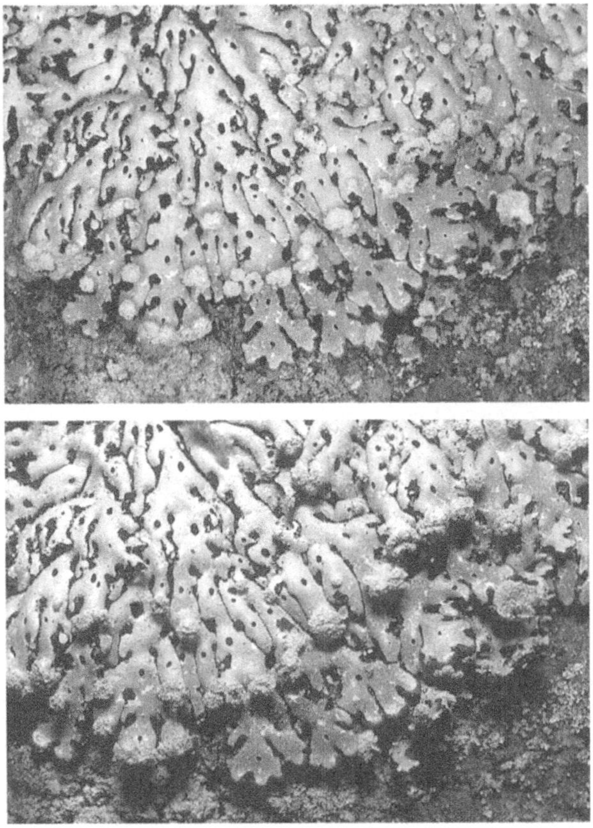

a

b

Fig. 3. Foliose lichen (*Menegazzia terebrata*). *a*: with ring flash lighting, the subject appearing unnaturally flat; *b*: with single flash unit from above left, the subject showing structures more naturally

Focussing and depth of field

Generally the subject plane should be parallel to the film plane. If you are photographing deep objects, it is usually desirable to get a large depth of field by using a small aperture. Position the important elements of the lichen in the foreground; it is usually better to have the foreground in sharp focus and the background out of focus. For example, keep the mushroom-like fruiting bodies of a large thallus of *Baeomyces* or the pseudopodetia of *Stereocaulon pileatum* in the front of your composition, and sharply in focus. You can use an out of focus background to document

a

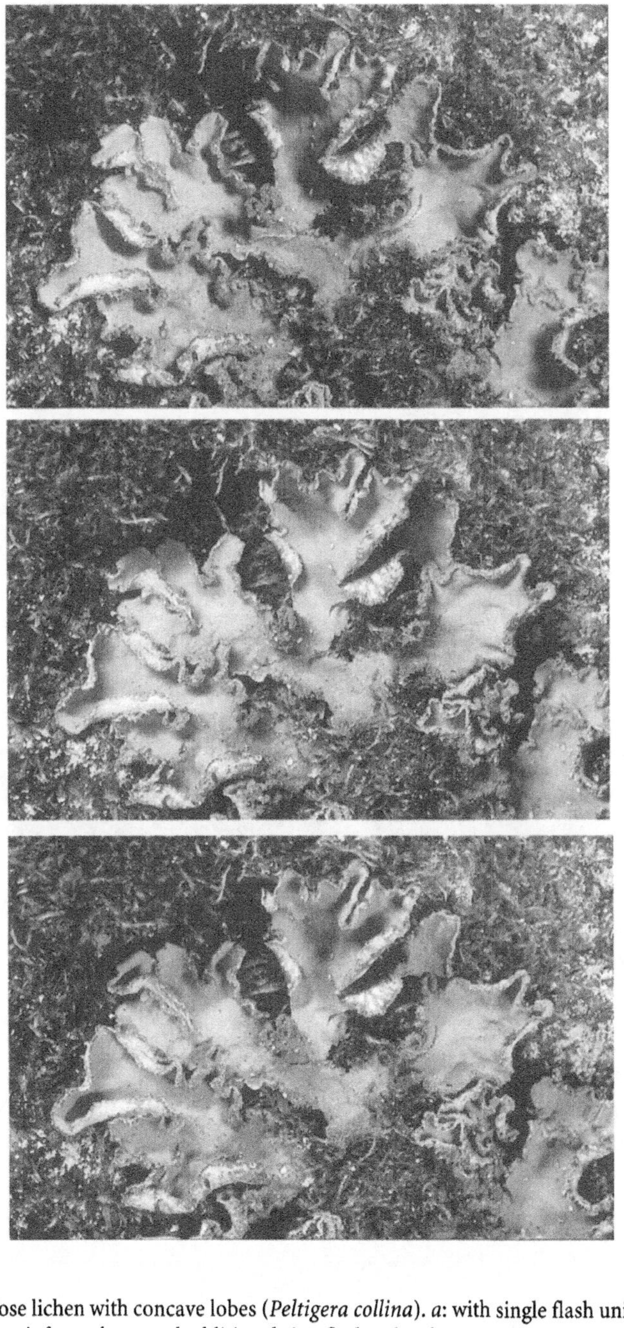

b

c

Fig. 4. Foliose lichen with concave lobes (*Peltigera collina*). *a*: with single flash unit; *b*: with single flash unit from above and additional ring flash to brighten up shadows; *c*: inadequate twin flash lighting, leaving unnatural shadows in the depth of concave structures

a

b

Fig. 5. Fruticose lichen (*Cladonia cenotea*). *a*: with ring flash lighting which offers for this type of growth form an adequate illumination; because of the depth of the subject no "flat" impression occurs; *b*: with natural light conditions producing too extended shadows

the extension of the thallus. Out of focus objects in the foreground of a photograph will spoil the overall appearance of the picture. Very occasionally it may be desirable to use a wider aperture, and thus a shallow depth of field. For example, you may need to simplify a background rich in disturbing details or emphasise the main object in focus.

Illumination

Illumination is much more important than supposed by inexperienced photographers. At low magnifications natural lighting is still an alternative to flash lighting. However, bright sunshine can produce surprisingly high contrast between the light and dark parts of a photograph (Figure 6). Conversely, diffuse light may produce uniform illumination and your subject may lack detail. To avoid unwelcome surprises, observe the distribution of light and shadow very carefully in the viewfinder. Flash lighting makes you independent of subjective assessment decisions of contrast and shadow pattern that occur when you are working with natural light. Experience will tell you which flash equipment is best suited for your particular shot and at which angle it should be positioned. Using a ring flash will give very different results to a spot light. Many photographers prefer a ring flash or a twin flash because they allow you to hold the camera with both hands. This is also possible with a single spot flash if you mount the flash unit directly on the camera or on a bracket, but you can only vary the flash angle if you have a flexible bracket. Mounting the flash unit directly on the camera (e.g. on the flash shoe above the view finder) gives you compact and convenient equipment for macrophotography but the results may be unsatisfactory. When photographing objects close to the camera the lens barrel may shade part of your subject.

The ring flash will guarantee a homogeneous lighting almost totally lacking shadows. This means that the surface structures appear flat (Figure 3), so that flash is therefore less suited to subjects like crustose and some foliose lichens. Areolation will appear indistinct, and isidia and soralia will lack contrast. On the other hand the ring flash is very useful for subjects rich in three-dimensional structures, e.g. *Cladonia*, *Cetraria* or *Stereocaulon* species (Figure 5). It is also the best kind of flash to use when working at very close distances. Working with twin flash gives similar effects but gives a little more shadow (Figure 1), and this often shows surface structures more distinctly than a ring flash.

In many situations you can get excellent results just with one single flash unit. Varying the angle of the flash unit on the subject allows you to control the length and direction of the shadows in your photograph. This is particularly useful for emphasising surface features of crustose or foliose lichens (Figure 1, 6). However, holding the flash at too high an angle will make the shadows too large, and they may hide important subject parts and also spoil the photograph from an aesthetic point of view.

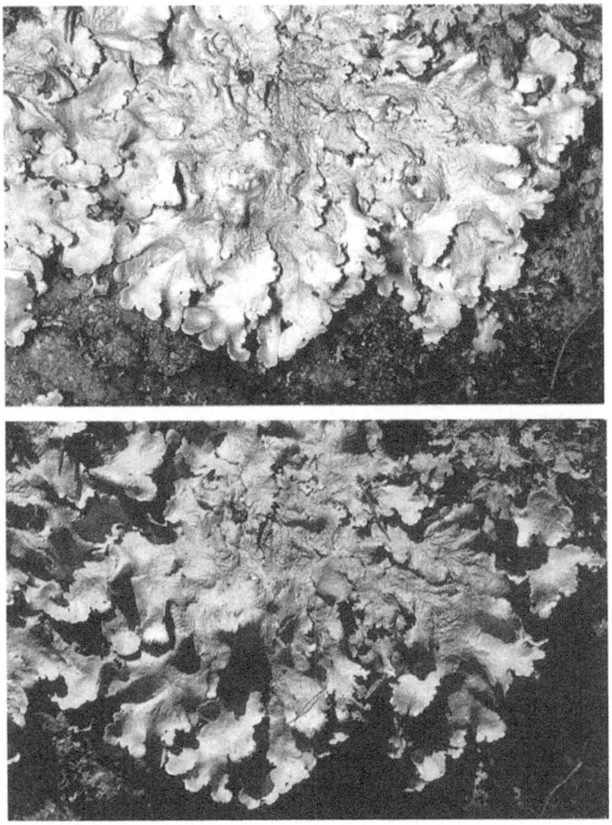

a

b

Fig. 6. Foliose lichen (*Parmelia caperata*). *a*: with single flash unit positioned at small/close angle to the camera, producing little shadow; *b*: natural lighting where you must accept structure hiding shadows

Microphotography

This chapter only documents the photography of lichens by macrophotography. For questions on microphotography I recommend that you look at the catalogues and user manuals of firms producing microscopes. You can use the same camera body that you use for macrophotography, and here TTL-measuring of exposure time with flash is very beneficial. At very high magnification, blur can occur not only as a result of vibration, even that caused by molecular motion, and so use of flash is often necessary. Microscope manufacturers may offer micro flash equipment, or you can construct your own using instructions published in special magazines for microscopy and microphotography e.g. Mikrokosmos.

Comments

Literature

Further up to date advice is provided by books on close-up and macrophotography, and there are also books dedicated to one specific manufacturer. I recommend that you read a standard book on macrophotography before buying any equipment. The decision on which camera you buy will to some extent determine the choice of accessories for macrophotography. Ask successful photographers and your local photo shop staff which equipment they can recommend for your intended purposes. You will have less choice if you want to take photographs at high magnifications. Before you actually start taking photographs, consider the advice given here. However, ultimately nothing can replace experimenting with your equipment, for example, by taking a wide range of photographs of the same object.

Acknowledgements. I am indebted to Mr. Klaus Paysan for discussing the manuscript.

Computer-Aided Identification Systems for Biology, with Particular Reference to Lichens

GERHARD RAMBOLD

Introduction

The process of attributing the correct name to objects or organisms is called "identification". For identifying and accessing desired names, descriptive data of morphological, anatomical, ecological, chemical, or molecular characters need to be checked and compared with the data of objects with known identity. Such information can be obtained from conserved specimens, illustrations, taxonomic descriptions, data tables, diagnostic keys, or computer-aided identification tools. Until recently, the most popular tools for identification were printed keys of di- or polychotomous structure, published in taxonomic revisions, monographs, and floristic or faunistic articles or books. Such keys mostly do not reflect the natural classification of organisms under consideration and are therefore called 'artificial' keys. The first published document including this type of key for plant identification, is Lamarck's *Flore Françoise* (Lamarck 1779). Arrangements of names of lichen genera, to some extent being equivalent to diagnostic keys, date back to E. Acharius, e. g., in the *'Dianome lichenum'* of his *Lichenographiae sueciae prodromus* or the *'Generum dianome'* in *Lichenographia universalis* (Acharius 1798, 1810). Due to the fixed sequence of character presentation, diagnostic keys enforce a stepwise process during identification, and neither allows free character selection or the option to handle unknown, rare, or ambiguous characters. Conventional keys may provide a more or less clear impression of the respective organismal group to the expert user, but their application is frequently considered inconvenient by non-taxonomists. Indeed, due to a lack or bad development of characters, even specialists may have difficulties when applying taxonomic keys. The introduction of synoptic or

Gerhard Rambold, Universität Bayreuth, Lehrstuhl für Pflanzensystematik, Universitätsstraße 30, 95447 Bayreuth, Germany (*phone* +49-921-55-2453; *e-mail* Gerhard.Rambold@uni-bayreuth.de)

tabular keys, and subsequently of edge- or body-punched key cards (Hall and Johnston 1954, Hansen and Rahn 1969), were promising attempts to overcome the shortcomings of conventional keys and facilitate the identification process by making free character selection possible. Punched card systems proved to be especially useful for identifying fragmentary material. However, for various reasons, these systems were set up for only a few taxonomic groups and published at rather low edition numbers.

The application of computers for identification in biology opened unforeseen possibilities. It started in the early sixties, shortly after the introduction of high-level computer languages. Boughey et al. (1968) and Morse (1974) were the first to use programs for interactive identification with a monitor as user interface. An early example of a taxonomic online identification system, the XPER program (Lebbe 1984), was applied to various organisms, including mushrooms and toadstools (see Pankhurst 1991). Now, a substantial number of highly sophisticated computer applications for specimen identification have been developed, and represent genuine alternatives to conventional identification keys. They are used in various biological disciplines as – for instance – in lichenology. The advantages of interactive keys over conventional keys are pointed out in detail by Dallwitz (1997 onwards), Dallwitz et al. (1998), and Pankhurst (1998). A relatively early example of computerized data for identification in lichenology is the key of Zeltyn and Pchelkin (1993) for epiphytic lichens in North-western and Central Asia (see Insarov et al. 1997). Data sets for various fungal groups, including foliose and fruticose lichen genera (*Bryoria*, *Cladonia*, *Hypogymnia*, and *Peltigera* of the North-western United States) and selected crustose lichen species, established some years ago, were distributed by F. Rhoades on diskette and now are available via the Internet (Pasarell 1999). These data are to be used with the programs 'Taxon' (for key creation, data editing and access), and 'Askataxa' (for identification and data retrieval). Descriptive information on lichens from the Negev Desert, suitable for import into most database applications, is published on diskette in dBase format by Insarov and co-workers (Insarov et al. 1997). The British Lichen Society published a CD-ROM with information on characters, a key and illustration of *Parmelia* and related genera on the British Islands (The British Lichen Society 1998). A recently distributed CD-ROM on epiphytic lichens from Central Europe includes data of about sixty species (Kirschbaum and Wirth 1998) and contains two interactive keys, one with free character selection for identification of about 150 species. Not only data collections for computer-aided identification of lichens and lichenicolous fungi, but also databases for identifying secondary compounds of lichens have been set up. Elix et al. (1988) developed a computer program that facilitates the identification of lichen substances using TLC data. The program uses R_f values and ranges of possible spot colours in standard solvent systems, and generates a list of fitting names of compounds.

Rather early on in its development, the community of lichenologists recognised the potential value of the Internet as a medium for publishing and exchanging taxonomic data. Recently, quite a number of WWW sites, exclusively dedicated to lichens, have been established. Web sites dedicated to checklists and databases with lichen names and records → exist (http://www.checklists.de) or the Mediterranean (Grube 1998-2001). Such checklists for defined operational geographical units may be used as tools for identification as well. HTML-formatted diagnostic keys with internal links as key markers for online use or printouts are accessible on various web pages as well. Their use, via display, may be considered as 'interactive' in the sense that the user needs to click internal links to move forwards and backwards within the key document, but these keys lack the advantages of true interactive identification. Very helpful keys of this kind exist for lichen genera of the Neotropics, the genera of the Guianas (tropical South America). Species of selected genera from various regions of the world are found on the 'Lichen determination keys available on Internet' page of Sipman (1998). Wetmore (1999) presents keys to some species groups of *Caloplaca*. A 'Key to the genus *Parmelia* in Great Britain and Ireland' (The British Lichen Society 1997) is also accessible via Internet and part of a CD-ROM distributed by the Society.

Interactive identification with server-sided databases offers the 'LichenLand' site (Pittam and Hanus 1997) with a 'Lichen Synoptic Key' which includes about 90 species of nearly 50 genera, and uses static HTML forms of the web-based graphic interface HyperSQL for character selection under a Unix operating system. The LIAS web site (Botanische Staatssammlung München 1995-2001) includes an online database with generic data of 820 lichenized and lichenicolous ascomycete genera (Rambold and Triebel 1997-2001) as well as species-level data for various lichen groups. A list of more than morphological, anatomical, chemical and ecological categorical and numerical characters is being extended under international participation. The database contents can be modified and completed by co-operating lichen experts via dynamically generated, HTML-formatted data revision and entry submission forms.

Structuring descriptive data for computer-aided use

The most common way for taxonomists to collect and store biological data has been for them to write their observations on paper or in form sheets in a more or less structured way. Today, most taxonomists store records in PC-based databases. Such electronic repositories potentially allow sophisticated data retrieval, the generation of data tables and production of various types of formatted reports. They may also be used as a starting point for traditionally established taxonomic descriptions or printed keys for identification.

The interchange format for descriptive data, DELTA (DEscription Language for TAxonomy), introduced by Dallwitz (1980), allows the use of data for a wide range of applications. DELTA describes taxonomic objects in a format that can be used for generating taxonomic descriptions in natural language and printed polychotomous keys, as well as for computer-aided interactive identification (Dallwitz and Paine 1993 onwards). The format is widely used today and was endorsed as an international data standard by the Taxonomic Database Working Group (TDWG), a subdivision of the IUBS. The coding of text, multi-state characters and numerical data follows rather simple principles and thus remains readable to some extent. Editors for DELTA exist for DOS platform, e.g., Taxasoft (Gouda 1998) and Dedit DOS which is part of Pankey program suite (Pankhurst 1988, 1999), and for Windows® 95/98/NT platforms, e.g., the editor of CSIRO-Delta program package (Dallwitz et al. 1993 onwards). These editors make working with this coding language data very convenient and do not confront users unnecessarily with the original coding language, but enable them to focus attention on the essentials of data capture. DELTA can be recommended as a most useful tool for storing descriptive data, and is used in the present article to exemplify the setting up of databases for interactive identification. Data can also be imported into software for statistic or cladistic analyses.

Procedure

Setting up a DELTA-based data collection

To set up a Delta database, it is recommended to create first a preliminary list of characters with a standard text editor or one of the CSIRO-Delta editors mentioned above.

Selecting and decomposing the characters

It depends on the purpose of the database as to which characters should be included. For determination work in the field, mainly ecological, geographical and morphological traits or chemical spot reactions may be relevant, while characters like micro-anatomical details or secondary metabolites, which can only be assessed with specific technical equipment, may not be desirable.

The properly structured decomposition of characters is a major precondition for the establishment of identification tools that work well. According to Diederich (1997), morphological and other kinds of characters may be decomposed into three major elements (Table 1):

- the structure
- the property, and
- the character states.

For example, the character statement 'ascospores large' includes 'ascospores' as a morphological/anatomical structure and the attribute 'large' as the character state, entailing the character property 'size'. It is evident, that the character property term is implicit and usually needs not to be present in the character statement itself. In the statement 'apothecia margin white', for instance, the property term 'colour' does not need to occur.

Structural units of morpho-anatomical and other characters may be represented by single terms, e.g. 'apothecia', or composed terms, e.g. 'ascospores wall'. While the number of morphological structures is rather infinite, the number of basic properties to be used for specifying structures is limited. Diederich (1997) provides a list of such basic properties for morphological and anatomical characters, grouped into four categories:

- **Appearance:** presence, shape, kind (distinctive trait), colour, texture, arrangement, symmetry;
- **Placement:** position relative to (different structure), orientation, angle with (different structure); distance to (different structure);
- **Measurement:** length, height, width, diameter, depth, weight, ratio of (structure), size;
- **Quantity:** number, quantity.

An extended list of properties is located on the LIAS web site of the Botanische Staatssammlung München, 1995-2001. Basic properties are determined by the kind of character states. For instance, the basic property 'shape' is entailed by character states like 'ellipsoid' or 'globose', but not 'green'.

Table 1. Example of character decomposition according to Diederich et al. (1997)

Structure	Substructure	(Basic) property	Character states
Apothecia	–	presence	1) absent – 2) present
Apothecia	disc	colour	1) black – 2) brown – 3) yellow

Arranging the characters

The characters recognised as useful for identification need to be grouped and arranged in a suitable order, for instance, according to the analysis techniques to be applied or the morphological topology (which may correspond with the first criterion). An appropriate character grouping would be: Taxonomy - Ecology - Geography - Morphology - Anatomy - Chemistry - Photobionts. If possible, substructures should be ordered hierarchically within the same character complex (in this example: 'ascospores'):
Ascospores <presence>
Ascospores wall <thickness>
Ascospores wall surface <structure>
Ascospores wall surface warts <kind>

If more than one property are applied to one structure (in this example: 'Ascospores wall surface warts'), the sequence of property names should always remain the same, for instance, following the list of Diederich (1997): presence, density, size, colour, etc.
Ascospores wall surface warts <presence>
Ascospores wall surface warts <density>
Ascospores wall surface warts <size>
Ascospores wall surface warts <colour>

DELTA distinguishes between five major character types:

- * **Unordered multistate characters** for categorical characters with two or more states, which may include two or more states and need not or cannot be brought in a natural order of sequence.
  Example: Ascospores wall surface <texture>/
  1) smooth/
  2) porate/
  3) rugose/

- **Ordered multistate characters** for categorical characters with more than two states and following a natural order of sequence.
  Example: Ascomata <exposure>/
  1) immersed/
  2) subimmersed/
  3) adnate/

4) substiptate/
5) stipitate/

Note: The best way to test, whether the sequence of character states is ordered or not, is to join pairs of states by the term 'to' and ask whether the statement implies the inclusion of the intermediate states or not. With respect to the selected example this is the case, because 'immersed to stipitate' includes the intermediate states 'subimmersed', 'adnate' and 'substiptate'.

- **Integers**, being numerical values resulting from counts. (Note, that mean values of integers may be real numerics!)
  Example: Ascospores <number>/
- **Real numerics**, being numerical values derived from measurements and are often measured by units, in the present example: 'µm'.
  Example: Ascospores <length>/ µm/
- Text characters may include any kind of textual data.
  Example: Phorophyte <name>/

It is possible, but **not** recommended to use characters with a combination of two or more basic properties:
Example: Ascospores septation <presence, kind>/
1) absent/
2) parallel 1-septate/
3) parallel 2-5-septate/
4) muriform/

Depending on the taxonomic group under consideration and the purpose of the identification key, the list of characters of a database project may comprise up to several hundred characters. Table 2 provides the example of a short character listing for crustose lichens in the DELTA format.

The short phrases proceeded by an asterisk symbol are so-called DELTA directives. The numbering symbol ('#') and the delimiters ('/') are obligatory formatting tags of the coding language. Basic property terms are enclosed in angular brackets.

Table 2. DELTA example file 'items' with characters of lichenized fungi

```
*SHOW Lichenized Fungi - Chars                    REVISED 990131_09:12

*CHARACTER LIST
#1. references:/
#2. notes:/
#3. occurrence: <name>/
   1. Africa/
   2. Antarctic/
   3. Asia-Temperate/
   4. Australasia/
   5. Europe/
   6. Northern America/
   7. Southern America/
   8. Pacific/
   9. Asia-Tropical/
#4. altitudal classification: <category>/
   1. coastal/
   2. colline/
   3. submontane/
   4. montane/
   5. subalpine/
   6. alpine/
#5. substratum: <kind>/
   1. soil/
   2. bryophytes/
   3. lichens/
   4. wood/
   5. bark/
   6. non-calciferous rock/
   7. calciferous rock/
   8. living leaves/
```

Table 2. Continuous

*SHOW Lichenized Fungi - Chars	REVISED 990131_09:12

#6. thallus <kind>/

1. endosubstratic/
2. episubstratic/

#7. thallus surface <structure>/

1. smooth/
2. rough/
3. byssoid/
4. pulverulaceous/
5. farinose/

#8. ascomata <diameter>/

mm in diam./

#9. ascomata <exposure>/

1. immersed/
2. semi-immersed/
3. sessile/
4. substipitate/
5. stipitate/

#10. ascomata margin <development>/

1. indistinct/
2. slightly prominent/
3. prominent/

#11. ascomata lower layers colour (HSB value) <range>/

#12. asci wall layers apex inner layers amyloid reactivity <pattern>/

1. with amyloid tube/
2. with amyloid cap/
3. entirely amyloid/

#13. ascospores <number>/

per ascus/

#14. ascospores <length>/

µm long/

Table 2. Continuous

*SHOW Lichenized Fungi - Chars	REVISED 990131_09:12
#15. ascospores septation <presence>/	
1. absent/	
2. present/	
#16. ascospores septation <kind>/	
1. parallel 1-septate/	
2. parallel 2-5-septate/	
3. muriform/	
#17. ascospores wall surface <texture>/	
1. smooth/	
2. porate/	
3. rugose/	
#18. primary photobiont <kind>/	
1. cyanobacterioid/	
2. chlorococcoid/	
3. trentepohlioid/	

Setting up the character specifications

For subsequent import in the Delta editor, database specifications, stored in a file named 'specs' are needed. (If the character list is entered directly into the editor, this file is generated automatically.) An example of a minimum specification is given here:

*SHOW *taxon* - SPECIFICATIONS	REVISED 990131_09:12
*NUMBER OF CHARACTERS 18	
*MAXIMUM NUMBER OF STATES 9	
*CHARACTER TYPES 1-2,TE 8,RN 9-10,OM 11,TE 13,IN 14,RN	
*NUMBERS OF STATES 3,9 4,6 5,8 7,5 9,5 10,3 12,3 16-17,3 18,3	

The first DELTA directive 'SHOW' (proceeded by an asterisk symbol) is optional but useful for labelling the file. The second one refers to the number of character present in the 'chars' file. The MAXIMUM NUMBER OF STATES directive is followed by the maximum number of states occurring in a character of the chars file. [*Example*: the maximum number of states is 9 (occurring in character #3).] Character types are coded as follows: unordered multi-state character: UM, ordered multi-state: OM, integer values: IN, real numerics: RN, text: TE. Character numbers are followed by comma and the relevant abbreviation or symbol. (*Example*: '9,OM' refers to character #9 which is an ordered multi-state character.) In the final directive of the file, the state numbers have to be specified for each character, being separated from the character numbers by a comma. (*Example*: '9,5' refers to character #9 with 5 character states.)

Note: Character type = UM and number of states = 2 are default values and have been omitted.

According to the DELTA convention, character and specification listings are stored in files named 'chars' and 'specs' in ANSI format (previously in ASCII format). It is recommended to store the two files in a separate project folder (e.g., named 'MyDeltaProject').

It should be stressed that it is possible to start a whole project using the Delta editor from the beginning. Because it allows the direct copying of whole parts from published character listings, e.g., from the listing for lichenized and lichenicolous Ascomycetes on the LIAS web site (Botanische Staatssammlung München 1995-2001), the external setting up of chars and specs files is nevertheless useful.

A very handy tool for data entry and maintenance is the editor distributed with the CSIRO-Delta program package. The package ('delt32.exe') can be downloaded from the Programs and Documentation page at http://biodiversity.uno.edu/delta/. When started, the self-extracting file proposes the creation of a folder called 'Delta' to contain various program, system and data files. The CSIRO-Delta package is shareware. A test period of 1 month is allowed, after which the programs must be registered as described in the file Register.txt in the 'doc' subfolder (this file is also available directly from the Programs and Documentation Web page).

A User's Guide to the Delta Editor is located in the subfolder 'docs'. The 'sample' subfolder contains a complete, working data set. These files (e.g. intkey.ini and toolbar.ini) can be useful as guides for setting up your own files. The relevant ones can be copied from the 'sample' folder to your own project folder ('MyDeltaProject'), using the DELTA 'sample' option (at the MS-DOS prompt), Windows Explorer, or the Delta Editor (as described in the documentation of the Editor).

Entering item-related data in the Delta editor

Having started the editor 'Delta.exe', the chars and specs files in the project folder first need to be imported. This is done by selecting in the menu bar: File → New and then File → Import directives.

Entry of taxon-related data (taxon name and data on characters) is possible in the major data entry form of the editor. Multi-state characters are selected by checkboxes of the lower right frame of the interface. Often more than one state per character needs to be selected, except in the case where states are defined as being exclusive and do not allow more than one state to be present per item. Numeric and text data are entered into the lower left frame. Frequently, for some reason, certain traits of an organism may not have been examined, and data therefore do not exist. In this case, the respective characters are considered as 'unknown' and can be left uncoded. (*Example*: a species only known from its type collection has not yet been analysed chemically. The compounds of the species are unknown; however, potentially, data could be gained by later chemical examination.) Due to *a priori* absence by character state dependency, inapplicable characters are different. Here, the presence of a structure may depend on the presence of another superior structure. In other words, absence of a controlling character state causes inapplicability of the dependent character. (*Example*: character #16 'ascospores septation <kind>' is not applicable, if character #15 'ascospores septation <presence>' is set to 'absent'. Thus, due to the lack of septation, data of the kind of septation are not relevant.)

If the inclusion of additional characters becomes necessary during a project, they may be added to the character form of the editor. Characters should be defined precisely. A reference book, like the Dictionary of Fungi (Hawksworth et al. 1995), supplies appropriate definitions for being adopted and stored in the character notes field of the character edit form of the editor. Options to integrate picture and sound files are provided as well.

For making stored data accessible to queries and interactive identification, re-export of the data contents into DELTA data files is necessary. This is done by selecting, in the menu bar: File → Export directives). The minimum set of needed export files are 'chars', 'items', 'specs' and 'cnotes' with character- and taxa-related data, specifications, and notes to the characters. Alternatively, the files may be exported automatically when other programs are run from within the editor (by selecting in the menu bar: View → Action sets).

Be aware that during DELTA export, backup files are made automatically from the original chars and specs files. It should be pointed out that the step of DELTA exporting will be superfluous in future versions of the CSIRO-Delta Editor, which will generate output directly from the database.

Providing descriptive data for local and online use via the WWW

Various programs for identification and data query are capable of reading data in DELTA format, for instance Intkey (Dallwitz et al. 1995 onwards, 1998), Pankey (Pankhurst 1999), or DeltaAccess (Hagedorn 1997).

DeltaAccess is a free application for MS Access® for Windows, which builds up a relational structure from DELTA-based data during import. It includes modules for project management, data maintenance, data analysis and interactive identification.

Intkey is an interactive key program that enables rapid identification of biological specimens and uses files generated from the DELTA-coded data by Confor, the compiler of the CSIRO-Delta program suite. Intkey calculates the order of characters according to their potential capacity to resolve sets of taxa after each decision step during the procedure of identification. The order of characters is dynamically generated and those with the highest capacity appear at top of the window. Apart from identification, Intkey exhibits sophisticated features that greatly exceed those of ordinary data base applications and provides, for instance, the option to compare selected taxa or to create diagnostic descriptions.

How to compile DELTA data for use with intkey

DELTA files can be imported directly into DeltaAccess and Pankey, but for use in Intkey, the translation of code with the compiler 'Conforqw.exe' (part of the Delta program package) is necessary. Two files, toint and Intkey.ini, which contain DELTA directives, need to be prepared in advance and saved in ANSI format in the project folder:

- Create the file 'toint' needed for the compilation by Confor:

```
*SHOW Translate into INTKEY format
*HEADING Example Lichen Data Set
*LISTING FILE TOINT.LST
*INPUT FILE SPECS
*DISABLE DELTA OUTPUT
*TRANSLATE INTO INTKEY FORMAT
*REPLACE ANGLE BRACKETS
*INTKEY OUTPUT FILE ICHARS
*INPUT FILE CHARS
*INTKEY OUTPUT FILE IITEMS
*INPUT FILE ITEMS
```

- Create a file 'Intkey.ini', needed as a start-up file for Intkey:

```
*FILE TAXA iitems
*FILE CHARS ichars
```

```
*FILE INPUT toolbar.inp
*SET IMAGEPATH images
*SET RBASE 1.2
*DISPLAY UNKNOWNS OFF
*DISPLAY INAPPLICABLES OFF
```

- Start Confor at the MS-DOS prompt in the project folder and enter 'toint' (without quotation marks) at the prompt line. The compilation process starts with entering the <return> button. [In cases where the data files (chars, items, specs, cnotes and toint) include no errors, the process will be terminated with the message 'normal termination'.] Alternatively, Confor can be run from the Windows 'Start' menu, or within the editor by selecting in the menu bar: View → Action sets → toint → Run.

The new data files 'ichars' and 'iitems', created by compilation, are ready now to be used by Intkey.

Delta Intkey files or DeltaAccess databases may be put on an Internet server for downloading and subsequent local use. However, more convenient for the user is an option for direct data query online. Via a web browser, Intkey data files can be directly loaded into the memory of a local computer, while Intkey is evoked as a 'helper application'.

How to install Intkey files for their direct use over the WWW

- Pack the following files (e.g., with the program WinZip®) into a zip-file called 'example.zip': ichars, iitems, intkey.ini and Toolbar.inp.
- Create a text file (ANSI) 'filename.ink' with the following contents ['filename' should be replaced by name of file, name of file, '_mydomain_/ _myproject_/_'by the web address (domain name and path) of the file]:

```
;To run the Intkey from this file, you need Intkey5 and Windows95/98/NT
;Name of this file
InkFile=http://_mydomain_/_myproject_/filename.ink
;Name of the Intkey initialization file within the compressed data file
InitializationFile=intkey.ini
;Name of the compressed data file
DataFile= http://_mydomain_/_myproject_/example.zip
```

- Copy these three files onto a server under the 'DataFile' address and add the link address 'filename.ink' to a HTML page on which the files can be loaded down (hypothetical example!):

```
<html>
<body>
<p>
<a href ="http://_mydomain_/_myproject_/example.ink">Intkey Example Data<a/><br>
<p>
```

<body>
</html>

Three free web interfaces link DeltaAccess databases to the Internet: 1) DeltaAccess Perl (Cross 1997, Findling 1998a) 2) DAWI (Findling 1998b), 3) Navikey (Bartley and Cross 1999). The parallel installation of the interfaces at the LIAS site (http://www.mycology.net/lias/) allows one to study the pros and cons of the different techniques. This web page also links to information on how to install the interfaces on a server.

Summarising, the following steps need to be exercised for creating databases for local use and online use over the Internet:

1. Text Editor: character selection and decomposition set up → character arrangement → character specifications set up (in DELTA format)

2. CSIRO-Delta Editor: character and specifications data import → item-related data entry → data export (in DELTA format)

3. ConforW: DELTA data compilation → local installation for use with Intkey → installation on WWW server for download and use with Intkey

4. DeltaAccess: DELTA data import → local installation for use as database → installation on WWW server together with user interfaces (like DAP, DAWI and Navikey) for online use over the Internet

Comments

The advantages of computer-aided applications for interactive identification are obvious. In monographic treatments of taxa, descriptive data have often to be collected and stored electronically, for instance, to establish taxonomic descriptions in natural language or to set up a matrix for cladistics. Even if the goal is to create printed diagnostic text and keys only, it is essential to structure data beforehand. The inclusion of only a few additional taxa often entails major structural changes within the key. This means that sizeable efforts are required to update existing key documents. This problem is virtually non-existent if computer-aided tools generate keys. Further, material under examination is often fragmentary or badly conserved, exhibiting only parts of characters needed for use in a diagnostic key. Due to multiple entry access, most computer applications for interactive identification (and data retrieval) work well in such situations because they tolerate unknown characters or character states. Consequently, databases do not have to be complete to work, and so might even serve as data management and identification system at the early stages of a project. Crosschecking with even highly incomplete digital collections of descriptive data may reveal errors in the characters observed.

During interactive identification, a closer examination of taxon groupings that appear repeatedly even when different identification paths were used, may be enlightening in so much as they may reflect possible natural groupings. Computer-based data storage, with options for interactive identification and character analysis, provide in a concise form the knowledge of experts to non-specialists. These arguments strongly favour the creation and use of structured digitised data. DELTA allows the use of descriptive data for a wide range of applications and is, therefore, recommended as the interchange format of choice in systematics.

Acknowledgements. I would like to thank M. Dallwitz (Canberra) and G. Hagedorn (Berlin) for valuable comments on the manuscript.

References

[All Internet sites listed were consulted on June 2001; an updated list of the links below is provided under www.mycology.net/links/ProtocolsInLichenologyChapter31Links.html]

Acharius E (1798) Lichenographiae sueciae prodromus: i-xxiv, 1-264. Linköping

Acharius E (1810) Lichenographia universalis: 1-696. 14 tab. Gottingae

Bartley M and Cross N 1999. Navikey v. 2.0 . [http://www.herbaria.harvard.edu/computerlab/web_keys/navikey].

Boughey AS, Bridges K W and Ikeda A G (1968) An automated biological identification key. Research series no.2. Museum of Systematic Biology, University of California, Irvine.

Botanische Staatssammlung München (ed.) (1995) LIAS – a DELTA-based information and data storage system for lichenised and lichenicolous Ascomycetes. [http://www.mycology.net/lias/] (First posted on 95-01-01, most recent update 2001-06-01).

The British Lichen Society (ed.) (1997) Database for the Identification of *Parmelia* Ach. of Great Britain and Ireland. The British Lichen Society. [http://www.argonet.co.uk/users/jmgray/cd/index.htm].

The British Lichen Society (ed.) (1998) Identification of *Parmelia* Ach. The British Lichen Society. [ISBN 0 9523049 4 5, CD-ROM].

Cross N (1997) Delta Access Perl FAQ. [http://www.herbaria.harvard.edu/computerlab/web_keys/ delta_access_perl.html]. Cambridge, Mass.

Dallwitz M J (1980) A general system for coding taxonomic descriptions. Taxon 29: 41-46

Dallwitz M J (1997) Desirable attributes for interactive identification programs. [http://biodiversity.uno.edu/delta/].

Dallwitz M J, Paine T A (1993 onwards) Definition of the DELTA format. [http://biodiversity.uno.edu/delta/].

Dallwitz M J, Paine T A, Zurcher, E. J (1993 onwards) 'User's Guide to the DELTA System: a General System for Processing Taxonomic Descriptions.' 4th edition. [http://biodiversity.uno.edu/delta/].

Dallwitz M J, Paine T A, Zurcher E J (1995 onwards) 'User's Guide to Intkey: a Program for Interactive Identification and Information Retrieval.' 1st edition. [http://biodiversity.uno.edu/delta/].
Dallwitz M J, Paine T A, Zurcher E J (1998) Interactive keys. In: Bridge P, Jeffries P, Morse D R, Scott P R (eds) Information Technology, Plant Pathology and Biodiversity. CAB International, Wallingford, pp 201-212.
Diederich J (1997) Basic properties for biological databases: character development and support. Mathl. Comput. Modelling 25: 109-127
Diederich J, Fortuner R, Milton J (1997) Construction and integration of large character sets for nematode morpho-anatomical data. Fundam Appl Nematol 20: 409-424
Elix J A, Johnston J, Parker J (1988) A computer program for the rapid identification of lichen substances. Mycotaxon 31: 89-99
Findling A (1998a) DAP Ein Web-Interface zu DeltaAccess. [http://www.axel-findling.de/programs/dap/] (First posted on 98-09-30)
Findling A (1998b) DAWI - Ein Web-Interface zu DeltaAccess. [http://www.axel-findling.de/programs/dawi/] (First posted on 98-09-30)
Gouda E J (1998) Taxasoft Delta Programs. [http://biodiversity.bio.uno.edu/delta/taxasoft/].
Grube M (1998) (onwards). Lichen Database of the Mediterranean. [http://biobase.kfunigraz.ac.at/medlichens.html]
Hagedorn G (1997) DeltaAccess - an SQL interface to DELTA ('Description Language for Taxonomy'), implemented in Microsoft Access. [http://www.bgbm.fu-berlin.de/projects/DeltaAccess/] (Most recent update 2000-06-30)
Hall A V, Johnston R D (1954) Card sorting key for the identification of eucalypts. Forestry and Timber Bureau, Canberra
Hansen B, Rahn K (1969) Determination of angiosperm families by means of a punched-card system. Dansk Bot Ark 26: 1-45 [+ 172 cards]
Hawksworth D L, Kirk P M, Sutton B C, Pegler D N (1995) Ainsworth and Bisby's dictionary of the fungi. 8th Edition. Egham
Insarov G, Insarova I, and Zeltyn S (1997) IDENT: a computer-aided multi-access key for the identification of lichens of the Negev Desert. Brit Lich Soc Bull 80: 42-45
Kirschbaum U, Wirth V (1998) Bestimmung epiphytischer Flechten. FH Gießen. [CD-ROM].
Lamarck J B P (1779) ("1778"). Flore Françoise. 1 ed. Imprimerie Royal, Paris
Lebbe J (1984) Manuel d'utilisation du logiciel XPER. Micro Application, Paris
Morse L E (1974) Computer programs for specimen identification, key construction and description printing. Publ. Mus. Michigan State University, Biology 5
Pankhurst R J (1988) An interactive program for the construction of identification keys. Taxon 37: 747-755
Pankhurst R J (1991) Practical taxonomic computing. Cambridge University Press, Cambridge
Pankhurst, R J (1998) A historical review of identification by computer. In: Bridge P, Jeffries P, Morse D R, Scott P R (eds) Information technology, plant pathology and biodiversity. CABI, Wallingford, Oxford, pp 289-303
Pankhurst R J (1999) PANKEY - A suite of key generation and identification programs. [http://www.rbge.org.uk/research/pankey.htm]
Pasarell F (1999) Computer assisted programs or references available for the mycologists or microbiologist. [www.doctorfungus.org]

Pittam S K, Hanus, F J (1997) Lichen Synoptic Key. [http://mgd.nacse.org/hyperSQL/lichenland/]
Rambold G, Triebel D (1997 onwards) The genera of lichenised and lichenicolous Ascomycetes. LIAS Online Database. Botanische Staatssammlung München. [http://www.mycology.net/lias/ (→ Identification)] (Most recent update 2001-05-01)
Pasarell L (1999) Computer assisted programs or references available for the mycologists or microbiologist. [http://www.doctorfungus.org/]
Sipman H J M (1998) Lichen determination keys available on Internet. [http://www.bgbm.fu-berlin.de/sipman/keys/default.htm] (Most recent update 2001-02-02)
Wetmore C M (1999) *Caloplaca* data. [http://www.tc.umn.edu/~wetmore/Calop.html] (Most recent update 2000-03-28)
Zeltyn S A, Pchelkin A V (1993) Application of a computerized lichen identifier in background ecological monitoring systems. In: Problems of Ecological Monitoring and Ecosystem Modelling 15: 264-273. Gidrometeoizdat Publisher, St. Petersburg.

On-Line Documentation of Lichen Biodiversity

Martin Grube

Introduction

The Internet has dramatically accelerated the exchange of scientific information during the last decade and biodiversity research in particular is a field that substantially benefits from the distribution of data on the World Wide Web (WWW). More specifically, this chapter discusses the various methods available for documenting lichen biodiversity on the Web. Such information is of interest for lichenologists involved both in basic research or in applied approaches, e.g. where biomonitoring with lichens is used in studies of ecological continuity, air pollution and its long term changes and in identifying regions of elevated risks to health (Löscher et al. 1990, Cislaghi and Nimis 1997).

Previously, the compilation of basic data on biodiversity was rather time-consuming, especially when references were scattered in literature that was old and not easily accessible. Furthermore, data on relevant specimens was hidden with the material in herbaria and had to be searched manually. Once such data are available in electronic form, the dissemination of this information is greatly improved by the Internet. Because names and concepts may change after taxonomic revisions or re-examination of the material by specialists, a well maintained on-line information system is also useful for rapid communication of scientific progress. Web pages are the most common way to make biodiversity information publicly available, and ideally, databases are used to generate the content of the pages. Beside this, there are also simpler ways for quick communication in use, e.g. the lichens-l email list, a forum of the International Association for Lichenology (maintained by Clifford Smith, Honolulu).

Martin Grube, Karl-Franzens-Universität Graz, Institut für Botanik, Holteigasse 6, Graz, 8010, Austria (*phone* +43-316-380-5655; *fax* +43-316-380-9883; *e-mail* martin.grube@kfunigraz.ac.at)

Already existing on-line information systems in lichenology include taxonomic information systems (LIAS, Rambold 1997), literature databases (Triebel and Rambold 1998-1999, Timdal 1999), herbarium databases (Obermayer, this volume) or on-line keys and identification systems (Sipman 1998, Bartley and Dickinson 1999, Rambold, this volume). Beside these, numerous checklists are now available via the WWW as HTML files, e.g. lists for the Mediterranean region, North America, South America, and other smaller geographic units (see Table 1). Although HTML is a very simple format, these lists already contain much useful information, and they can be searched, printed or distributed. Clearly, more flexible access and evaluation of data is possible with databases that can be queried on-line (e.g. Grube and Nimis 1997, 2000).

There are many possibilities available to workers who want to develop on-line information systems, and it is important to think about some general points in advance. The functionality of the system should be adapted to the needs of the potential users. This can be simply on a technical level. For example, if many users have slow internet connections, it is better to avoid the excessive use of images. If most of the potential users are lichenologically inexperienced, more explanations and basic information will have to be included. It is also important to take into consideration the choice of software, hardware and financial resources of the users. In the case of complicated soft- or hardware, technical support may need to be funded. Also, it should be considered how much information will be available publicly and whether users will have on-line updating options, which could make password protection necessary. Once agreement about such and similar general points is achieved, the database-driven information system can be developed. Each of the major steps will be discussed in greater detail below.

Outline

The major steps in the development of an online biodiversity information system are as follows:

1. Compilation of data
2. Set-up of a database
3. Formatting of the data
4. Connection to the Internet
5. Development of application software

Table 1. Some Internet sites related to biodiversity of lichens.

Topic	Address
Biodiversity of Mediterranean Lichens	http://biobase.kfunigraz.ac.at/medlichens.html
Biomonitoring in the Pacific Northwest	http://mgd.NACSE.ORG/qml/lichenair/
Checklist of Bavarian Lichens	http://www.rrz.uni-hamburg.de/biologie/ialb/herbar/bay_f2.htm
Checklist of Foliicolous Lichens	http://www.uni-bayreuth.de/departments/planta2/research/lichens/checklis.htm
Checklist of the Lichenized Fungi of the Guianas (U.S. National Herbarium)	gopher://nmnhgoph.si.edu:70/11/.botany/.guianas/lichlist
Checklist of Japanese Lichens (Kochi University)	http://www.is.kochi-u.ac.jp/Bio/lichens/fljapan.html
Index Nominum Genericorum	http://www.nmnh.si.edu/cgi-bin/wdb/ing/names/form
ITALIC	http://dbiodbs.univ.trieste.it
Lichen checklists worldwide	http://www.rrz.uni-hamburg.de/biologie/ialb/herbar/lichenw.htm
North American Lichen Checklist Online	http://www.ndsu.nodak.edu/instruct/esslinge/chcklst/chcklst7.shtml
South American lichens Online	http://www.rrz.uni-hamburg.de/biologie/ialb/herbar/sa_f2.htm
Key to Lichens in the Neotropics	http://www.bgbm.fu-berlin.de/bgbm/staff/wiss/Sipman+H/keys/neokeyA.htm
Key to Lichens of the Pacific Northwest	http://mgd.nacse.org/cgi-bin/hyperSQL_gateway/?synoptic_key
LIAS	http://www.botanik.biologie.uni-muenchen.de/botsamml/lias/lias.html
Search Recent Literature on Lichens	http://www.toyen.uio.no/botanisk/bot-mus/lav/sok_rll.htm

Procedure

Data compilation

There are diverse sources of original biodiversity information. This includes data from herbaria, literature records, or field observations. The format and accuracy of these original data is usually variable. This will have a substantial impact on the applicability of the information system (e.g. if the original sources of the geographic information are inaccurate). More detailed data will be required if the database is used in biomonitoring projects, whereas in national checklist projects that normally have taxon names as their base units resolution may be lower. Some limitations of geographic resolution are common for the latter as these are at least partly based on references to old literature with poor information on the collecting localities. As a long-term goal, it is advisable to base such information systems on specimens being the basic information units, and stored in herbaria. These can be accessed for re-evaluation, which becomes a necessity when taxonomic concepts change at the species level. Moreover, a specimen-based approach is also important to implement database models incorporating multiple classifications (Pullan et al. 2000). A good example of a herbarium information system is available at the Botanical Museum in Oslo (Timdal 1999) and further examples are found in Obermayer (this volume).

Data collection and improvement in biodiversity projects is an open-ended, and complex process as is shown by the regular updates of checklists and floras. The compilation involves several main steps, which do not necessarily follow a sequential order.

1. Compile all available information for a geographic unit (revise literature, revise herbarium collections, carry out fieldwork to complement literature and herbarium data).
2. Consult experts in taxonomic groups.
3. Standardize and complete missing information (e.g. refine geographic information).
4. Store data in a common format.

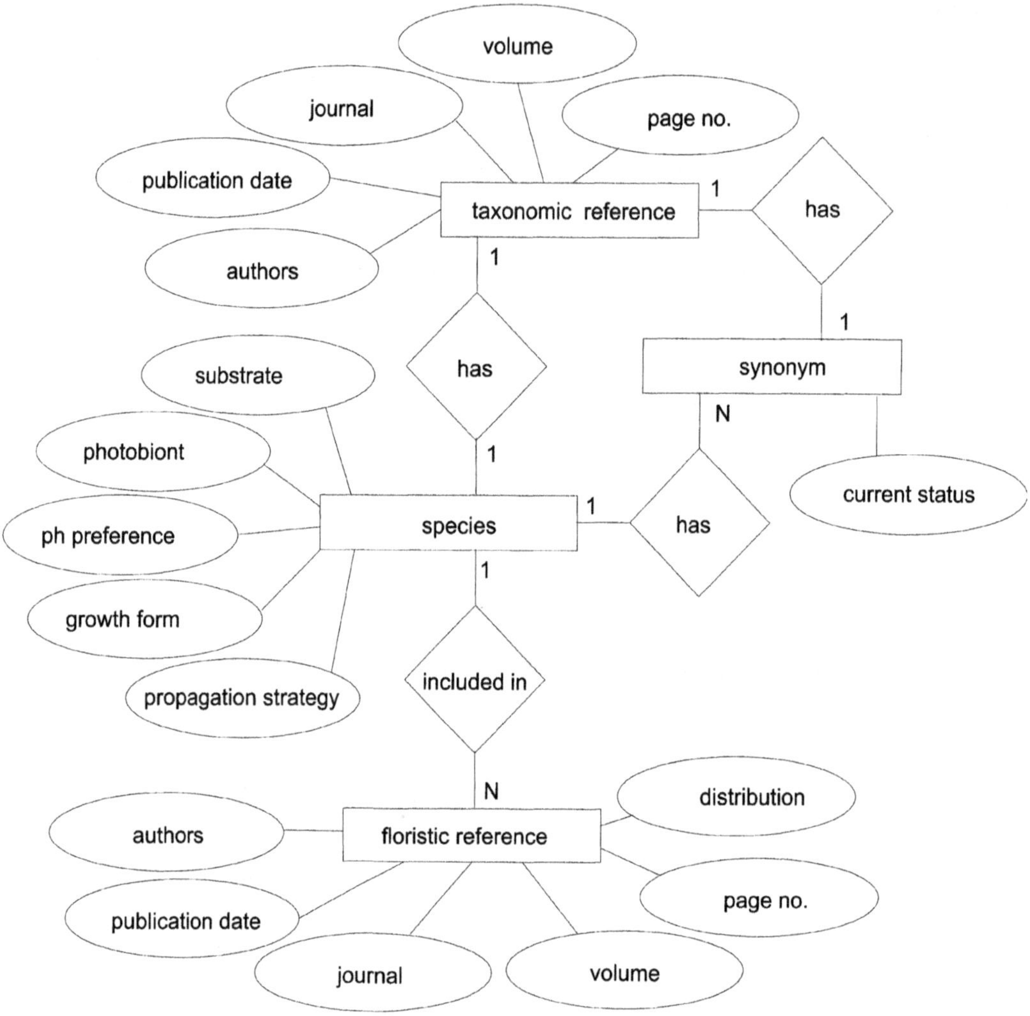

Fig. 1. Example for an entity relationship-diagram. Ovals represent attributes, boxes are entities, and facets are relations.

Database design

The design of a database should be carefully planned and involves an initial conceptual phase after which a logical data model has to be developed. Entity-relationship diagrams may facilitate this work and a simple example of such a diagram is shown in Figure 1. This article cannot explore further theoretical details of database design and the reader may wish to consult more detailed literature such as Conolly et al. (1998) for more information. For a comprehensive model on biological collections and surveys, see Behrendsohn et al. (1999). After conceptual and logical design considerations, individual tables are physically created which will later contain the data.

1. Physical representation of basic relations.
2. Design of constraints.
3. Design of transactions.
4. Disk storage.

Step 1 is the translation of a data model into a form that can be implemented in the target database system; practically it is the creation of tables, variables, etc. Step 2 can include various rules that have to be followed in the database, e.g. that only correctly written names can be entered, or that the same record is entered only once. This is very important to maintain the consistency of the database. The design and analysis of transactions is more important in larger and intensely used databases to keep a high level of performance, because complex transactions (queries and updates) via the Web must be processed in reasonable time. As the fourth step suggests, consideration should be given to the disk space that will be used, because sufficient storage space must be allocated. If the information system is maintained on different computers, portions of the database can be kept separately. This could either be tables or rows of a table (=horizontal splitting). Alternatively, and to avoid time-consuming queries across remote sites, read-only copies of all tables, so-called snapshots, can be maintained on different servers, which are regularly updated.

If several independent centres of research or authors contribute to a joint project, inconsistencies and standardisation of taxonomic concepts will become important issues, because different concepts are still followed by the lichenological community. This will lead to difficulties in searching the database unless the inconsistencies can be translated or managed by a thesaurus of synonymy. The synonymy can be incorporated as a separate table in the database including taxonomic information on each taxon, i.e. name, author, original reference, taxonomic status, and, if the name is no longer used, the currently accepted name. This is one possible implementation of the "potential taxon concept" proposed by Berendsohn (1995, 1997).

Once a database is established, an administrator should be responsible for its maintenance. Also, it should be clear how further flow of data is organised (Fig. 2). Direct writing access to the database via the Internet may accelerate the data flow and it is a feasible option, but must avoid inconsistency and should maintain high standards with respect to quality of data. Ideally, direct update access should be possible by very few persons and protected by a password. Alternatively, it might be a more appropriate approach if incoming data are collected in a separate file until they have been evaluated by the administrator and by experts for the taxonomic groups. It can also be useful to provide a public forum, i.e. a discussion page, which leaves space for comments on literature, determinations, or opinions. Any contribution may of course be stored in a separate table and retrieved on-line.

Another important aspect is the hardware. Complex database systems are usually difficult to maintain on outdated computer technology, and before choosing a database server software product, it should be confirmed that it will run on the available platform. Several software products are available, each with its own advantages and disadvantages. Which database system is best will depend mostly on financial resources, admin-

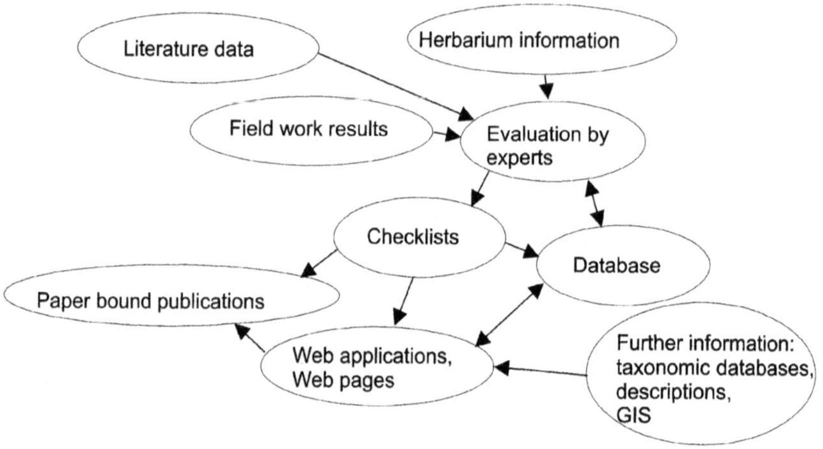

Fig. 2. Simple schema of information flow in an on-line biodiversity project.

```
<?XML version = "1.0" ?>
<!DOCTYPE DOCUMENT [
<!ELEMENT DOCUMENT (NAME)*>
<!ELEMENT NAME (GENUS,EPITHETON,OGU*)>
<!ELEMENT GENUS (#PCDATA)>
<!ELEMENT EPITHETON (#PCDATA)>
<!ELEMENT OGU (#PCDATA)>
]>
<DOCUMENT>
<NAME>
  <GENUS>Caloplaca</GENUS>
  <EPITHETON>aurantia</EPITHETON>
        <OGU>
          ITALY
        </OGU>
        <OGU>
          SPAIN
        </OGU>
        <OGU>
          TURKEY
        </OGU>
</NAME>
<NAME>
  <GENUS>Lecanora</GENUS>
  <EPITHETON>allophana</EPITHETON>
        <OGU>
          ITALY
        </OGU>
        <OGU>
          SPAIN
        </OGU>
</NAME>
<NAME>
  <GENUS>Lecanora</GENUS>
  <EPITHETON>rupicola</EPITHETON>
        <OGU>
         ITALY
        </OGU>
</NAME>
</DOCUMENT>
```

Fig. 3. An example of a simple XML file containing data about the distribution of lichens.

istrative capacity, and practical requirements. A relatively expensive, though powerful solution is Oracle 7 or 8. Further possibilities are provided with a SQL server, which can be integrated completely in NT and Office, or by Informix. Other products include DB2, Sybase, or Interbase, of which the latter is relatively easy to manage. Also, various desktop databases can now be linked to the web, e.g. Microsoft Access and Filemaker Pro.

Data formatting and input

The physical representation of basic relations of the database as structure of tables determines how the data needs to be formatted before they can be included into the database. The records of the database can then be entered manually, but if large amounts of updates are available as files (e.g. existing checklists as text files which follow a consistent format), they can then easily be converted into a suitable format using separators, for direct import into a database table.

For convenient exchange between different database systems and interoperability, data may be formatted as DELTA or XML files. DeltaAccess allows users to import DELTA formatted data into a relational database (Hagedorn 2001), while XML can be used with existing web protocols (HTTP and MIME). XML, as a simplified form of SGML, is designed specifically for Web documents. XML documents can be logically separated into two parts: one part defines the structure (Document Type Definition, DTD) and the other contains the data. A simple example of the structure of an XML file is shown in Fig. 3. XML documents must be parsed explicitly, which may be done in different ways: e.g. using JavaScript (XML is only supported by Internet Explorer at present) or using converting programs (e.g. Java applets) that translate to HTML. For DELTA files, which are imported in an Access database, convenient Web interfaces exist with DAP and DAWI (Cross 1997, Findling 1998a, b).

Connection to the Internet

There are numerous ways to make database information publicly accessible and therefore it is difficult to formulate a uniform and sequential protocol. The primary prerequisite is a server connected to the Internet, which accesses the stored information. This server provides downloadable files, or HTML pages, which may be generated from a database using application software. The necessary programs can be created individually or are provided commercially for certain products. Alice software offers an example: the tool AliceWeb allows structured and searchable web pages to be generated from Alice databases (Dransfield 1998, Grant et al. 1998). Also, DeltaAccess has a utility to generate HTML forms from databased delta files (Hagedorn 2001).

During the past few years the most common approach has been to directly connect to databases via the Internet using the common gateway interface (CGI). However, the CGI architecture is poorly scalable, and as a new CGI process has to be created to service each request, this may slow down the information system. Vendors solved these shortcomings by different proprietary architectures, tools, and application programming interfaces (see Gutierrez 1999). For example, the Microsoft Active Platform offers a range of tools to connect databases with the WWW. As a programming model, Active Server Pages (ASP) were introduced with the Microsoft Information Server (IIS) 3.0. This model contains so-called Active Data Objects as a programming extension for connecting various databases, e.g. desktop databases as Access or SQL server databases. HyperSQL is a scripting language to query remote Sybase and Oracle databases (Newsome et al. 1996). This language has been used to link to various lichenological databases, including an identification system and a herbarium database (Pittam and Hanus 1997). The programming language Java (Gosling et al. 1996) provides an alternative way to develop applications. A major advantage is the machine independence of the language. Code written once, can be used on any computer which supports the Java Virtual Machine. Database connectivity in Java is mediated by the JDBC package, an application-programming interface that supports basic SQL functionality and enables access to a range of relational database management system products. The list can certainly be extended.

A common model for Web database projects is the three-tier architecture, where the client, the Application Server, and the Data Server form separate layers (Figure 4). In such a model, the Application server forms the middle tier that provides transaction support across multiple HTTP requests. A three-layer architecture is implemented for example in the database project on Mediterranean lichens (Grube and Nimis 1997, 2001). This is done with Oracle software including the Oracle Web Application server 4.0 as the middle layer. The latter manages the accesses to the database, which contains the tables and PL/SQL programs that produce dynamic HTML output, based on the contents of the database. In practice, when a user sends a request, the system works as follows: within the Application server, the Web Listener receives an URL from the web browser

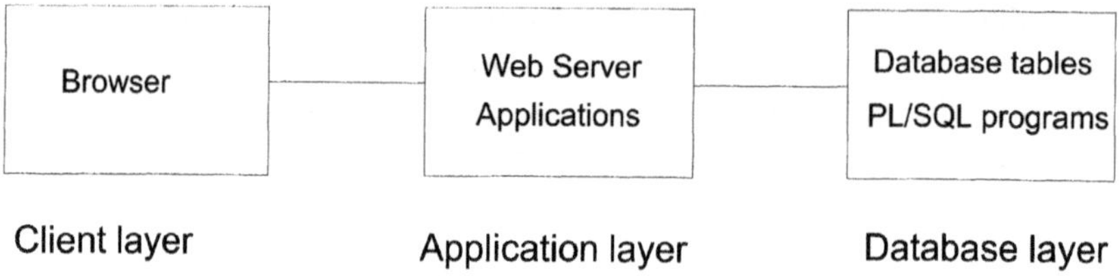

Fig. 4. A three-tier model for on-line databases.

and determines whether the request is for a static HTML page that can be accessed from the file system or whether the request requires the use of a Web Request Broker (WRB) cartridge. Such software cartridges execute the requests, e.g. by processing PL/SQL routines. Cartridges exist also for other languages and APIs (application programming interfaces), such as Java, Perl, LiveHTML and ODBC. As it is not possible here to discuss the Oracle Web Application Server in detail, the reader is referred to Harvey and Beitler (1998).

Development of application software

Once an on-line database is established, appropriate software must handle the requests by the clients. Forms are provided on the Web that allow the user to specify a query to the database. The completed form is submitted to the server that handles the request. The content of the form invokes a program which produces a response in the form of dynamically generated Web pages. Depending on the complexity of the query this may also involve other procedures and functions which are invisible to the user.

Programming applications may become a time-consuming and complex process.

Some basic steps involved in the development of applications for on-line databases are as follows:

1. Establishment of a general design of your web pages using an appropriate editor.
2. Creation of forms to handle requests and connect to the database.
3. Generation of programs for HTML output based on information in the database.

4. Generation of tutorial pages (help-desk) to make the system user-friendly.
5. Inclusion of an error handling system.
6. Inclusion of relevant links to related pages.
7. Incorporation of programs for graphical output (optional).

The listings in Figure 5 and 6 are examples for underlying PL/SQL code of a query form and an output page.

There are numerous options that can improve the design of pages presented to the client: for example, text areas are used in the input forms to let the user enter a search name, and checkboxes, radio buttons, menus, or font size controls can be used to refine queries and to improve the appearance of the form pages. Error handling is most useful to handle wrong input or to indicate that the queried data are not present in the database. It is also appropriate to inform the user about the reasons for a failure. This may be useful for example when the user enters an outdated name. Then, a taxonomic thesaurus may be invoked to look up a possible current name, which may be used to search the database again. Graphical output

```
/* a simple form to enter a query */
PROCEDURE Ask_database
IS

BEGIN
       htp.HtmlOpen;  /* this command opens a HTML page */
       htp.BodyOpen;  /* this opens the body of a HTML page */
       htp.FormOpen('Database_response','POST'); /* here, the procedure is called which
                                                    will handle the request */
       htp.p('Enter species:');
       htp.FormText('the_name','30'); /* creates an empty text field for entering the
                                          request */
       htp.p('With synonyms: '||htf.formcheckbox('Synonymy')); /* with this, the user
                              can specify whether the synonyms should be listed */
       htp.FormSubmit;
                     /* creates a submit button */
       htp.FormReset('Another species');
       htp.FormClose;
       htp.Mailto ('martin.grube@kfunigraz.ac.at', 'my email address'</P>');
       htp.BodyClose;
       htp.HtmlClose;
END;
```

Fig. 5. PL/SQL program generating an HTML form to query an Oracle database

```
/* in this simple case we assume that the entire distribution of a taxon is represented in a single data field */
PROCEDURE Database_response
(the_name in varchar2, Synonymy varchar2)
IS

cur_name VARCHAR2(100);

/* the following cursors are used to navigate through the floristic and taxonomic tables; they contain a select
statement to find matching entries */

CURSOR row_cursor IS
SELECT name, dist FROM italflo where UPPER(name) like UPPER(the_name);
dba_rec row_cursor%ROWTYPE;

CURSOR syn_cursor IS
SELECT name, syn FROM synonyms where UPPER(name) like UPPER(the_name);
syn_rec syn_cursor%ROWTYPE;

BEGIN
htp.HtmlOpen;
htp.BodyOpen;
OPEN row_cursor;
FETCH row_cursor INTO dba_rec;
   IF row_cursor%NOTFOUND THEN
                    /* If the name is not found, the following code looks in the synonym table */
     BEGIN
     SELECT DISTINCT name INTO cur_name FROM synonyms WHERE syn LIKE
           the_name;
     EXCEPTION
     WHEN NO_DATA_FOUND THEN
            htp.p(the_name||' is not found in the database. <BR>'); EXIT;
            END;
        htp.p('<TR>In this database, '||cur_name||' is the current name for
             the taxon you asked for ('||the_name||'). </TR>');
   END IF;
/*******************************************************/
EXIT WHEN row_cursor%NOTFOUND;
htp.p('<B>'||dba_rec.name||'</B>');
htp.p('Is present in the following regions: <BR>');
htp.p(dba_rec.dist||'<BR>');
htp.nl;

IF (Synonymy = 'on') THEN
    OPEN syn_cursor;
      LOOP
      htp.p('<B>Synonyms: </B>');
      FETCH syn_cursor INTO syn_rec;
      EXIT WHEN syn_cursor%NOTFOUND;
      htp.p(syn_rec.syn);
      ENDLOOP;
    CLOSE syn_cursor;
htp.nl;
end if;
htp.bodyClose;
htp.htmlClose;
END;
```

Fig. 6. PL/SQL program generating dynamic HTML output from an Oracle database

can make an information system more attractive. Java routines could be used to plot distributional data on a map. This can be achieved by submitting distributional data retrieved from the database as parameters to a Java program that is invoked at the client layer. More sophistically, the geographic distribution of a species can be represented using geographic information system (GIS) software.

Links to other electronic information resources

Various kinds of additional information can be linked. This may for example involve the following types of data:

- Ecological parameters for a species
- Literature references
- Taxonomic data
- Descriptions of species
- Images of species

Except for the latter two, these additional links are also integrated in the Lichen Database of Italy (Nimis 1999). Such links to other electronic resources are optional and can be managed by the application software. The latter can automatically format input properly to invoke external query routines. One example for this is implemented in the database of Mediterranean lichens, where the original query string is reformatted to retrieve information from a server in Trieste on specimens in the herbarium in Trieste (TSB), on specimens of seven herbaria databased in Oslo, or on recent literature and taxonomic information on the queried genus. When ecological data and descriptions of species, e.g. in DELTA format (Rambold, this volume), are available, this could lead to a completely new type of information resource. Interactive keys can then individually be generated for lichens occurring under certain ecological conditions in the operational geographic unit.

Further extensions are possible if the database is linked with a GIS. Complex geographic information can be stored as so-called themes in such a geographic database, and can be linked to other databases via open database connectivity (ODBC). For example, a theme may be geological data; another is altitude or vegetational data. The link with a GIS can be used to draw simple distribution maps of lichens or to carry out complex analyses of OGUs, for example, to compare the distribution of lichens with geological information. Also, as ecological data on lichens are available, predictive distribution maps for lichens can be generated (Nimis and Martellos 2000). A geographic information system can conveniently be linked to other programs via ODBC. With the InternetMapper by ESRI, it is also possible to connect a GIS directly to the WWW.

Comments

Copyright and citation of on-line information, which is continuously updated, are important issues and were discussed by Grube and Nimis (1997). It will be appropriate to regularly produce publicly accessible CD-Roms or paperbound versions of the stored information.

On-line databases offer another interesting possibility to store large amounts of data on specimens. Many journals do not accept detailed lists of studied specimens, e.g. in taxonomic revisions. However, these lists contain useful information and it might be interesting to store this in a publicly available database, similar to sequence data stored in Genbank (Bilovsky and Burke 1987). Such data may of course also be incorporated in biodiversity information systems.

Acknowledgements. Sabine Rostek and Guenther Berthold (EDV-Zentrum Graz) are acknowledged for technical support. I also thank Stefano Martellos, Pier Luigi Nimis (both Trieste), and Helmut Kammerer (Graz) for discussions.

References

Bartley M and Dickinson T (1999) Web-based interactive identification using Navikey and Pollyclave. XVI International Botanical Congress. Abstracts. St.Louis: 94

Berendsohn WG (1995) The concept of "potential taxa" in databases. Taxon 44: 207-212

Berendsohn WG (1997) A taxonomic information model for botanical databases: the IOPI model. Taxon 46: 283-309

Berendsohn WG, Anagnostopoulos A, Hagedorn G, Jakupovic J, Nimis PL, Valdés B, Güntsch A, Pankhurst R & White RJ 1999: A comprehensive reference model for biological collections and surveys. Taxon 48: 511-562

Bilovsky HS and Burke C (1987) The Genbank genetic sequence data bank. Nuc Acids Res 16: 1861-1863

Cislaghi C and Nimis PL (1997) Lichens, air pollution and lung cancer. Nature 387: 463-464

Conolly T, Begg C and Strachan A (1998) Database systems. A practical approach to design, implementation, and management. 2nd ed. Addison-Wesley, Harlow

Cross N (1997) Delta Access Perl FAQ. - Hyperlink: http://www.herbaria.harvard.edu/computerlab/web_keys/delta^accesss_perl.html
Dransfield J (1998) A database of Rattan palms. Alice software news 6: 14-15
Findling A (1998a) DAP - Ein Web-Interface zu DeltaAccess. Hyperlink: http://www.lrz.muenchen.de/~findling/programs/dap/
Findling A (1998b) DAWI - Ein Web-Interface zu DeltaAccess. Hyperlink: http://www.lrz.muenchen.de/~findling/programs/dawi/
Gosling J, Joy B and Steele G (1996) The Java Language specification. Addison-Wesley, Harlow
Grant R, Turnbull D and Winfield P (1998) Conservation of potato genetic resources in Europe. Alice software news 6: 8-9
Grube M and Nimis PL (1997) Mediterranean lichens on-line. Taxon 46: 487-493
Grube M and Nimis PL (2001) The Med-checklist of Mediterranean lichens on-line: towards a synthesis. Bocconea 13:291-297
Gutierrez D (1999) Web Database Development for Windows Platforms. Prentice Hall, New York
Hagedorn G [in press]. Making DELTA accessible: datasing descriptive data. Bocconea
Harvey D and Beitler S (1998) The Developer's Guide to Oracle Web Application Server 3. Addison Wesley. Reading, Massachusetts
Löscher W, Einspieler C, Holzer-Sutter A, Grill D, Moser M, Haidmayer R, Kurz R and Kenner T (1990) Zusammenhänge zwischen dem "Plötzlichen Säuglingstod (SIDS)" und der durch Bioindikation dargestellten Luftverschmutzung in Graz. Mitt. Naturwiss Vereins Steiermark 120: 287-290
Newsome M, Pancake C, Hanus J and Moore L (1996) HyperSQL User's Guide and Language Reference. Hyperlink: http://www.NACSE.ORG/hsql/docs/manual/
Nimis PL (1999) Lichen Database of Italy 1.0. Univ. of Trieste, Dept. of Biology, IN1.0/99. Hyperlink: http://biobase.kfunigraz.ac.at/flechte/owa/askitalflo4
Nimis PL and Martellos S (2000) Predictive distributional maps of Italian lichens. The fourth IAL Symposium. Abstracts:121
Obermayer W (2001) Management of a lichen herbarium. In: Kranner I, Beckett R and Varma A (Eds) Protocols in Lichenology (Springer Lab Manual), Springer Berlin Heidelber New York
Pittam SK and Hanus FJ (1997) Lichen Synoptic Key. Hyperlink: http://mgd.nacse.org/hyperSQL/lichenland/
Pullan MR, Watson MF, Kennedy JB, Raguenaud C & Hyam R (2000) The Prometheus Taxonomic Model: a practical approach to representing multiple classification. Taxon 49: 55-75
Rambold G (1997) LIAS - the concept of an identification system for lichenized and lichenicoous ascomycetes. - In: Türk, R. and Zorer, R. (ed.), Progress and problems in lichenology in the Ninties -IAL 3. Biblioth. Lichenol. 68: 67-72
Rambold G (2001) Computer-aided identification systems of lichens. In: Kranner I, Beckett R and Varma A (Eds) Protocols in Lichenology (Springer Lab Manual), Springer Berlin Heidelberg New York
Sipman H (1998, July). Identification key and literature guide to the genera of Lichenized Fungi (Lichens) in the Neotropics. Provisional version. Hyperlink: http://www.bgbm.fuberlin.de/bgbm/staff/wiss/Sipman+H/keys/neokeyA.htm
Timdal E (1999) Herbarium database. Hyperlink: http://www.toyen.uio.no/botanisk/bot-mus/lav/soklavhb.htm

Triebel D and Rambold G (1998-99, first posted on May 19th, 1998; most recent update: January 2nd, 1999). DALI: Database of ascomycete literature. Botanische Staatssammlung München, Hyperlink: http://dbsys.botanik.biologie.uni-muenchen.de/botsamml/mycology/dali/home.html

Glossary

Definitions

Most definitions are taken from Hawksworth et al (1995), others from Tootill (1984), Wittig (1993) or from authors themselves.

- **Akinete**, a thick-walled resting spore that is formed during unfavourable conditions by certain cyanobacteria. On germination a hormogonium may be formed.
- **Amerospore**, a one-celled (non-septate) spore with a length to width ratio of less than 15:1.
- **Apothecium**, a cup- or saucer-like ascoma in which the hymenium is exposed at maturity.
- **Appressorium** (of lichens), a swelling on a hypha, for attachment to a photobiont cell.
- **Areolate**, thallus formed of small areas (=areoles).
- **Ascocarp**, see ascoma.
- **Ascoma**, an ascus-containing structure.
- **Ascospore**, a spore produced in an ascus by free cell formation.
- **Ascus**, typically sac-like cell diagnostic for ascomycetes in which (after karyogamy and meiosis) ascospores (generally eight) are produced by free cell formation.
- **Axenic** (of cultures), consisting of one organism; uncontaminated; a pure culture.
- **Basidioma**, a basidium-producing organ.
- **Basidiospore**, a propagative cell produced after meiosis on a basidium.

- **Basidium**, the cell or organ, diagnostic for basidiomycetes, from which (after karyogamy and meiosis) basidiospores (generally four) are produced.
- **Biatorine** (of lichen apothecia), lacking photobionts in a pale coloured margin.
- **Bioindicator**, an organism, or a part of an organism or a society of organisms, that gives information on the quality of all, or a part, of its environment.
- **Biological half-life (T_{bio})**, the time required for a 50% reduction of the environmental radioactivity in living organisms.
- **Biomonitor**, an organism, or a part of an organism or a society of organisms, that quantifies the quality of all, or a part, of its environment.
- **Byssoid**, cotton-like; made up of delicate threads; floccose.
- **Caducous** (of spores), readily falling off.
- **Cartilaginous**, firm and thoug , but readily bent.
- **Central cord**, see chondroid axis.
- **Cephalodium**, a delimited region within, or a warty, squamulose, or fruticose structure on the surface of, a lichen thallus containing a photobiont different to that characteristic of the rest of a thallus. Generally, cephalodia contain cyanobacteria (e.g. *Nostoc*) whereas the rest of the thallus contains green algae.
- **Chondroid axis**, cartilaginous axis occupying the central portion of the medulla in *Neuropogon* and *Usnea.*
- **Conidiomata**, specialised, multi-hyphal conidia-bearing structure.
- **Conidiophore**, simple or branched hypha bearing conidiogenous cells from which conidia are produced.
- **Conidium**, a specialised non-motile asexual spore usually caducous, not developed by cytoplasmic cleavage or free cell formation.
- **Cortex** (of lichens), a more or less thick outer covering of the thallus.
- **Corticolous**, living on bark.
- **Crustose**, crust-like; used for lichens having a thallus stretching over and firmly fixed to or inside the substratum; such thalli generally lack rhizinae and a lower cortex.

- **Cultivation**, technique that proliferates individuals in glass houses, growth chambers or in the field, non-aseptically.
- **Culture**, *in vitro* technique that proliferates cells, organs or individuals in sterilised tubes, dishes, flasks or fermentors, aseptically.
- **Cyanobiont**, a cyanobacterial photosynthetic symbiont of a lichen.
- **Cyanomorph**, morphotype of a photosymbiodeme in which the major photosynthetic symbiont is a cyanobacterium.
- **Diaspore**, any unit of dissemination, e.g. a spore or fragment of mycelium; in lichens particularly applied to vegetative propagules; see hormocyst, isidium, soredium.
- **Endolithic**, living beneath the surface of stones.
- **Epilithic**, living on the surface of stones.
- **Epiphytic**, living above the ground surface, growing on plants or other objects, but not as a parasite.
- **Excipulum** (of an ascoma), marginal mycelium delimiting or enclosing the fertile structures in an ascoma; -proprium, ascomatal margins without photobionts; -thallinum, ascomatal margins containing photobionts.
- **Farinose**, like flour.
- **Foliose**, leaf-like, having a stratose thallus, usually with a lower cortex and attached to the substratum either by rhizinae or at the base but not by the whole lower surface.
- **Free cell formation**, the process by which the eight nuclei, each with some adjacent cytoplasm, are cut off by walls in the immature ascus to become ascospores.
- **Fruticose**, having an upright, brush-like, or hanging, beard-like thallus.
- **Globose**, spherical or almost so.
- **Heterocysts**, any of the large cells that occur at intervals in the filaments of certain species of cyanobacteria, probably involved in nitrogen fixation.
- **Homoiohydric**, maintaining the water content at relatively constant level; unable to survive in the desiccated state.

- **Hormocyst,** a resting stage formed by certain filamentous cyanobacteria from side branches of the filament.
- **Hormogonium,** a short filament of more or less spherical cells that may be formed on germination of an akinete in certain filamentous cyanobacteria. When it comes to rest on a suitable surface, it gives rise to filaments.
- **Hymenium,** the spore-baring layer of a fruit-body.
- **Isidium,** a photobiont-containing, corticate protuberance of the thallus in lichens which may be warty, cylindrical, scale-like, simple, or branched; unit of dissemination.
- **Lecanorine** (of a lichen apothecium), having an excipulum thallinium.
- **Lecideine** (of a lichen apothecium), having an excipulum proprium; usually applied when the excipulum is dark pigmented.
- **Leprose,** having the surface or the whole thallus entirely dissolved into soredia.
- **Lirella,** a long, narrow apothecium.
- **Lobate,** lobed.
- **Lobulate,** having small lobes.
- **Macrolichen,** usually applied to large lichens of squamulose, foliose, or fruticose habit.
- **Mazaedium,** a dry spore mass formed in an ascoma (e.g., in Caliciales), in which the spores become passively free from the asci.
- **Medulla** (of lichen thalli), the loose layer of hyphae below the cortex and algal layer.
- **Microlichen,** mainly applied to small crustose lichens.
- **Morphotype,** a group of morphologically differentiated individuals of a species.
- **Muriform** (of ascospores), with transverse and longitudinal septation.
- **Perithecium,** a subglobose or flask-like ascoma.
- **Phorophyte,** the host tree of an epiphyte.
- **Photomorph,** an organism whose form is determined by the nature of the photobiont.

- **Photosymbiodeme,** lichen thallus with different possible photobionts.
- **Phototype,** each of the morphologically distinct structures derived by symbiosis between a single mycobiont and different photobionts.
- **Placodioid; placoid** (of a lichen thallus), disc-shaped, with plicate lobes at the circumference.
- **Plectenchyma,** aggregation of hyphae becoming twisted and fixed together.
- **Plicate,** folded into pleats.
- **Podetium,** lichenized, stem-like portion bearing the hymenial discs and sometimes conidiomata in a fruticose apothecium.
- **Poikilohydric,** unable to regulate water content which fluctuates with environmental water availability; able to survive in the desiccated state for longer or shorter periods without losing the ability to regenerate and grow.
- **Primordium,** the earliest stages of development of an organ.
- **Pseudopodetium,** a lichenized, podetium-like structure of vegetative origin.
- **Pulverulaceous,** powdered.
- **Redifferentiation,** thallus reformation from undifferentiated cell-aggregates where mycobionts and photobionts co-exist in the culture, as fragments or soredia.
- **Resynthesis,** recombination of previously separated lichen symbionts to reform a thallus.
- **Rhizina,** a root-like hair or thread formed by several hyphae; the attachment organs of many foliose lichens.
- **Rugose** (of ascospores), wrinkled.
- **Saxicolous,** growing on rocks.
- **Sclerotium,** a firm, frequently rounded, mass of hyphae with or without the addition of substratum, normally having no spores in or on it; may give a fruit-body, a stroma or mycelium.

- **Soralium**, decorticate portions of a thallus where soredia are located, usually formed from meduallary tissues thrusting upwards through the cortical layers.
- **Soredium**, non-corticate combination of photobiont cells and fungal hyphae having the appearance of a powdery granule, and capable of reproducing a lichen vegetatively.
- **Spathulate**, spoon-like in form.
- **Sporodochium**, conidioma in which the spore mass is supported by a superficial cushion-like mass of short conidiophores.
- **Squamulose**, having small scales.
- **Stipitate**, stalked.
- **Stratose**, lichen thallus having tissue in horizontal layers.
- **Stroma** (of lichens), a dense mass of plectenchyma, often dark pigmented.
- **Terricolous**, growing on the ground.
- **Thallus fragment culture** (Yamamoto method, "Lichen tissue culture"), *in vitro* culture of undifferentiated cell-aggregates derived from thallus fragments, containing algal and fungal symbionts.
- **Umbilicate**, growth form of lichens where the thallus is fixed to the substratum with a central holdfast.

References

Hawksworth DL, Kirk BM, Sutton BC, Pegler DN (1995) Ainsworth & Bisby's dictionary of the fungi. 8ed. CAB International, Wallingford, UK

Tootill E (1984) Dictionary of Botany, Penguin, London, UK

Wittig R (1993) Generals aspects of biomonitoring heavy metals by plants. In: Markert B. (ed). Plants as biomonitors. Indicators for heavy metals in the terrestrial environment. VCH Weinheim, New York, pp 4-27

Subject Index

GPSR Compliance

The European Union's (EU) General Product Safety Regulation (GPSR) is a set of rules that requires consumer products to be safe and our obligations to ensure this.

If you have any concerns about our products, you can contact us on ProductSafety@springernature.com

In case Publisher is established outside the EU, the EU authorized representative is:

Springer Nature Customer Service Center GmbH
Europaplatz 3
69115 Heidelberg, Germany

Batch number: 10755698

Printed by Printforce, the Netherlands